Biomaterials, Bioengineering and Sustainability

Volume 1

There is an urgent need to address the current paradigm shifts in bioengineering for Human Health aiming the creation of breakthrough tissue engineered products, manufacturing technologies and effective regenerative treatments for tackling different diseases/disorders in a personalized manner. Yet, the excessive costs and wastes related to the development and production of biomaterials and advanced therapy medicinal products, and the increasing use of plastics in cell culture methods and animal derived reagents has recognized the importance of decreasing the direct carbon footprint and thus, implement sustainable principles and solutions in the innovation ecosystem. The main goal of the volumes in Biomaterials, Bioengineering and Sustainability series is to catapult and consolidate new concepts and solutions towards the development of the next-generation of sustainable and eco-friendly biomaterials and tissue engineering and regenerative medicine approaches. Each volume will focus on the latest developments dealing with the identification of new sources of sustainable or recycled biomaterials, providing ideas for green technologies and methods that can be applied for biomaterials advanced processing and scaffolding strategies, and applications in biofabrication, tissue engineering, regenerative medicine, and drug delivery systems. It also aims to include the exploitation of renewable and sustainable source of human cells applied for cell therapies or in combination with sustainable biomaterials. The develop complex in vitro 3D/4D models and dynamic cell culture systems will be other subjects to be further explored from a sustainable perspective. This series aims to attract the contributions of leading experts in bioengineering, cell biology, materials engineering, and environmental sciences.

Rishabha Malviya • Sonali Sundram
Editors

Sustainable Green Biomaterials As Drug Delivery Systems

Springer

Editors
Rishabha Malviya
Department of Pharmacy
Galgotias University
Greater Noida, Uttar Pradesh, India

Sonali Sundram
Department of Pharmacy
Galgotias University
Greater Noida, Uttar Pradesh, India

ISSN 2731-751X ISSN 2731-7528 (electronic)
Biomaterials, Bioengineering and Sustainability
ISBN 978-3-031-79061-4 ISBN 978-3-031-79062-1 (eBook)
https://doi.org/10.1007/978-3-031-79062-1

This Springer imprint is published by the registered company Springer Nature Switzerland AG
The registered company address is: Gewerbestrasse 11, 6330 Cham, Switzerland

Foreword

The development and application of sustainable green biomaterials in drug delivery systems is the need of the hour. This book, "Sustainable Green Biomaterials as Drug Delivery Systems: Synthesis and Properties," edited by Dr. Rishabha Malviya attempts to fill this gap by demonstrating the potential of green biomaterials toward reshaping the drug delivery systems.

The book is organized into several chapters that demonstrate the understanding of numerous green techniques for obtaining sustainable biomaterials, analyze the prospects of biodegradable materials, and discuss the approaches for synthesizing biodegradable polymers. The book also explores the development of probiotic bacterial cellulose as a bio-mediated nanomaterial for therapeutic uses. It includes the revolutionary impact of 3D and 4D printing technology for the sustainable manufacturing of green biomaterials, and addresses the role of proteins as biocompatible materials for biomedical applications. The book explores applications of graphene-based carbonaceous compounds and green chemistry for synthesizing silk fibroin biomaterial scaffolds. The book also highlights the function of green catalysts in the synthesis of biomaterials and includes a case study for the green synthesis of metallic nanoparticles using plant extracts. The book discusses the applications of metal frameworks as biosensors and elaborates the role of cellulose, chitin, and chitosan in the formulation of composite-based biomaterials. It also focuses on the sustainable synthesis of cellulose-derived hydrogels for tissue engineering applications and highlights hydroxyapatite-starch-based biomaterials for healthcare applications and surfactant-free approaches for synthesizing metal and metal oxide nanomaterials for biomedical applications.

The book offers a complete and insightful discussion of the synthesis and application of sustainable green biomaterials in drug delivery systems, thus bringing research and development done in this critical area of biomedical engineering at a single platform in the form of this book. Dr. Rishabha Malviya and the team meticulously organized the chapters which resulted in a book that advances our understanding of green biomaterials while emphasizing the importance of sustainability in scientific research and healthcare. Happy reading and best wishes.

Professor (HAG Scale) and Former Head,
Department of Pharmaceutical
Engineering & Technology, IIT(BHU)
Varanasi, Uttar Pradesh, India

Brahmeshwar Mishra

Preface

The book, "Sustainable Green Biomaterials as Drug Delivery Systems: Synthesis and Properties," explores the synthesis, properties, and potential applications of sustainable biomaterials in drug delivery systems, aiming to provide a valuable resource for researchers, academicians, industrial experts, and students in biomedical engineering, materials science, and environmental science. The book addresses the urgent need for sustainable and biocompatible materials that can enhance drug delivery efficiency while minimizing environmental impact.

The book discusses innovative and sustainable approaches in the synthesis and application of green biomaterials. It focuses on the sustainable synthesis of cellulose-derived hydrogels for tissue engineering and regenerative medicine, the synthesis and properties of hydroxyapatite-starch-based biomaterials, and the surfactant-free synthesis of metal and metal oxide nanomaterials. The book aims to integrate environmental sustainability with cutting-edge biomedical engineering, paving the way for future advancements in drug delivery systems. It highlights green methods for obtaining sustainable biomaterials, the prospects of biodegradable materials, and their importance in creating sustainable and patient-centric solutions in biomedical engineering; presents strategies for synthesizing biodegradable polymers; and highlights their significance in developing sustainable drug delivery systems. It also explores the synthesis of probiotic bacterial cellulose, examining its potential as a bio-mediated nanomaterial for healthcare applications. The book also addresses the innovative use of 3D and 4D printing technologies in the sustainable manufacturing of green biomaterials, and utilization of proteins as biocompatible materials with vast applications in biomedical engineering.

The book integrates environmental sustainability with biomedical engineering, aiming to advance drug delivery systems. The book aims to inspire further innovation and development in this emerging field. Editors thank contributors for their support and readers for their interest in sustainable biomedical solutions, aiming to create a more sustainable and health-conscious future.

Uttar Pradesh, India

Rishabha Malviya
Sonali Sundram

About the Book

The book, "Sustainable Green Biomaterials as Drug Delivery Systems: Synthesis and Properties," explores the critical need for sustainable and environmentally friendly materials in biomedical applications. The book focuses extensively on the production, characteristics, and uses of green biomaterials in drug delivery systems. Key qualities include in-depth investigation, cutting-edge research, diverse viewpoints, clinical applications, and environmentally friendly solutions.

The book covers various topics, such as green methods for obtaining sustainable biomaterials, prospects of biodegradable materials, strategies in synthesizing biodegradable polymers, probiotic bacterial cellulose, 3D and 4D printable biomaterials, proteins as biocompatible materials, graphene-based carbonaceous materials, silk fibroin biomaterial scaffolds, green catalysts in biomaterial synthesis, utilization of plant extracts for the synthesis metallic nanoparticles, metal frameworks for biosensing applications, composite-based biomaterials using cellulose, chitin, and chitosan, sustainable synthesis of cellulose-derived hydrogels for tissue engineering applications, hydroxyapatite-starch-based biomaterials, and surfactant-free synthesis of metal and metal oxide nanomaterials.

The book is an invaluable resource for researchers and scholars in biomedical engineering, materials science, environmental science, and related subjects. It also provides information for industry professionals, students, and clinical practitioners interested in sustainable biomedical engineering. Practitioners in the biomedical and pharmaceutical industries will find useful applications and case studies to supplement their professional endeavors. Graduate and postgraduate students will learn about the principles, techniques, and applications of green biomaterials, preparing them for service in sustainable biomedical engineering.

This book is an essential read for anybody interested in the future of sustainable biomedical engineering. This book brings up the opportunities for eco-friendly and effective drug delivery systems by bringing together the most recent research and novel approaches, showcasing the revolutionary potential of green biomaterials in healthcare while promoting environmental sustainability.

Contents

Sustainable Green Biomaterials in Drug Delivery

1

Ruby Srivastava

Abstract

The recent breakthrough of biomaterials in drug delivery has created a new beginning of personalised medicare with smart "sustainable green" biomaterials. These biomaterials are versatile, adaptable and have enriched therapeutic outcomes with reduced side effects. Recent advancements in nanomaterials and gene therapeutics have enhanced the precision drug discovery. As biomaterials can be tuned according to the individual genomic profiles and disease characteristics, they are used for optimum benefits and minimum side effects. Biodegradable polymers have adaptable features, managable degenerative rates, customized mechanics, and surface properties that coordinate well with the human body. Yet biomaterials have not enough potential to change healthcare sector, as the treatment is customized to the individual patient's needs. Novel biomaterials led to the discovery of absorbable stents, used to replace the permanent metallic stents in cardiology. Metamaterials also play a vital role due to their exceptionally designed properties. They are used for medical applications as they have excellent potential to exploit sound and electromagnetic energy. These metamaterials are used in radiology, optical and masking devices. Metamaterials are used to improve diagnostics for better resolution and susceptibility. In this chapter we will discuss various types of sustainable biomaterials, their significant role in drug delivery, advantages and their applications in oncology, dentistry, cardiovascular diseases and neurological diseases. The challenges are also highlighted which include the toxicity issues and the complexity of manufacturing processes.

R. Srivastava (✉)
Bioinformatics, CSIR-CCMB, Hyderabad, India

R. Malviya, S. Sundram (eds.), *Sustainable Green Biomaterials As Drug Delivery Systems*, Biomaterials, Bioengineering and Sustainability 1,
https://doi.org/10.1007/978-3-031-79062-1_1

Keywords

Biomaterials · Sustainable · Drug delivery · Cardiovascular diseases · Side effects

Abbreviations

E. coli	Escherichia coli
DOX	*Doxorubicin*
B-Mori	Bombyx mori
HER2	Human Epidermal growth factor Receptor 2
In Situ	In the normal location
PVC	polyvinyl chloride
PEHD	polyethylene high density
PTFE	polytetrafluroethylene
PMMA	polymethylmethacrylate
UHMWPE	Ultra-High Molecular Weight Polyethylene
PLA	Polylactic acid
NFC	nanofibrillated cellulose
BNC	bacterial nanocellulose
CNC	cellulose nanocrystals
ELRs	elastin-like recombinamers
RLPs	Recombinant resilin-like peptides
ECM	extracellular matrix
FDA	Food and Drug Administration
Cel/Ch/Chs	Cellulose, Chitosan, and Chitin
GO	Graphene Oxide
LCA	Life Cycle Management
in vivo	within the living
in vitro	within the glass

1.1 Introduction

Biomaterials are natural or artificial substances, which interact with an organ, tissue or any other system for the diagnostic and treatment purposes. Biomaterials are used to replace the damaged tissue, send and receive signals from the surrounding cells and adjust the behaviour of immune cell (Colombani et al. 2021a, b), or the functioning of biological materials (Rodrigo-Navarro et al. 2021; Whitaker et al. 2021). Green materials have started a new need based era that is used to reform the system and serve the society at a larger scale. The optimization of green materials is used to utilize the systems that relate to the environment and physiology in a better way. The development of green sustainable biomaterials have increased due to the global importance as these green sustainable materials have less harmful effect to the environment. These biomaterials are developed with natural and renewable resources. Green sustainable materials are produced with sensible properties and it covers the whole cycle from production to disposal with reduced environmental hazards. These materials are extracted from the manufacturing plants and used for diagnostic and therapeutic and then as a medical waste. During this process, these materials are receptive to the impurity and successive degradation through different (thermal, hydro-mechanical, or biochemical) processes in air, clay and water. It plays a major role for the biological safety. The complexities arise due to the fusion of similar or different type of biomaterials. Biomaterials affect the surface chemistry, structural strength and porosity and other factors which effect the repurposing of biomaterials. These green materials emphasize on the accurate configuration of concepts which are used to coordinate the sustainable green biomaterials with a new dimension. Green sustainable biomaterials are based on ten fundamental principles which are as follows: avoid the toxic wastes, applications for recycling reactions, use of safe and green solvents, use of sustainable and safe feedstocks, use of biomimic protocols and approaches, use of green sustainable reagents, reducing the carbon emission, sustaining the life cycle, conservation of resources and production of efficient products. These principles are used to design the applications oriented materials with non-toxic, biodegradable, biocompatible, multifunctional and recyclable materials. The real-time and *in situ* analysis should be carried out to avoid the toxic reactions and to design sustainable materials with ease and fast clearance from functional systems. The assessment of the sustainable biomaterials needs to be assessed from production to disposal and to maintain and improve the durability of green sustainable biomaterials. The advanced biomaterials are used to regulate the macrophage polarization (Mao et al. 2022), drug delivery (Khan et al. 2019; Servatan et al. 2020), and regenerative therapies (Chen et al. 2020; Rana et al. 2023) See Fig. 1.1. Other biomaterials for stimuli-responsive systems are also

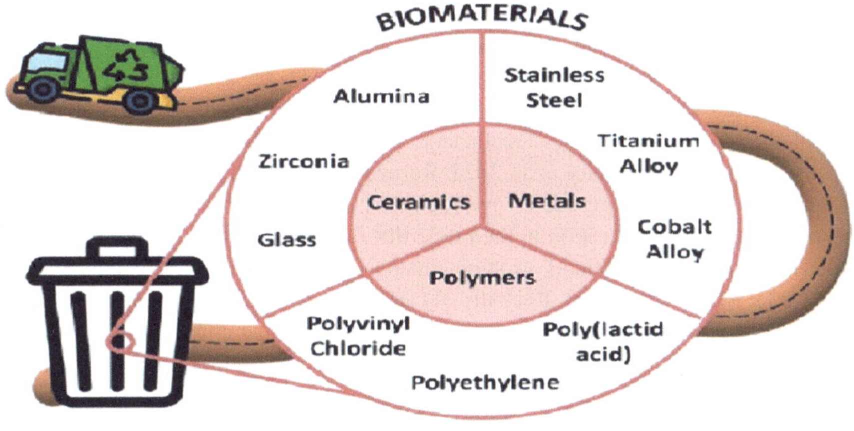

Fig. 1.1 Composition of biomaterials used for sustainable recycling. (Reference Wiśniewska P, et al. Front. Biomater. Sci. 2023, 2, 1260402)

designed for biological applications (Rogers et al. 2020; Yazdi et al. 2022). There are advanced approaches to shape and regulate the viability of cells of these biomaterials (Brokesh and Gaharwar 2020; Chua et al. 2021; Fu et al. 2022). 3D printing is used to develop function-shaped implants (Jafari et al. 2022) from titanium and its alloys, based on patients need.

These biomaterials are considered as an interdisciplinary work of materials science, biology, chemistry, engineering, and fabrication process. For example: Ti, stainless steel, polyvinyl chloride (PVC), polystyrene, polyurethane, polyethylene, polyesters, Si, Al_2O_3. The life cycle of material includes sensible and substantial properties, with ecological effect on production, refining, degradation, reusing, or final discard after its utility (Lefèvre and Auger 2016; Biswal et al. 2020; Zhu et al. 2016). Based on the properties these biomaterials are classified as materials produced from sustainable sources, environment friendly manufacturing, safety, repurposing after proper utilization. The other salient features should be economic attainability and usefulness for the society. The environmental impact of a sustainable biomaterial should be considered and justified for the societal needs (Kalirajan et al. 2021).

1.2 Classification

Biomaterials are composed of ceramics, polymers, glass, metals, and composites. These biomaterials are classified based on their function, compatibility, response and acceptability. If we look from the chemical point of view, these biomaterials are identified by composition, structure, backbone, and surface chemistry (Kargozar et al. 2019). Bioceramics are inert, bioresorbable, or bioactive materials. Bioceramics has wider applications in non-adherent fibrous tissue, resorbed and included into the local tissue and forming bond between the implant and the tissues (Punj et al. 2021). Biometals are biocompatible and have load-bearing capacity. Few biometals are biodegradable (Saini et al. 2015), and few other biometals have corrosive nature. The reaction mechanism (anode, cathode) of biomaterials can control the corrosion rate of implants (Eliaz 2019). The flexibility of polymers for molecular design can provide a larger application for tissue engineering, nanoplatforms, engrafts, stents, catheters, bone setting support, and prosthesis (Kalirajan et al. 2021; Rabiee et al. 2022; Rezaeeyazdi et al. 2022). These polymeric biomaterials are biodegradable, hydrophilic, surface chemistry, and erosion mechanisms. The modification in the properties of biomaterials can be significantly carried out with nanoparticles or supplementary biodegradable polymers (Saeedi et al. 2022; Seidi et al. 2021). Recently 3D and 4D polymeric biomaterials are used due to reduced cost and flexibility in design for biomedical applications (Pugliese et al. 2021; Shokrani et al. 2022). After serving the purpose, these biomaterials are replaced or withdrawn. The applications of biomaterials are wide, but the generated medical waste has become a huge problem.

Right now these biomaterials are either dumped or combust (Joseph et al. 2021). Since these biomaterials are infectious waste, discarding them can produce the environmental hazards and pollution (soil, water). Also the other disposal method is the inadequate incineration which if not properly done can lead to the release of harmful chemicals in the surroundings (Joseph et al. 2021). The wide applications of biomaterials have improved the quality of life and time span for humans. These biomaterials are hydrogels, natural polymers (Arif et al. 2022a, 2023; Khalid and Arif 2022) and the synthetically manufactured materials. These biomaterials are used to replace the parts of the body such as: Heart valve, shoulder, knee, elbow and hip joints, ears, orthodental system, and implants (Pesode and Barve 2021a; Barbin et al. 2020). These biomaterials are classified as follows:

1.2.1 Metallic Biomaterials

Metallic biomaterials are used mainly for medical applications. These materials are corrosion resistance and have good mechanical features. Such as CrCo amalgam, stainless steel, coinage metal alloys, Ti alloys, Mg and related alloys, and NiTi (Pesode and Barve 2021b, 2022). The titanium implants are used for its bony combination with the jaw bones.

1.2.2 Polymeric Biomaterials

Polymeric biomaterials are used for biomedicine applications. These polymeric biomaterials are chemically stable, biocompatible, environmental friendly, viable and biodegradable (Rana et al. 2023). For example: polyethylene high-density (PEHD), polytetrafuroethylene (PTFE), polymethylmethacrylate (PMMA). These biomaterials are affordable, biocompatible and easily available. Other polymer biomaterials include acrylic, polyamide, polyester, polyethylenes, polysiloxanes, and polyurethanes. These polymers are used for the pacemakers, Arthroplasty, artificial skin, sutures, skin and maxillofacial implants, synthetic blood vessels. These biomaterials are widely used for pancreas, artificial hearts, livers, kidneys, and bladders (Burg et al. 2000; Tariq et al. 1970; Arif et al. 2022b) related artificial devices.

1.2.3 Ceramic Biomaterials

Ceramic biomaterials include HAp ceramic particles into a biocompatible HDPE as a replacement for bone. The ceramic biomaterials include graphite, calcium phosphate, apatites, aluminium oxide and glass which is used for medicinal purposes (Pesode and Barve 2021b; Burg et al. 2000). Ceramics are nontoxic, inert and create variety of shapes for the body. These include bone grafting, artificial heart valves, hip prostheses, knees replacement, prosthesis and dental products. Since the ceramics have inferior mechanical properties, it is advisable to use ceramics carefully for load-bearing and stress related applications. Ceramics should be used carefully and safely for implant devices that bear large tensile stresses (Arif et al. 2022b).

1.2.4 Composite Biomaterials

Composite biomaterials are widely used for designing artificial parts of the body and dental products. Carbon fibres are used to strengthen polyethylene (UHMWPE) matrix of high molecular weight. Fabricated from pyrolyzing acrylic fibres, these structures are created with high young modulus and great durability. These carbon fibres have 6–15 nm of diameter and they are randomly aligned in the matrix. The core properties of these composite biomaterials are tuned for applications in intra medullary rods and artificial joints (Pesode and Barve 2022; Burg et al. 2000). It is possible to change the flexibility of the materials design. These biomaterials are unique and stronger than any other homogeneous substances. They are used for tissue in-growth as per the requirements. For example: Al_2O_3 deposit on carbon, carbon and PTFE, Al_2O_3 and PLA coated with carbon fibres (Burg et al. 2000; Takur et al. 2022).

1.2.5 Nanocellulose

Nanocellulose is novel biomaterials with unique characteristics and bio sustainability. Cellulose, the natural polymers are found in plant cell walls, is purified to the nanoscale. Cellulose is taken out as cellulose nanofibers or nanocrystals. These novel and artificially created biomaterials from natural polymers and biocellulose are used for clinical applications (Lin and Dufresne 2014; Mehanny et al. 2021). Nanocellulose made from natural cellulose fibres has a diameter (100 nm approximately) and a length (in micrometres).

Nanocellulose can be classified as:

(a) nanofibrillated cellulose (NFC),
(b) bacterial nanocellulose (BNC), and
(c) cellulose nanocrystals (CNC).

Nanocellulose has high elasticity (110–220 GPa), hardness (7.5–7.7 GPa), large specific surface area, personalized surface functionality, adjustable crystallinity, excellent chemical resistance and polymerizability (Fagone et al. 2017). Because of the multifacet properties of these sustainable nanocelluloses, these biomaterials are used for various interdisciplinary applications (Kumar et al. 2019). As the precision in dimensions of nanocellulose is important, it is essential to understand the dimensions of nanocellulose retrieve by different plants (Khalid et al. 2021).

Let's discuss the applications of different types of sustainable biomaterials in next section.

1.3 Different Types of Green Biomaterials for Biomedical Applications

1.3.1 Biocompatible Proteins

Proteins are produced by plants and animals (Stie et al. 2022; Liang et al. 2006), which are abundant, biodegradable and used to made the environmental supportive products (Grigsby et al. 2020). Proteins are sustainable source of polymers produced in large quantities. Proteins rely on renewable sources and they have low greenhouse gas emissions (Eshel et al. 2019). Plants derived proteins are not related to animal transmitted diseases and become an alternative source for animal derived proteins (Jones et al. 2013). Animal proteins are mostly expensive compared to the plant proteins. These proteins can serve for osseous tissue engineering and osteoinductive biocomposite scaffolds (Râpă et al. 2020; Arslan et al. 2017). The collagen from marine wastes is used as a substitution for natural collagen (Barros et al. 2015). We get proteins from human hair and biological throwaway for the domestic fowls (Eshel et al. 2019; He et al. 2017). Protein-based biomaterials have varied applications provided that these protein sources should have low energy and environmentally sustainable (Zuin and Ramin 2018). Proteins from livestock excrement or food feeder provide a sustainable, low cost, and dependable feedstock (Sozer et al. 2017). Several fibrous proteins, such as keratin, silk, resilin, and elastin are also used as green biomaterials (Costa et al. 2017). Collagen, gelatin, keratin, and silk are used for various eco-friendly purposes and numerous applications, such as freeform fabrication, micro and tissue engineering. See Table 1.1.

Protein-based biomaterials are plenty and renewable (Bonduelle 2018; DeFrates et al. 2018), easily modified and have chemical heterogeneity. These features of

Table 1.1 Biomaterials developed from traditional and alternative protein sources

Protein	Traditional source	Alternative source
Keratin	Wool fibres, horns, nails and feathers from butchery	Biomimetic recombinant hagfish thread keratin, recombinant human hair keratins K31 and K81
Collagen	Collagen type I: Mammalian skin, and tendon tissues (porcine, bovine and ovine in origin); collagen type II: bovine, porcine and chicken cartilaginous tissues; marine collagen	Recombinant human collagen in different prokaryotic, eukaryotic, plant and mammalian expression systems
Silk	Domestic silkworms: Bombyx mori (B. mori); wild silkworms: Antheraea pernyi and Samia cynthia ricini; Spiders: Nephila clavipes and Araneus diadematus.	Recombinant silk proteins in different host systems
Elastin and resilin	Animal-derived tropoelastin, recombinant production.	Recombinant elastin-like recombinamers (ELRs); recombinant resilin-like peptides (RLPs) proteins

Reference Agnieray et al. Biochemical Society Transactions (2021) 49953–964 protein biomaterials are due to the variable amino acid composition which a protein opt from various resources (Shadish and DeForest 2020; Wang et al. 2019). As the proteins are water-soluble, the production is limited for organic solvents (Capezza et al. 2019). Protein biomaterials have excellent physical properties, and they are intrinsically biodegradable (Abascal and Regan 2018). So it is necessary to develop novel protein-based materials and sourcing proteins in various manners (Costa et al. 2017; Bonduelle 2018; DeFrates et al. 2018; Shadish and DeForest 2020). 3D printing, a sustainable manufacturing allows the flexible designs and print of complex structures. 3D printing associate biopolymers such as proteins, polysaccharides and other green materials for novel biomaterial inks (DeFrates et al. 2018). The physicochemical and biological properties of proteins are tuned to design novel biomaterials for tissue engineering. For example: structural scaffolds for medical devices, tissue mimics and biosensors. Bioengineered silk are used for wider applications. Silk spidroin proteins are associated to the H2.1 or DOX peptide (Shamshina et al. 2019), which binds to Human Epidermal growth factor Receptor 2 (HER2), responsible for breast cancer. DOX shows affinity for doxorubicin, a cancer drug. Composite silk spidroin spheres offer controlled, low cost and targeted drug delivery system (Shamshina et al. 2019) which is effective for the patients. Sustainable bioengineered silk is designed for prokaryotic and eukaryotic models which included *E. coli*, rice, tobacco, and kidney cells (Galanakis 2015).

The improvement of insulin delivery in patients become effective by inserting subcutaneous silk fibroin hydrogels injections encapsulated with insulin for the control of blood glucose in the studies of diabetic rats (Shamshina et al. 2018). Silk fibroin microneedle patches are used for drug delivery to preserve tissue integrity (Miculescu et al. 2017a). Collagen biomaterials are used to deliver drug with natural biocompatibility and ECM protein (Fagone et al. 2017). Sponge scaffolds and other materials are produced by the extracted collagen from fowl industry. Gelatin, which is produced by partial hydrolysis of collagen, is used for drug delivery (Fagone et al. 2017). Hydrogels are used in the form of injections to transport anti-inflammatory drugs for osteoarthrosis patients (Dorozhkin 2011). Gelatin/salecan along with an antibiotics (vancomycin) is given in intravenous form to deliver drugs (Bouler et al. 2017). Delivery vehicles made with sustainable protein-based scaffolds such as keratin form rigid, yet porous, scaffold which can be used for osseous matrix applications.Wool keratin scaffolds showed good physicochemical properties and can be produced by liquid casting (Abutalib and Yahia 2017) and electrohydrodynamic processes (Miculescu et al. 2017b). The biocompatibility of keratin scaffolds does not obtain intrinsic or antibody mediated safe responses in standard large animal model, which is important for implantation (Takur et al. 2017). Elastin, an ECM protein sourced from the ligaments of animal neck show flexibility and durability to tissue. The incorporation of elastin cross-linking into collagen scaffolds for vascular system improve the strength and increases survival for overall muscle cell (Takur and Voicu 2016). Hydrogels and films created from elastin expressed in *E. coli* are used for improvement of vascular tissue engineering. Elastin improves the mechanical stress of protein-based materials, so that these biomaterials can be used in skin tissue engineering (Patel et al. 2015). Resilin with high elasticity and flexibility is used to repeat mechanical loading (Eshel et al. 2019) and bacterial cultivation (Eshel et al. 2019; Jones et al. 2013; Râpă et al. 2020). Resilin-like composite proteins are used to imitate tissue and contribute to cell differentiation (Eshel et al. 2019; Jones et al. 2013; Râpă et al. 2020). Recombine peptides based on human type I collagen is processes by fermentation of yeast (Khalid et al. 2022). Hydrogels formed from animal-derived methacrylated gelatins imitate ECM and help enclosed stem cells (Vanapalli et al. 2021). The bacterial fermentation is put to use with human-like collagen expressed in bacterial host (*E. coli* BL21) in addition to chitosan to produce hydrogels which is used to remove the full-thickness skin defects (Vanapalli et al. 2021). Poultry skin collagen composites are used to develop engineered skin substitutes. Marine Products 'blue biomaterials', and elastin hydrolysate, human tropoelastin and α-elastin are used to make or modify scaffolds such as collagen for skin substitutes. Collagen extracted from eel (Govindharaj et al. 2019), codfish (Carvalho et al. 2018), and tilapia (Li et al. 2019) skin show sustainable structures as compared to conventional, mammalian collagens. Even with these advantages, there are limitations and challenges in designing effective viable protein-based biomaterials with collected variation and complications of translating them for clinical purposes.

Sustainable technologies are used for promoting greener, safer, and environmentally safe biomaterials (Tipping and Wolfe 2016). There is a need to reduce the ecological track by 2030, by restructuring life process of raw materials and production (Khalid et al. 2023; Keshipour and Hadidi 2023). Protein biomaterials in green chemistry use decomposable materials, minimum waste generation, and minimal use of insufficient raw substances. Animal sourced proteins biomaterials show more mechanical strength than proteins from plants resources (Fiorentini et al. 2021). Another type of composite biomaterials is made from animal and plant proteins, (Karri et al. 2016; Samadian et al. 2020), polycarbohydrates or other proteins mixed with biopolymers (natural/synthetic) (Alam and Shubhra 2015; Yao et al. 2017; Khabbaz et al. 2019) with a selection of right solvent (Al Kayal et al. 2020; Chong et al. 2019). Proteins are biocompatible due to its intrinsic source; and they are used for clinical applications. Proteolytic enzymes degrade proteins, and the rate of breakdown can affect the functioning of protein biomaterials *in vivo*. Proteases have diverse affinity for particular recognition motifs which depend on primary amino acid sequence, made some proteins unsafe to proteolytic degeneration by certain peptidase (Shamshina et al. 2018). The secondary and tertiary structures of some proteins are less manageble to proteolytic cleavage. A protein is resisted for being break down by certain proteases if it is no able to recognize the motifs. So the behaviour of materials and degradation process of protein is determined by the type of protein (Müller-Herrmann and Scheibel 2015). Since there is crosslinking of the constituents of proteins in biomaterials, it is possible to increase the responsiveness of proteolytic disintegration and tune the behaviour of proteins (Müller-Herrmann and Scheibel 2015). The ossification and modification of the secondary structure of proteins is modified by exposing these proteins to solvents (CH_3OH, HCOOH) during the manufacturing of a biomaterial. The sensitivity of protein-based materials is also reduced due to degradation (Müller-Herrmann and Scheibel 2015). It is easy to predict the properties of novel protein-based materials *in vivo* by evaluating degeneracy. Biodegradable proteins can be used as substitution to synthetic polymers due to ecological green biodegradation (Maeda et al. 2001). Peptides and proteins are approved safe for the ecosystem. Protein-based delivery devices are FDA approved and these devices are used for drug delivery and treatment. Neurontin brand gebapentin capsules are used for neurological disorders (Shamshina et al. 2018). By modification of the structures, robust and flexible plant protein biomaterials are used for coatings, films, microcapsules, and personal care products. GelaCellTM using gelatin and corn-based zein is used to fabricate 3D nanoscaffolds for the treatment of injury and tissue engineering.

1.3.2 Composites (Cellulose, Chitosan, and Chitin)

The biocomposites made of Cel/Ch/Chs possess sustainable biological activity and used for biomedical applications. Cellulose, Chitin, and Chitosan Composite have antibacterial activity against many bacteria and fungi (Madni et al. 2021). Chemical modification of Cel/Ch/Chs enhanced the biological interactions, whereas protonation of chitosan increases its solubility. Adding substituents, the microbial activity of chitin and chitosan is decreased (Ma et al. 2008). The high content of chitosan causes sufficient reduction in weight (leaching). As cellulose has high pH, cellulose does not protonate for the application of pH range (3–8) (Yang et al. 2019). The same explanation can be applied to the modified anionic and cationic chitin/chitosan and cellulose derivatives. Natural polymers include polysaccharides like Cel/Ch/, and it's partly diacetylated derivatives. Chitosan (Bassas-Galia et al. 2017) with functional groups ($-NH_2$ and –OH) have high biosustainability and cytocompatibility. These biomaterials are nontoxic, and biodegradable. Due to the Chitin and cellulose inter-relations, these biomaterials give structural stability and protection to plants and some animals. Synthetic polymers have low cost and compositional consistency, but the low reconcilability and toxicity of their disintegrated product limits its uses. So it is of major concern to look into the environmental friendly production and disposal as well as biocompatible production. Such as: personal protective equipment made of fossil fuels (Liu et al. 2019). New composites of various polymers can be created with customized properties. Also advanced techniques are used to fabricate high quality tissue engineering fibres and materials. For instance, pure chitosan is not used for the industrial purpose due to its poor mechanical properties; but cellulose can be used to enhance its activities. These composites are made from mixing of biopolymers. These biomaterials are multifunctional and biodegradable and used for tissue engineering (Kucharczyk et al. 2020). Cellulose and cellulose acetate are used in food and beverage industries for ultra and nanofiltration (Whittall et al. 2020). The Cel/Ch and Cel/Chs biocomposites are extracted into desired "shapes". The one-pot method solution blending is used to reduce the complexity of process and expenditure (Maity et al. 2020). The heterogeneous synthesis of Cel/Ch and Cel/Chs composites in various combinations are used for biomedical applications.

1.3.3 Hydroxyapatite-Starch Based Biomaterials

These sustainable biomaterials are used for bone replacement therapies. There is great demand of using artificial materials for substituting the bone tissues. These synthetic materials are used in large collections, unlike the traditional transplant methods which are used in practise. These biomaterials are used for artificial implants and other biomedical applications. With proper usage, the risk of biological infection and immunologic inconsistency is reduced (Zhu et al. 2020; Kuo et al. 2018). These biomaterials mimic the natural osseous tissue with composites which includes an organic matrix with a ceramic component. Since these biomaterials are used for bone implantation so it is vital to use functionalised and recycled biomaterials to nurture the regrowth of host tissue and made it easier for newly generated bone tissue to replace the implanted materials (Fan et al. 2020). The biceramics made of calcium phosphates have good osteoconductive properties, which made them suitable candidates for bone regeneration provided that vascularization supported porosity is provided (Bloise et al. 2020). Calcium phosphates is brittle and molding it to a complex geometry is difficult, yet it is useful for bone interaction. Mixing with natural polymers, these composite materials are used for tissue engineering (Naderi et al. 2020). Natural polymers are both reconcilable and may increase the protein-based ceramics (Dias et al. 2020; Ghazanfari et al. 2019). Various other biomaterials such as hydroxyapatite and starch mixtures are used for medical applications. Plant and animal based polymers are abundant in nature. Some of these biomaterials are biocompatible, and some are even biodegradable. These biomaterials use fast, simple and low cost processes to produce sustainable properties (Khalili et al. 2019). 3D cell-laden nanohydroxyapatite/protein hydrogels are used for bone regeneracy (Sadat-Shojai et al. 2015). The 3D cell-filled HAp/hydrogel nanocomposites integrated with gelatin hydrogel provide strength and bioactivity, as these biocomposites are active and osteoconductivive (Hing et al. 2004). These biomaterials absorb proteins (vitronectin and fibronectin), so that its bioactive capabilities are improved (Lee et al. 2006; Dorozhkin 2010). The limitations of using hydroxyapatite are that these are fragile and have low tensile strength, which cause poor mechanical properties. Hydroxyapatite also increased post implantation infection risks. Starch is also used in clinical applications as starch is low cost material, renewable, decomposable in multienvironments without generating harmful substances. Starch can thicken easily, swell, and gel, which is used to change the properties in water at high temperatures (Wong and Louie 2017). Amylose and amylopectin are found in starch. Amylopectin is responsible for the transparency of the materials. Starch grains are also useful for wider applications and it is present in plants (wheat, rice, and potatoes) (Gregorova et al. 2009). The properties of starch depend on which source they are derived (Salgado et al. 2004). The composite materials (polymeric films and starches) increase the tensile strength, fracture toughness, and abrasion resistance with increased amylose concentrations. Starch is used for many applications, but still it has some restrictions as poor mechanical properties, water sensitivity and processing. Though these restrictions is removed by modifying the appropriate additions and chemical changes, so that the biomaterials can become stable. The properties are also tuned by using the appropriate blending of starch, phyllosilicates, clays, and other polysaccharides for useful applications.

1.3.4 Carbonaceous Materials

Carbonaceous materials are abundant in nature. Biomaterials made from renewable carbon materials are used for various applications (Ma et al. 2014). These carboneous materials are made from eutectic solvents, graphitic nanostructures and cellulose fibres and carbon nanotubes. For example: cellulosic fibres, agricultural biomass (such as pomelo skins and maize stalks). The experimental results showed that pyrolysis-based synthesis of cellulose carbon fibres contained branched carbon nanotubes. Carbonaceous materials are used in catalysis, energy generation, production of polyester, photo and electrochemical applications. The surface area of carbon fibres are increased by adding the metal containing salt solution (Zhao et al. 2013). The composite multiwalled carbon nanotubes with heavy metal ions are used to manage the nuclear waste (Wang et al. 2005). GO nanosheets and its sulphonated derivatives worked as biosorbents (Zhao et al. 2011a, b) to separate Cd (II) or Co (II) ions, naphthalene, and 1-naphthol, respectively. Carbonaceous gels processed via hydrothermal treatment are a cheap and environmentally safe biomaterial. As aerogels have poor mechanical and/or thermal stability, 3D carbonaceous flexible hydrogel and aerogel is developed. The entire substance include carbon source (crude biomass) and the network inserted by Fe_3O_4 nanoparticles, (Wu et al. 2013) so that it can be used for electrochemical applications due to enhanced mechanical properties and strong chemical activity. These properties are used for catalytic supports, adsorbents, materials, and biomedical applications. These biomaterials are chemically inactive, have specified surface area, sizeable volumes, and high mechanical strength. Carbon compounds made from waste biomass is the harmful dye constituent malachite green, which when released causes hazards for aquatic life. Recent researches showed that sulphuric acid-stimulated carbon produce waste biomass palm flowers to absorb malachite green (Nethaji et al. 2010). Banana peels are also used to create sponge-like functional carbons with magnificent properties to remove the methylene blue (Liu et al. 2014a). Also, nitrogen-doped porous carbon nanosheets are used to enhance energy storage (Liu et al. 2014b). By adding doped materials, the electrical and electrochemical properties are tuned for varied applications (Wang et al. 2014; Asghari and Keshipour 2023; Keshipour and Asghari 2022; Keshipour and Eyvari-Ashnak 2023; Gao et al. 2014). Nitrogen-doped carbon composites showed good electrocatalytic activities due to high porosity and large surface areas. Despite the usefulness of carbon materials at certain level, their industrial applications are hindered due to low cost, environment, time, and complex routes. The disadvantages of these materials are excess energy requirement, solvent revival, tools decomposition, and toxicity. The environmentally safe techniques can use the wastes as a source for creating novel biomaterials. Soot, xylan, biofuel, polysaccharides, and turf are used for these purposes (Kumar Gupta et al. 2015).

1.4 Perspective

In recent years, the design, production, and industrial use of biomaterials have captured the industry. Further, demand for 3Rs (repair, replacement, recycling) of biomaterials is also increased. These biomaterials are not fully degradable in case of oral implantation or within the body. It addresses great concern regarding the sustainability and circularity. The Life Cycle Assessment (LCA) reported ecological effects and carbon footmark outcomes of biomaterials (Hjuler and Hansen 2018; Soman and Ajitha 2018). Now these recycling and environmental safety has become a major concern. See Fig. 1.2.

1.4.1 Current Recycling Strategies

The green biomaterials show specific and relevant purpose *in vivo* and/or *in vitro*. Biomaterials are continuously been used in regenerative medicine and various medical devices. Once their applications are served, they are replaced or withdrawn. As biomaterials have prominent role in medicine, their collection and management in medical waste is major concern. These biowastes are often landfilled or incinerated. Since they are infectious waste, they are very harmful for the environmental. The waste degradation causes oil and water pollution. Also, inadequate incineration release toxic chemical compounds into the environment, which need to be addressed urgently.

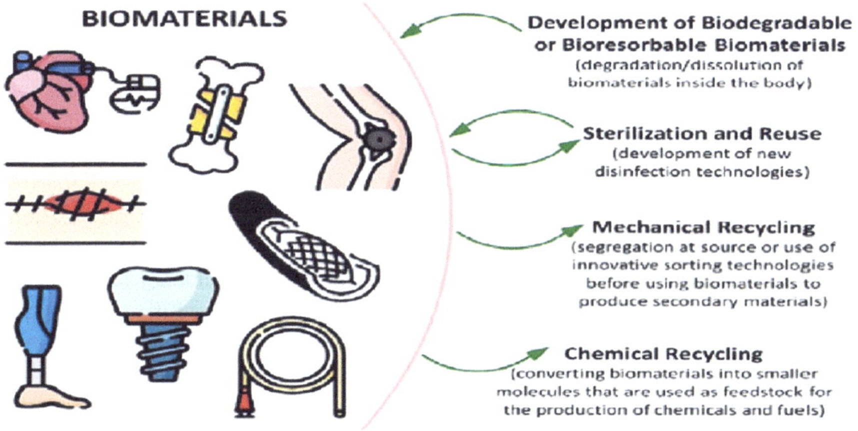

Fig. 1.2 Various strategies with actions taken for biomaterials recycling. (Reference Wiśniewska P, et al. Front. Biomater. Sci. 2023, 2, 1260402)

1.4.2 Dental and Orthopedic Implants

Since biomaterials are used in dental and orthopedics, proper recycling of metal amalgam used in dental fixing and silver composites is needed (Thopegowda et al. 2013). Studies showed (Yadav et al. 2020) many efficient ways to recycle dental and orthopedic implants. It is proposed that polymeric implants can be recycled by extrusion, while metallurgical powder or liquid resin recycling can be used for metallic and ceramic-based implants. The recycled products can be used to manufacture the other subsequent implants and biomaterials. It is vital to see the importance of the biomaterials and to decide whether to reuse or remanufacture or recycle it. Clinical research showed that thrice recasting of Pd-Cu-Ga and Au-Pt metal ceramic dental alloys showed no significant difference in its interconnected structures. Though in some cases these structures showed different morphologies. Such as recasting of some metallic alloys change the corrosiveness and cytotoxic properties of these alloys. The recycled Ni-Cr dental casting alloys caused genotoxicity when it is recycled for larger materials. It is also complicated to recycle mixed polymeric biomaterials during production and processing (Tripuraneni and Namburi 2008).

1.4.3 Medical Plastic Waste

Mechanical and chemical methods are adopted for secondary and tertiary recycling of polymeric medical waste biomaterials (Kheirabadi and Sheikhi 2022). Sterilization is necessary step for the waste biomaterials before including it with plastic wastes. The recycled polymer biomaterials have weak connected adherence and strength. Though advanced techniques and machine learning (ML) tools are used to overcome these challenges, many other possible strategies are also identified to recycle these biomaterials (Kheirabadi and Sheikhi 2022).

1.4.4 Sterilization and Reusability

It is necessary to reuse the biomaterials after proper treatment and sterilization. There are successful reports on this strategy which showed that the material properties of the biomaterials can be maintained (Yadav et al. 2020; Kheirabadi and Sheikhi 2022) without compromise. The mechanical strength of unutilized, inserted, removed smaller implants from pig iliac bone with and without supplementary treatment showed that the distortion of screws was not changed. But in some other reports, the sterilization affects the final behavior of the biomaterial. Such as the orthodontic archwires composed of Ni-Ti and Cu-Ni-Ti alloys, which experience deterioration and surface irregularities after two cycles of uses. In other studies the stainless steel orthodontic brackets receive reduction in shear bond strength by flaming with ultrasonic cleaning (Sayed et al. 2018). The flame treatment with abrasive blasting showed adequate strength for its use for clinical applications.

Now it is the demand of time to discover novel, environmentally friendly disinfection technologies to reuse biomaterials without deteriorating their properties (Attrah et al. 2022).

1.4.5 Waste Management for Recycling

Though the above mentioned process is used for the waste management, it needs major changes in the collection of waste biocompatible materials. The health sector requires trained staff with awareness of recycling and disposal of the waste-collection used in the hospitals. These biomaterials need to be separated and categorised, and if possible it should be recycle after sterilization. These biomaterials need not to be an environmental burden; rather they are decomposed or dissolve inside the body after a preset time (Thopegowda et al. 2013; Kheirabadi and Sheikhi 2022; Attrah et al. 2022). The biomaterials have low tensile strength which limits the use to these biomaterials in bone transplant applications (Yadav et al. 2020).

1.5 Conclusion and Future Challenges

There are remarkable researches in the field of biomaterials, which are focussing on recycling or disposal of toxic chemical industry products to sustain environmental safety (Rodrigo-Navarro et al. 2021; Whitaker et al. 2021). Since the biomaterials associated with various harmful chemical products overshadow its benefits for biomedical and clinical applications, developing vigorous measures, categorize, and processing methods that bond to the rigid standard control is vital. The contamination and degradation are the responsible factors for the biological safety. Additionally, including biomaterials into other materials can affect surface chemistry, strength and chemical properties of the biomaterials. This effect the reusability and recyclability of the biomaterials. These biomaterials are used for various therapeutic applications. 4D printing has great influence on the biomedical industry. New sustainable materials are fabricated, and high-resolution, economical printers are also made. 5D printing is also used for products with twisted surfaced support system that will improve the speed and strength by proper adjustments. These methods are used to decrease waste materials, reduce transportation costs, improve the manufacturing costs, simplifying the supply chain, and improve the sustainability of ecosystem. These technologies are used by the variety of industries to use the recycled and recovered materials when necessary. Numerous applications of tissue engineering include appropriate natural and synthetic biomaterials and biocomposites.

Solution blending is also possible, which is used to dissolve both biopolymers simultaneously or singly in suitable solvents before refining them into the proper "shapes." Solution blending use one-pot approach to remarkably decrease the complex processing and expenses. Hydroxyapatite and starch combinations are used for the recent biomedical uses with extensive potential of biomaterials. There are many plants and animal based biocompatible and biodegradable polymers which improve the bioactivity of ceramics. The potential of sustainable and recycled biomaterials transform the healthcare by combined efforts of policymakers, scientists and industry stakeholders. We can create sustainable ecosystem by addressing the challenges that allows the efficient reuse of biomaterials for therapeutic and diagnostic applications. Until now, few researches have been carried out for the improvement of these biomaterials. A considerable amount of work with innovative and emerging green design technologies can lead to a better reasonable and supportable society in future.

Acknowledgements DST WOSA project (SR/WOS-A/CS-69/2018) scheme. Author is thankful to Prof. Shrish Tiwari, Bioinformatics, CCMB-CSIR, and Prof. G. Narahari Sastry, Director, CSIR-NEIST for the support.

References

Abascal NC, Regan L (2018) The past, present and future of protein-based materials. Open Biol 8: 180113. https://doi.org/10.1098/rsob.180113

Abutalib MM, Yahia IS (2017) Novel and facile microwaveassisted synthesis of Mo-doped hydroxyapatite nanorods: characterization, gamma absorption coefcient, and bioactivity. Mater Sci Eng C 78:1093–1100

Al Kayal T, Losi P, Pierozzi S, Soldani G (2020) A new method for fibrin-based electrospun/sprayed scaffold fabrication. Sci Rep 10(1):5111–5114

Alam AKMM, Shubhra QT (2015) Surface modified thin film from silk and gelatin for sustained drug release to heal wound. J Mater Chem B 3(31):6473–6479

Arif ZU, Khalid MY, Sheikh MF, Zolfagharian A, Bodaghi M (2022a) Biopolymeric sustainable materials and their emerging applications. J Environ Chem Eng 10(4):Article ID 108159

Arif ZU, Khalid MY, Zolfagharian A, Bodaghi M (2022b) 4D bioprinting of smart polymers for biomedical applications: recent progress, challenges, and future perspectives. React Funct Polym 179:Article ID 105374

Arif ZU, Khalid MY, Noroozi R et al (2023) Additive manufacturing of sustainable biomaterials for biomedical applications. Asian J Pharm Sci 18(3):Article ID 100812

Arslan YE, Sezgin Arslan T, Derkus B, Emregul E, Emregul KC (2017) Fabrication of human hair keratin/jellyfsh collagen/eggshell-derived hydroxyapatite osteoinductive biocomposite scafolds for bone tissue engineering: from waste to regenerative medicine products. Colloids Surf B: Biointerfaces 154:160–170

Asghari A, Keshipour S (2023) Green and one-pot synthesis of Co (II) citrate complex nanoparticles/graphene oxide nanocomposites towards photocatalytic hydrogen evolution. Int J Hydrog Energy 89:34750–34765

Attrah M, Elmanadely A, Akter D, Rene ER (2022) A review on medical waste management: treatment, recycling, and disposal options. Environments 9:146. https://doi.org/10.3390/environments9110146

Barbin T, Veloso DV, Del Rio Silva L et al (2020) 3D metal printing in dentistry: an in vitro biomechanical comparative study of two additive manufacturing technologies for full-arch implant-supported prostheses. J Mech Behav Biomed Mater 108:Article ID 103821

Barros AA, Aroso IM, Silva TH, Mano JF, Duarte ARC, Reis RL (2015) Water and carbon dioxide: green solvents for the extraction of collagen/gelatin from marine sponges. ACS Sustain Chem Eng 3(2):254–260

Bassas-Galia M, Follonier S, Pusnik M, Zinn M (2017) Natural polymers: a source of inspiration. In: Bioresorbable Polymers for Biomedical Applications. Woodhead Publishing, Sawston, pp 31–64

Biswal T, BadJena SK, Pradhan D (2020) Sustainable biomaterials and their applications: a short review. Mater Today Proc 30:274–282. https://doi.org/10.1016/j.matpr.2020.01.437

Bloise N, Patrucco A, Bruni G, Montagna G, Caringella R, Fassina L et al (2020) In vitro production of calcified bone matrix onto wool keratin scaffolds via osteogenic factors and electromagnetic stimulus. Materials (Basel) 13:3052. https://doi.org/10.3390/ma13143052

Bonduelle C (2018) Secondary structures of synthetic polypeptide polymers. Polym Chem 9:1517–1529. https://doi.org/10.1039/C7PY01725A

Bouler JM, Pilet P, Gauthier O, Verron E (2017) Biphasic calcium phosphate ceramics for bone reconstruction: a review of biological response. Acta Biomater 53:1–12

Brokesh AM, Gaharwar AK (2020) Inorganic biomaterials for regenerative medicine. ACS Appl Mater Interfaces 12:5319–5344. https://doi.org/10.1021/acsami.9b17801

Burg KJ, Porter S, Kellam JF (2000) Biomaterial developments for bone tissue engineering. Biomaterials 21(23):2347–2359

Capezza AJ, Newson WR, Olsson RT, Hedenqvist MS, Johansson E (2019) Advances in the use of protein-based materials: toward sustainable naturally sourced absorbent materials. ACS Sustain Chem Eng 7:4532–4547. https://doi.org/10.1021/acssuschemeng.8b05400

Carvalho A, Marques A, Silva T, Reis R (2018) Evaluation of the potential of collagen from codfish skin as a biomaterial for biomedical applications. Mar Drugs 16:495. https://doi.org/10.3390/md16120495

Chen B, Xiang H, Pan S, Yu L, Xu T, Chen Y (2020) Advanced theragenerative biomaterials with therapeutic and regeneration multifunctionality. Adv Funct Mater 30:2002621. https://doi.org/10.1002/adfm.202002621

Chong C, Wang Y, Fathi A, Parungao R, Maitz PK, Li Z (2019) Skin wound repair: results of a pre-clinical study to evaluate electrospun collagen–elastin–PCL scaffolds as dermal substitutes. Burns 45(7):1639–1648

Chua K, Khan I, Malhotra R, Zhu D (2021) Additive manufacturing and 3D printing of metallic biomaterials. Eng Regen 2:288–299. https://doi.org/10.1016/j.engreg.2021.11.002

Colombani T, Eggermont LJ, Hatfield SM, Rogers ZJ, Rezaeeyazdi M, Memic A et al (2021a) Oxygen-generating cryogels restore T cell mediated cytotoxicity in hypoxic tumors. Adv Funct Mater 31:2102234. https://doi.org/10.1002/adfm.202102234

Colombani T, Eggermont LJ, Rogers ZJ, McKay LG, Avena LE, Johnson RI et al (2021b) Biomaterials and oxygen join forces to shape the immune response and boost COVID-19 vaccines. Adv Sci 8:2100316. https://doi.org/10.1002/advs.202100316

Costa F, Silva R, Boccaccini AR (2017) Fibrous protein-based biomaterials (silk, keratin, elastin, and resilin proteins) for tissue regeneration and repair. Pept Proteins Biomater Tissue Regen Repair:175–204. https://doi.org/10.1016/B978-0-08-100803-4.00007-3

DeFrates K, Moore R, Borgesi J, Lin G, Mulderig T, Beachley V et al (2018) Protein-based fiber materials in medicine: a review. Nanomaterials 8:457. https://doi.org/10.3390/nano8070457

Dias GJ, Ramesh N, Neilson L, Cornwall J, Kelly RJ, Anderson GM (2020) The adaptive immune response to porous regenerated keratin as a bone graft substitute in an ovine model. Int J Biol Macromol 165:100–106. https://doi.org/10.1016/j.ijbiomac.2020.09.133

Dorozhkin SV (2010) Calcium orthophosphates as bioceramics: state of the art. J Funct Biomater 1(1):22–107

Dorozhkin SV (2011) Biocomposites and hybrid biomaterials based on calcium orthophosphates. Biomatter 1(1):3–56

Eliaz N (2019) Corrosion of metallic biomaterials: a review. Materials 12:407. https://doi.org/10.3390/ma12030407

Eshel G, Stainier P, Shepon A, Swaminathan A (2019) Environmentally optimal, nutritionally sound, protein and energy conserving plant based alternatives to US meat. Scien Rep 9(1): 10345–10411

Fagone M, Loccarini F, Ranocchiai G (2017) Strength evaluation of jute fabric for the reinforcement of rammed earth structures. Compos Part B 113:1–13
Fan Z, Cheng P, Yin G, Wang Z, Han J (2020) In situ forming oxidized salecan/gelatin injectable hydrogels for vancomycin delivery and 3D cell culture. J Biomater Sci Polym Ed 31:762–780. https://doi.org/10.1080/09205063.2020.1717739
Fiorentini F, Suarato G, Grisoli P, Zych A, Bertorelli R, Athanassiou A (2021) Plant-based biocomposite flms as potential antibacterial patches for skin wound healing. Eur Polym J 150: Article ID 110414
Fu Z, Ouyang L, Xu R, Yang Y, Sun W (2022) Responsive biomaterials for 3D bioprinting: a review. Mater Today 52:112–132. https://doi.org/10.1016/j.mattod.2022.01.001
Galanakis CM (2015) Separation of functional macromolecules and micromolecules: from ultrafltration to the border of nanofltration. Trends Food Sci Technol 42(1):44–63
Gao S, Chen Y, Fan H et al (2014) Large scale production of biomass-derived N-doped porous carbon spheres for oxygen reduction and supercapacitors. J Mater Chem A 2(10):3317–3324
Ghazanfari S, Alberti KA, Xu Q, Khademhosseini A (2019) Evaluation of an elastic decellularized tendon-derived scaffold for the vascular tissue engineering application. J Biomed Mater Res A 107:1225–1234. https://doi.org/10.1002/jbm.a.36622
Govindharaj M, Roopavath UK, Rath SN (2019) Valorization of discarded marine Eel fish skin for collagen extraction as a 3D printable blue biomaterial for tissue engineering. J Clean Prod 230: 412–419. https://doi.org/10.1016/j.jclepro.2019.05.082
Gregorova E, Živcova Z, Pabst W (2009) Starch as a pore-forming and body-forming agent in ceramic technology. Starch/Staerke 61(9):495–502
Grigsby WJ, Scott SM, Plowman-Holmes MI, Middlewood PG, Recabar K (2020) Combination and processing keratin with lignin as biocomposite materials for additive manufacturing technology. Acta Biomater 104:95–103
He M, Zhang B, Dou Y, Yin G, Cui Y, Chen X (2017) Fabrication and characterization of electrospun feather keratin/poly (vinyl alcohol) composite nanofbers. RSC Adv 7(16): 9854–9861
Hing KA, Best SM, Tanner KE, Bonfeld W, Revell PA (2004) Mediation of bone ingrowth in porous hydroxyapatite bone graft substitutes. J Biomed Mater Res A 68(1):187–200
Hjuler SV, Hansen SB (2018) LCA of biofuels and biomaterials. Life cycle assessment: theory and practice. Springer, Berlin, pp 755–782
Jafari A, Ajji Z, Mousavi A, Naghieh S, Bencherif SA, Savoji H (2022) Latest advances in 3D bioprinting of cardiac tissues. Adv Mater Technol 7:2101636. https://doi.org/10.1002/admt.202101636
Jones BA, Grace D, Kock R et al (2013) Zoonosis emergence linked to agricultural intensifcation and environmental change. Proc Natl Acad Sci 110(21):8399–8404
Joseph B, James J, Kalarikkal N, Thomas S (2021) Recycling of medical plastics. Adv Ind Eng Polym Res 4:199–208. https://doi.org/10.1016/j.aiepr.2021.06.003
Kalirajan C, Dukle A, Nathanael AJ, Oh T-H, Manivasagam G (2021) A critical review on polymeric biomaterials for biomedical applications. Polymers 13:3015. https://doi.org/10.3390/polym13173015
Kargozar S, Ramakrishna S, Mozafari M (2019) Chemistry of biomaterials: future prospects. Curr Opin Biomed Eng 10:181–190. https://doi.org/10.1016/j.cobme.2019.07.003
Karri VVSR, Kuppusamy G, Talluri SV et al (2016) Curcumin loaded chitosan nanoparticles impregnated into collagen-alginate scafolds for diabetic wound healing. Int J Biol Macromol 93:1519–1529
Keshipour S, Asghari A (2022) A review on hydrogen generation by phthalocyanines. Int J Hydrog Energy 47(26):12865–12881
Keshipour S, Eyvari-Ashnak F (2023) Nitrogen-doped electrocatalysts, and photocatalyst in water splitting: efects, and doping protocols. ChemElectroChem 10(7):Article ID e202201153
Keshipour S, Hadidi M (2023) Palladium, and platinum nanoparticles supported on chitosan aerogel as a photocatalyst of formic acid degradation. J Polym Environ 32:1–12

Khabbaz B, Solouk A, Mirzadeh H (2019) Polyvinyl alcohol/soy protein isolate nanofibrous patch for wound-healing applications. Prog Biomater 8(3):185–196

Khalid MY, Arif ZU (2022) Novel biopolymer-based sustainable composites for food packaging applications: a narrative review. Food Packag Shelf Life 33:Article ID 100892

Khalid MY, Al Rashid A, Arif ZU, Ahmed W, Arshad H (2021) Recent advances in nanocellulose-based different biomaterials: types, properties, and emerging applications. J Mater Res Technol 14:2601–2623

Khalid MY, Arif ZU, Ahmed W, Arshad H (2022) Recent trends in recycling and reusing techniques of different plastic polymers and their composite materials. Sustain Mater Technol 31:Article ID e00382

Khalid MY, Arif ZU, Hossain M, Umer R (2023) Recycling of wind turbine blade through modern recycling technologies: road to zero waste. Renew Energy Focus 44:373

Khalili S, Khorasani SN, Razavi SM, Hashemibeni B, Tamayol A (2019) Nanofibrous scaffolds with biomimetic composition for skin regeneration. Appl Biochem Biotechnol 187:1193–1203. https://doi.org/10.1007/s12010-018-2871-7

Khan MSJ, Khan SB, Kamal T, Asiri AM (2019) Agarose biopolymer coating on polyurethane sponge as host for catalytic silver metal nanoparticles. Polym Test 78:105983. https://doi.org/10.1016/j.polymertesting.2019.105983

Kheirabadi S, Sheikhi A (2022) Recent advances and challenges in recycling and reusing biomedical materials. Curr Opin Green Sustain Chem 38:100695. https://doi.org/10.1016/j.cogsc.2022.100695

Kucharczyk K, Florczak A, Deptuch T, Penderecka K, Jastrzebska K, Mackiewicz A et al (2020) Drug affinity and targeted delivery: double functionalization of silk spheres for controlled doxorubicin delivery into Her2-positive cancer cells. J Nanobiotechnol 18:56. https://doi.org/10.1186/s12951-020-00609-2

Kumar Gupta G, De S, Franco A, Balu A, Luque R (2015) Sustainable biomaterials: current trends, challenges and applications. Molecules 21(1):48

Kumar R, Ul Haq MI, Raina A, Anand A (2019) Industrial applications of natural fibre-reinforced polymer composites–challenges and opportunities. Int J Sustain Eng 12(3):212–220

Kuo CY, Guo T, Cabrera-Luque J, Arumugasaamy N, Bracaglia L, Garcia-Vivas A et al (2018) Placental basement membrane proteins are required for effective cytotrophoblast invasion in a three-dimensional bioprinted placenta model. J Biomed Mater Res A 106:1476–1487. https://doi.org/10.1002/jbm.a.36350

Lee KY, Park M, Kim HM et al (2006) Ceramic bioactivity: progresses, challenges and perspectives. Biomed Mater 1(2):R31–R37

Lefèvre T, Auger M (2016) Spider silk inspired materials and sustainability: perspective. Mater Technol 31:384–399. https://doi.org/10.1179/1753555715Y.0000000065

Li D, Gao Y, Wang Y, Yang X, He C, Zhu M et al (2019) Evaluation of biocompatibility and immunogenicity of micro/nanofiber materials based on tilapia skin collagen. J Biomater Appl 33:1118–1127. https://doi.org/10.1177/0885328218820180

Liang M, Chen VY, Chen HL, Chen W (2006) A simple and direct isolation of whey components from raw milk by gel fltration chromatography and structural characterization by Fourier transform Raman spectroscopy. Talanta 69(5):1269–1277

Lin N, Dufresne A (2014) Nanocellulose in biomedicine: current status and future prospect. Eur Polym J 59:302–325

Liu RL, Liu Y, Zhou XY, Zhang ZQ, Zhang J, Dang FQ (2014a) Biomass-derived highly porous functional carbon fabricated by using a free-standing template for efficient removal of methylene blue. Bioresour Technol 154:138–147

Liu Q, Duan Y, Zhao Q, Pan F, Zhang B, Zhang J (2014b) Direct synthesis of nitrogen-doped carbon nanosheets with high surface area and excellent oxygen reduction performance. Langmuir 30(27):8238–8245

Liu J, Sun L, Xu W, Wang Q, Yu S, Sun J (2019) Current advances and future perspectives of 3D printing natural-derived biopolymers. Carbohydr Polym 207:297–316. https://doi.org/10.1016/j.carbpol.2018.11.077

Ma G, Yang D, Zhou Y, Xiao M, Kennedy JF, Nie J (2008) Preparation and characterization of water-soluble N-alkylated chitosan. Carbohydr Polym 74(1):121–126

Ma M, Dai Y, Zou JL, Wang L, Pan K, Fu HG (2014) Synthesis of iron oxide/partly graphitized carbon composites as a high-efficiency and low-cost cathode catalyst for microbial fuel cells. ACS Appl Mater Interfaces 6(16):13438–13447

Madni A, Kousar R, Naeem N, Wahid F (2021) Recent advancements in applications of chitosan-based biomaterials for skin tissue engineering. J Bioresour Bioprod 6(1):11–25

Maeda M, Kadota K, Kajihara M, Sano A, Fujioka K (2001) Sustained release of human growth hormone (hGH) from collagen film and evaluation of effect on wound healing in db/db mice. J Control Release 77(3):261–272

Maity B, Samanta S, Sarkar S, Alam S, Govindaraju T (2020) Injectable silk fibroin-based hydrogel for sustained insulin delivery in diabetic rats. ACS Appl Bio Mater 3:3544–3552. https://doi.org/10.1021/acsabm.0c00152

Mao J, Chen L, Cai Z, Qian S, Liu Z, Zhao B et al (2022) Advanced biomaterials for regulating polarization of macrophages in wound healing. Adv Funct Mater 32:2111003. https://doi.org/10.1002/adfm.202111003

Mehanny S, Abu-El Magd EE, Ibrahim M et al (2021) Extraction and characterization of nanocellulose from three types of palm residues. J Mater Res Technol 10:526–537

Miculescu F, Maidaniuc A, Voicu SI, Takur VK, Stan GE, Ciocan LT (2017a) Progress in hydroxyapatite– starch based sustainable biomaterials for biomedical bone substitution applications. ACS Sustain Chem Eng 5(10):8491–8512

Miculescu F, Mocanu AC, Dascalu CA et al (2017b) Facile ˇ synthesis and characterization of hydroxyapatite particles for high value nanocomposites and biomaterials. Vacuum 146:614–622

Müller-Herrmann S, Scheibel T (2015) Enzymatic degradation of films, particles, and nonwoven meshes made of a recombinant spider silk protein. ACS Biomater Sci Eng 1(4):247–259

Naderi P, Zarei M, Karbasi S, Salehi H (2020) Evaluation of the effects of keratin on physical, mechanical and biological properties of poly (3-hydroxybutyrate) electrospun scaffold: potential application in bone tissue engineering. Eur Polym J 124:109502. https://doi.org/10.1016/j.eurpolymj.2020.109502

Nethaji S, Sivasamy A, Tennarasu G, Saravanan S (2010) Adsorption of Malachite Green dye onto activated carbon derived from Borassus aethiopum fower biomass. J Hazard Mater 181(1–3): 271–280

Patel AD, Telalovic S, Bitter JH, Worrell E, Patel MK (2015) Analysis of sustainability metrics and application to the catalytic production of higher alcohols from ethanol. Catal Today 239:56–79

Pesode P, Barve S (2021a) Surface modifcation of titanium and titanium alloy by plasma electrolytic oxidation process for biomedical applications: a review. Mater Today Proc 46:594–602

Pesode PA, Barve SB (2021b) Recent advances on the antibacterial coating on titanium implant by micro-arc oxidation process. Mater Today Proc 47:5652–5662

Pesode P, Barve S (2022) Magnesium alloy for biomedical applications. In: Advanced materials for biomechanical applications. CRC Press, Boca Raton, pp 133–158

Pugliese R, Beltrami B, Regondi S, Lunetta C (2021) Polymeric biomaterials for 3D printing in medicine: an overview. Ann 3D Print Med 2:100011. https://doi.org/10.1016/j.stlm.2021.100011

Punj S, Singh J, Singh K (2021) Ceramic biomaterials: properties, state of the art and future prospective. Ceram Int 47:28059–28074. https://doi.org/10.1016/j.ceramint.2021.06.238

Rabiee N, Atarod M, Tavakolizadeh M, Asgari S, Rezaei M, Akhavan O et al (2022) Green metal-organic frameworks (MOFs) for biomedical applications. Microporous Mesoporous Mater 335: 111670. https://doi.org/10.1016/j.micromeso.2021.111670

Rana D, Colombani T, Saleh B, Mohammed HS, Annabi N, Bencherif SA (2023) Engineering injectable, biocompatible, and highly elastic bioadhesive cryogels. Mater Today Bio 19:100572. https://doi.org/10.1016/j.mtbio.2023.100572

Râpă M, Gaidău C, Stefan LM et al (2020) New nanofbers ˘ based on protein by-products with bioactive potential for tissue engineering. Materials 13(14):3149

Rezaeeyazdi M, Colombani T, Eggermont LJ, Bencherif SA (2022) Engineering hyaluronic acid-based cryogels for CD44-mediated breast tumor reconstruction. Mater Today Bio 13:100207. https://doi.org/10.1016/j.mtbio.2022.100207

Rodrigo-Navarro A, Sankaran S, Dalby MJ, del Campo A, SalmeronSanchez M (2021) Engineered living biomaterials. Nat Rev Mater 6:1175–1190. https://doi.org/10.1038/s41578-021-00350-8

Rogers ZJ, Zeevi MP, Koppes R, Bencherif SA (2020) Electroconductive hydrogels for tissue engineering: current status and future perspectives. Bioelectricity 2:279–292. https://doi.org/10.1089/bioe.2020.0025

Sadat-Shojai M, Khorasani MT, Jamshidi A (2015) 3 – Dimensional cell-laden nano-hydroxyapatite/protein hydrogels for bone regeneration applications. Mater Sci Eng C 49:835–843

Saeedi M, Vahidi O, Moghbeli M, Ahmadi S, Asadnia M, Akhavan O et al (2022) Customizing nano-chitosan for sustainable drug delivery. J Control Release 350:175–192. https://doi.org/10.1016/j.jconrel.2022.07.038

Saini M, Singh Y, Arora P, Arora V, Jain K (2015) Implant biomaterials: a comprehensive review. World J Clin Cases WJCC 3:52. https://doi.org/10.12998/wjcc.v3.i1.52

Salgado AJ, Coutinho OP, Reis RL (2004) Novel starch-based scaffolds for bone tissue engineering: cytotoxicity, cell culture, and protein expression. Tissue Eng 10(3–4):465–474

Samadian H, Zamiri S, Ehterami A et al (2020) Electrospun cellulose acetate/gelatin nanofbrous wound dressing containing berberine for diabetic foot ulcer healing: in vitro and in vivo studies. Sci Rep 10(1):8312–8312

Sayed Y, Aboulazm K, Fahmy A (2018) Evaluation of ultimate tensile strength and surface roughness of NiTi and CuNiTi orthodontic archwires after recycling procedures. Egypt Orthod J 54:47–57. https://doi.org/10.21608/eos.2018.77146

Seidi F, Yazdi MK, Jouyandeh M, Dominic M, Naeim H, Nezhad MN et al (2021) Chitosan-based blends for biomedical applications. Int J Biol Macromol 183:1818–1850. https://doi.org/10.1016/j.ijbiomac.2021.05.003

Servatan M, Zarrintaj P, Mahmodi G, Kim S-J, Ganjali MR, Saeb MR et al (2020) Zeolites in drug delivery: progress, challenges and opportunities. Drug Discov Today 25:642–656. https://doi.org/10.1016/j.drudis.2020.02.005

Shadish JA, DeForest CA (2020) Site-selective protein modification: from functionalized proteins to functional biomaterials. Matter 2:50–77. https://doi.org/10.1016/j.matt.2019.11.011

Shamshina JL, Zavgorodnya O, Choudhary H, Frye B, Newbury N, Rogers RD (2018) In search of stronger/cheaper chitin nanofbers through electrospinning of chitin–cellulose composites using an ionic liquid platform. ACS Sustain Chem Eng 6(11):14713–14722

Shamshina JL, Berton P, Rogers RD (2019) Advances in functional chitin materials: a review. ACS Sustain Chem Eng 7(7):6444–6457

Shokrani H, Shokrani A, Saeb MR (2022) Methods for biomaterials printing: a short review and perspective. Methods 206:1–7. https://doi.org/10.1016/j.ymeth.2022.07.016

Soman S, Ajitha A (2018) Life cycle assessment of metallic biomaterials. In: Fundamental biomaterials: metals. Woodhead Publishing, Duxford, pp 411–423

Sozer N, Nordlund E, Ercili-Cura D, Poutanen K (2017) Cereal side-streams as alternative protein sources. Cereal Foods World 62:132–137. https://doi.org/10.1094/CFW-62-4-0132

Stie MB, Kalouta K, Vetri V, Fodera V (2022) Protein materials as sustainable non-and minimally invasive strategies for biomedical applications. J Control Release 344:12–25

Takur VK, Voicu SI (2016) Recent advances in cellulose and chitosan based membranes for water purifcation: a concise review. Carbohydr Polym 146:148–165

Takur S, Govender PP, Mamo MA, Tamulevicius S, Takur VK (2017) Recent progress in gelatin hydrogel nanocomposites for water purifcation and beyond. Vacuum 146:396–408

Takur B, Barve S, Pesode P (2022) Magnesium-based nanocomposites for biomedical applications. In: Advanced materials for biomechanical applications. CRC Press, Boca Raton, pp 113–131

Tariq A, Arif ZU, Khalid MY et al (1970) Recent advances in the additive manufacturing of stimuli-responsive soft polymers. Adv Eng Mater

Thopegowda NB, Shenoy K, Shankarnarayana RK, Kukkila J, Vaddya SB, Gingipalli K (2013) Recycling of materials used in dentistry with reference to its economical and environmental aspects. Int J Health Rehabil Sci 2:140–145. https://doi.org/10.6049/kjjbydc.2016020160

Tipping A, Wolfe IR (2016) Options for follow-up and review of the trade-related elements of the 2030 Agenda for Sustainable Development. International Centre for Trade and Advances in Materials Science and Engineering 19 Sustainable Development, Winnipeg, Manitoba, Canada

Tripuraneni S, Namburi S (2008) Evaluation of genotoxicity of recycled Ni-Cr dental casting alloys: an in vitro study. J Appl Biomater Biomech 6:47–54

Vanapalli KR, Sharma HB, Ranjan VP et al (2021) Challenges and strategies for effective plastic waste management during and post COVID-19 pandemic. Sci Total Environ 750:Article ID 141514

Wang X, Chen C, Hu W, Ding A, Xu D, Zhou X (2005) Sorption of 243Am (III) to multiwall carbon nanotubes. Environ Sci Technol 39(8):2856–2860

Wang L, Zhang Q, Chen S et al (2014) Electrochemical sensing and biosensing platform based on biomass-derived macroporous carbon materials. Anal Chem 86(3):1414–1421

Wang Y, Katyal P, Montclare JK (2019) Protein-engineered functional materials. Adv Healthc Mater 8:1801374. https://doi.org/10.1002/adhm.201801374

Whitaker R, Hernaez-Estrada B, Hernandez RM, Santos-Vizcaino E, Spiller KL (2021) Immunomodulatory biomaterials for tissue repair. Chem Rev 121:11305–11335. https://doi.org/10.1021/acs.chemrev.0c00895

Whittall DR, Baker KV, Breitling R, Takano E (2020) Host systems for the production of recombinant spider silk. Trends Biotechnol S0167–7799(20):30244–30244. https://doi.org/10.1016/j.tibtech.2020.09.007

Wong THT, Louie JCY (2017) Te relationship between resistant starch and glycemic control: a review on current evidence and possible mechanisms. Starch/Staerke 69(7–8):Article ID 1600205

Wu XL, Wen T, Guo HL, Yang S, Wang X, Xu AW (2013) Biomass-derived sponge-like carbonaceous hydrogels and aerogels for supercapacitors. ACS Nano 7(4):3589–3597

Yadav D, Garg RK, Ahlawat A, Chhabra D (2020) 3D printable biomaterials for orthopedic implants: solution for sustainable and circular economy. Resour Policy 68:101767. https://doi.org/10.1016/j.resourpol.2020.101767

Yang J, Dahlstrom C, Edlund H, Lindman B, Norgren M (2019) pH-responsive cellulose–chitosan nanocomposite films with slow release of chitosan. Cellulose 26(6):3763–3776

Yao CH, Lee CY, Huang CH, Chen YS, Chen KY (2017) Novel bilayer wound dressing based on electrospun gelatin/keratin nanofibrous mats for skin wound repair. Mater Sci Eng C 79:533–540

Yazdi MK, Sajadi SM, Seidi F, Rabiee N, Fatahi Y, Rabiee M et al (2022) Clickable polysaccharides for biomedical applications: a comprehensive review. Prog Polym Sci 133:101590. https://doi.org/10.1016/j.progpolymsci.2022.101590

Zhao G, Li J, Ren X, Chen C, Wang X (2011a) Few-layered graphene oxide nanosheets as superior sorbents for heavy metal ion pollution management. Environ Sci Technol 45(24):10454–10462

Zhao G, Jiang L, He Y et al (2011b) Sulfonated graphene for persistent aromatic pollutant management. Adv Mater 23(34):3959–3963

Zhao X, Lu X, Tze WTY, Kim J, Wang P (2013) Cellulosic carbon fibers with branching carbon nanotubes for enhanced electrochemical activities for bioprocessing applications. ACS Appl Mater Interfaces 5(18):8853–8856

Zhu Y, Romain C, Williams CK (2016) Sustainable polymers from renewable resources. Nature 540:354–362. https://doi.org/10.1038/nature21001

Zhu M, Liu Y, Jiang F, Cao J, Kundu SC, Lu S (2020) Combined silk fibroin microneedles for insulin delivery. ACS Biomater Sci Eng 6:3422–3429. https://doi.org/10.1021/acsbiomaterials.0c00273

Zuin VG, Ramin LZ (2018) Green and sustainable separation of natural products from agro-industrial waste: challenges, potentialities, and perspectives on emerging approaches. Top Curr Chem Collect 376:229–282. https://doi.org/10.1007/978-3-319-90653-9_8

Dr. Ruby Srivastava, a theoretical physicist, is the principal investigator at CSIR-CCMB, Hyderabad, India. After fifteen years in teaching, she pursued her PhD in 2010 and started her research career with CSIR-IICT, Hyderabad, India. She has published several solo authored papers, solo review articles, fourteen book chapters, and edited many books. She has solo authored two books and co-authored one book project. One of her book chapter is selected in the Web of Science™ Core Collection (BKCI) by the Book Citation Index. Her research interests involve Artificial Intelligence (AI) applications in Drugs and Diseases and Reverse Vaccinology in mRNA Vaccines.

Prospects of Biodegradable Material: Sustainable and Patient-Centric Approach in the Realm of Biomedical Engineering

2

Dhanalekshmi Unnikrishnan Meenakshi, Alka Ahuja, Selvasudha Nandakumar, Lekshmi Salim, Chilaka Baburao, and Shah Alam Khan

Abstract

The creation of biodegradable materials that are both sustainable and patient-friendly has become essential in the current era of rapid industrialization and increasing environmental awareness. In the domain of healthcare, science and technology, collaborations have led to significant medical science advancements, resulting in the development of numerous life-improving, life-enhancing, and life-saving biomedical engineering innovations. Biodegradable materials are required for fostering environmentally friendly options and therefore biodegradable materials have a promising sustainable future. Biodegradable materials hold great promise as environmentally friendly materials with a wide range of applications. They are particularly advantageous in patient-centered implanted biomedical devices because of their exceptional and distinctive qualities, which include hydrophilicity, bioactivity, renewability, biocompatibility, and biodegradability. This chapter explores a few of the main fields, such as tissue engineering and drug delivery systems, where biodegradable materials are having an enormous impact. It also highlights the technological breakthroughs in biodegradable materials, which combine stimuli responsive materials with 3D and 4D printing to create dynamic, patient-specific biodegradable materials. It also briefly addresses the current trend and instances of biodegradable materials-based sensors for medical therapeutic, diagnostic, and body monitoring applications. The major technological obstacles to the clinical use of biodegradable materials have also been covered. Despite the promising prospects, the clinical utilization

D. U. Meenakshi (✉) · A. Ahuja · L. Salim · S. A. Khan (✉)
College of Pharmacy, National University of Science and Technology, Muscat, Oman
e-mail: dhanalekshmi@nu.edu.om; shahalam@nu.edu.om

S. Nandakumar · C. Baburao
Department of Biotechnology, Pondicherry University, Puducherry, India

R. Malviya, S. Sundram (eds.), *Sustainable Green Biomaterials As Drug Delivery Systems*, Biomaterials, Bioengineering and Sustainability 1,
https://doi.org/10.1007/978-3-031-79062-1_2

of biodegradable materials faces significant challenges such as rising prices, aging populations, and a high burden of chronic illness linked to health-related activities. To address these challenges, healthcare must completely change to become patient-centered, preventive, evidence based, and proactive with an emphasis on enhancing quality of life. Overall, this chapter provides a comprehensive overview of the scientific advancement, challenges, and promising directions for the development of next generations of patient-centered and sustainable biodegradable materials.

Keywords

Biodegradable · Drug delivery · Biomedical engineering · Patient centric · Sustainable

2.1 Introduction

Rapid industrialization and heightened environmental consciousness in recent years lead to the development of tailored biodegradable/sustainable materials that prioritize patient needs, ensuring optimal therapeutic outcomes while minimizing side effects.

Pharmaceutical industries and health workers are focusing on the advancement of patient oriented biomedical materials (Mohd et al. 2023). In recent years, biomaterials have shown great potential to revolutionize the biomedical field by promoting sustainability and patient centered care. Additionally, the biodegradable materials help in patient centric approach by reducing the adverse effects with long term implantation. Unlike biodegradable implants, permanent implants need to be removed through surgery. This surgical protocol is not convenient to the patients, and it also carries the risk of infection or rejection with multiple unpredictable complications. Hence, biodegradable materials provide a convenient and comfortable experience to the patients without the need for follow up surgeries and related complications arising out of it (Abyzova et al. 2023). Physiologically, biodegradable implants interact together with the process of tissue repair and regeneration, offering the best possible support for the healing process and perhaps shortening recovery periods. Biodegradable scaffolds play a major role in the field of tissue engineering and regenerative medicine by providing a basic framework for cell adherence, proliferation, differentiation, and regeneration, as most of their components resemble extracellular matrix (ECM) components, as discussed elsewhere in this chapter (Olteanu et al. 2024). Biodegradable materials offer a promising solution to reduce our environmental impact and have a bright future ahead if the demand for sustainable biomaterials continues to rise. Their exceptional and distinguished properties, such as hydrophilicity, renewability, bioactivity, biodegradability, and biocompatibility, make them appropriate for patient-centered implanted biomedical devices. Some polymers, such as polyesters, polyanhydrides, and polyurethanes, are engineered to break down drugs in a controlled manner by interacting with biological matters. Some encompass copolymers of poly (lactic acid), poly(-ε-caprolactone), and poly (glycolic acid) (Duoyi et al. 2021). As their derivation is from petroleum and other non-renewable resources, such synthetic polymers are non-biodegradable. Biodegradable materials have also seen scientific and technological advancements, permitting for the design of patient-specific, dynamic materials through the integration of stimulus responsive materials with 3D and 4D printing and is briefly covered in this chapter. From orthopedic implants to cardiovascular stents, the use of biodegradable polymers is revolutionizing the field of medical device engineering. These polymers offer a biologically compatible alternative to traditional materials such as metals and ceramics, reducing the risk of device-related complications and enabling better long-term outcomes for patients (Fig. 2.1).

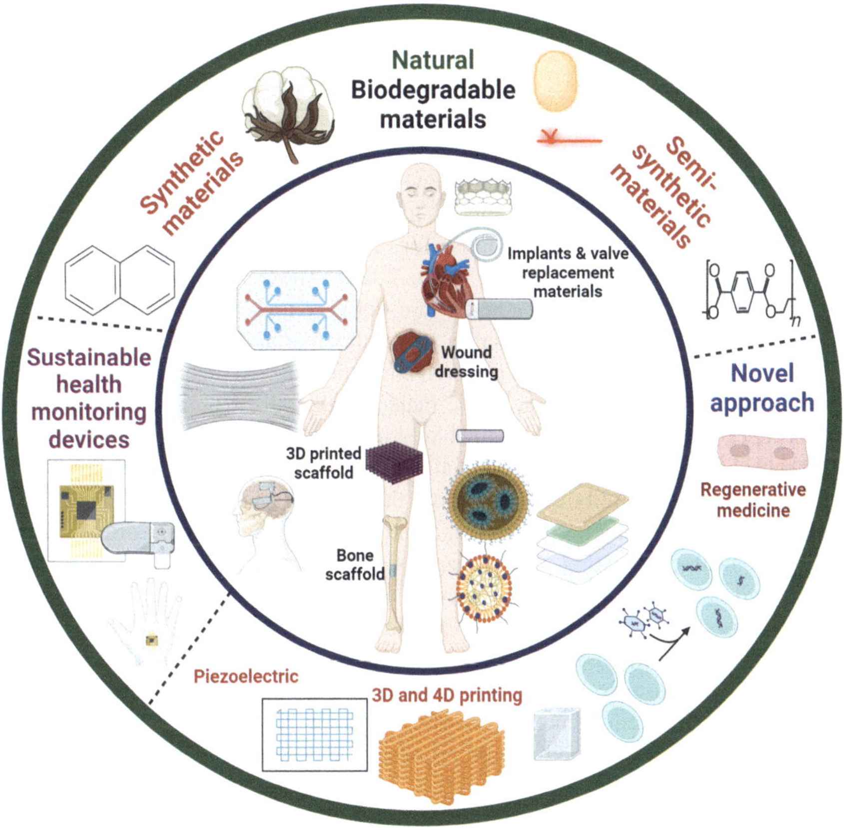

Fig. 2.1 A patient-centric approach utilizing biodegradable polymers involves the development of sustainable health monitoring/medical devices or drug delivery systems that prioritize patient comfort, safety, and convenience. By incorporating biodegradable polymers, these devices can be designed to gradually degrade within the body, reducing the need for additional procedures for removal. This approach not only enhances patient compliance and overall experience but also aligns with sustainable and environmentally friendly practices in healthcare

This chapter also discusses the prominent impact of biodegradable materials in the field of tissue engineering and drug delivery systems. Researchers can tailor the composition, degradation kinetics, and mechanical properties of these materials to optimize performance for various therapeutic purposes. From drug delivery systems capable of controlled release to implantable devices engineered for specific anatomical sites, the adaptability of biodegradable materials opens new avenues for personalized medicine and targeted interventions (Trucillo 2024). Nano particles with large surface-area-to-volume ratio and small sizes can be engineered to tailor drug administration, like matrix stability, molecular solubility, and specific functionalities (like optical, electrical, magnetic). Novel smart device designs in biomedical applications like *in vivo* imaging, molecular targeting, bio-sensing, gene delivery, artificial implants, and cancer therapy, offer considerable benefits for environment, industry, and society via eco-friendly, sustainable approaches, promoting healthier lives and a cleaner world.

In the field of tissue engineering, biodegradable polymers serve as scaffolds for the regeneration of functional tissues and organs as mentioned before. They also mimic the cellular environment and provide support for cell adhesion, multiplication, growth, and differentiation. After some days, these scaffolds degrade by internalization and are replaced by the newly formed cells without the mark of any differentiation and thus help biologically to restore the organ function effectively. This approach holds significant promise for organ transplantation, wound healing, and cartilage regeneration (Olteanu et al. 2024). These materials also hold the potential to significantly change healthcare delivery and medical practice in the coming years because they offer safer, more effective and eco-friendly treatment options to the patients. Continued research and innovation in this field could unlock the limitless potential of biodegradable materials to address unmet medical needs and elevate patient outcomes. Sustainable biomaterials are always compatible with tissue environment, biodegradable and allow faster resorption (Trinath et al. 2020).

This chapter also briefly explains about the benefits and features of sustainable biomaterials based on their composition, sources and applications in drug delivery and biomedical engineering across various medical specialties, such as wound healing and cancer treatment. The present trend and examples of biodegradable materials-based on sensors for therapeutic, diagnostic, and body monitoring applications is also discussed. The potential impact of sustainable biomaterials on healthcare is immense, as they hold the promise of personalized treatments that cater to individual patient requirements, providing hope for improved treatment effectiveness, reduced side effects, and a more promising outlook for medical care.

2.2 Sustainable Green Biomaterials

Due to the unique properties sustainable biomaterial are being efficiently employed in technologies like tissue engineering and drug delivery. Tissue engineering is a discipline which work with combination of engineering, materials, and cells to restore, maintain, improve or replace different type of biological tissues. In short, tissue engineering develops biological substitutes. Scientists have investigated renewable and organic material alternatives to embark upon this problem. In recent years, "bio-sustainability" has gained attention in pharmaceutical, and biomedical research. Developing "green," sustainable materials that perform as well as a traditional ones is being the great challenge. During novel product designing, green biomaterial ensures stability, quality, efficacy, and safety. Reducing the polymer sector's environmental effect can aid accomplish today's greater sustainable development goals. Scientists are constantly searching for better ways to create sustainable polymers, which opens the door to green materials derived from biomass or other natural sources.

2.2.1 Naturally Derived Polymers and Polymer Substrates

Scientists have focused on both animal and plant sources of biopolymers, incorporating chitosan, hyaluronic acid, collagen, silk, alginates, arabic gum, and dextrans (Liu et al. 2022a; Hong et al. 2020; Sun et al. 2020). However, additional processing methods like extraction and purification are generally required for these materials, which can augment costs and decrease yield. Cellulose, starch, hemicellulose, keratin, lignin, and silk fibroin are few examples of renewable resource-based bio-sustainable polymers that have demonstrated capabilities based on biodegradability, affordability, and biocompatibility. Such biopolymers are quite cost-effective, and eco-friendly as they are made from biomasses that are derived from industrial or agricultural wastes (Liu et al. 2022a). Biomaterials are used in areas like skin tissue engineering (skin burns, wound healing etc.), bladder tissue engineering, cardiovascular tissue engineering and scaffolding. Natural biomaterials are of two types: polysaccharide nature (chitosan, alginate, carrageen, cellulose, pectin, agarose, and hyaluronic acid) and protein nature (collagen, gelatin, silk, keratin, and fibrin). Due to their enhanced performance, functionality, and reproducibility compared to natural biomaterials, synthetic biomaterials have attracted researchers for use in various technologies. Examples of such materials are Poly lactic acid (PLA), Poly lactic-co-glycolic acid (PLGA), Poly-p-dioxanone (PDS), Poly glycolic acid (PGA), Poly (ethylene succinate) (PES), Poly (butylene succinate) (PBS). Utilizing supercritical fluids for polymer dissolution without using organic solvents is a greener technology with considerable environmental and medicinal benefits. Different research groups are evaluating supercritical fluid methods (SCF) to create pharmacological carriers with nutraceutical (e.g., vitamins, CoQ10) or inorganic compounds of growing interest (Kankala et al. 2021). These methods are ideal for alginates, and chitosan, because they do not negatively influence biocompatibility, and hydrophilicity, which are significant to control molecular release processes. Diffusion processes mediated by matrix characteristics can specifically target drugs like doxorubicin into exact tissue sites, reducing collateral effects during chemotherapy (Martău et al. 2019). Plant-derived extracts suggest anti-inflammatory characteristics and are a greener approach for synthesizing copper, silver, and gold nano particles (Thiruvengadam et al. 2019).

2.2.1.1 Protein Based Sustainable Biomaterials

Collagen, gelatin and keratin are protein-based biomaterials used in drug delivery and tissue engineering. Collagen molecule constitutes three polypeptide chains, entwined like a three-stranded rope. There is a unique twist in the opposite direction in every chain. The main factor influencing the development of a helix is the amount of glycine and amino acid residues. The primary forces holding the strands together are covalent bonds and hydrogen bonding between neighbouring CO and NH molecules. The fundamental collagen molecule is a rod-like molecule with a molecular weight of around 300 kDa and dimensions of roughly 3000 and 15 A, respectively. Glycine, proline, and hydroxyproline are the three main amino acids found in the triple helix structure. Glycine, proline, and X—any amino acid—make up the pattern. Amino acids cause the collagen to perform certain tasks. The helical structure is held together by hydrogen bonds, which connect peptide bonds (Shenoy et al. 2022). In the field of drug delivery to encourage the creation of bones, collagen film and matrix were used as gene delivery vehicles. To track the growth of bones and changes in carrier collagen's absorbent state, a composite of collagen and recombinant human bone morphogenetic protein 2 (rhBMP-2) was developed as a patient centric approach. Whereas collagen alone did not result in any bone development, the rhBMP-2/collagen only implant caused active bone production. Collagen matrix that was loaded with bone morphogenetic protein (BMP) and near osteogenic cells resulted in direct osteo induction, however cartilage formation was not triggered and hence raised the concern towards patient benefits (Khan and Khan 2013). In the field of tissue engineering hydrogel was made with recombinant human collagen and carboxylated chitosan using 1 ethyl-3 (3-dimetyl) aminopropyl carbodiimide as a crosslinking agent. Hydrogel tested on Sprague Dawley rats for gene expression on extracellular membrane- associated proteins were measured by qRT-PCR and for wound healing capacity as a prospect of sustainable biomaterials in the clinical practice. This study revealed the expression of collagen 1, collagen 3 and integrin and hence the effectiveness of sustainable biomaterial. That means developed hydrogel has good wound healing capacity with creation of environment for adhesion, spreading, migration and proliferation of fibroblast cells (Yang et al. 2021). Recently, technology suppliers have created a variety of customized (patient centric approach) wound dressings intended especially for the treatment of chronic wounds, demonstrating the medicinal application of various extracted sustainable collagens. They hope that rate of chronic wound closure will rise and clinical results will be much improved by these customised dressings (Zhao et al. 2023).

Bioengineered silk, synthesized using recombinant systems in several organisms, recommends a sustainable alternative to traditional silk harvesting. Functioning of silk spidroin proteins with peptides such as DOX and H2.1 are used to design targeted drug delivery systems. For instance, composite silk spidroin spheres can deliver cancer medication especially to breast cancer cells expressing the HER2 receptor (Maity et al. 2020). Turkey tendon collagen, a squander product from the chicken industry, can be processed like sponge scaffolds that could release medicines such as prilocaine hydrochloride. Gelatin, a derivative from collagen, can be synthesized as drug delivery hydrogels, such as those utilized to deliver anti-inflammatory drugs to knee joints to treat osteoarthritis. Vancomycin-loaded salecan/gelatin composites have also shown promise in reducing bacterial co-culture populations like *Staphylococcus aureus*, and *E. coli* (Zhu et al. 2020).

Zein, a protein present in maize, is another biodegradable material that can be used for encapsulating active substances incorporating micronutrients or propolis (Jaski et al. 2022). Silk fibroin, derived from arachnids and insects, is also of great interest because of its mechanical strength, and biocompatibility. Byssal threads derived from shellfish, and mussel-inspired polymers have been utilized in several applications like tissue engineering, nanomedicine, and adhesives (Vineis et al. 2021). Byssal threads, like silkworm silk, are collagen fibres engrossed in polyphenolic resin (3,4-dihydroxy-L-phenylalanine or L-DOPA). Several molecules, including polydopamine, necessitate L-Dopa as an intermediary. Pheomelanin, and emulelanin can be obtained by mimicking its oxidative melanin polymerization (Cavallini et al. 2020). Recombinant peptides synthesized from methacrylated gelatins, and human type I collagen have been utilized to design hydrogels that take off the extracellular matrix and carry stem cells. Marine-derived collagen from codfish, eel, and tilapia skin is structurally akin to mammalian collagen and functions as a promising protein-based biomaterial (Dias et al. 2020).

Due to the use of mammalian gelatin as a protein source for agri food and pharmaceutical applications, protein-based nano fibers are gaining attention. Gelatin-based nano fibers activate cell adhesion leading to binding motifs including Arg-Gly-Asp (RGD) to create tissue-native integrins, used as a biocompatible in a few *in vitro* studies (Sanchez et al. 2022). However, there are challenges in designing efficient and sustainable protein-based biomaterials, including intricacies in clinical translation, and batch-to-batch variance. Future research should focus on addressing these challenges, delving into additional protein sources, and establishing novel material production approaches. Overall, protein biomaterials are the probable materials to improve tissue engineering, and drug delivery, leading to better patient outcomes and reduced medical expenses.

2.2.1.2 Polysaccharides Based Sustainable Biomaterials

Both chitin as such and its derivative, chitosan, represent a unique class of polysaccharides due to their distinctive properties and application potential. Chitin, of which chitosan is derived, is a polysaccharide found in the exoskeleton of most crustaceans and the cuticles of insects as well as in the cell wall of algae and fungi. Chitosan, a viable polysaccharide, is soluble in an acidic environment owing to the protonation of primary amine groups. The chitin and its deacetylated form, chitosan, are linear polysaccharides formed from various linear segments of linked ($\beta 1 \rightarrow 4$) residue units of N-acetyl-2-amino-2-deoxy-D-glucose and 2-amino- 2-deoxy-D-glucose, as such. In contrast, they have a high number of acetylated units, and chitin presents itself as insoluble in acid (Aranaz et al. 2021).

Being the only polycation in nature, chitosan's charge density is dependent on the degree of acetylation and pH of the surrounding solution. The former also remains one of the factors determining chitosan's solubility alongside the other attached molecular ion, namely, the molecular weight. Chitosan oligomers are soluble in a wide range of pH. Regardless, voluminous chitosan exhibits suitable solubility only in acidic solutions, restricting its applications to such; the majority of physiological media are either weakly acidic or neutral (Aranaz et al. 2021). Furthermore, while most chitosan currently produced has been created through chitosan deacetylation, there's been a growing interest in various other types. Chitosan from fungi and insects has attracted attention due to the increased demand for vegan friendly products as well as the ability to control the molecule's viscosity via the use of a different biosynthesis method (Huq et al. 2022). Biological and technological properties of chitin and chitosan involves the following: mucoadhesive activity, an anti-inflammatory, antioxidant, anti-microbial, anti-fungal, anti-hyperglycemic, anti-tumoral, wound healing effect, and others; some of these properties largely depend on physicochemical features of polymers such as molecular weight and acetylation degree. Chitosan science is a rapidly expanding area with increased research opportunities and commercial potential. It may represent one of the most promising research areas for further development of biotechnological polymers with implications in bio catalysis, drug delivery systems, and other fields (Harugade et al. 2023). A simple freeze-drying method designed a 3D macro porous from bacterial cellulose, used in drug screening and in-vitro cancer biology.

Good biocompatibility, minimal immunogenicity, and biomimetic scaffolding were demonstrated in a study of composite constructed from nanocellulose and nanochitin hydrogels for bone tissue engineering (Torres et al. 2015). The most universal biopolymer, cellulose, has multiple uses and may form the basis of nano-structured materials like nanocrystals and nanofibers (Zuppolini et al. 2022). The second most general biopolymer, lignin, is convenient in polymer and composite materials and is recurrently available in fiber waste products from natural processes. The overarching objective is to spot sustainable biomaterials that do not impair humans or other animals, are generally commercially feasible, and do not exhaust natural resources. The use of these materials may appreciably vary the landscape of drugs in numerous ways, including drug delivery methods. Electrospun cellulose fibre meshes were manufactured for use in bone tissue engineering. These meshes have a high ability to retain human recombinant bone morphogenic protein-2 and maintain their structural and mechanical integrity for a period of seven days (Filion et al. 2011). Another study demonstrated the controlled release of antibacterial medicines along with good biocompatibility and non-toxicity when ethyl hydroxy ethyl cellulose (vinyl alcohol) nanofibers were created for tissue engineering and drug delivery (Wali et al. 2018). With the development of gelatin carboxy methyl-cellulose hydrogels, scientists worked to apply their application on engineering vascularized and cell dense 3D tissues organs. This resulted in properties that are a suitable candidate for the quick preparation of perfusable vascular networks, cytocompatibility, and a sustainable microenvironment for angiogenesis (Kageyama et al. 2016). By imitating natural tissues and structures, biological materials are greatly advancing in medical technology and improving their compatibility with the human body for a patient centric approach (Trucillo 2024).

2.2.1.3 Hydroxyapatite Based Sustainable Biomaterials

Hydroxyapatite is a calcium phosphate derivative. It is considered most suitable component for bone engineering as more than half of the bone itself is made of compounds resembling hydroxyapatite (Fernando et al. 2016). Hydroxyapatite ceramics infused composites show physical properties like that of bone (Ohtsuki 2008). Hence many studies reported improved bone formation and repair. A randomized clinical trial by Resende et al. (2019), carbonated hydroxyapatite was shown to be a better alveolar socket preservation before the dental implantation, with a higher degradation rate compared to current commercial xenografts. The hydroxyapatite nanocomposites are even shown to improve the bone tumor therapy, even the malignant (Zhang et al. 2023). The spin coating method was used to create the starch-hydroxyapatite thin film components, which have demonstrated exceptional tensile strength, elongation, and modulus when applied to skin. Additionally, the film is biodegradable and exhibits good biocompatibility with human blood lymphocyte cells (Pramanik et al. 2019).

2.2.1.4 Carbonaceous Sustainable Biomaterials

Carbon materials are gaining a lot of interest as a rapidly emerging class of high-performance materials in various scientific and technical domains for a variety of biomedical uses, including prosthetic joints, scaffold building, bioimaging, implantation, tissue engineering, and bone engineering. They are ideal because of their exceptional biological characteristics. Carbon quantum dots (CQDs) and their composites, carbon nanotubes (CNTs) and their composites, carbon nanofiber (CNF) reinforced composites, and graphene and its composites are examples of carbonaceous materials (Kim et al. 2024). Doxorubicin has been reported to be able to bind to SWCNT surfaces. This work demonstrated that 1 g of SWCNTs could encapsulate 4 g of doxorubicin (DOX) on their surface; the drug delivery method including loaded DOX is dependent on pH balance. Subsequent research by the same group has introduced DOX loaded on SWCNT-PEG, which is less harmful to mice and has good antitumor effects when administered intravenously (Liu et al. 2007, 2009). These strategies have multiple benefits, such as increased medication efficacy, less adverse effects, and extended release for long-term illnesses thereby increasing health outcomes (Trucillo 2024).

2.2.2 Synthetic Polymer Substrate

Even though natural polymers provide good scaffolds promoting cell proliferation and adhesion, they still do not check all the criteria like the immunogenicity or pathogen transmission and henceforth the clinical outcomes. Synthetic polymers were derived with no immunogenicity, controlled structure and to mimic the extra cellular matrix (Magnusson et al. 2011). Different types of synthetic polymers are used according to their properties: Aliphatic polymers like PLA (Poly lactic acid), PGA (Poly glycolide), Polyanhydrides, Polyphosphazenes, Polyurethanes, Polyglycerol sebacate. PLGA is the polymerised product of PLA and PGA, widely used for artificial bone-substitution and drug delivery (Duoyi et al. 2021). Though it is advantageous than other polymers, its hydrophobicity limited the cell adhesion. Therefore, substance like fibronectin, vitronectin were added to enhance the cell adhesion (Rentsch et al. 2012). The capacity to modify drug-release patterns has changed patient care by boosting therapeutic efficacy, decreasing adverse effects, and enhancing patient compliance. Synthetic polymer substrate will surely continue to be at the forefront of innovation in pharmaceutical research as it develops, offering adaptable answers to challenging drug delivery problems and advancing medical science. Bio-sustainable materials, like polyhydroxyalkanoates (PHAs), and polylactides (PLA) are obtained from renewable sources and have characteristics that make them appropriate for nanomedicine, packaging, controlled drug delivery and tissue engineering scaffolds (Coiai et al. 2021). PLA, is recyclable, cheap, and non-toxic. PHAs, in general, are fully biodegradable, making them epitome for the circular economy (Elmowafy et al. 2019). Among the commercially available PHAs, poly(3-hydroxybutyrate-co-3-hydroxyvalerate) (PHBV) is particularly versatile and is utilized in several implant devices (like grafts, stents, patches, pins, etc.,), tissue engineering scaffolds, and controlled drug release. In general, these biodegradable materials offer a range of possibilities for controlled release systems, and sustainability as well as usages in various industries (Pramanik 2023). In a patient-centered approach, the combination of quercetin (QCT) and gemcitabine (GMC) has shown a synergistic effect in inhibiting the migration of pancreatic cancer cells. This treatment strategy involves loading GMC and QCT together into biodegradable nanoparticles made of poly (lactic-co-glycolic acid) and enhanced with hyaluronic acid (HA; PPHA NPs). These nanoparticles are designed to specifically target cancer cells by interacting with the overexpressed CD44 receptor. By utilizing this targeted approach, patients may benefit from a more effective and precise treatment that aims to combat pancreatic cancer cell migration while minimizing potential side effects on healthy tissues. (Serri et al. 2019).

2.2.3 Biodegradable Metal Substrates

Metals and their alloys are used to make bone implants and using biodegradable metals will reduce the invasiveness of the procedure since after implantation they degrade in the body and does not require removal (Dorozhkin 2015). They must be biocompatible, light and corrosion resistance mainly. Magnesium, Iron, Manganese, Calcium, Zinc are most used and these aid in metabolic functions inside the body and has much advantageous properties than stainless steel, which is currently used in used medicine for implantable devices (Patricia et al. 2023). In considering patient care, the development of alloys such as Zn-Cu-Mg, which exhibit enhanced biocompatibility and controlled biodegradability, is crucial. It is essential to monitor the degradation of these alloys carefully, as rapid degradation could potentially elevate metal concentrations above safe levels, leading to toxicity risks for patients. This highlights the importance of selecting materials that not only support patient health but also ensure their safety throughout the treatment process (Zibo et al. 2017).

2.2.4 Encapsulants and Adhesives Biodegradable Electronic and Sensing Devices

In vivo sensors and monitoring devices have been developed over decades, but their biocompatibility remains a concern. Biodegradable materials like dielectrics, semiconductors and conductors are being currently employed to fabricate devices (Hosseini et al. 2021). Green electronics are being developed to promote environmental sustainability using carbon-based frames (Min et al. 2023). Recently, elasticity has emerged a major criterion to make electronics. Polyester elastomers are mainly used in tissue engineering and drug delivery which can mimic the elastin properties. Its biodegradability is different than other biopolymers which undergo bulk degradation.

Conventional biomedical methods are fast evolving to slot in green technology, resulting in bio-inspired materials in alternative to animal-based goods for bio-based leather and food (Zuppolini et al. 2022). Research has focused on creating micro- or nanofibers for cell growth, studying the effects of released chemicals (Malviya et al. 2021). Pre-clinical studies used FDA-approved biodegradable nanofibers to look at the influence of 5-azacytidine (5-AZA) on muscle gene up regulation (Fasolino et al. 2017). Grafting of cellulose nano fibers (T-CNF) with soy protein were designed using a TEMPO-mediated oxidation process, a biodegradable one (Abdul Khodir et al. 2022).

The polyester elastomers undergo linear mass loss through surface degradation leading to retain its strength while degradability and achieving sustained release (Chen et al. 2022). Furthermore, they improve adherence and decrease the frequency of dose, which helps with patient compliance issues, especially in chronic illnesses where frequent dosing can be stressful. Therefore, by providing tunable features that allow for exact control over drug encapsulation and release, these biomaterials may change the landscape of drug delivery in near future. By creating new fields such as scaffolds, thin films, hydrogels, nanocomposites, cryogels, drug delivery systems, etc., and using them to treat patients in different ways, biomaterials transformed tissue engineering and drug delivery towards patient centric approach. The ease of administration and painlessness of biomaterials may soon cause them to opt for these novel drug delivery systems. Another promising development in biomaterials is the development of 3D printing technologies as discussed elsewhere in the chapter, will enable the production of biological substitutes for patient application in a variety of treatments, including ready-made biological tissues. The less procedure time and the need for personalization in patient care can be met with these new strategies.

2.3 Regenerative Medicine Applications of Smart Materials Towards Patient-Oriented Approach

Biodegradable polymers are now acknowledged as a sustainable and versatile group of materials, with important implications for regenerative medicine when integrated into 4D printing technology. The use of these polymers, designed to naturally degrade, offers numerous applications that are greatly reshaping the realms of tissue engineering, regenerative treatments, and patient-focused healthcare. By utilizing the body's innate healing abilities, regenerative medicine has enormous potential to treat a wide range of diseases. Smart materials are essential for the advancement of regenerative medicine applications because they provide novel approaches to tissue creation, medication delivery, and individualized treatment plans (Zhang et al. 2022). Sustainable biomaterials 3D printing has become a game changer, providing a patient centric approach to tailored treatment plans. Environmentally friendly and sustainable biomaterials promote tissue regeneration by adhering to the guidelines of patient centered care approach (Murphy and Atala 2014). 3D printing scaffolds and implants are designed by using sustainable biomaterials to tailor each patient's needs. Better patient outcome and therapeutic response can be achieved by creating these specialized concepts that are tailored to the patient therapeutic and anatomical requirements.

Additionally, the use of sustainable biomaterials in 3D printing reduces the reliance on traditional materials that may have adverse effects on patient health and the environment. 3D printing has revolutionized design and manufacturing sector by facilitating unprecedented design freedom and efficiency (Zhou et al. 2015). The commencement of 4D printing has made this method a step forward by letting the user to design program in a specific way for the thing to function. Stimulus-responsive materials and shape memory polymers allow 4D printing a promising one to create adaptive equipment, and buildings, and biomedical devices. 4D printing incorporates self-assembling smart materials that can change according to their surroundings and combine the best of 3D printing along with them. Various industries, including regenerative medicine and healthcare, stand to gain largely from this 4D printing technology (Sheikh et al. 2023).

In healthcare, 3D printing has previously made noteworthy progressions in tissue engineering, and medical device manufacturing. Nevertheless, the integration of 4D printing takes this evolution to a whole new level. Through the addition of a temporal dimension to 3D-printed structures, smart materials can act in response to stimuli like humidity, temperature, light, and mechanical forces, permitting the printed structures to change shape, self-organize, and carry out dynamic activities (Kantaros 2022).

In the field of medical device manufacturing and tissue engineering, 3D printing has achieved a noteworthy progression. Integrating 4D printing into 3D technique get hold of these evolution to a next level. Adding a temporal dimensional structure to 3D-printed materials, and smart materials will result in a greater response to external stimuli such as temperature, humidity, mechanical forces, and light, will make possible the printed structures to self-organize, alter shape, and perform dynamic activities (Kantaros 2022).

Shape-modifying characteristics is the major in 4D printing process, achieved through designing and establishing intelligent smart materials. The designed smart materials can either be shape memory or stimulus-responsive, in which the smart materials can reinstate to a specific shape during stimulation or undergo reversible alterations in their qualities or form during external trigger, respectively (Sheikh et al. 2023). Shape memory polymers (SMPs), liquid crystal elastomers (LCEs), magnetic shape memory alloys (MSMAs), biodegradable polymers, and cell-laden biomaterials are among the most commonly used raw materials in the 4D printing process. They are also employed in the production of complex 3D structures that function in response to external stimuli and they have shape modifying properties (Wu et al. 2018). When 3D printed temporary implants come into touch with heat stimuli or additives like SMPs, they can help them regain their original shape (Prevot et al. 2018). LCEs can be utilized to print complex structures that replicate real tissue procedures, while MSMAs can be utilized in magnetic field-responsive devices, and implants (Zhang et al. 2022).

In addition, the employment of biodegradable polymers in 3D and 4D printing lets for the design of drug delivery systems, and tissue scaffolds, and temporary supports, that liquefy as new tissue grows, reducing the requirement for additional measures to take out implants (Wu et al. 2016). The usage of cell-laden materials in 4D printing grasp enormous promise for designing bio-mimetic tissue structures, highly functional that can progress and change over time, mimicking the human body's natural curative processes. In general, 4D printing usage in regenerative medicine recommends highly developed alternatives for tissue scaffolds that strongly imitate real tissue dynamics and support cell growth, proliferation, and differentiation. It also facilitates the exact release of therapeutic drug compounds during physiological changes, resulting in enhanced regeneration therapy and optimized outcomes. Also, 4D printing lets for the customization of patient-specific implant, plummeting complications and enhancing long-term results. Employing smart materials in 4D printing progresses adaptability, precision, and personalization of printed structures, thereby boosting the field of regenerative medicine (Kantaros and Piromalis 2021; Mathur et al. 2023; Qu et al. 2023). Sustainable 3D printed biomaterials have enormous potential to transform healthcare applications towards a patient-centric model. Eco friendly plant-based polymers are used to produce customized 3D printing implants and medical equipment. This approach not only improves treatment results but also complies with environmental responsibility and sustainability standards (Goyanes et al. 2015).

2.4 Bio-degradable Piezoelectrics for Medical Implants

Advancements in implanted medical devices and the internet of things (IoT) in recent years lead diagnostics to patient centric approach rather than hospital-based approach. Piezoelectric materials were initially used in electromechanical systems due to their ability to generate electrical charges and strain when subjected to mechanical stress. These same characteristics have since made them influential in the field of biomedical engineering. Nebulizers, implanted therapeutic devices, hand pieces and ultrasonic actuators are examples of piezoelectric materials that work under the same principle (Zaszczyńska et al. 2020). Piezoelectric materials were then influenced by the integration of biodegradable materials and this integration was considered as a technological revolution by biomedical scientists (Kamel 2022).

Biodegradable piezoelectric materials hold the advantage of mimicking the tissue environmental characteristics as per the requirements of individual patient and hence they are considered better for therapeutic applications with less adverse reactions. This feature enables bio-sensing capabilities, real-time energy harvesting, and therapeutic effects for a range of disorders. By leveraging these materials, healthcare providers can tailor treatment strategies to individual patients, promoting personalized care and targeted interventions. The application of piezoelectric materials in biomedical implants also influences various physiological cell functions, including proliferation, differentiation, migration, and apoptosis. These adjustments play a crucial role in tissue regeneration and repair, ultimately enhancing the overall well-being and recovery of patients. This patient-centric approach underscores the importance of utilizing cutting-edge technologies to deliver more effective and personalized healthcare solutions.

Piezoelectric reactions occur in different tissues like tendon, bone, cartilage, ligament, dentin, skin, cornea, and sclera. Living cells own intrinsic piezo electricity because of proteins. Comparison to quartz, bone has a higher piezoelectric constant. Collagen elimination removes bone piezoelectricity, while demineralization conserves it. Flexoelectricity of hydroxyapatite (HA) ceramic surfaces enhances bone healing (Vasquez et al. 2018). Electrical stimulation has enduring osteogenic effects on mesenchymal stem cells. Low-intensity electrical current promotes collagen organization and angiogenesis in bone grafts, enhancing osteo-conductivity and graft growth (Fonseca et al. 2019). Electrical stimulation influences pre-osteoblasts and macrophages, rising osteogenesis (Srirussamee et al. 2019). Piezoelectricity has a significant role in bone regeneration by triggering signaling pathways, re-organizing dipoles, opening calcium channels, and enhancing bone healing at different stages (Jacob et al. 2018). Piezoelectricity triggers calcineurin, which de-phosphorylates nuclear factor of activated T-cells (NF-AT). The promotion of bone healing is done using translation of growth factors like bone morphogenetic protein (BMP) and transforming growth factor β (TGF β) (Goonoo and Luximon 2022).

PLLA is a biodegradable piezoelectric polymer, that is utilized in tissue engineering. A fiber-based scaffold based on piezoelectric PLLA enhances stem cell differentiation into osteoblasts in *in vitro* and bone production *in vivo*. The piezoelectric charge on the scaffold's surface can be provoked by ultrasonic waves to augment bone repair and osteogenesis. PLLA piezoelectric nanofibers promote cartilage regeneration. These nano-fiber tissue scaffolds in *in-vivo* self-stimulate TGF-signaling pathways, cell migration, and chondrogenesis required for cartilage repair (Liu et al. 2022b).

Electrospun hybrid membrane comprising glass-reinforced hydroxyapatite granules (gHA) and PLLA polymer is used for bone regeneration (Santos et al. 2017). It showed enhanced osteoblastic cell activity and bone-bonding than pure PLLA substrates. PLLA nanotubes functioned as a bio-soft piezoelectric interface in cell culture studies uncovers that electro-mechanical contact with human cells enhances cell differentiation and proliferation (Tai et al. 2021). PHB's piezoelectric properties and cellular sensitivity can be progressed by mixing CNTs, HA, and barium titanate. Piezoelectric PHBV scaffolds enhance osteoblast development and calcium deposition (Gorodzha et al. 2017). In a patient-centric approach to healthcare, the utilization of innovative technologies such as Au-coated ZnO nanorods and piezoelectric materials holds significant promise for improving treatment outcomes and enhancing patient well-being (Hoop et al. 2017).

2.5 Promising Areas of Biopolymer Applications and Patient Centric Approaches

2.5.1 Wound Healing

The role of biodegradable biomaterials is contemporary in wound treatment protocols, offering sustainable, compatible, and efficient healing (Fig. 2.2). Biodegradable biomaterials are especially promising in this field because of their efficient compatibility and ability to support the natural healing process. Most of these polymers as discussed are obtained from natural sources including plants, animals, and microorganisms. Polysaccharide based biomaterials like chitosan, alginate and protein derived biomaterials like collagen and gelatin are getting significant attention in this field because of their unique characteristics to link in with the physiological environment (Boateng et al. 2008). Peptides and proteins components (endogenous) are vital for the process of wound healing because they support tissue repair, infiltration, and regeneration cascades in the healing process. These natural biodegradable polymers because of their chemical composition (proteins, polysaccharides) act as scaffolds for cell adhesion, proliferation, regeneration as they have the capacity to mimic the extracellular matrix and thereby promotes tissue regeneration cascades in the healing process. Gelatin derived from collagen is frequently utilized as bandages for wounds. Another chemical component, keratin, also plays an important role in tissue regeneration and it's upregulated during wound healing process. Hence, they are incorporated as one of the biomaterials in several wound dressings eg. Keragel® and Keraderm® marketed products. Novel drug delivery systems based on polysaccharides and keratin have been researched for their potential role in tissue regeneration, cell migration and proliferation (Moholkar et al. 2021). Alginate and chitosan promote wound healing because of their effective antimicrobial property in addition to other biological, physical, and chemical characteristics (Bano et al. 2017). The structural and biological qualities of chitosan are supportive to the process of gelation, electrospinning and 3D printing that's important characteristics for the production and function of scaffolds for wound healing. It also can form films, hydrogels, fibers, and sponges and hence widely used for tissue regeneration and repair process as it enables the biomaterials tissue interface or receiving tissue to be shaped in the same way (Singh et al. 2022). Furthermore, it permits the regeneration of primary tissue cells and stem cells, and the chemicals are like the components of extra cellular matrix, which raises the prospects of chemical modification to adjust and adaptable to the individual patient requirements. Hence the application of chitosan in biomedical engineering is

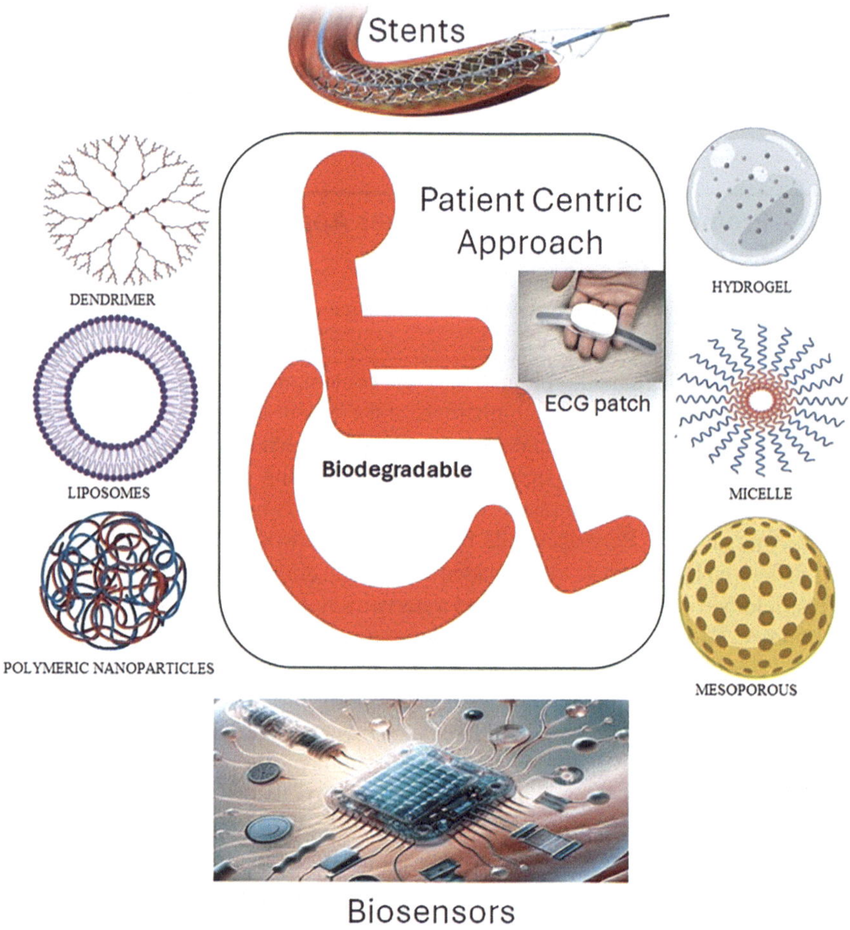

Fig. 2.2 Biodegradable biomaterials in wound healing

outstanding. Chitosan is used in intra-abdominal surgery and healing procedure as it has good surface bonding to the lesion as well as able to maintain hemostasis (Koumentakou et al. 2020). Meanwhile, seaweed derived alginate also possesses gel like consistency like chitosan and hence holds the moisture content effectively that supports one of the ideal requirements of the wound healing scaffolds. Dextran, another component, also plays a major role in the application of tissue engineering because of its anti-thrombotic properties which stimulate angiogenesis and skin regeneration. Starch, a natural biopolymer, is also the desirable component in wound dressings because of its physical, chemical, and biodegradable properties like chitosan. It also enhances epithelial tissue regeneration that has an immense therapeutic role in wound healing. It's also attractive because of its high compatibility to the natural cellular environment. Several research works are going on to modify the mechanical characteristics and moisture sensitivity of this biopolymer by the incorporation of some chemicals and carriers to make it par with some costly commercial polymers (Waghmare et al. 2017). According to reports, cutaneous cell growth and multiplication were enabled by the nanofibrous mats of starch which may find use in wound healing application. Another biopolymer is hyaluronic acid (HA) and shows promise component in tissue engineering because of its physiological activity. Improved effectiveness of HA hydrogel scaffolds may augment tissue regeneration, repair, and growth. Pectin, when combined with other materials/chemicals, generated gel that facilitate wound gealing by acting as a barrier and encouraging moisture content with effective antimicrobial property (Yang et al. 2022).

Biodegradable polymers such as poly (lactic acid) PLA and poly (glycolic acid) (PGA) are commonly used in the fabrication of wound closure materials and sutures. With time, these polymers degrade naturally by the physiological stimuli hence eliminating the necessity of removal process from the body. These dressings can also be customized to meet the specific demands of the patient by processing biopolymers into a variety of forms including films, gels, and nanofibers (Gobi et al. 2021). The therapeutic potential of biopolymers are further increases when they combined with bioactive substances like growth factors, cytokines, antibacterial and other therapeutics. These bioactive substances facilitate blood vessel creation, cell multiplication, lowering inflammation, scavenges free radicals and thus speeds up healing process. Biomedical engineering concepts utilize their newer ideas to incorporate/combine these beneficial molecules without any adverse interaction to promote the beneficial effects as per the requirement of the patient (Alven et al. 2022). These combinations address the various issues of tissue repair and regeneration because of their unique characteristics. By utilizing these characteristics of mixed biopolymers, scientists are creating innovative wound dressings and medication delivery systems that promote healing and enhance patient outcomes. There are new advanced hydrogel dressing materials employed for wound healing: mechanically active sticky hydrogels and self-adapting hydrogels. These smart hydrogel systems can actively encourage the wound healing process in comparison to traditional hydrogel dressings. These mechanically active and adhesive hydrogels can stick to the wounds effectively and actively shrink them in reaction to outside stimuli such as changes in temperature, moisture content and pH etc. Self-adapting hydrogels may change their shape, size, and thickness to perfectly fit wounds while also adjusting to the local physiological environment. This property may be considered positive in several aspects and negative in some respects (Zhang et al. 2022). The advancement in tissue engineering and skin biology of scarless wound healing led to the development of skin counterparts to regenerate whole skin by accessing the intrinsic regeneration potential with the use of pro-regenerative scaffolds. Full skin regeneration appears to be greatly enhanced by pro-regenerative scaffolds that trigger regenerative immune response (Krishani et al. 2023). These scaffolds provide a supportive framework for cell attachment, proliferation, and tissue regeneration. They can mimic extracellular matrix natural architecture and hence promote wound healing.

Moreover, using α-cyclodextrin chitosan modified into hydrophobic demonstrated good hemostatic properties and reduced bleeding time in a rat model by 90%. Chitosan-Kaolin porous microsphere had noteworthy hemostatic properties in two rat models (tail amputation and liver laceration). A commercial chitosan-based hemostat powder and standard medical gauze were compared, and the chitosan sponges were demonstrated to have good characteristics such as easy handling, hemocompatibility, biocompatibility, mechanical stability, and absorption capacity to be applied as wound dressing material clinically (Gheorghiță et al. 2023). Commercially available chitosan-based wound care products are effectively applied clinically. Axiobio(sponge), ChitoSAM™ 100 (nonwoven), ChitoClot Bandage (nonwoven), ChitoClot Pad (sponge), and ChitoClot (Gauze) consist of 100% chitosan. Gel-type products like ChitoClear® and Opticell® are some martketetd chitosan-based dressing used for patient care (Thambiliyagodage et al. 2023).

Recent studies have explored innovative applications of collagen-based drug delivery in various biomedical contexts. Kumar et al. (2014) developed a collagen-mimetic peptide for thrombosis treatment, demonstrating its potential as a hemostat through in vitro assays on platelet behavior and clotting kinetics. Another study used a solvent-based approach to create a collagen/chitin biomaterial with superior properties, showing promising results in vitro and in vivo models of hemostasis. Furthermore, combining collagen with γ-polyglutamic acid resulted in a gel with enhanced viscosity and mechanical strength. Implanted in rat skin incisions, this gel promoted cell adhesion, fibroblast migration, and collagen fibril maintenance, suggesting its potential as a scaffold for tissue repair. Pad-type dressing materials like Biopad (100% collagen), Puracol (collgen+honey), Promogran (collagen+cellulose) 100%, gel-type wound care materials like Stimulen (collgen+glycerine) Mesh like products like ColActive (Gelatin, sodium alginate, carboxymethylcellulose, EDTA, plasticizers) are some of collagen-based dressing applied clinically (Tronci 2019). Results from complex animal and cellular research of these natural biopolymers utilizing natural material and less process costs, can make the finished products a cost effective, patient-oriented treatment option for wound healing in patients especially beneficial to the patients from lower socioeconomic backgrounds.

2.5.2 Drug Delivery Systems

Biopolymers are used for the fabrication of various drug delivery systems, which have advantages over the application of conventional dosage forms. Some drug delivery systems in which biopolymers are used are discussed in this section.

2.5.2.1 Nano-based Drug Delivery Systems

Nano-based drug delivery systems have revolutionized therapeutic studies, particularly treating various cancers and chronic diseases. The drug delivery mechanism offers several pharmacokinetic and pharmacodynamic advantages, including improved bioavailability, controlled or targeted delivery, and enhanced stability for long-term storage (Sudhakar et al. 2021).

2.5.2.2 Polymeric Nanoparticles

Polymeric nanoparticles are effective at delivering nutraceuticals and boosting antigens antimicrobial/immune related action. For instance, PLGA nanoparticles loaded with the vaccine antigen against acute *P. aeruginosa* demonstrated encouraging results and tobramycin loaded alginate/surface modified chitosan nanoparticles showed high bioavailability and effective therapeutic efficacy against *P. aeruginosa* (Wibowo et al. 2021). Nanogel, a form of polymeric nanoparticle showed promising therapeutic adherence because of its sustained and regulated drug release characteristics. Polymers and drug molecules have three-dimensional network structures that can hold small amounts of chemicals or other nutraceuticals. Hence, they allow for precise control over the release characteristics, physical properties and efficacy of the active molecule and reduce side/adverse effects. These characteristics holds promise for the safe bio distribution and delivery methos for vaccine (Trucillo 2024).

2.5.2.3 Solid-Lipid Nanoparticles (SLNs)

The application of solid lipid nanoparticle (SLNs) formulation in the management of cutaneous diseases and related issues has demonstrated notable success in the field of biomedical engineering. This formulation approach provides targeted delivery, sustained/controlled release, and protect the stability of the active constituent because of their biodegradable nature and biocompatible chemical composition with the physiological environment. They can encapsulate different types of therapeutic agents including genes, cytokines, and several proteins. Physiologically, intact skin inhibits topical therapy permeation. However, SLNs rearranges and fluidize the stratum corneum lipid matrix because of its physiological lipid composition and thus enhances drug permeability. They also provide a sealing effect that help to minimize water loss from the skin and leaving skin more moisturised. Innovations in lipid based nano formulations have reduced the cost, frequency of administration toxicity while increasing the half-life and bioavailability of several therapeutic agents especially related to skin wound and diseases. Most membranes and materials used as regeneration agents are hydrophilic, polar macromolecules. Thus, the therapeutic effect is limited because of its inability to approach the target wound region through various natural barriers such as the vascular endothelium. On the contrary liposomes, SLNs, nanoemulsions and micelles are examples of nanoscale lipid formulations. Targeting can be accomplished by altering the surface of the medicine such that it accumulates in a particular location, which lowers toxicity and increases efficiency. Numerous studies that used lipid nano formulation bilayer and scaffolds to repair tissue damage have shown promising outcomes in recent years (Fig. 2.3). To produce concentrated drug carriers for biomedical applications or signal intensity diagnostic probes, both hydrophilic and hydrophobic species can be effectively entrapped with phospholipid matrix. The application of SLN towards targeted chemotherapy and diagnostics as well as surface modification for tissue engineering applications reported to be successful (Filipczak et al. 2021). SLNs are a viable therapy option for burn and chronic wounds because of their sustainable biological characteristics. Additionally, SLNs, liposomes and niosomes have been

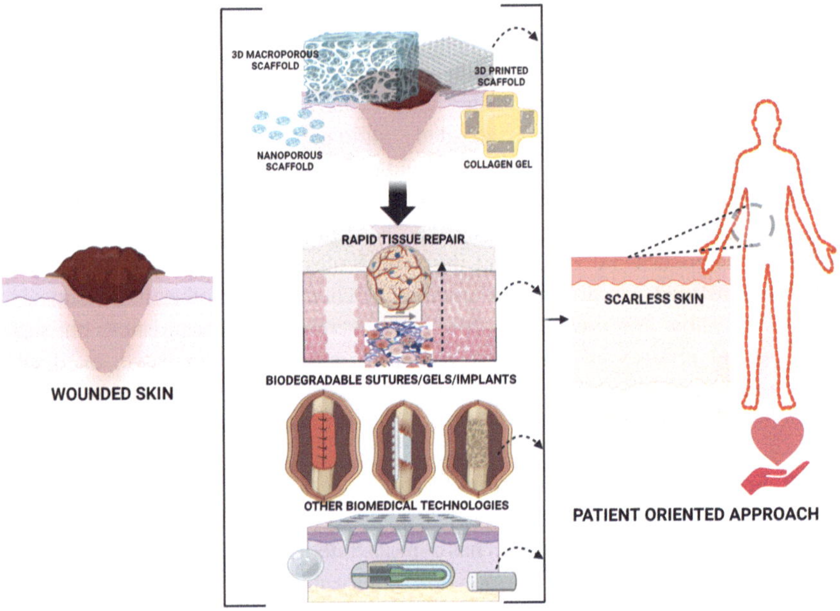

Fig. 2.3 The biodegradable nature of ECG patch ensures comfort and convenience for the user, as well as environmentally friendly disposal once the device is no longer needed. In a similar way sustainable biomaterials in drug delivery minimize the risk of adverse reactions and improve patient compliance by providing a more convenient and efficient method of drug administration

demonstrated to efficiently transport medication to wound sites and biodegradable polymers like PLA and PEG have been utilised as scaffolds to encourage tissue regeneration. When combined, these two systems may have a beneficial influence on wound healing. They can encapsulate a variety of drugs including hydrophilic and hydrophobic substances and are biocompatible. They also could target tissues and cells that plays important role in the process of wound healing including macrophages, release medications over an extended period and offer physical support for tissue regeneration. Maintaining ideal therapeutic concentration at the wound site can also be aided by the prolonged release of medication from the polymers. There are still certain issues that need to be resolved despite the beneficial effects of combinational systems. The absence of a consistent process for assessing these systems efficacy is one of the main obstacles. Another challenge is the absence of long-term data on the effectiveness and safety of these combinational formulations. Further detailed research is needed to ascertain the ideal dosage, course of treatment, as well as to validate the safety and effectiveness of these system for chronic use in humans and for tailoring the effectiveness as per the requirements of the individual patient (Farasati et al. 2023).

Micelles, known for their stability and acid degradability, are favored carriers for anticancer agents, with hydrotropic polymer micelles showcasing versatility in delivering poorly water-soluble drugs. Zhang et al. (2017) and Kim et al. (2011)

investigated the solubilizing abilities of polymeric forms of N, N-diethyl nicotinamide (DENA) and N, N-dimethyl benzamide (DMBA) for various pharmaceuticals.

In medical applications, three micelles are typically studied: regular micelles, reverse micelles, and unimolecular micelles. As exemplified by PEG-PLGA and PEG-PLA, regular micelles self-assemble with a hydrophobic core and a hydrophilic exterior. They have been explored for in vivo targeted drug delivery, such as aptamer-conjugated nanoparticle systems targeting prostate cancer. Multiple branched polymeric molecules called dendrimers offer exciting biomedical applications owing to their hyper branching, compact nature, and biodegradability. They find use in targeted cancer therapy, bone regeneration and controlled therapeutics delivery (Chauhan 2018).

Dendrimers have the potential in receptor- mediated targeting and enhance therapeutics uptake while reducing toxicity, although challenges related to the dendrimers size and kinetics persist. The toxicity of dendrimers, especially in relation to surface charge characteristics, is a concern despite their potential and hence several researchers are fabricating the surface modified dendrimers to get better therapeutic efficiency with less adverse consequences to the patients in the clinical settings (Simanek 2021).

2.5.2.4 Liposomes

Phospholipid bilayers are the main component of the skin and most of the physiological membranes. These bilayers play a major role in biomedical engineering as they are the main targets for drug delivery, diagnostics, implant applications etc. Phosphatidylcholine (Lecithin), phosphatidyl ethanolamine, phosphatidyl serine, phosphatidyl inositol, phosphatidyl glycerol, and cholesterol are lipids used for liposome preparation (Fathi and Oyelere 2016; Laouini et al. 2012). Therefore, liposomes are majorly made of lipid biomaterials. These biomaterials are thought to be utilized more frequently as tools for regenerative medicine, reconstructive surgery, and sustained drug release as they can mimic the physiological environment (Monteiro et al. 2014). In clinical settings, liposome-based therapy is commonly employed in the treatment of systemic fungal infections and cancer. However, the application of liposomes has gone beyond these clinical uses and now they are prominently used in the vaccine development, imaging, and cosmetics because of its adaptable characteristics with physiological environment. Furthermore, several tissue engineering scaffolds manufactured with liposome technology have been approved for human applications. Magnetic cationic liposomes (MCLs) are positively charged magnetic nanoparticles encapsulated/bounded/fabricated with liposomes and they can easily interact with negatively charged cell membranes. They are commonly used in tissue engineering, regenerative medicine, drug delivery formulations, and scaffolds fabrication. The trend of employing magnetic force-based tissue engineering is hype. They are easily accepted by the patients because of their ability to interfere with the tissue angiogenesis, tissue regeneration, proliferation, and bone remodeling in a tailored manner. Cell seeding into the core spaces of the scaffolds was successfully supported by MCLs (Monteiro et al. 2014).

For tissue engineering applications related to gene delivery, liposomal scaffold systems are commonly employed as biodegradable biomaterials because of their efficiency in localized delivery in the target space. Some of the clinically significant patient centric approaches in gene therapy includes bone and cartilage regeneration. These modified genes encode specific molecular components in the bone and cartilage, which play a major role in the bone differentiation, maturation, and ossification process. Some of these genes transfer through bonne marrow stromal cells and were utilized to take advantage of tissue vascularization. Liposomes loaded with genes that promote wound healing showed beneficial results in animal models. Hence, to combine the formulation advantages of liposomes with the physical and mechanical strength of scaffolds, liposomal biomaterial hybrid systems are developed in the biomedical engineering field (Nsairat et al. 2022). This development in the field of biomedical engineering ultimately resulted in the common usage and reputation of novel systems that are appropriate for patient centric approach.

2.5.3 Medical Devices

2.5.3.1 Implants

Implantable devices play a crucial role in modern medicine, aiming to replicate or replace damaged body parts to maintain normal bodily functions. While traditional materials like metals, ceramics, and synthetic polymers have been widely used, they have drawbacks such as immunological rejection and concerns about biodegradation products. Biopolymers from living organisms offer promising alternatives due to their well-defined structures, degradability, and renewability (Bhatt and Jaffe 2015). PLA, derived from natural sources like corn starch and rice, presents properties like synthetic polymers while being bioabsorbable. The property of biodegradation makes PLA a suitable material to be used in applications like stents as it internalizes with the body fluids and eliminates the needs for additional removal surgeries. Preclinical and clinical research demonstrated that PLA based stents work efficiently compared to other non-biodegradable/biomaterial implants. Furthermore, this polymer has been utilized in composite scaffolds for tissue engineering and demonstrated promising results in inducing bone formation when used as a carrier for proteins (Mukherjee and Kao 2011). Another commonly employed biodegradable polymer in implants is chitosan. It has diverse applications in the field of tissue regeneration including bone, skin, liver, and nerves. Research reports indicate that chitosan is efficacious in fostering hemocompatibility and endothelial cell compatibility in coronary stents and augmenting mechanical properties when combined with other natural biopolymers like alginate, gellan gum, etc. (Meng et al. 2009).

Biopolymer silk has exceptional mechanical strength and biocompatibility mainly due to its distinct beta sheet structure and hence it is used in several implanted devices, drug delivery formulations and tissue engineering scaffolds. 3D porous silk fibroin scaffolds offering ideal conditions for cell adhesion and proliferation in recent years and used for patient centric approach (Meinel and Kaplan 2012).

2.5.3.2 Other Applications

Biopolymers are employed to develop diagnostic devices such as biosensors, biochips, and immunoassays. Functionalized biopolymers, including antibodies, enzymes, and DNA probes, are immobilized onto sensor surfaces to detect specific biomolecules or analytes in biological samples with high sensitivity and specificity. They are also used in dental and orthopedic applications to fabricate dental implants, bone graft substitutes, and orthopedic implants. Materials like PLA, PGA, PLGA, and hydroxyapatite-based biopolymers provide structural support, promote osseointegration, and facilitate bone regeneration in dental and orthopedic surgeries (Rebelo et al. 2017).

2.6 Challenges and Constraints to the Clinical Use of Biodegradable Materials

For the benefit of human applications or patient centric approach, it is crucial to comprehend the properties and behavior of biomaterials derived from both lab-based and living tissue (Mohd et al. 2023). Metal alloys, ceramics, polymers, and composites are a few examples of biodegradable materials for human usage and the difficult aspect is determining which biodegradable material is appropriate for a given patient. Since a very smaller number of studies determined that biomaterials including implants had long-term negative effects, there is also concern regarding the chronic consequences of these biomaterials in the human body. Although effective creations of muscle, cartilage, skin, and liver tissue have been made in the biomedical area, tissue engineering is still considered as a relatively new technique. However, it is accepted that the method holds considerable promise for the future development of medical operations (Deperlioglu 2019). Present biodegradable piezoelectric materials have limited medical implant uses owing to their somewhat modest output. Optimizing piezoelectric output is critical to implement new-Health systems for *in-situ* diagnosis and monitoring. The second restriction is that the application direction is not properly aligned with the maximal piezoelectric constant mode. Overall, all biodegradable materials, particularly biopolymers, have shear modes with maximal piezoelectric constants, that are lesser than longitudinal d33. This quality may limit the global usability of certain items. Even if various research has revealed that piezoelectric components may be bio-degraded, getting to "true" biodegradability is even now complicated. Current implants comprise non-degradable components like transmitters, signal amplifiers, rectifiers, substrates, batteries, and wireless consoles. Scanty research is available on the correlation that exists between device functioning, piezoelectric performance, and deterioration rates. Proper design is significant to uphold component degradation rates and thwart premature device faults within the stipulated lifespan. The challenges emphasized include: (i) Limited global usability: The quality of certain items being limited may hinder their widespread use on a global scale, potentially restricting access to beneficial technologies. (ii) Achieving "True" biodegradability: Despite research indicating that piezoelectric components can be bio-degraded, achieving complete biodegradability remains a complex task, posing a challenge in the development of sustainable medical devices. (iii) Presence of non-degradable components: Current implants contain non-degradable elements such as transmitters, signal amplifiers, rectifiers, substrates, batteries, and wireless consoles, which can impede the overall biodegradability of the device. (iv) Lack of research on correlation: There is a scarcity of research on the relationship between device functionality, piezoelectric performance, and degradation rates, making it difficult to optimize the design and performance of biodegradable implants. (v) Importance of proper design: Proper design is crucial to maintain component degradation rates and prevent premature device failures within the expected lifespan, highlighting the need for meticulous planning and execution in the development of biodegradable medical devices.

2.7 Prospects and Conclusion

Biomedical engineering and patient centric approach will be wedged by novel engineering concepts in the future. A few examples of this are artificial limbs and organs, the genetic engineering of new species, computer-simulated surgery, and tissue/organ regeneration. With the progress of biomedical research, cell treatments and neural probes are now being used to treat disability in people with uncommon disorders. Antimicrobial resistance is treated via phage therapy, which has been recognized as a potentially catastrophic worldwide concern. Many of the people who have limited mobility will benefit from this. Given the increasing demand for alternative treatments for chronic illnesses, such as organ failure and severe tissue damage, tissue engineering holds enormous promise for the future of healthcare and will address several issues. Therefore, it makes economic sense for biomedical engineering to incorporate sustainable and biodegradable polymers into the patient-centric approach.

The present state of drug delivery systems is highly dependent on synthetic polymer materials, which do not offer optimal long-term performance. Natural polymers have been used in recent times to make biomimetic or bioactive carriers, which suggests both merits and demerits. Discoveries in green chemistry are paving the way for the creation of low-cost, environmentally friendly, readily available, and minimally hazardous drug delivery systems. To design novel, highly functioning drug delivery systems, it is necessary to optimize green nanotechnologies for the use of nano scale biomaterials. This opens up new potential for smart devices in biology and medicine.

In the realm of therapeutic and medical diagnostics, the development of biodegradable piezoelectric materials hold incredible promise for the future. Updates on biodegradable piezoelectrics for its usage in medical implants are discussed in detail. The overview and the recent discoveries including crystal controls, compositing, and new structural design are discussed. Bio-sensing for real-time health condition monitoring, energy harvesting for powering implantable electronics, and therapeutic applications are just a few of the novel biomedical uses made possible by biodegradable piezoelectric implants. These innovations will aid to light up the next generation patient-centered advanced medical devices. Potential applications of biodegradable polymers promise several advantages for people and the environment, signaling the arrival of a patient-centric approach in healthcare. These polymers provide a sustainable substitute for conventional materials since they are made from renewable resources or via environmentally acceptable methods of synthesis. They can also be removed from the biological system without interfering with invasive and costly treatment protocols and consequently they reduce the biological and economical risk of the patients. Hence, the application of biodegradable/sustainable biomaterials tailored to patient requirements seems promising in the biomedical field. Their compliance abilities with the physiological environment guarantee the best therapeutic outcomes and beneficial in the patient centric approach. They also reduce the side effects by enabling precise control of medication release especially in medical implants and drug delivery systems and in tissue engineering by promoting cell proliferation, repair, and integration. By using biodegradable polymers, medical personnel can treat patients more safely and effectively while reducing environmental harm and focusing on the needs of the patient. These sustainable biomaterials will open up new avenues towards patient centered care in the fields of biomedical engineering from orthopedic implants to drug eluting stents and other medical devices.

References

Abdul Khodir WKW, Abd Hamid S, Yusof MR et al (2022) Electrospun Sulfonatocalix[4]arene loaded blended nanofibers: process optimization and in vitro studies. Pharmaceutics 14(9):1912. https://doi.org/10.3390/pharmaceutics14091912

Abyzova E, Dogadina E, Rodriguez RD et al (2023) Beyond tissue replacement: the emerging role of smart implants in healthcare. Mater Today Bio 29(22):100784. https://doi.org/10.1016/j.mtbio.2023.100784

Alven S, Peter S, Mbese Z et al (2022) Polymer-based wound dressing materials loaded with bioactive agents: potential materials for the treatment of diabetic wounds. Polymers (Basel) 14(4):724. https://doi.org/10.3390/polym14040724

Aranaz I, Alcántara AR, Civera MC et al (2021) Chitosan: an overview of its properties and applications. Polymers 13(19):3256

Bano I, Arshad M, Yasin T et al (2017) Chitosan: a potential biopolymer for wound management. Int J Biol Macromol 102:380–383. https://doi.org/10.1016/j.ijbiomac.2017.04.047

Bhatt R, Jaffe M (2015) Biopolymers in medical implants. In: Excipient applications in formulation design and drug delivery. Springer, Cham, pp 311–348

Boateng JS, Matthews KH, Stevens HNE et al (2008) Wound healing dressings and drug delivery systems: a review. J Pharm Sci 97(8):2892–2923. https://doi.org/10.1002/jps.21210. PMID: 17963217

Cavallini C, Vitiello G, Adinolfi B et al (2020) Melanin and melanin-like hybrid materials in regenerative medicine. Nanomaterials (Basel) 10(8):1518. https://doi.org/10.3390/nano10081518

Chauhan AS (2018) Dendrimers for drug delivery. Molecules 23(4):938

Chen Z, Wu C, Chu Y et al (2022) Biodegradable elastomers and gels for elastic electronics. Adv Sci 13(9):2105146

Coiai S, Di Lorenzo ML, Cinelli P et al (2021) Binary green blends of Poly(lactic acid) with Poly (butylene adipate-co-butylene terephthalate) and Poly(butylene succinate-co-butylene adipate)

and their nanocomposites. Polymers (Basel) 13(15):2489. https://doi.org/10.3390/polym13152489
Deperlioglu O (2019) Intelligent techniques inspired by nature and used in biomedical engineering. In: Biotechnology: concepts, methodologies, tools, and applications. IGI Global, Hershey, pp 666–692
Dias GJ, Ramesh N, Neilson L et al (2020) The adaptive immune response to porous regenerated keratin as a bone graft substitute in an ovine model. Int J Biol Macromol 165:100–106
Dorozhkin SV (2015) 7 – Surface modification of magnesium and its biodegradable alloys by calcium orthophosphate coatings to improve corrosion resistance and biocompatibility. In: Surface modification of magnesium and its alloys for biomedical applications, Woodhead Publishing series in biomaterials. Woodhead Publishing, Duxford, pp 151–191
Duoyi Z, Tongtong Z, Jie L et al (2021) Poly(lactic-co-glycolic acid)-based composite bone-substitute materials. Bioact Mater 2(6):346–360
Elmowafy E, Abdal HA, Skouras A et al (2019) Polyhydroxyalkanoate (PHA): applications in drug delivery and tissue engineering. Expert Rev Med Devices 16(6):467–482. https://doi.org/10.1080/17434440.2019.1615439
Farasati FB, Naimi-Jamal MR, Sedaghat M et al (2023) Combinational system of lipid-based nanocarriers and biodegradable polymers for wound healing: an updated review. J Funct Biomater 14(2):115
Fasolino I, Guarino V, Cirillo V et al (2017) 5-Azacytidine-mediated HMSC behavior on electrospun scaffolds for skeletal muscle regeneration: behavior of 5-azacytidine-mediated HMSC. J Biomed Mater Res A 105(9):2551–2561. https://doi.org/10.1002/jbm.a.36111
Fathi S, Oyelere AK (2016) Liposomal drug delivery systems for targeted cancer therapy: is active targeting the best choice? Future Med Chem 8(17):2091–2112
Fernando S, Mcenery, Madison et al (2016) Polyurethanes for bone tissue engineering, 1st edn, pp 481–501. https://doi.org/10.1016/B978-0-08-100614-6.00016-0
Filion TM, Kutikov A, Song J (2011) Chemically modified cellulose fibrous meshes for use as tissue engineering scaffolds. Bioorg Med Chem Lett 21(17):5067–5070. https://doi.org/10.1016/j.bmcl.2011.04.032
Filipczak N, Yalamarty SSK, Li X et al (2021) Lipid-based drug delivery systems in regenerative medicine. Materials (Basel) 14(18):5371. https://doi.org/10.3390/ma14185371
Fonseca JJH, Bagne L, Meneghetti DH et al (2019) Electrical stimulation: complementary therapy to improve the performance of grafts in bone defects? J Biomed Mater Res Part B Appl Biomater 10(1):75. https://doi.org/10.3390/bioengineering10010075
Gheorghiță D, Moldovan H, Robu A et al (2023) Chitosan-based biomaterials for hemostatic applications: a review of recent advances. Int J Mol Sci 24(13):10540. https://doi.org/10.3390/ijms241310540
Gobi R, Ravichandiran P, Babu RS et al (2021) Biopolymer and synthetic polymer-based nanocomposites in wound dressing applications: a review. Polymers 13(12):1962. https://doi.org/10.3390/polym13121962
Goonoo N, Luximon AB (2022) Piezoelectric polymeric scaffold materials as biomechanical cellular stimuli to enhance tissue regeneration. Mater Today Commun 31:103491
Gorodzha AR, Muslimov DS, Syromotina AS et al (2017) A comparison study between electrospun polycaprolactone and piezoelectric poly (3- hydroxybutyrate -co-3 hydroxy valerate) scaffolds for bone tissue engineering. Colloids Surf B: Biointerfaces 160(2017):48–59
Goyanes A, Buanz AB, Hatton GB et al (2015) 3D printing of modified-release aminosalicylate (4-ASA and 5-ASA) tablets. Eur J Pharm Biopharm 89:157–162. https://doi.org/10.1016/j.ejpb.2014.12.003
Harugade A, Sherje A, Pethe A (2023) Chitosan: a review on properties, biological activities and recent progress in biomedical applications. React Funct Polym 191:105634. https://doi.org/10.1016/j.reactfunctpolym.2023.105634
Hong S, Choi DW, Kim HN et al (2020) Protein-based nanoparticles as drug delivery systems. Pharmaceutics 12(7):604. https://doi.org/10.3390/pharmaceutics12070604

Hoop M, Chen XY, Ferrari A et al (2017) Ultrasound-mediated piezoelectric differentiation of neuron-like PC12 cells on PVDF membranes. Sci Rep 7(1):4028. https://doi.org/10.1038/s41598-017-03992-3

Hosseini ES, Dervin S, Ganguly P et al (2021) Biodegradable materials for sustainable health monitoring devices. ACS Appl Bio Mater 4(1):163–194. https://doi.org/10.1021/acsabm.0c01139

Huq T, Khan A et al (2022) Sources, production and commercial applications of fungal chitosan: a review. J Bioresour Bioprod 7:85–98. https://doi.org/10.1016/j.jobab.2022.01.002

Jacob J, More N, Kalia K et al (2018) Piezoelectric smart biomaterials for bone and cartilage tissue engineering. Inflamm Regen 38:2

Jaski AC, Schmitz F, Horta RP et al (2022) Zein-a plant-based material of growing importance: new perspectives for innovative uses. Ind Crop Prod 186:115250

Kageyama T, Osaki T, Enomoto J et al (2016) In situ cross-linkable gelatin-CMC hydrogels designed for rapid engineering of Perfusable vasculatures. ACS Biomater Sci Eng 2(6):1059–1066

Kamel NA (2022) Bio-piezoelectricity: fundamentals and applications in tissue engineering and regenerative medicine. Biophys Rev 14(3):717–733. https://doi.org/10.1007/s12551-022-00969-z

Kankala RK, Xu PY, Chen BQ et al (2021) Supercritical fluid (SCF)-assisted fabrication of carrier-free drugs: an eco-friendly welcome to active pharmaceutical ingredients (APIs). Adv Drug Deliv Rev 176:113846. https://doi.org/10.1016/j.addr.2021.113846

Kantaros A (2022) 3D printing in regenerative medicine: technologies and resources utilized. Int J Mol Sci 23(23):14621

Kantaros A, Piromalis D (2021) Fabricating lattice structures via 3D printing: the case of porous bio-engineered scaffolds. Appl Mech 2(2):289–302

Khan R, Khan MH (2013) Use of collagen as a biomaterial: an update. J Indian Soc Periodontol 17(4):539–542. https://doi.org/10.4103/0972-124X.118333

Kim JY, Kim S, Pinal R (2011) Hydrotropic polymer micelles as versatile vehicles for delivery of poorly water-soluble drugs. J Control Release 152(1):13–20

Kim CH, Lee SY, Rhee KY et al (2024) Carbon-based composites in biomedical applications: a comprehensive review of properties, applications, and future directions. Adv Compos Hybrid Mater 7:55

Koumentakou I, Terzopoulou Z, Michopoulou A (2020) Chitosan dressings containing inorganic additives and levofloxacin as potential wound care products with enhanced hemostatic properties. Int J Biol Macromol 162:693–703. https://doi.org/10.1016/j.ijbiomac.2020.06.187

Krishani M, Shin WY, Suhaimi H et al (2023) Development of scaffolds from bio-based natural materials for tissue regeneration applications: a review. Gels 9(2):100

Kumar VA, Taylor NL, Jalan AA et al (2014) A nanostructured synthetic collagen mimic for hemostasis. Biomacromolecules 15(4):1484–1490. https://doi.org/10.1021/bm500091e

Laouini A, Jaafar MC, Limayem BI (2012) Preparation, characterization, and applications of liposomes: state of the art. J Colloid Sci Biotechnol 1(2):147–168

Liu Z, Sun X, Nakayama RN et al (2007) Supramolecular chemistry on water-soluble carbon nanotubes for drug loading and delivery. ACS Nano 1:50–56. https://doi.org/10.1021/nn700040t

Liu Z, Fan AC, Rakhra K et al (2009) Supramolecular stacking of doxorubicin on carbon nanotubes for in vivo cancer therapy. Angew Chem Int Ed 48:7668–7672. https://doi.org/10.1002/anie.200902612

Liu P, Chen G, Zhang J (2022a) A review of liposomes as a drug delivery system: current status of approved products, regulatory environments, and future perspectives. Molecules 27(4):1372. https://doi.org/10.3390/molecules27041372

Liu Y, Dzidotor G, Le TT et al (2022b) Advances in biodegradable piezoelectrics for medical implants. Sci Transl Med 14(627):eabi7282. https://doi.org/10.1126/scitranslmed.abi7282

Magnusson JP, Saeed AO, Fernandez TF et al (2011) Synthetic polymers for biopharmaceutical delivery. Polym Chem 2:48–59
Maity B, Samanta S, Sarkar S et al (2020) Injectable silk fibroin-based hydrogel for sustained insulin delivery in diabetic rats. ACS Appl Bio Mater 3:3544–3552
Malviya R, Raj S, Fuloria S et al (2021) Evaluation of antitumor efficacy of chitosan-tamarind gum polysaccharide polyelectrolyte complex stabilized nanoparticles of simvastatin. Int J Nanomedicine 16:2533–2553. https://doi.org/10.2147/IJN.S300991
Martău GA, Mihai M, Vodnar DC (2019) The use of chitosan, alginate, and pectin in the biomedical and food sector—biocompatibility, bioadhesiveness, and biodegradability. Polymers (Basel) 11: 1837. https://doi.org/10.3390/polym11111837
Mathur V, Agarwal P, Srinivasan V et al (2023) Facet of 4D printing in biomedicine. J Mater Res 38:2–18
Meinel L, Kaplan DL (2012) Silk constructs for delivery of musculoskeletal therapeutics. Adv Drug Deliv Rev 64(12):1111–1122
Meng S, Liu Z, Shen L (2009) The effect of a layer-by-layer chitosan–heparin coating on the endothelialization and coagulation properties of a coronary stent system. Biomaterials 30(12): 2276–2283
Min J, Jung Y, Ahn J et al (2023) Recent advances in biodegradable green electronic materials and sensor applications. Adv Mater 35(52):e2211273. https://doi.org/10.1002/adma.202211273
Mohd J, Abid H, Ravi PS et al (2023) Sustaining the healthcare systems through the conceptual of biomedical engineering: a study with recent and future potentials. Biomed Technol 1(2):39–47. https://doi.org/10.1016/j.bmt.2022.11.004
Moholkar DN, Sadalag PS, Peixoto D et al (2021) Recent advances in biopolymer-based formulations for wound healing applications. Eur Polym J 160(8):110784. https://doi.org/10.1016/j.eurpolymj.2021.110784
Monteiro N, Martins A, Reis RL et al (2014) Liposomes in tissue engineering and regenerative medicine. J R Soc Interface 11(101):20140459
Mukherjee T, Kao N (2011) PLA based biopolymer reinforced with natural fibre: a review. J Polym Environ 19:714–725
Murphy SV, Atala A (2014) 3D bioprinting of tissues and organs. Nat Biotechnol 32(8):773–785
Nsairat H, Khater D, Sayed U (2022) Liposomes: structure, composition, types, and clinical applications. Heliyon 8(5):e09394. https://doi.org/10.1016/j.heliyon.2022.e09394
Ohtsuki C (2008) 22 – Ceramic–polymer composites. In: Bioceramics and their clinical applications, Woodhead Publishing series in biomaterials. Woodhead Publishing, Duxford, pp 501–517. https://doi.org/10.1533/9781845694227.2.501
Olteanu G, Neacşu SM, Joiţa FA et al (2024) Advancements in regenerative hydrogels in skin wound treatment: a comprehensive review. Int J Mol Sci 29;25(7):3849. https://doi.org/10.3390/ijms25073849
Patricia SB, Gowsihan P, Amy NN et al (2023) 3 – Materials for 3D printing in medicine: metals, polymers, ceramics, and hydrogels. In: 3D printing in medicine, Woodhead Publishing series in biomaterials, 2nd edn. Woodhead Publishing, Duxford, pp 59–103
Pramanik N (2023) A tool for biomedical application: synthesis and modification of Polyhydroxyalkanoates. Sustain Chem Pharm 32(5):101041. https://doi.org/10.1016/j.scp.2023.101041
Pramanik S, Agarwala P et al (2019) Human-lymphocyte cell friendly starch–hydroxyapatite biodegradable composites: hydrophilic mechanism, mechanical, and structural impact. J Appl Polym Sci 137(1). https://doi.org/10.1002/app.4891
Prevot M, Ustunel S, Hegmann E (2018) Liquid crystal elastomers—a path to biocompatible and biodegradable 3D-LCE scaffolds for tissue regeneration. Materials 11(3):377
Qu G, Huang J, Gu G et al (2023) Smart implants: 4D-printed shape-morphing scaffolds for medical implantation. Int J Bioprint 9(5):764. https://doi.org/10.18063/ijb.764
Rebelo R, Fernandes M, Fangueiro R (2017) Biopolymers in medical implants: a brief review. Procedia Eng 200:236–243

Rentsch R, Bernhardt D, Scharnweber WS et al (2012) Embroidered and surface coated poly (caprolactone-co-lactide) scaffolds: a potential graft for bone tissue engineering. Biomatter 2(3):158–165
Resende RFB, Sartoretto SC, Uzeda MJ et al (2019) Randomized controlled clinical trial of nanostructured carbonated hydroxyapatite for alveolar bone repair. Materials (Basel) 12(22): 3645. https://doi.org/10.3390/ma12223645
Sanchez RDO, Vineis C, Cruz Maya I et al (2022) Wool keratin nanofibers for bioinspired and sustainable use in biomedical field. J Funct Biomater 14(1):5
Santos D, Silva DM, Gomes PS (2017) Multifunctional PLLA-ceramic fiber membranes for bone regeneration applications. J Colloid Interface Sci 504:101–110. https://doi.org/10.1016/j.jcis.2017.05.032
Serri C, Quagliariello V, Iaffaioli RV et al (2019) Combination therapy for the treatment of pancreatic cancer through hyaluronic acid-decorated nanoparticles loaded with quercetin and gemcitabine: a preliminary in vitro study. J Cell Physiol 234(4):4959–4969. https://doi.org/10.1002/jcp.27297
Sheikh A, Abourehab MAS, Kesharwani P (2023) The clinical significance of 4D printing. Drug Discov Today 28(1):103391. https://doi.org/10.1016/j.drudis.2022.103391
Shenoy M, Abdul NS, Qamar Z et al (2022) Collagen structure, synthesis, and its applications: a systematic review. Cureus 14(5):e24856. https://doi.org/10.7759/cureus.24856
Simanek EE (2021) Two decades of triazine dendrimers. Molecules 26(16):4774
Singh V, Marimuthu T, Makatini MM et al (2022) Biopolymer-based wound dressings with biochemical cues for cell-instructive wound repair. Polymers 14(24):5371
Srirussamee S, Mobini NJ, Cassidy SH et al (2019) Direct electrical stimulation enhances osteogenesis by inducing Bmp2 and Spp1 expressions from macrophages and preosteoblasts. Biotechnol Bioeng 116(12):3421–3432. https://doi.org/10.1002/bit.27142
Sudhakar K, Fuloria S, Subramaniyan V et al (2021) Ultraflexible liposome nanocargo as a dermal and transdermal drug delivery system. Nanomaterials (Basel) 11(10):2557
Sun Y, Jing X, Ma X et al (2020) Versatile types of polysaccharide-based drug delivery systems: from strategic design to cancer therapy. Int J Mol Sci 21(23):9159. https://doi.org/10.3390/ijms21239159
Tai S, Yang S, Yu A (2021) Modulation of piezoelectric properties in electrospun PLLA nanofibers for application-specific self-powered stem cell culture platforms. Nano Energy 89(4):106444. https://doi.org/10.1016/j.nanoen.2021.106444
Thambiliyagodage C, Jayanetti M, Mendis A et al (2023) Recent advances in chitosan-based applications—a review. Materials 16(5):2073
Thiruvengadam M, Chung IM, Gomathi T et al (2019) Synthesis, characterization and pharmacological potential of green synthesized copper nanoparticles. Bioprocess Biosyst Eng 42(11): 1769–1777. https://doi.org/10.1007/s00449-019-02173-y
Torres RJG, Femmer T, De Laporte L et al (2015) Bioactive Gyroid scaffolds formed by sacrificial Templating of Nanocellulose and Nanochitin hydrogels as instructive platforms for biomimetic tissue engineering. Adv Mater 27(19):2989–2995. https://doi.org/10.1002/adma.201405873
Trinath B, Badjena SK, Pradhan D (2020) Sustainable biomaterials and their applications: a short review. Mater Today Proc 30:274–282. https://doi.org/10.1016/j.matpr.2020.01.437
Tronci G (2019) The application of collagen in advanced wound dressings. In: Advanced textiles for wound care. Woodhead Publishing, Duxford, pp 363–389. https://doi.org/10.1016/B978-0-08-102192-7.00013-8
Trucillo P (2024) Biomaterials for drug delivery and human applications. Materials (Basel) 17(2): 456. https://doi.org/10.3390/ma17020456
Vasquez SF, Abdollahi A, Damjanovic D et al (2018) Flexoelectricity in bones. Adv Mater 30(9). https://doi.org/10.1002/adma.201705316
Vineis C, Cruz MI, Mowafi S et al (2021) Synergistic effect of Sericin and keratin in gelatin based nanofibers for in vitro applications. Int J Biol Macromol 190:375–381
Waghmare VS, Wadke PR, Dyawanapelly S et al (2017) Starch based nanofibrous scaffolds for wound healing applications. Bioact Mater 3(3):255–266. https://doi.org/10.1016/j.bioactmat.2017.11.006

Wali A, Zhang Y, Sengupta P et al (2018) Electrospinning of non-ionic cellulose ethers/polyvinyl alcohol nanofibers: characterization and applications. Carbohydr Polym 181:175–182. https://doi.org/10.1016/j.carbpol.2017.10.070
Wibowo D, Jorritsma SHT, Gonzaga ZJ et al (2021) Polymeric nanoparticle vaccines to combat emerging and pandemic threats. Biomaterials 268:120597. https://doi.org/10.1016/j.biomaterials.2020.120597
Wu Z, Su X, Xu Y et al (2016) Bioprinting three-dimensional cell-laden tissue constructs with controllable degradation. Sci Rep 6:24474
Wu JJ, Huang LM, Zhao Q et al (2018) 4D printing: history and recent Progress. Chin J Polym Sci 36:563–575
Yang Y, Alastair CR, Nicola ME (2021) Recombinant human collagen/chitosan-based soft hydrogels as biomaterials for soft tissue engineering. Mater Sci Eng C Mater Biol Appl 121:111846. https://doi.org/10.1016/j.msec.2020.111846
Yang Y, Xu L, Wang J et al (2022) Recent advances in polysaccharide-based self-healing hydrogels for biomedical applications. Carbohydr Polym 283:119161. https://doi.org/10.1016/j.carbpol.2022.119161
Zaszczyńska A, Gradys A, Sajkiewicz P (2020) Progress in the applications of smart piezoelectric materials for medical devices. Polymers (Basel) 12(11):2754. https://doi.org/10.3390/polym12112754
Zhang S, Xu J, Chen H et al (2017) Acid-cleavable unimolecular micelles from amphiphilic star copolymers for triggered release of anticancer drugs. Macromol Biosci 17(3):1600258
Zhang W, Nan Y, Wu Z et al (2022) Photothermal-driven liquid crystal elastomers: materials. Align Appl Mol 27(14):4330
Zhang Q, Qiang L, Liu Y et al (2023) Biomaterial-assisted tumor therapy: a brief review of hydroxyapatite nanoparticles and its composites used in bone tumors therapy. Front Bioeng Biotechnol 11:1167474. https://doi.org/10.3389/fbioe.2023.1167474
Zhao Z, Deng J, Fan D (2023) Green biomanufacturing in recombinant collagen biosynthesis: trends and selection in various expression systems. Biomater Sci 11(16):5439–5461. https://doi.org/10.1039/d3bm00724c
Zhou Y, Huang WM, Kang SF et al (2015) From 3D to 4D printing: approaches and typical applications. J Mech Sci Technol 29(10):4281–4288
Zhu M, Liu Y, Jiang F et al (2020) Combined silk fibroin microneedles for insulin delivery. ACS Biomater Sci Eng 6:3422–3429
Zibo T, Hua H, Jialin N et al (2017) Design and characterizations of novel biodegradable Zn-Cu-Mg alloys for potential biodegradable implants. Mater Des 117:84–94. https://doi.org/10.1016/j.matdes.2016.12.075
Zuppolini S, Salama A, Cruz MI et al (2022) Cellulose amphiphilic materials: chemistry. Process Appl Pharm 14(2):386

Dr. Dhanalekshmi Unnikrishnan Meenakshi holds a doctorate in Pharmacology (specialization in Nanomedicine). She has over 13 years of research and teaching experience with leading National and International Organizations. She worked as a Scientist at the Council of Scientific and Industrial Research, NEIST, Assam, India. She was involved in government-funded projects for the Northeast Exploration of Pharmaceuticals from natural sources. She has been working on various projects relating to pharmacology, nanotechnology and cancer therapy. Her research vicinity focuses on preclinical and clinical trials. She uses state-of-the-art technology to systematically evaluate the efficiency of novel polymeric nanoparticles encapsulated with biologically active agents. She is a faculty member in the College of Pharmacy, National University of Science and Technology, Muscat, Sultanate of Oman. She published extensive research papers related to pharmacology, nanomedicine, drug delivery, and formulation technology in peer-reviewed reputed journals, books, and international conferences. She received several grants for carrying out her research. Her active membership of several professional bodies, which demonstrates her dedication and commitment to the field, is a testament to her professional standing. She has been conferred many awards nationally and internationally for her excellent scientific contributions.

Dr. Alka Ahuja, Professor (Pharmaceutics) and Chair of Pharmacy Program Chair of Research and Innovation Committee College of Pharmacy, National University of Science and Technology, Muscat Sultanate of Oman (In academic partnership with West Virginia University, USA). She was a chair/member of various academic and research committees for over 35 years. Her research work has been presented at various international conferences in India, Oman and abroad (Controlled Release Society CRS conference in Paris, FIP, Spain SanDiego, Utah, The Netherlands, Scotland, Sydney, Boston, Hawaii, Miami etc.). Her research work has citations in the famous book Bioadhesive Drug Delivery Systems: Fundamentals, Novel Approaches and Development edited by Edith Mathiowetz, Marcel Dekker Series, New York, U.S.A. besides in high-impact journal publications. She worked on major research projects financed and supported by ICMR, UGC, AICTE, and MOHERI, Oman. She is an active member of editorial boards for national/international professional journals. She has won several awards for her research presentations and publications. Her best publication awards include the Motan Devi Dandiya Prize by P.C Dandiya Endowment Trust for notable research publications and Prof. M.L. Khorana Memorial Prize for best paper in the field of Pharmaceutics and Biopharmaceutics, published in IJPS. She published more than 250 publications in journals of repute with close to 11,000 plus citations and 200 plus papers presented at Seminars and Conferences. She worked on consultancy projects financed and supported by a Scottish company and a drug manufacturing company in Muscat. She authored 11 books in different areas—Pharmaceutical Technology, Pharmaceutical Management, Drug regulatory and Pharmacokinetics and worked on 20 funded research projects by UGC, AICTE and MOHERI, Oman.

Dr. Selvasudha Nandakumar received her B. Pharmacy and M. Pharmacy degree from The TN Dr.M.G.R Medical University, Tamil Nadu, India and obtained her Ph.D. at PRIST University, Puducherry, India. She has additional qualifications in patent law and drafting. She has fifteen years of teaching and research experience. She has published articles in reputable journals with a cumulative impact factor of more than 50. She has over ten book chapters in her credits and has attended several conferences and workshops. She received several patents for her innovative ideas. She has been awarded as a woman scientist by funding agencies of Govt. of India and obtained several innovation awards. She is registered at an MSME start-up company. Drug delivery and biomedical devices are her areas of interest. Her passion for inventing cost-effective healthcare products, which reflects her empathy and concern for society, is a driving force in her work. She has a passion for inventing cost-effective healthcare products that benefit society.

Dr. Lekshmi Salim, a distinguished academician and researcher in Pharmacy Practice, brings a unique perspective to the field. With over five years of teaching experience, she has honed her skills and knowledge. Her academic journey includes a B. Pharmacy from Dr. MGR Medical University, Chennai, India, and a Pharm. D and PhD from Annamalai University Chidambaram, India. Her research prowess is evident in her numerous presentations at National and International conferences, including the 'Best Poster Award' at the 7th Medication Safety Conference in Abu Dhabi, UAE. Her contributions extend to the publishing of several research articles in reputed national and international journals, further solidifying her position as a leading figure in the field.

Mr. Chilaka Baburao completed B.Sc. in Silver Jubilee Government College, Affiliated to Rayalaseema University. He received a master's in biotechnology from Pondicherry University. During masters, he was recipient of prestigious GAT-B fellowship under DBT scheme. In his first publication, he worked on developing a 3D printed transdermal patch and published in peer reviewed journal with impact factor 8.2. He also received the best poster presentation award in the National seminar on Smart Drug Delivery systems. He is interested in biomaterials, nanoparticles and drug delivery research.

Dr. Shah Alam Khan is currently working as a Professor of Pharmaceutical Medicinal Chemistry at the College of Pharmacy, National University of Science and Technology, Muscat, Oman (In academic partnership with West Virginia University, USA). He earned his Ph.D. from Jamia Hamdard University, New Delhi, India in 2003 for studying the chemistry and role of *Silybum marianum* (Milk thistle), silymarin and its synthetic analogues in the hepatoprotection. Dr. Khan is a recipient of the Junior Research Fellowship (JRF-UGC, GATE qualified) and Senior Research fellowship (SRF, CSIR) from the Government of India for pursuing master's and doctorate degrees. With over 20 years of teaching and research experience, Dr. Khan's field of research interests include the synthesis and biological evaluation of natural products and synthetic heterocyclic compounds. He has published 190+ full length scientific articles in the Scopus/WoS/ISI indexed national and international journals. In addition to this, he has contributed over 15 book chapters to publications by renowned publishers such as Elsevier, CRC Press, Springer Nature, Nova, and Wiley. He is serving as an editorial board member of several international journals and has reviewed scientific papers for over 50 journals in the field of pharmaceutical sciences. Dr. Khan has supervised 40+ undergraduate and postgraduate research projects, many of which have won awards in poster/oral presentation category at the international conferences. He has successfully completed seven funded research projects and a few more are ongoing. He has been invited as a guest speaker at numerous national and international conferences. His achievements have been recognized with the 'NU Best Researcher award' in 2020–21, 'NU Best academic excellence award in 2021-22' and Best Professor in Pharmacy in July 2023. He is a life member of professional bodies including the IPA, ATINER, IPGA and APTI.

3 Strategies in Synthesis of Biodegradable Polymers

Alper Durmaz, Erdi Can Aytar, İbrahim Mizan Kahyaoğlu, and Selcan Karakuş

Abstract

Biodegradable polymers present a viable substitute for their petroleum-derived equivalents due to their eco-friendly characteristics and inherent degradation process. This chapter explains the various uses of biodegradable polymers or composites, such as food additives, biomedical products, packaging, and environmental solutions, and describes how they are classified into three categories: biomass-based, animal-derived, and microorganism-derived biodegradable polymers. Due to their special qualities, which are invaluable in a variety of industries, thorough study on the synthesis of biopolymers has become more focused. Researchers are focusing on natural, chemically produced, and microbially derived biopolymers in an effort to reduce plastic pollution. Its broad acceptance is hampered, therefore, by ongoing challenges with production and purification process optimization. To fully utilize biodegradable polymers in enhancing sustainable development and global well-being, several challenges must be addressed.

A. Durmaz
Artvin Coruh University, Ali Nihat Gokyigit Botanical Garden Application and Research Center, Artvin, Türkiye
e-mail: alperdurmaz@artvin.edu.tr

E. C. Aytar
Department of Horticulture, Usak University, Faculty of Agriculture, Usak, Türkiye

İ. M. Kahyaoğlu
Faculty of Science, Department of Chemistry, Ondokuz Mayıs University, Samsun, Türkiye

S. Karakuş (✉)
Department of Chemistry, Istanbul University-Cerrahpaşa, Faculty of Engineering, Istanbul, Türkiye
e-mail: selcan@iuc.edu.tr

R. Malviya, S. Sundram (eds.), *Sustainable Green Biomaterials As Drug Delivery Systems*, Biomaterials, Bioengineering and Sustainability 1,
https://doi.org/10.1007/978-3-031-79062-1_3

Keywords

Biopolymer · Biodegradability · Natural biopolymers · Synthesis methods · Extraction · Sonochemical

Abbreviation

DNA	Deoxyribonucleic acid
PCL	Polycaprolactone
PHAs	Polyhydroxyalkanoates
PHBV	Polyhydroxybutyrate-co-hydroxyvalerate
PLA	Polylactic acid
PVA	Polyvinyl alcohol
RNA	Ribonucleic acid
SPPS	Solid-phase polypeptide synthesis

3.1 Introduction

With significant advantages including eco-friendliness and biodegradability that set them apart from petroleum-based polymers, biodegradable polymers are poised to challenge the presence of synthetic polymers in the market. Three primary types of biopolymers can be distinguished: microbe-derived biopolymers (cellulose, polyhydroxyalkanoate and xanthan), animal-based biopolymers (silk, collagen, casein, and gelatine), and plant-based biopolymers (starch, cellulose, zein and gluten) (Prabu Udayakumar et al. 2021). These biopolymers exhibit a wide range of potential uses across various issues, including food additives, drug delivery systems, biomedical products, packaging, environmental solutions, biosensors, wound healing, and skin tissue engineering. Their unique chemical, biological, physical, and mechanical characteristics make them invaluable in these applications. Due to their extraordinary features, there has been significant emphasis on studying biopolymer synthesis.

To promote global wellbeing and reduce pollution in the environment, researchers are concentrating more and more on biodegradable and biocompatible materials as a means of countering the manufacture of non-biodegradable plastic waste. Three types of biopolymer synthesis can be distinguished: natural biopolymers, chemically synthesized biopolymers, and microbially derived biopolymers. Many biodegradable and biocompatible polymers, including biopolymers-based hydrogels, biopolymers-based composites, and biopolymers-based nanoparticles, have emerged in sectors such as agriculture, pharmaceuticals, water treatment, petroleum-based products, and cosmetics. However, a significant hindrance in advancing biopolymer production lies in the inefficiencies and associated costs of manufacturing and purification processes.

3.2 Natural Biopolymers

Natural biopolymers are a desirable alternative in a variety of uses in commerce, healthcare, and biotechnology due to their widespread availability, biodegradability, low toxicity, versatility, efficiency, sustainability, and biocompatibility. In terms of their chemical structures, natural polymers including polysaccharides, polynucleotides, and polypeptides, are divided into three distinct groups (Bashir et al. 2020). Polysaccharide-derived polymers and proteins, sourced from both animals and plants, are distinct from natural biopolymers isolated from biomass. Examples of protein-based biopolymers include keratin, zein, collagen, soy protein, fibrinogen, gelatine, wheat gluten, microfibrillar proteins, alginate, egg protein, whey protein, and milk protein. Carbohydrate-derived polymers include agar, cellulose, starch, pectin, kappa carrageenan, alginate, galactans, chitosan, and natural gums. Synthetic polymers are produced chemically from petroleum-based materials such as polycaprolactone and polyvinyl alcohol (PVA), as well as biomass-based materials like polylactic acid (PLA). Biopolymers synthesized by microbes, such as levan, gellan gum, curdlan, xanthan gum, dextran, polyhydroxybutyrate-co-hydroxyvalerate (PHBV), and polyhydroxyalkanoates (PHAs), constitute the third group.

Condensation dehydration processes, which release water to join discrete and small groups of monomers, are the standard procedure that forms biodegradable polymers. The condensation of 20 different kinds of amino acids results in proteins. Four different types of nucleotides condense to form polynucleotides. The primary polymer in the the natural world, cellulose, is produced when one kind of monomer, glucose, condenses. Less than 20 distinct monosaccharides are present in complex cell-surface polysaccharides (Runnels et al. 2018).

3.2.1 Polysaccharides

These high-molecular-weight polysaccharides often contain more than one hundred monosaccharide molecules; in certain cases, they may even contain more than 100,000. They can have either a linear or branching structure; the latter might be made up of branch-on-branch or linear branches that are arranged differently on a linear backbone. Based on their chemical structure, polysaccharides can be classified into two main groups: homopolysaccharides and heteropolysaccharides. Based on their charge, polysaccharides can also be categorized into cationic polysaccharides (e.g. chitosan, chitin), anionic polysaccharides (e.g. hyaluronic acid, chondroitin sulfate, alginic acid, and heparin), and nonionic polysaccharides (e.g. starch, cellulose, and dextran) types.

To achieve specific biological activity, polysaccharides including chitosan, alginate, cellulose, xanthan gum, guar gum, collagen, fibrinogen, keratin, hydroxyethylcellulose, carboxymethylcellulose, gelatine, scleroglucan, welan gum, schizophyllan, lignin, zein, starch, wheat gluten, microfibrillar proteins, soy protein, egg protein, whey protein, and milk protein, are frequently utilized either directly or in combinations with other polymers by copolymerization. Polysaccharides and components with regenerated cellulose are frequently used as essential ingredients in products for packaging. Polysaccharide-based materials possess free hydroxyl groups, which facilitate their simple modification for a range of purposes. For example, in the presence of glycerol, xanthate groups (–O–(S)C–S–) or ester groups (–O–(O)C–) can be added, offering flexibility in their modification. The various functional groups that compose polysaccharides—such as the carboxyl group (–COOH), amino group (–NH_2), and hydroxyl groups (–OH)—all play a role in how well the molecules bond to tissues in the body. The increased adhesiveness to mucous layers, which enhances medication absorption, is explained by this chemical reaction. Consequently, polysaccharide-based nanostructures have become popular and are employed for carrying a variety of pharmaceuticals.

Polymers based on polysaccharides from marine organisms exhibit potential as wound dressings because of their accessibility, low cost, and biological properties. Porous sodium alginate/chitosan composites containing green antibacterial agents were developed by Jiang et al. to assist in the healing of infected wounds. A suitable dressing for a wound should be highly antibacterial and foster an environment that is conducive to the regeneration of skin cells. Using sericin, they synthesized silver nanoparticles, which they subsequently combined with curcumin to produce an antibacterial agent. Following that, a double cross-linked three-dimensional network constructed from chitosan and sodium alginate encapsulated this hybrid material. Excellent retention of moisture, permeability, durability, and strong antimicrobial activity against Pseudomonas aeruginosa and Staphylococcus aureus were all showed by the resultant biocomposite materials (Jiang et al. 2023).

3.2.2 Polynucleotide

All living organisms use polynucleotides, such as deoxyribonucleic acid (DNA) and ribonucleic acid (RNA), as biopolymers that carry genetic data. In terms of structure, RNA is primarily single-stranded, but DNA is composed of up of two spiral chains of polynucleotide that are organized in a helical form. The backbone, which is composed of up of ribose and deoxyribose in DNA and RNA, respectively, is another distinction. Guanine, adenine, cytosine, and thymine are the primary nucleic acids found in DNA, whereas uracil, is found in RNA. Furthermore, the distinct structure of the DNA biomolecule offers an array of attributes absent from other biopolymers. In addition to organic DNA being mentioned as a possible substrate for data storage uses, certain molecular reactions have been shown to result in higher light emission and higher energy conversion efficiency.

In recent years, researchers worldwide have been concerned with improving the long-term viability and excellence of environmentally friendly products to protect biodiversity and the natural world. Organic dyes have been the subject of extensive research recently because of their intriguing photosensitive qualities. These are becoming a significant class of biodegradable optical compounds for contemporary uses in optical and photonics, protection, and health. The DNA modified with various synthetic dyes, such as thiazole orange, Disperse Red 1, Rhodamine 6G, Rhodamine B, and cyanine dyes, provides novel compounds with advantageous uses in photonics. A novel class of luminescent materials with intriguing photo-sensitive capabilities is developing biopolymers loaded with organic dyes (Kawabe and Kato 2012).

3.2.3 Polypeptides

Recently, proteins from many different sources have been extensively studied as potentially biodegradable polymers due to their dramatically unique physical, chemical, and biological features. Using approaches from self-assembled biodegradable materials and biochemical technology, polypeptide-based materials are deeply investigated. Recent studies address the basic concepts and applications of polypeptide-based biodegradable polymers, as well as the current state of the field's bioresearch and potential future approaches. The process of chemical synthesis is one way of synthesizing peptide-based biodegradable polymers, which can also include biologically ordered polypeptides such as poly(amino acid) and elastin-based polypeptides, as well as biodegradable polymers with recurring peptide units. The biopolymers are currently being employed for a wide range of motives, particularly in the pharmaceutical and food sectors. These include drug delivery systems, therapeutic targets, anticancer therapeutic agents, protein purification, injectable scaffolds bio-gels, and tissue engineering applications. Improved uses are also being investigated at the same time. While, in comparison to other synthetic materials, polypeptide-based biopolymers have several advantages and drawbacks. These consist of:

- The drawback of having few structural components.
- The directed and self-assembled polypeptide-based biopolymers as suitable approaches, which have drawn a lot of research attention for the development of therapeutic biomaterials; and.
- The incapacity to use them as platforms for tissue engineering bio-based materials and drug delivery systems owing to their rapid biodegradation in the circulatory system (Nasrollahzadeh et al. 2021).

Two common methods used to design polypeptides are the recombinant DNA approach and stepwise solid-phase polypeptide fabrication. Recombinant DNA technological advances: this technique generates specific polypeptides by modifying DNA materials. Researchers can introduce genes containing codes for certain polypeptides into host living things, such fungi, or bacterial cells, and use them as manufacturers for synthesizing proteins. Finding the genetic material that codes for the required polypeptide and introducing it into a vector, is the first step in the procedure. After being inserted into the host's cell, the viral genome replicates alongside the DNA of the living body. Upon growth and division, the environment that supports the species generates the essential polypeptide. The chemical method of stepwise solid-phase polypeptide synthesis (SPPS) entails constructing the polypeptide one amino acid at a time, beginning at the C-terminus, and working your way up to the N-terminus. Usually, SPPS attaches the initial amino acid to a stable polymer or resin.

3.3 Chemically Synthesized Biodegradable Polymers

Biodegradable polymers can be produced using various methods. For example, they can be synthesized through fermentation, extraction, bulk synthesis of biopolymers, as well as endo- and exo-biopolymer production. Chemical synthesis of monomer units, which can be broken down by enzymes, microorganisms, or biomass, is an additional production technique. The following is an overview of some widely used methods to produce biodegradable polymers from both synthetic and organic materials.

For tissue engineering applications, Ragothaman et al. developed porous 3D scaffolds by combining collagen and poly(dialdehyde)guar gum and adding a growth agent (Ragothaman et al. 2014). The sodium periodate technique was utilized to produce biopolymer blends. These scaffolds included hybrids of covalently linked collagen-based biopolymer blends, in which collagen-based biopolymer blends were necessary for bonding biopolymers by creating imine groups and stabilizing collagen. Up to 13 days of continuous biopolymer blends release from the scaffolds has been found in in vitro tests. Moreover, the 3D hybrid scaffolds showed impressive proliferation of fibroblast cell populations, highlighting the capacity to facilitate tissue regeneration and recovery. In another study, Wang et al. investigated the effect of porcine fibrinogen biopolymers wound dressing and poly(l-lactide-co-caprolactone) on a preclinical animal wound model (Ragothaman et al. 2014). According to their studies, biopolymers significantly altered the diabetes-related wound microenvironment, re-epithelialization, the development of mature arteries, and supporting dermal repair. In addition, the external governance of biopolymers patch exhibited exceptional performance to encourage wound healing independent of the wound's diabetic or physiological state. According to this, there is a lot of opportunity for biopolymers dressing as a therapeutic tool that can help in the healing of wounds. In 2023, Zheng and colleagues prepared a new film consisting of xanthan gum, flaxseed gum, carboxymethyl cellulose, and polygonatum cyrtonema extracts. Because polyphenols, flavonoids, and polysaccharides, are included in extracts, adding them reduced light absorption while improving UV block efficiency, antioxidant capacity, mechanical properties (Zheng et al. 2023).

3.3.1 Extraction Methods of Biodegradable Polymers

The selection of appropriate solvents and the design of green processes are of paramount importance for the benefit of the environment and future generations. Consequently, the use of solvents in the production of green and sustainable polymers derived from renewable sources will be pivotal in processing industries as the 'solvents of the future'. Nowadays, it's common knowledge that new solvent structures, often known as "neoteric solvents," are environmentally friendly methods of digesting biomass and producing novel advanced materials. Green solvent systems provide an alternative for dissolving and processing biopolymers, even though they tend to dissolve in water-based solutions and treated with a variety of chemicals. The ionic liquids and deep eutectic solvents are two of the few green solvent types that are gaining popularity as alternate solvent options for processing and dissolving various biopolymers. These solvent systems have several special qualities, including low vapour pressure, low volatility, thermal resistance, high recycling capacity, high selectivity, high biodegradability, and more (Prasad and Sharma 2019).

Especially in biomedical studies and technology, chitin-based materials are one of the most attractive abundant polysaccharide polymers for creating structures with sophisticated capability due to their distinctive characteristics. Owing to their biodegradable and biocompatible characteristics, they have a wide range of uses in biomedical and therapeutics. Chitin can make up to 30% of the number of crustacean shells produced by the seafood industry. This is a significant source of waste that comes with a high cost of pollution. Chitin is primarily extracted chemical method from shrimp shells on a commercial basis. Chitin is a polysaccharide derived from N-acetyl-D-glucosamine groups that is mostly found in shrimp shells, insect species, and microbes including algaes, fungus, and yeast cells. It is the second most abundant biopolymer on Earth after cellulose. Demineralization in an acid medium (such as HCl) and deproteinization in a basic medium (such as NaOH) are the two steps in the extraction of chitin. On the other hand, acidic mediums (e.g., HNO_3, H_2SO_4, CH_3COOH, CH_2O_2) can cause the deproteinization process. Usually, industrial extraction uses NaOH. The method of deacetylation into chitosan usually uses alkaline solutions at high temperatures. Organic solvents may be utilized for decolorization. Although practical, commercial chemical techniques have drawbacks for the environment and product performance, such as reduced viscosity and molecular weight. The lengthy nature of the procedure calls for the development of more effective techniques. Due to its ability to reduce response times, boost yields, and conserve energy, microwave irradiation has become a more effective heating technique than traditional techniques. Chemical modifications that take hours can be performed in a matter of minutes by using microwave radiation for three-dimensional heating. Numerous studies have exhibited the beneficial effects of microwave irradiation in several phases of chitin extraction, such as deacetylation, demineralization, and deproteinization. Microwave-assisted extraction takes much less time—typically 20–30 min—than traditional procedures and provides greater quantities and higher degrees of deacetylation. This makes industrial-scale microwave extraction financially viable. The molecular weight, degree of deacetylation, and crystallinity of chitin and chitosan can all be affected by microwave irradiation, which may influence their practical characteristics like dissolution, viscosity, and durability. As a result, careful adjustment of the conditions of extraction is essential to guarantee the preservation of the intended properties. These days, ultrasonic-assisted extraction is another cutting-edge technique that is widely employed due to its advantages, which include decreased extraction times, economic benefits, lower consumption of energy, and increased extraction productivity. Typically, high-intensity ultrasound is the basis of the ultrasonic technology process, which speeds up the extraction of solid substance from a liquid solvent by creating cavitations in the liquid solution. The studies that have been provided make it clear that ultrasonication is a viable and alternative method of extracting chitosan with high yields. Up to 50 kHz, ultrasonic radiation was used in conjunction with capacities ranging from 50 W to 400 W. As a result, ultrasonic irradiation proved to be an effective method for chitin deacetylation, requiring less reaction time than the traditional procedure and consuming less energy. Overall, research is needed to completely understand the impacts of ultrasonic extraction on the qualities and attributes of the extracted components, as well as to optimize the extraction conditions, even if this technology has shown itself to be a feasible one for chitin and chitosan extraction. The advancement of effective and sustainable extraction techniques is crucial for the further acceptance and integration of chitin and chitosan across diverse sectors. The use of bacteria and proteolytic enzymes in biological techniques has become more popular since it is more secure, fewer resources and solvent-free, and more sustainable. Proteases such as papain, alcalase, pancreatin, pepsin, trypsin, and devolvase, are among the proteases that can be employed by enzyme-based extraction to deproteinize a variety of invertebrate shells, such as shrimp, crab, acetes shrimp, mealworms, and squid gladius. In summary, different methods are used to extract chitin and chitosan, each with its own benefits and applications. Chemical extraction typically involves the use of bases, acids, or organic solvents to perform demineralization, deproteinization, decolorization, and deacetylation. Ultrasonic-assisted extraction utilizes high-intensity ultrasonic waves to expedite the extraction process, while microwave-assisted extraction employs microwave energy for quicker and more effective procedures. Biological extraction utilizes proteolytic enzymes and bacteria, providing a more environmentally friendly and safe option compared to chemical techniques. These methods are essential for extracting chitin and chitosan, facilitating their integration into various industries while considering performance and sustainability factors (Ben Aoun et al. 2024).

3.3.2 Polymerization of Biodegradable Polymers

Biodegradable polymers termed biopolymers are present in living things. A biodegradable polymer has a long chain molecule composed of monomers joined by covalent bonds to form a molecule that is biodegradable. Biopolymers primarily originate from organic materials such as bacteria, plants, and trees. Compared to biopolymers, which are intricate molecules with clearly defined three-dimensional structures, synthetic polymers are more random and simpler. Small molecules known monomers—the majority of which are organic—may unite with other monomers to form polymer compounds, which are larger molecules in size. All monomers can form chemical links with two or more additional subunits. Synthetic compounds called polymers are composed of multiple smaller units known as monomers. Biopolymers compounds contain chains with an arbitrary number of monomer units. Polymerization, a chemical process, creates biopolymers by combining many monomer units together. Biopolymers with a linear or branching morphology can be produced by the polymerization process. Biodegradable polymers may also resemble a complex, three-dimensional network. Biopolymers, which include proteins (polymerization of amino acids), nucleic acids (polymerization of nucleotides), and polysaccharides (polymerization of sugars), are biological macromolecules created through the polymerization of monomeric units. A sequence of distinct reactions that match to the activities of the reacting substances and their steric impacts come after the polymerization stage.

The commonly used poly(lactic acid) (PLA), a linear aliphatic polyester, holds great promise for global sustainable development. PLA is derived from a thermoplastic monomer and possesses superior properties. For example, it offers high strength, chemically inert, low heat distortion temperature, low melting point, low thermal resistance, poor toughness, high hydrophobicity, low rate of crystallization, and excellent layer adhesion. The largest amount of PLA synthesized today originates from the polymerization of lactide, which is formed from lactic acid through ring-opening. The ability to produce high quality lactic acid serves as the foundation for the development of PLA bioplastics. Furthermore, lactide, a crucial intermediate for PLA fabrication, has a significant influence on the entire PLA commercial network. Ring-opening polymerization and polycondensation have been utilized in PLA fabrication thus far. Azeotropic dehydration, Direct polycondensation, second order melt polycondensation, and third-order solid phase polycondensation processes are all included in polymerization. Under comparable circumstances, ring-opening polymerization usually produces PLA with a high molecular weight than polycondensation. The dominance of ring-opening polymerization may be challenged to some extent by recent developments in azeotropic dehydration, second-order melt polycondensation, and third order solid phase polycondensation approaches, which have demonstrated the ability to generate high molecular weight polymers. Nevertheless, the most efficient method for producing high molecular weight PLA on a wide scale is still ring-opening polymerization of lactide (Yu et al. 2023).

3.3.3 Fermentation Method of Biodegradable Polymers

Nevertheless, the significant energy requirements of these two procedures will further increase manufacturing expenses. Recently, scientists have been focusing more on the search for innovative, sustainable extraction and filtration methods. From this perspective, additional studies are needed to determine the ideal recovery approach that will be least expensive, most efficient, and have minimal adverse impacts on the environment when applied in a commercial setting. This includes exploring the use of safe surfactants, green solvents, ultrasound-based extraction, and floating air dissolved methods.

Anaerobic metabolism is utilized to generate renewable energy, such as natural gas and bio-hydrogen, while aerobic fermentation has primarily been used to produce biopolymers. Researchers have reported a variety of microorganisms in their experimental studies, including recombinant *Bacillus species* (spp.), *Escherichia coli (E. coli)*, *Pseudomonas species* (spp.), *Alcaligenes species* (spp.), *Nocardia species* (spp.), *methylotrophs, Rhizobium species* (spp.), and *Azotobacter species* (spp.) (Yadav et al. 2023).

Microbial biodegradable polymers are biopolymers formed through fermentation using microorganisms that have certain sources of nitrogen, carbon, mineral substances, and salts. These microbial biopolymers are produced mostly because of their defence system or storage component. Generally, fermentation is used by microbes, fungi, and algal to generate biopolymers. For instance, Bacillus sp. produces polyhydroxyalkanoates (PHAs), whereas Xanthomonas campestris produces xanthan gum. For the purpose of producing biopolymers, various biological reactors can be constructed for both experimental and continuous fermentation procedures. The solvent composition, product yield, waste management strategy, and manufacturing expenses should all be thoroughly assessed before commencing operations. The primary factor adversely impacting the overall finances of large-scale polymer fabrication is the cost of raw materials. Consequently, several low-cost carbon sources are typically utilized, including syrup, milk protein, glucose, active sludge, and maple syrup. To produce bacteria biomass in the fermenter, a great deal of these sources must first be examined, prepared, or evaluated which raises the prior expenses for polymer synthesis. The downstream stage is crucial and significantly affects the total cost of synthesis. The recovery of biopolymers from bacterial culture is now accomplished mostly using solvent-based extraction approaches; nevertheless, the use of these halogenated solutions compromises the preservation of the environment. Two important processes in the solvent extraction and chemical decomposition method that result in pure biopolymers are sedimentation and filtration (Das et al. 2023).

3.3.4 Sonosynthesis of Biodegradable Polymers

An increasing amount of research endeavors have concentrated on the sonochemical approach because of its many applications in recent years. Because it does not harm the natural environment, ultrasonic cavitation is extremely valuable in producing novel biopolymers, especially when it comes to changing naturally formed biodegradable polymers. According to examinations, ultrasonic effect is consistent with modern techniques of biomaterial preparation since it improves the mechanical characteristics, morphological properties, agglomeration, and crystal structure of materials (Cai et al. 2022).

Ultrasonic waves, with frequencies ranging between 2×10^4 and 10^7 Hz, are categorized as a type of mechanical wave. The development of bio-based materials aided by sound is linked to the sonosynthesis approach, which is the chemical consequence of the phenomenon known as acoustic/ultrasonic cavitation. When

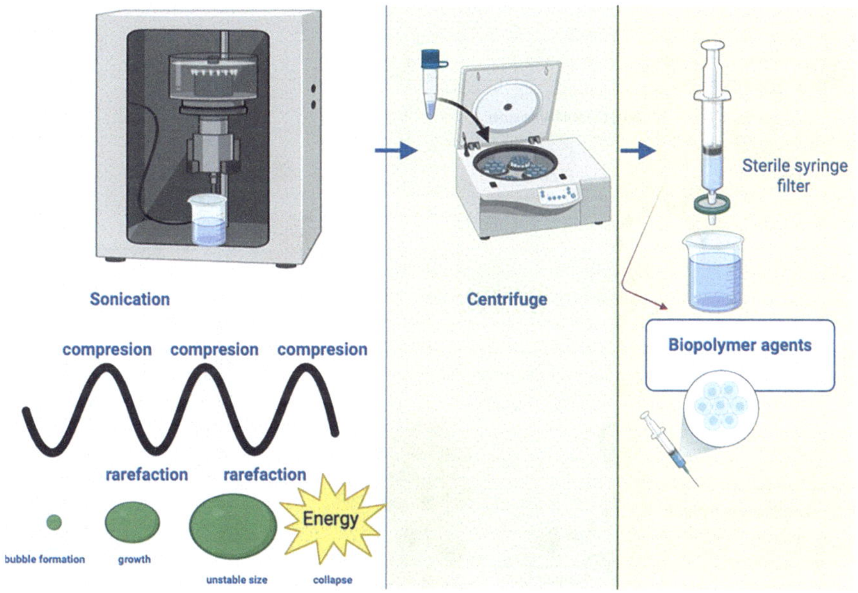

Fig. 3.1 Sonosynthesis of biopolymer agents

sufficient acoustic waves pass through a liquid, bubbles begin to form, expand, and quickly collapse a process known as sonochemical synthesis. In Fig. 3.1, the sonosynthesis of biopolymers are presented. Up to 5000 K of temperature and 1000 atm of pressure are produced by the combined mechanical power of the acoustic bubbles before they collapse, transforming into heat as well as additional types of energy. The heat produced by the high temperatures and high pressure in the region is sufficient to break all bonds between molecules and form many free radicals, which are then used to interconnect biodegradable polymers that are useful. The chemical reaction is divided into primary sonochemical reaction, and secondary sonochemical reaction categories based on the reaction processes involved. Secondary sonochemical reaction originates when radical species spread into the aqueous solution from the acustic air pockets, whereas primary sonochemical reaction is based on the sono-reaction started by the molecules of gases inside the collapse of acoustic bubbles. There are numerous variables that influence the synthesis of biopolymers, including surface tension, sonochemical frequency, acoustic power, sonochemical time, viscosity, probe types, solvent volume, and the type of free radicals. Cavitation significantly impacts the yield in sonochemical processes. A two-phase system can generally be effectively sonicated at the interfaces to prepare biopolymers such as bovine serum albumin, chitosan, xanthan gum, egg white protein, gum acacia, lysozyme, sodium alginate, and starch, among others. Usually, this sono-process operates at frequencies of about 20 kHz, with durations of between 30 s and 5 min and power densities of between 23 W/cm^2 and 500 W/cm^2. In the aqueous medium, it consists of proteins, polysaccharides, or their bio-composites to

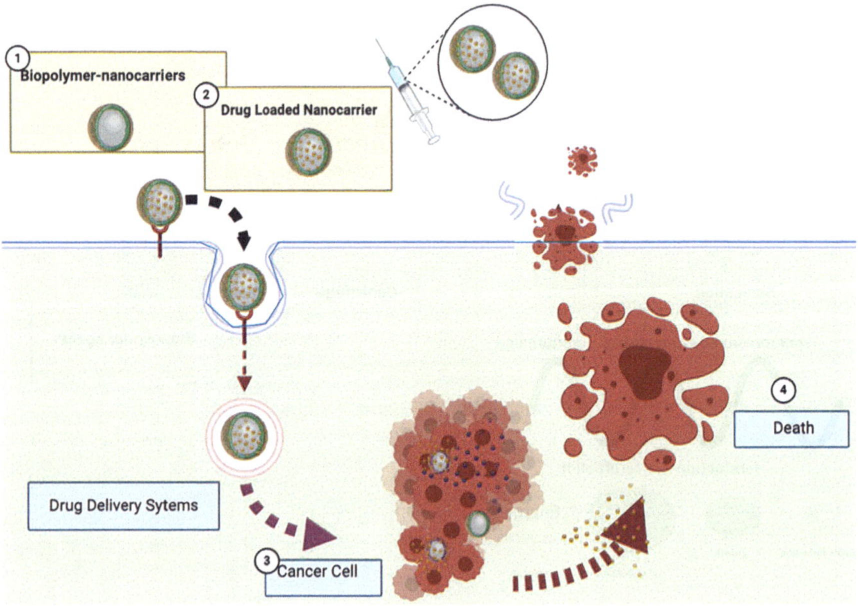

Fig. 3.2 Drug delivery systems of drug-loaded biopolymer nanocarriers for cancer therapy

form biopolymeric formations; in the oil medium, it contains hydrophobic bioactive components. The sonochemically manufactured aqueous core-shell biopolymer based nanocapsules that are in great demand, especially in clinical settings where they are used for producing cross-linked water/oil (W/O) formulations. The ideal pH must be reached for each carbohydrate and protein mixture to achieve a negative charge during the W/O synthesis process. This is a crucial phase since it helps to reduce the electrostatic interaction that forms after mixing these biopolymer-based mixtures. The blend of biopolymeric solutions comprising hydrophilic bioactive and oil components is subjected to sonication under specific conditions in the next phase of producing W/O emulsions. The outcome of the filtration procedure is a liquid cross-linked encapsulated bio-based structure. Generally, core-shell biopolymer composites or nanocarriers are formed through covalent bonds and electrostatic interactions and exhibit redox/pH sensitive properties. The drug delivery mechanisms of drug-loaded biopolymer nanocarriers intended for cancer treatment are illustrated in Fig. 3.2. These extraordinary properties ensure effective and highly efficient drug release in the cellular environment and gastrointestinal tract in drug release studies, thus facilitating controlled release. In the light of this information, it becomes clear that double-effect structures resulting from the combination of pH-sensitive strategies with a simple and environmentally friendly sonochemistry method can be produced and multifunctional biopolymer composites can play an important role in smart redox/pH sensitive drug delivery systems (Tan et al. 2022). While sonochemical synthesis is an attractive method for creating a wide range of structures, the delivery of biomass-derived chemicals using this approach has only recently gained attention. Previous studies have addressed the primary types of cross-linked biopolymer agents available on the market for biomedical applications, including aqueous core-shell nanocarriers, surface-modified biopolymer composites or blends, biopolymer composites or blends, cross-linked biopolymer composites or blends, and aqueous core-shell-shell nano/microcapsules.

In summary, the sonosynthesis process of manufacturing these vehicles largely relies on the chemical changes that occur during ultrasonic cavitation, initiating the cross-linking of biopolymers at liquid interfaces. Within ultrasonic cavitation, various polysaccharides and proteins can bind together. Resultantly, the modified vehicles exhibit high encapsulation performance, robust tolerance to environmental stresses, and pH-responsive release characteristics.

3.3.5 Solvent Casting Method of Biodegradable Polymers

The main advantages of the solvent casting process over other processes such as spin coating, microfluidic spinning, salt leaching, and three-dimensional printing are its low cost, ease of usage, and ability to create robust biofilms with superior physical characteristics and uniformity. External factors, however, can also cause significant differences in the features of the components created by the solvent casting technique between production batches, which might impede manufacture (Borbolla-Jiménez et al. 2023). The achievement of the objectives of manufacturing biodegradable polymers films depends on addressing the difficulty of maintaining a sterile environment during the whole production process in biomedical applications. The quantity of standardized commercial products is noticeably lacking, despite the method's great promise for clinical use and industrial scalability in the development of biodegradable polymers or composites (Anbukarasu et al. 2015). One of the most frequently employed and popular ways of manufacturing biodegradable materials based on polysaccharides is solvent casting. In order to develop environmentally friendly polymers, firms have additionally adopted electrospinning method, layer-by-layer assembled method, phase immersion method, injection molding method and extrusion method, as polymer synthesis methods. A more attractive approach to developing biopolymer-based films with extremely high quality parameters is solvent casting. The biopolymer was properly mixed into the solution; following the homogenization process the highly viscous solution should have a small amount of solid material; and the film should form and be simply removed from the casting support. These were the primary principles of this synthesis process. The solvent casting method has higher transparency, more uniform thickness dispersion, and higher quality compared to extrusion. Nevertheless, this method has more drawbacks than benefits. These include the need for a lengthy drying period, film shrinkage after drying, low fracture resistance, uneven distribution of drugs, and various unfavorable storage conditions. A disadvantage of this method is the "rippling effect," which causes the film's shear stress to become problematic and ends up resulting in a lack of uniformity. Nucleate boiling, a different term for bubble creation, can occasionally be attributed to over-drying. Plasticizing agents can be added in conformity with the polymer mixture in order in order to alleviate these limitations. This will decrease the cohesion intermolecular tension within the polymers, which leads to decreased tension strength and increased elasticity over the films (Das Adhikary 2023).

3.3.6 Electrospinning Method

The research on the use of bioresorbable cardiovascular implants has been stimulated, in part, by the current rise in fascination with electrospinning. One of its primary attractions is the fact that electrospinning is a straightforward technique requiring minimal professional supplies. An ordinary electrospinning apparatus typically consists of a spinneret, needle pump, grounded target, and high voltage power source usually an instrument with a blunt tip needle. The method itself is effectively conducted out in an experimental setting and may be readily scaled up for production on a large scale with little changes. To put it briefly, a biopolymer solution is electrically charged with a high voltage, often several kilovolts, and carried into the spinneret. To produce a static electric field, a grounded target is set up at a specific distance away from the charged polymer solution. At the extremity of the spinneret, a fine jet of entangled polymer networks is pulled out when the surface tension is overcome by the electrostatic repulsion of the biopolymers, reaching a critical point when the electrical potential achieves a critical importance. The solvent in the biomaterial evaporates as this jet hurtles through the air in the direction of the grounded target, forming a dry bio-based fiber. Depending on how the biopolymer mixture or solvent is treated, the electrospinning method will produce fibers with a diameter ranging from 50 nm to 10 μm that are gathered on the grounded target (Sell et al. 2009). In 2020, Gutierrez-Gonzalez et al. manufactured alginate/polyethylene oxide biodegradable polymers-based nanofibers with curcumin (Gutierrez-Gonzalez et al. 2020). The mechanical characteristics of the nanofibers were shown to be greatly improved by crosslinking with trifluoroacetic acid and curcumin, with the latter method significantly exceeding the benefits of the former. They also found that electrospinning at high humidity levels did not require a regulated environment when a spinning drum was employed. With an emphasis on protein delivery uses, Rafiei et al. developed 3D fibrous biodegradable polymers-based scaffolds composed of polycaprolactone (PCL) by coaxial electrospinning method (Rafiei et al. 2020). A scaffold containing biodegradable core-shell fibers was successfully created by wet electrospinning; it has a greater porosity and hydrophilic properties than traditional two-dimensional (2D) biodegradable biopolymer mats. Proteins could be efficiently loaded into the fiber core thanks to this biodegradable structure design, which also provided controlled release properties that were better than those achieved through blended electrospinning approaches. Cell cytotoxicity tests were used to evaluate the biodegradable polymers-based scaffold's biological compatibility and determine its suitability for a range of biomedical uses. Overall, the 3D fibrous biodegradable polymers-based scaffold that was created showed potential as an adaptable structure for tissue engineering and regenerative drug applications including regulated protein delivery systems. In 2022, Mosayebi et al. developed a needleless electrospinning approach to create biodegradable polymers nanofibers composed of gelatine and Spirulina protein concentration. Because biodegradable polymers were added, the resulting biodegradable nanofibers had higher antioxidant activity than pure gelatine nanofibers. Furthermore, the electrospun biodegradable nanofibers demonstrated ultrafast dissolving properties in an aqueous media, highlighting their potential use in applications requiring rapid dissolution. Wound dressings, drug delivery systems, and food packaging materials are just a few of the biomedical and medicinal uses that could benefit from this novel manufacture technique and the improved features of the biodegradable composite nanofibers (Mosayebi et al. 2022). In 2023, a novel method of covering caffeine in multilayered nanofibers formed from gelatin, polyvinyl alcohol (PVA), and anionic hydroxyethyl starch was developed by Kazemianrad et al. (Kazemianrad et al. 2023). Caffeine's thermal resistance was enhanced through encapsulation when compared to its natural form. Moreover, the multilayered nanofibers' reduced caffeine release rate suggests regulated release features. A non-Fickian diffusion model described the release process, indicating a complicated interaction between factors affecting the release kinetic models. This novel encapsulating method has a chance of application in many different kinds of applications, including cosmetics, drugs, and functional foods, that call for controlled caffeine delivery. In summary, a needleless electrospinning technique developed enabled the creation of biodegradable micro/nanofibers. These biodegradable materials exhibited enhanced biological activity and ultrafast dissolving properties, showing promise for applications in wound dressings, drug delivery systems, and food packaging materials.

References

Anbukarasu P, Sauvageau D, Elias A (2015) Tuning the properties of polyhydroxybutyrate films using acetic acid via solvent casting. Sci Rep 5(1 5):1–14. https://doi.org/10.1038/srep17884

Bashir S, Hina M, Iqbal J et al (2020) Polymers fundamental concepts of hydrogels: synthesis, properties, and their applications. Polymers 12(11):2702. https://doi.org/10.3390/polym12112702

Ben Aoun R, Trabelsi N, Abdallah M et al (2024) Towards a greener future: exploring the challenges of extraction of chitin and chitosan as bioactive polysaccharides. Mater Today Commun 39:108761. https://doi.org/10.1016/J.MTCOMM.2024.108761

Borbolla-Jiménez FV, Peña-Corona SI, Farah SJ et al (2023) Films for wound healing fabricated using a solvent casting technique. Pharmaceutics 15(7):1914

Cai B, Mazahreh J, Ma Q et al (2022) Ultrasound-assisted fabrication of biopolymer materials: a review. Int J Biol Macromol 209:1613–1628. https://doi.org/10.1016/j.ijbiomac.2022.04.055

Das Adhikary N (2023) Recent advances in different fabrication techniques to formulate biodegradable plastic and their effects on functional properties of the film: a mini-review Nibedita Das Adhikary. Pharma Innov J 12:1760–1766

Das S, Das P, Lal B (2023) Development of microbial biopolymers: the eco-friendly sustainable products for environmental applications. In: Novel polymeric materials for environmental applications. World Scientific Publishing Company. https://doi.org/10.1142/9789811265938_0001

Gutierrez-Gonzalez J, Garcia-Cela E, Magan N, Sameer Rahatekar S (2020) Electrospinning alginate/polyethylene oxide and curcumin composite nanofibers. Mater Lett 270:127662. https://doi.org/10.1016/j.matlet.2020.127662

Jiang M, Li S, Ming P et al (2023) Rational design of porous structure-based sodium alginate/ chitosan sponges loaded with green synthesized hybrid antibacterial agents for infected wound healing. Int J Biol Macromol 237:123944. https://doi.org/10.1016/j.ijbiomac.2023.123944

Kawabe Y, Kato S (2012) Spectroscopic study of cyanine dyes interacting with the biopolymer, DNA. Dyes Pigm 95(3):614–618. https://doi.org/10.1016/j.dyepig.2012.07.002

Kazemianrad F, Koocheki A, Ghorani B (2023) Encapsulation of caffeine in sandwich structured Alyssum homolocarpum seed gum/PVA/gelatin nanofibers using electrospinning technique. Food Hydrocoll 140:108604. https://doi.org/10.1016/j.foodhyd.2023.108604

Mosayebi V, Fathi M, Shahedi M et al (2022) Fast-dissolving antioxidant nanofibers based on Spirulina protein concentrate and gelatin developed using needleless electrospinning. Food Biosci 47:101759. https://doi.org/10.1016/j.fbio.2022.101759

Nasrollahzadeh M, Maham M, Nezafat Z, Shafiei N (2021) Protein and polypeptide biopolymer chemistry. In: Biopolymer-based metal nanoparticle chemistry for sustainable applications, pp 107–144. https://doi.org/10.1016/B978-0-12-822108-2.00005-3

Prabu Udayakumar G, Muthusamy S, Selvaganesh B et al (2021) Ecofriendly biopolymers and composites: preparation and their applications in water-treatment. Biotechnol Adv 52:107815. https://doi.org/10.1016/j.biotechadv.2021.107815

Prasad K, Sharma M (2019) Green solvents for the dissolution and processing of biopolymers. Curr. Opin. Green Sustainable Chem. 18:72–78. https://doi.org/10.1016/j.cogsc.2019.02.005

Rafiei M, Jooybar E, Abdekhodaie MJ, Alvi M (2020) Construction of 3D fibrous PCL scaffolds by coaxial electrospinning for protein delivery. Mater Sci Eng C Mater Biol Appl. 113:110913. https://doi.org/10.1016/j.msec.2020.110913

Ragothaman M, Palanisamy T, Kalirajan C (2014) Collagen-poly(dialdehyde) guar gum based porous 3D scaffolds immobilized with growth factor for tissue engineering applications. Carbohydr Polym 114:399–406. https://doi.org/10.1016/j.carbpol.2014.08.045

Runnels CM, Lanier KA, Krish Williams J et al (2018) Folding, assembly, and persistence: the essential nature and origins of biopolymers. J Mol Evol 86:598–610. https://doi.org/10.1007/s00239-018-9876-2

Sell SA, McClure MJ, Garg K et al (2009) Electrospinning of collagen/biopolymers for regenerative medicine and cardiovascular tissue engineering. Adv Drug Deliv Rev 61:1007–1019. https://doi.org/10.1016/J.ADDR.2009.07.012

Tan C, Zhu Y, Ahari H et al (2022) Historical Perspective Sonochemistry: An emerging approach to fabricate biopolymer cross-linked emulsions for the delivery of bioactive compounds. Adv. Colloid Interface Sci. 311:102825. https://doi.org/10.1016/j.cis.2022.102825

Yadav J, Ray S, Soni M, Kashyap BK (2023) Microbial fermentation system for the production of biopolymers and bioenergy from various organic wastes and by-products. In: Kashyap BK, Solanki MK (eds) Current research trends and applications in waste management. Springer, Singapore, pp 307–339. https://doi.org/10.1007/978-981-99-3106-4_12

Yu J, Xu S, Liu B et al (2023) PLA bioplastic production: from monomer to the polymer. Eur Polym J 193:112076. https://doi.org/10.1016/j.eurpolymj.2023.112076

Zheng M, Su H, Xiao R et al (2023) Effects of Polygonatum cyrtonema extracts on the antioxidant ability, physical and structure properties of carboxymethyl cellulose-xanthan gum-flaxseed gum active packaging films. Food Chem 403:134320. https://doi.org/10.1016/j.foodchem.2022.134320

Alper Durmaz received his Master of Science degree in botany from Ondokuz Mayis University in 2016, and his Ph.D in botany from Ondokuz Mayis University in 2021. He has currently an Research Assistant at Artvin Çoruh University, Turkiye. He has research experience in plant systematic and plant ecology. He has worked on a variety of projects funded by Ondokuz Mayis University and has published several review and research articles.

Erdi Can Aytar is currently a Research Assistant in the Department of Horticulture, at Usak University, Türkiye. He received her Master of Science degree in biology from Gazi University in 2016, and her Ph.D. in biology from Ondokuz Mayıs University in 2024. He has research experience in Antioxidant and antimicrobial activity, cancer, isolation of organic compounds, plant physiology and germination. He has worked on national and international projects. He has published several research articles.

İbrahim Mizan Kahyaoğlu is currently an Research Assistant in the Department of Chemistry, at Ondokuz Mayis University, Türkiye. He received her Master of Science degree in physical chemistry from Ondokuz Mayis University in 2023, and he is in Ph.D. program in analytical chemistry at Ondokuz Mayis University now. He has research experience in drug carrier systems, nanoparticles, nanocomposites, copolymer blends, heavy metal detecting, analysis of organic and inorganic compounds. He has published several research articles and book chapters.

Selcan Karakuş is currently an Associate Professor in the Department of Chemistry, at Istanbul University-Cerrahpaşa (IUC), Türkiye. She received her Master of Science degree in physical chemistry from Istanbul University (IU) in 2006, and her Ph.D. in physical chemistry from IU in 2011. She has been a visiting researcher at the University of Massachusetts, Department of Polymer Science and Engineering. She has research experience in drug carrier systems, nanoparticles, nanocomposites, nanoemulsion self-assembled polymeric nanostructures, and copolymer blends. She has worked on a variety of projects funded by IUC and has published several research articles and book chapters.

Probiotic Bacterial Cellulose: A Bio-mediated Nanomaterial for Health Care Applications

4

Mainak Chaudhuri, Nabanita Saha, and Petr Saha

Abstract

Nanomaterials have been used for various applications in recent years due to their interesting physiochemical properties. Nevertheless, conventional physical and chemical ways of synthesizing nanomaterials are expensive and easily result in hazardous byproducts that are hazardous to the environment and public health. One of the very prominent applications of bio-mediated nanomaterials is in the field of healthcare applications. These nanomaterials are ideal candidates for drug delivery, imaging, and therapeutic applications. Additionally, bio-mediated nanomaterials are promising in antimicrobial therapy. Bacterial cellulose (BC) shows every characteristic of a biopolymer for healthcare applications, having a high surface area-to-volume ratio, tunable surface chemistry, great water-holding capacity, biocompatibility and biodegradability. The biosynthesis of BC using *Komagataeibacter xylinus* is very cost-effective and eco-friendly. BC itself is a very good biopolymer with wound-healing properties. There are various probiotic bacteria that have beneficial effects on our health, such as several *Lactobacillus* sp. have antimicrobial properties against pathogenic bacteria; *Pediococcus pentosaceus* have antimicrobial, antioxidant and antitumor activity. Combining the physiochemical properties of BC as a biopolymer and the medicinal properties of probiotic bacteria, probiotic bacterial cellulose (PBC) can be derived as a bio-mediated nanomaterial for different healthcare applications. This chapter focuses on different probiotic bacteria and their beneficial effects, derivatives of probiotic bacteria responsible for those effects, different methodologies of synthesizing PBC by combining BC and probiotic

M. Chaudhuri · N. Saha (✉) · P. Saha
Centre of Polymer Systems, University Institute, Tomas Bata University in Zlin, Zlín, Czech Republic
e-mail: chaudhuri@utb.cz; nabanita@utb.cz; saha@utb.cz

R. Malviya, S. Sundram (eds.), *Sustainable Green Biomaterials As Drug Delivery Systems*, Biomaterials, Bioengineering and Sustainability 1,
https://doi.org/10.1007/978-3-031-79062-1_4

bacteria and their derivatives, and the applications of PBC in different biomedical applications.

Keywords

Biosynthesis · Nanomaterials · Probiotic bacterial cellulose · Antimicrobial · Healthcare

Abbreviations

BC	Bacterial cellulose
PBc	Probiotic bacterial cellulose
AAD	Antibiotic-associated diarrhoea
RCTS	Randomized controlled trials
IBS	Irritable bowel syndrome
IBD	Inflammatory bowel diseases
LGG	*Lactobacillus rhamnosus* GG
SCFAs	Short-chain fatty acids
EPS	Exopolysaccharides
BSH	Bile salt hydrolases
GABA	Gamma-aminobutyric acid
GI	Gastrointestinal

4.1 Introduction

In recent years, there has been a growing interest in developing novel biomaterials with multifaceted applications in healthcare. Among these, probiotic bacterial cellulose (PBC) has emerged as a promising bio-mediated nanomaterial with immense potential for various healthcare applications. Bacterial cellulose (BC) is a biopolymer produced by different acetic acid bacteria like *Komagataeibacter xyninus*, where the biopolymer is synthesized through the fermentation process. It possess a number of unique characteristics as it is very pure (more than plant cellulose, as BC does not contain lignin and hemicellulose in its structure), has good mechanical properties and it is easily biodegradable and it is also very biocompatible with living cells, making BC to be one of the most preferred polymers to be used in biomedical and healthcare applications. It has been demonstrated that when BC co-cultured with probiotic bacteria PBC is obtained, which is heralded as the dawn of an entirely new protocol to utilize the synergism properties of probiotic bacteria and of BC, which is an innovative avenue in the scope of advanced pharmaceutical and medical fields.

Several researches have been done on synthesising BC by acetic acid bacteria, mainly belonging to different *Komagataeibacter* and *Gluconacetobacter* strains. The synthesis process has been extensively optimized for industrial-scale production in various fields (Gopu and Govindan 2018). These bacteria produce cellulose as a metabolic byproduct extracellularly in micro or nanofiber form, which then self-assembles into a network-like three-dimensional structure that results in a mechanically sturdy structure with high porosity (Florea et al. 2016). BC become a perfect polymer for drug delivery systems, wound dressing, tissue engineering, and other biomedical applications because of its intrinsic properties like high water-holding capacity (98%), oxygen permeability and its porous structure (Gorgieva and Trček 2019). When probiotic bacteria are introduced into the BC mat, the mat shows additional therapeutic functions by increasing its capability for various healthcare applications.

Probiotics are living microorganisms that upon food intake have many health kick that they provide in the body. Probiotics are well researched and shown to have positive effects on gut health, prevention of disease, and immune system support (Hill et al. 2014). This BC merged with probiotics will generate a synergistic delivery system possessing beneficial properties of two partners, leading to enhanced therapeuticsay and targeted delivery of bioactive compounds.

PBC has a wide spectrum of medical applications in areas such as tissue engineering, drug delivery, gut health and wound healing. Compared with those of traditional medical materials, PBC-based wound dressings show advantages in gas permeability, moisture-affinity, and biocompatibility (Jayani et al. 2020). Additionally, the dressings with probiotic bacteria build-positive surrounding for wound healing by bringing in some changes in the local microenvironment to maintain tissue regeneration and minimize the threaten of infection (Savitskaya et al. 2019). Resent researches found that PBC derived dressings are effective for increasing wound healing, reducing inflammation and improving healing and other outcomes (Sabio et al. 2021).

PBC has bright prospects as a probiotic delivery vehicle in the genre of gastrointestinal health. BC as a defensive barrier protects probiotic bacteria to get harsh gastrointestinal environment, so probiotic bacteria can easy to alive and accurately arrive in intestines(Singh et al. 2018; Jayani et al. 2020). This is of particular benefit in enhancing the effectiveness of probiotic interventions in conditions such inflammatory bowel disease, irritable bowel syndrome and antibiotic-associated diarrhoea (Wang et al. 2022). In addition, the sustained release of probiotics from PBC matrices enables the longer residence of probiotics in the intestine consequently, enhancing the therapeutic effect of probiotics and promoting the homeostasis of the gut microbiota (Savitskaya et al. 2024).

Together with its applications in wound healing and gastrointestinal treatments, PBC has gained interest as a prospective substrate for controlled drug delivery and tissue repair. The inherently porous nature or BC serves as a ready-made vehicle for containing and prolonging the release of any number of bioactive payload, including pharmaceutical agents, growth factors, and signalling molecules. By incorporating probiotic bacteria into those framework, PBC-based drug delivery strategies can be sculpted to exhibit major imploring functionalities like site-specific delivery to tissues or disease loci (Yang et al. 2024). In addition, the good biocompatibility and mechanical properties of PBC also make it an ideal scaffold material for tissue-engineering applications by providing a micro-environment that supports cell proliferation, growth, and differentiation (Pang et al. 2020). Inclusion of probiotic bacteria in PBC substrates offers an added advantage for enhancing their healing competence, by promoting epithelial regeneration, balancing the local immune response, and preventing infections (Gregory et al. 2021).

Known as a probiotic bacterial cellulose (PBC), this multifunctional biofabricated nanomaterial presents profound opportunities for their use in healthcare. The integration of unique properties of bacterial cellulose with the beneficial features of probiotic bacteria makes PBC an integrated platform for advanced therapeutic approaches for wound healing, gastrointestinal health, drug delivery, and tissue engineering. The realm of exploration and improvisation in this area seems advantageous such that eventually, the PBC-centric discoveries could be moved towards clinical application inspiring novel interventions for multiple health demands.

Bacterial cellulose (BC) as well as probiotic cellulose exhibit significant advantages with respect to cost-effectiveness and environmental sustainability for medical application. So it is very important to their ecological sustainability that they are produced through controlled bacterial fermentation. Probiotic cellulose and BC are eco-friendly since they are produced from natural sources rather than the petrochemicals used in the production of synthetic materials (Ma et al. 2020).

Compared to conventional production techniques, BC and probiotic cellulose require less resources and are low-waste production. Probiotic bacteria could use various agricultural waste streams or by-products as a substrate in turn reducing the ecological burden of those materials. We are thus providing an ideal solution for environmentally sustainable cellulose synthesis, employing probiotic bacteria (Abdelraof et al. 2019).

Due to the relatively easy process of production as well as the abundance of raw materials, BC and PBC can be produced in large amount, facilitating their application as cost-effective alternatives compared to synthetic biomaterials in many cases. If synthetic media is not available BC and PBC can be produced through fermentation process, using cheap alternative nutrient sources and equipment (Kamal et al. 2022).

Both BC and probiotic bacterial cellulose have appeared as well-suited materials for a variety of healthcare applications due to their ecofriendly and cheaper nature. These materials offer long-term solutions, focusing on environmental and budget concerns, which are an ever-increasing frenzy.

The next chapter describes the physical and chemical properties of BC and Probiotic BC, the bacteria cellulose producer matrix and the production protocols, as well as probiotic bacteria and their benefits. It further delineates different techniques of Probiotic BC synthesis as well as Healthcare applications of Probiotic BC.

4.2 Probiotic Bacterial Cellulose and Bacterial Cellulose

Bacterial cellulose (BC), a natural biomaterial, is produced by some specific bacterial species, mostly from some *Komagataeibacter* and *Gluconacetobacter* strains (Ullah et al. 2016, 2017). It can be distinguished by its specific structure and chemical composition from the plant-derived cellulose. BC is made up of linear chains made up of β-(1 → 4)-linked D-glucose units same as plant cellulose. Due to the absence of hemicelluloses, lignin, or other non-cellulosic compounds in BC, it is purer and crystalline compared to plant cellulose (Naomi et al. 2020).

The cellulose synthase complexes embedded within the bacterial cell membrane facilitate the extracellular synthesis of BC. These complexes catalyze the polymerization of glucose molecules from UDP-glucose precursors, resulting in the formation of cellulose chains that assemble into microfibrils. BC is discharged into the surroundings as a gel-like membrane, where it self-assembles into a network of nanofibrils. This process differs from plant cellulose production, which occurs internally within the cell and is subsequently deposited in the cell wall (Mensah et al. 2022).

BC is composed of nanofibrous chains in a regularly arranged highly crystalline microfibril morphology. The chains repack and form crystallites, which are the crystalline regions of BC with high order and alignment, leading to strong intermolecular hydrogen bonds and mechanical properties. The amorphous areas, represent the more disordered areas interspersed into these structure and are likely to be responsible for the elasticity and the ability of the BC to deform (Jia et al. 2017).

Several characteristics of BC for instance, structure and properties, are vastly influenced by different factors such as bacterial strain, culture conditions, and post-synthesis treatments. Typically, things like the type of carbon source (glucose or glycerol) culture medium affect BC's molecular weight, crystallinity, and porosity. During fermentation, the temperature, pH alteration, and agitation can also induce modifications in the morphology and order of arrangement of BC nanofibrils.

The chemical integrity of BC could be enhanced with chemical or enzyme treatment to introduce functional group besides to improve its compatibility to the foreign material. Periodate and TEMPO-mediated oxidation introduce aldehyde or carboxyl groups on BC that could allow for the chemical modification or polymer cross-linking (Luo et al. 2013). However, the amorphous fraction of BC can be removed by an enzymatic degradation via cellulases or hemicellulases followed by improving its crystallinity and mechanical properties.

The nanofibrous structure is the most characteristic physical property of BC. Compared to that it is composed of a highly organized nanocellulose network of entangled fibers with diameters in the range of 20–100 nm (Park et al. 2023). This nanofibrous structure gives BC superior mechanical properties and tensile strength over plant cellulose and synthetic polymers (Yao et al. 2017). Moreover, the high aspect ratio of BC nanofibres gives rise to a number of interesting optical properties, including transparency and birefringence, which in turn is useful for optical applications, e. g., photonics and displays.

In addition, BC shows excellent water holding capacity and liquid absorption properties due to its high surface area and hydrophilicity. It was considered that BC is an excellent candidate for use as a polymer in wound dressings and skin care preparations because of its ability to hold large quantities of water and absorb it when necessary, such as to maintain a moist wound environment and promote wound healing (Swingler et al. 2021). The porous structure of BC simply provides room for the flow of gases, as well as transfer of nutrients, which allows for many cell adhesion, growth, and tissue regeneration properties in biomedical scaffolds and implants.

A very important physical property of BC that can be applied to various heat-resistant and fireproof products is its heat resistance and flame retardant property. The high decomposition temperature and low flammability of BC becanse its dense structure and cellulose content (Torgbo and Sukyai 2020).

Bacterial cellulose can be used in healthcare industry for wound healing (Deng et al. 2024), tissue engineering (Wang et al. 2023), drug delivery (Ye et al. 2024), etc. Due to its non-toxicity, biodegradability and biocompatibility (Nayak et al. 2024) this biopolymer is an excellent candidate to accelerate tissue regeneration and to promote wound healing. During the usage of BC as a wound dressing material, a fluid environment is provided which assist in fast healing along with prevents the wound from a potential pathogen infection (Palem et al. 2024).

Also, the incorporation of probiotic bacteria into BC matrices will offer further therapeutic advantages which will therefore enhance the potential of BC in healthcare applications. Probiotics are live microorganisms that provide similar health benefits when an adequate amount of them are ingested. Several studies have been conducted to see how they can help maintain gut health and strengthen immunity as well as prevent against a wide array of diseases (Varela-Trinidad et al. 2022).

The immobilization of a broad range of probiotic bacteria in BC mats offers several advantages. The cellulose matrix protects the probiotic bacteria mainly in two ways. First, by providing a protective environment which isolates the probiotic bacteria from adverse conditions (e.g. from the effect of enzymes in the gut or the low pH in the stomach), so they can survive and maintain their stability. This entrapment allows the PBC mat to precisely release probiotics, creating a sustained delivery system to the intestines across a long time frame due to the porous structure of bacterial cellulose (Jayani et al. 2020; Oliveira-Alcântara et al. 2020). Probiotic bacteria also release molecules, such as bacteriocins, short-chain fatty acids, butyrate and polyphosphate with health beneficial properties of antibacterial and immuno-stimulatory and anti-inflammatory and anti-cancerogenic activities (George Kerry et al. 2018). The capability of those compounds to be entrapped in the BC network enables the development of functional material with selective health benefits by performing with specific probiotic bacteria.

This synergy of bacterial cellulose and probiotic bacteria provides new horizons for healthcare applications. For example, some probiotic-incorporated BC dressing may help not only in the wound healing process but also may fight against the pathogenic bacteria in the wound bed, and therefore mitigate the risk of infection in the wound. Likewise, bacterial cellulose formulations along with probiotic bacteria, may provide greater efficiency in the treatment of intestinal disorders due to the capacity to enhance probiotic survival and colonization in the gastrointestinal tract after oral administration (Martín et al. 2013).

4.3 Producers of Bacterial Cellulose

The production of bacterial cellulose involves several specialized bacterial strains, cultured under controlled conditions to yield this valuable biomaterial. Producers of bacterial cellulose play a very important role in the fermentation process to synthesize this green biomaterial, useful in various industries. Among several other BC producing bacteria, the following bacterial strains are mainly involved in the synthesis of BC:

Komagataeibacter xylinus The previously named *Gluconacetobacter xylinum*, now called *Komagataeibacter xylinus*, is mainly used for its ability for the production of BC, a unique biopolymer applied in different fields of technology. BC is synthesized by *K. xylinus* through the fermentation process. In the process, it converts the carbon sources, that are present in the culture media (mainly glucose) into cellulose chains. The cellulose chains are secreted out of the bacterial cells as a metabolic byproduct and spontaneously assemble among themselves in a special architecture as nanofibers are organized in dense networks. Factors such as pH, temperature, oxygen availability and nutrient composition, have been shown to influence to what extent *K. xylinus* is able to synthesize BC. These and other related parameters might be optimized in order to determine the possibility of yield and quality of BC. It has various important properties like mechanical strength, high purity, porosity and water retention capacity, which makes it an essential biopolymer in various industries (Bilgi et al. 2016; Volova et al. 2018).

Acetobacter pasteurianus *Acetobacter pasteurianus* bacteria are also a well-established strain for producing bacterial cellulose. The remarkable cellulose-producing abilities of this bacteria (high cellulose yield) actually makes it a good candidate for industrial scale production of BC - promoting the commercialization of good bacterial cellulose based products in general (Kumar et al. 2019).

Gluconacetobacter hansenii This bacterial strain is renowned for its outstanding ability to grow in different environments and the high-level production of cellulose. The use of this bacteria for BC production will enable the growth of BC under different circumstances enabling BC to be a factory-level biopolymer for the usage in industries (Gromovykh et al. 2017).

Komagataeibacter sucrofermentans *Komagataeibacter sucrofermentansis* one of the BC producing bacteria that synthesizes BC by using different carbon source such as sucrose, one of the main part element in horticultural loss. Therefore, this microbe is one of the key players in the dissemination of the BC biosynthesis. This substrate adaptability enables this bacteria to utilize a wide variety of inexpensive clean feedstocks which complements the synthesis route more flexible and sustainable for BC production (Adamopoulou et al. 2024).

Other Producers In addition to the above bacterial strains, several other bacterial strains are involved in bacterial cellulose synthesis. Each strain have special characteristics that influence the quality and characteristics of the produced BC. Those bacterial species include *Komagataeibacter rhaeticus, Komagataeibacter europaeus*, and others (He et al. 2020; Tseng et al. 2023).

4.3.1 Process of Bacterial Cellulose Synthesis

BC is synthesized by acetic acid bacterial strains by a controlled fermentation process from *Gluconacetobacter* and *Komagataeibacter* genera. Normally, this fermentation process occurs in a culture medium containing several minerals, organic vitamins, and different sources of carbon and nitrogen bred in a controlled atmosphere.

Normally, the Hestrin-Schramm (HS) medium is considered as the conventional media for the growth BC producing of bacterial strains viz. *Gluconacetobacter* and *Komagataeibacter* for BC synthesis. Several studies suggest that different alternate media can be used to produce BC through bacterial fermentation. Researchers showed that *Acetobacter xylinum* grown in pineapple juice produced BC and has better mechanical properties than conventional glucose-based media, i.e., HS media. The resulting BC has higher tensile strength and crystallinity, due to the special ingredients in the composition of pineapple juice (Mahsuli et al. 2019). Vitamin C, which is found in orange juice, acts as an antioxidant and may have an influence on BC structure and synthesis kinetics. Research by Andritsou et al. (2018) found that BC synthesis by using orange juice as culture media showed better thermal stability and water-holding capacity compared to BC synthesised using standard HS media (Andritsou et al. 2018). Pectinolytic enzymes are usually present in apple juice. Those enzymes can influence the microstructure and porosity of BC, which can affect its mechanical properties. Previous research reported that BC synthesized by using apple juice has better biocompatibility and water absorption compared to using synthetic media (He et al. 2020), which makes it ideal for biomedical applications such as wound dressings and tissue engineering. Polyphenols and antioxidants have shown promise as a supplement in the bacterial growth medium for BC production with antimicrobial properties. Research by Xu et al. (2014) found that BC synthesized in grape juice has inhibitory effects against gram-positive bacteria *Staphylococcus aureus* and gram-negative bacteria *Escherichia coli* (Xu et al. 2014).

The process of BC synthesis consists of several essential phases, viz. inoculation, fermentation and purification. Each of these phases is very crutial for the characteristics, yield, and purity of the final synthesized BC.

1. Inoculation:
   Bacterial inoculum with the desired strain of BC is first inoculated in the designated culture medium prior to the synthesis of BC. Components of the culture medium together with inoculation conditions are adjusted to optimize the bacterial growth and cellulose synthesis. It is then inoculated and the culture is incubated under controlled conditions at a given temperature (varies for different bacterial strain, usually 30 °C), pH, and oxygen, under which the bacteria grow and hence produce cellulose. Fermentation usually takes several days to weeks, contingent upon the bacterium and fermentation conditions.
2. Fermentation:
   This happens post inoculation when the BC producing bacteria release cellulose fibrils in the culture medium as part of their metabolism. The cellulose nanofibrils then self-assemble outside the bacterial cell into a 3D network structure. During the fermentation process, a thick BC membrane is formed on the surface of the culture medium. BC quality and purification can be performed on the produced BC mat after the conclusion of fermentation. The ability to control the media composition as well as the pH, temperature and agitation during fermentation is essential for the successful production and it may vary according to some of the utilized cultivation methods.
3. Purification:
   After the completion of the fermentation, the BC mat is usually collected from the culture medium and purification is done to remove the bacterial cells, residual nutrients, and other impurities present in the culture media. The BC mat is purified usually by washing with NaOH solution to remove soluble components (bacterial cells and other nutrients) and then again withholding. The purified BC pellicle can further be processed into required shapes and forms (i.e., sheets, membranes or fibers) by means of cutting or spinning techniques.

Figure 4.1 shows a schematic diagram of the process of BC synthesis and purification, from inoculation of *Komagataeibacter xylinus* to purification of produces BC.

Ultimately, the dehydrated BC product is cleaned, packed, and stored for further use in culinary, pharmaceutical, and industrial applications. Due to its special qualities, which include high purity, mechanical strength, water-holding capacity, and biocompatibility, the synthesised BC may be used in a variety of ways. Researchers can modify the characteristics of BC to satisfy certain application needs by adjusting the synthesis process parameters and enhancing culture conditions (Blanco Parte et al. 2020).

4.4 Probiotic Bacteria and Their Beneficial Effects

Probiotics from the genera *Lactobacillus, Streptococcus and Pediococcus* are among those bacteria that have been recognised for their beneficial effects on human health. Due to their contribution in boosting the immune system and maintaining gut microbiota makes them a particularly good choice for many of the different

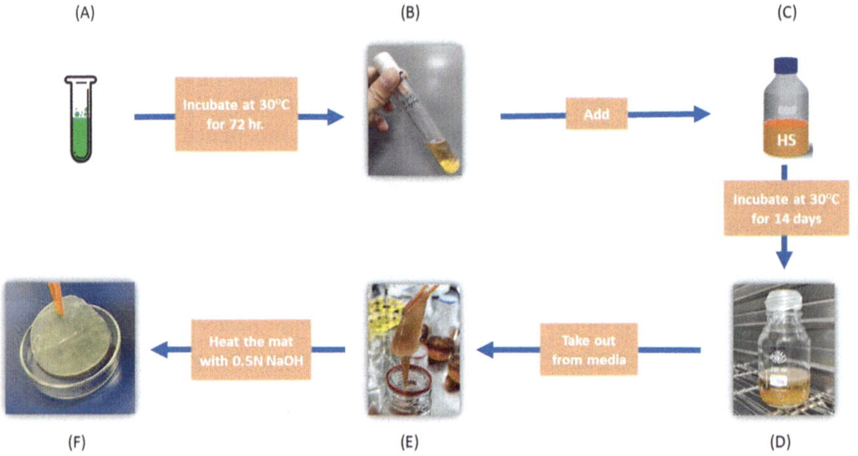

Fig. 4.1 Synthesis of bacterial cellulose. (**a**–**b**) inoculation, (**c**–**d**) fermentation, (**e**–**f**) Purification

physiological systems. Several studies have been done on probiotic bacteria for their beneficial effects on human health.

Probiotic bacteria have shown more promise when it comes to digestive health. *Lactobacillus acidophilus*, a genus commonly found in yoghurt and various other fermented foods, has shown benefits in protecting and handling antibiotic-associated diarrhoea (AAD). The study by Xie et al. (2015) involved a meta-analysis of sixty-three randomized controlled trials (RCTs) that determined that supplementation with *L. acidophilus* could greatly reduce the risk of AAD (Xie et al. 2015). Also, María Remes-Troche et al. (2020) comprehensive review and meta-analysis proved that *L. acidophilus* supplementation for adults and children reduced the incidence of AAD (María Remes-Troche et al. 2020).

In addition to AAD, probiotics have emerged as a treatment for irritable bowel syndrome (IBS), a chronic functional gastrointestinal disease characterised by bloating and abdominal distension, altered bowel habits, and abdominal discomfort. Recently, increasing interest has turned to a probiotic strain, *Bifidobacterium infantis*, naturally found in the gut of the healthy infant, for its potential therapeutic effects in the management of IBS. In the study by Ringel-Kulka et al. (2017) patients with IBS were provided supplementation of *B. infantis* resulted in better clinical implications of IBS, stomach distress scores to placebo when compared with management during the experiment (Ringel-Kulka et al. 2017). The research of Yuan et al. (2017) further found in the study of that IBS symptom severity was significantly reduced and bowel habits improved satisfactorily in *B. infantis*-treated patients (Yuan et al. 2017).

In addition, probiotics may be a useful tool for the treatment of inflammatory bowel diseases (IBD) such as Crohn's disease and ulcerative colitis. Some studies demonstrated the ability of certain probiotics to modulate the intestinal microbiota, and decrease inflammation in patients with IBD. For example, Vakadaris et al.

(2023) conducted a randomized controlled experiment to test the ability of a multispecies probiotic containing *Bifidobacterium breve, L. acidophilus, and L. plantarum* in sustained remission of ulcerative colitis patients (Vakadaris et al. 2023). Patients, treated with probiotics had a significantly lower disease prevalence and improved health conditions than those in the placebo group. Also, Prantera and Scribano (2002) observed positive outcomes with probiotic supplementation in Crohn's disease patients a reduction in inflammatory markers and overall symptoms (Prantera and Scribano 2002).

In addition to gastrointestinal disorders, probiotics have been studied for their potential use in preventing and treating allergic disorders, such as eczema and allergic rhinitis. One extensively studied probiotic strain is *Lactobacillus rhamnosus* GG (LGG), which has shown promising results in regulating the immune system and preventing the onset of various allergic disorders. A comprehensive review and meta-analysis by Cabana et al. (2017) of RCTs evaluating the effectiveness of LGG in preventing eczema (Cabana et al. 2017) found a noticeable decrease in the risk of eczema in babies who were supplemented with LGG. Also, another study reported favourable outcomes with probiotic supplementation while used in the case of Crohn's disease patients, including reduced inflammatory markers and better clinical symptoms.

Also, the potential of probiotic application for metabolic health has been reported in glycemic control, lipid profiles, and reduction in metabolic syndrome. It indicates the preventive effect of probiotics for metabolic disorders such as cardiovascular diseases and type 2 diabetes by improving metabolic health. In a systematic review and meta-analysis of randomized controlled experiment by Li et al. (2023) was performed for examining the effects of probiotics on glycemic control in patients with type 2 diabetes found that supplementation with probiotic significantly reduced blood glucose and HbA1c (Li et al. 2023). Similarly, Razmpoosh et al. (2019) reported that after probiotic supplementation supplementation in both healthy individuals as well as in patients with metabolic disorders, lipid profiles was improved, which includes reductions in triglycerides total and triglycerides (Razmpoosh et al. 2019).

Studies have shown that *Lactobacillus* plantarum is able to suppress the growth of an array of pathogens, making it suitable for use both in food preservation as well as supplementation in health care applications. Arena et al. (2016) reported that some specific strains of *L. plantarum* can show activity against different foodborne pathogens like *Salmonella spp., Listeria monocytogenes, and E. coli* (Arena et al. 2016). Another study conducted by Zeng et al. (2020) proved *Lactobacillus plantarum* collected from traditional Chinese fermented foods showed significant antibacterial activity against several bacteria, including *Bacillus cereus* and *Staphylococcus aureus* (Zeng et al. 2020). These results prove that *L. plantarum* could be used as an alternative to food preservation as well as its usability in various therapeutic strategies are based on novel antimicrobial treatment tools. Moreover, the antimicrobial activity of *L. plantarum* is not limited only to food preservation applications. Studies have shown several beneficial effects of *L. plantarum* against different infections and maintaining gut health in humans. More researches are required to identify the mechanisms of the antimicrobial activity of *L. plantarum* and to explore their potential in therapeutic and clinical settings.

Pediococcus pentosaceus, a lactic acid bacterium commonly isolated from different fermented foods and the human gastrointestinal tract, has drawn attention for its interesting ability to be used as a probiotic and for the variety of health benefits it may provide. Recent investigations have also highly associated that a few strains of *P. pentosaceus* may contain antitumoral and anti-cholesterol properties. Hence, these findings may further increase the potential of the drug in the biomedical area. For example, Ilavenil et al. (2016), performed a laboratory study of cholesterol-lowering properties of *P. pentosaceus*, isolated from fermented dairy product. The studies showed that supplementing this bacterium decrease serum cholesterol (Ilavenil et al. 2016). Furthermore, several reports demonstrated that *P. pentosaceus* possibly have anti-tumor effects. Jiang et al. found that *P. pentosaceus* isolated from traditional fermented foods show anti-proliferative effects on cancer cells in vitro. According to the research it has potential as a new probiotic having therapeutic effects for the treatment of cancer. These results describes the vast applications of this bacteria for several medicinal applications and focuses on the need of more research in future to study the modes of action of the probiotic and assess its effectiveness in pharmaceutical contexts (Jiang et al. 2021).

Apart from having different medicinal applications, probiotic bacteria also have a positive effect via different other pathways, likewise, improvement of the intestinal barrier function, the competitive exclusion of different pathogenic microorganisms, immune response alteration, and synthesis and secretion of bioactive metabolites like organic acids, bacteriocins and short-chain fatty acids (SCFAs). These systems support the preservation of both general host health and intestinal homeostasis. Table 4.1 demonstrates various beneficial effects of different probiotic bacteria.

Probiotic bacteria are a potentially effective treatment option for a variety of illnesses, including metabolic problems, allergic diseases, gastrointestinal disorders, and more. The growing body of data from current studies emphasises their numerous health benefits and emphasises the need to take probiotic supplements into account as part of an all-encompassing strategy for managing health and illness. To better understand the intricate relationships between probiotics and the host microbiota and immune system, as well as to determine the best strains, doses, and treatment durations for certain illnesses, further study is necessary. However, an increasing amount of research backs up the use of probiotics in clinical settings and public health campaigns that enhance people's health and well-being.

Probiotic bacteria have drawn the attention of researchers for a long time back due to their beneficial health effects. Apart from they are present in the gut, probiotic bacteria secrete several compounds which have various therapeutic effects. Some of the most important compounds are briefly described below with their acting principle and producing probiotic bacteria.

1. **Short-Chain Fatty Acids (SCFAs):**
   SCFAs, primarily acetate, propionate and butyrate, which form the main class of metabolites discharged by probiotics. These SCFAs are primarily produced in the colonic fermentation pathway of some non-digestible dietary fibers and resistant starches by saccharolytic probiotic bacteria, especially species of *Lactobacillus*

Table 4.1 The beneficial effects of different probiotic bacteria on human health

Probiotic bacteria	Beneficial effect	References
Lactobacillus plantarum	Inhibitory properties to *Streptococcus mutants* and *Candida albicans*, which are the main responsible bacteria for oral infection	Zeng et al. (2022)
Lactobacillus rhamnosus	Reduces *Streptococcus mutans*	Pahumunto et al. (2020)
Lactobacillus. rhamnosus from human breast milk	Strong antibacterial activity against pathogenic *E. coli, Salmonella typhimurium* Produce biosurfactants Inhibition of bacterial adhesion and biofilm eradication by cell-bound biosurfactants Anti-biofilm action	Riaz Rajoka et al. (2018)
Lactobacillus plantarum	Antibacterial activity *E. coli*	Patel et al. (2012)
Weisella confus	Antibacterial activity *E. coli*	Patel et al. (2012)
Pediococcus parvulus	Antibacterial activity *E. coli*	Patel et al. (2012)
Weisella cibaria	Antibacterial activity *E. coli*	Patel et al. (2012)
Lactobacillus strains from Pulp of the durian (*Durio zibethinus*) fruit	Antimicrobial, Antioxidant Reduces cholesterol	Khalil et al. (2018)
Pediococcus pentosaceus	Antimicrobial activity Antioxidant activity Antitumor activity, α-amylase and α-glucosidase inhibitions	Ayyash et al. (2020)
Lactobacillus plantarum KX041	Possessed the immune activity DPPH/ABTS radicals scavenging Activities and DNA damage productive effect	Xu et al. (2019)
Lactobacillus gasseri FR4	In-vitro antioxidant, antibacterial against food borne pathogens And anti-biofilm activity of EPS	Rani et al. (2018)
Lactobacillus sp. Ca6	Antimicrobial activity and sensitive to several antibiotics The wound healing activity of EPS-Ca6 was assessed using excision wound model in rats	Trabelsi et al. (2015)

and *Bifidobacterium* (Marasco et al. 2020). SCFAs have been recognized as critical elements essential for gut mucosal function in several ways, among which is the main energy source for colonocytes and their fundamental role in maintaining gut immune homeostasis and protection against intestinal inflammatory pathophysiology. In particular, butyrate has received the most attention because of its anti-inflammatory and anti-cancer bioactivity, which may represent a potential therapeutic window for the probiotic-derived SCFA.

2. **Bacteriocins:**
   Bacteriocins are peptides which are secreted by probiotic bacteria and can inhibit the growth of a number of pathogens. As indicated by the experiment of Heilbronner et al. 2021, these are small peptides capable of providing very limited spectrum of activity against closely related bacterial species, at the same time sparing the gut normal flora. Bacteriocins are mainly produced in large a large extent by various lactic acid bacteria, including *Lactobacillus* and *Streptococcus* species (Heilbronner et al. 2021). Bacteriocins can help regulate gut-microbiota balance by competitive inhibition and through their direct antimicrobial effects.
3. **Exopolysaccharides (EPS):**
   Exopolysaccharides are complex carbohydrates that is secreted by several bacteria as well as probiotic bacteria that having numerous benefitial effect to health. According to Castro-Bravo et al. (2018), different EPS play very important roles in biofilm formation, enhanceing the colonization of probiotics and their persistence in the gut (Castro-Bravo et al. 2018). Furthermore, according to the research of Kaur et al. (2023), EPS have prebiotic effects as they selectively promote the growth of different beneficial bacteria such as *Bifidobacterium* and *Lactobacillus* species (Kaur et al. 2023). Also, EPS have been observed for modulating host immune responses and to protect against intestinal pathogens (Rahbar Saadat et al. 2019).
4. **Bile Salt Hydrolases (BSH):**
   Probiotic bacteria secrete some enzymes called bile salt hydrolases which can catalyze the hydrolysis of conjugated bile salts. BSH enzymes influence bile salt metabolism and cholesterol homeostasis in cholesterol in the host by disconjugating of bile salts. Furthermore, BSH activity supports the endurance and persistence of probiotic bacteria in the harsh environment of the gut (Yang et al. 2019). Through bile salt hydrolysis, probiotic-derived BSH enzymes help in lipid metabolism, maintainance of gut microbiota, and overall gut health.
5. **Neurotransmitters and Metabolites:**
   Neurotransmitters are chemicals that mostly act directly on the physiology and behaviour of the host. Probiotic bacteria such as *Bifidobacterium* and *Lactobacillus* can produce neurotransmitters as dopamine, serotonin and gamma-aminobutyric acid (GABA) that affect mood, stress and cognition of the host (Tette et al. 2022). Probiotic bacteria also produce a range of metabolites including tryptophan and polyamines. Gut microbiota and the immune system are inextricably linked by these metabolites with varied effects on immune regulation and cross-talk with the gut-brain axis (Bosi et al. 2020). These researches open doors for the use as well as investigation of probiotics-based therapeutics, such as PBC, for neuropsychiatric conditions and demonstrates interactions of probiotic bacteria with the human neurological system.

Probiotic bacteria secrete an array of molecules with an impact that encompasses many aspects of host health and welfare, accessed via the oral cavity into the gut. These bioactive compounds (short-chain fatty acids, bacteriocins, exopolysaccharides, bile salt hydrolases, and even neurotransmitters) exert potential

Table 4.2 Different probiotic bacteria and their secreted molecules

Probiotic bacteria	Secreted material	References
Lactobacillus acidophilus	Bacteriocins, SCFAs	Wang et al. (2020)
Bifidobacterium breve	Exopolysaccharides (EPS)	Wang et al. (2019)
Streptococcus thermophilus	Lactocepin, Exopolysaccharides	Wang et al. (2021)
Lactobacillus rhamnosus	Bacteriocins, Lipoteichoic acids	dos Santos et al. (2015)
Bifidobacterium animalis	Short-Chain Fatty Acids (SCFAs)	Lee et al. (2017)
Enterococcus faecium	Aggregation Substance (Agg)	Boltz et al. (2019)
Bacillus coagulans	Spores, Heat-Stable Enzymes	Konuray and Erginkaya (2018)
Bacillus coagulans	Microcins, Outer Membrane Vesicles	Juturu and Wu (2018)
Propionibacterium freudenreichii	Microcins, Outer Membrane Vesicles	Parizzi et al. (2012)

benefits on the gut physiology, immune function, and even neurological processes. Different probiotic bacteria and their secreted materials are mentioned in Table 4.2.

4.5 Methods of Synthesizing Probiotic Bacterial Cellulose

Bacterial cellulose (BC), as a biopolymer that is highly biocompatible and pure and possess special structure, is considered a good green polymer in the field of biomedicine. One tempting route to enhance BC-based material performance for healthcare uses, such as drug-release, tissue engineering, and wound healing, is to produce BC in the presence of probiotic bacteria, which has not been previously realised. The probiotic BC can be in several ways produced, such as:

1. **Static Culture Method:** Through this process, a bacterial inoculum was soaked in a culture medium, BC then formed at the interface between air and media. Due to their own advantages of high scalability and simplicity, they are now utilized heavily for large-scale BC synthesis. In that approach, various culture conditions and nutrient supplements as well as pH adjustment have been investigated in recent research to improve BC production and purity (Öz and Kalender 2023).
2. **Agitated Culture Method:** In this technique agitation is generally used to increase the distribution of nutrients and oxygen in the culture medium to boost the synthesis of beta-carotene. Agitation can be achieved by several ways, such as shaking or stirring the culture vessel with different instruments. To increase the efficiency of BC synthesis, researchers have investigated at different parameter like agitation speed and duration (Bi et al. 2014).
3. **Immobilisation Techniques:** Cross-linking and matrix immobilization might enhance BC production by maintaining a micro-niche of constant conditions required for biopolymer (PBC) production. Earlier methods of immobilisation such as encapsulation and biofilm formation have been reported in the literature

for co-culture systems of probiotic bacteria and BC-producing bacteria for the development of composite materials with improved functionalities (Jayani et al. 2020).

4. **Co-cultivation Technique:** To create BC with improved qualities, co-cultivating BC-producing bacteria, such as *Komagataeibacter* xylinus, in the presence of probiotic bacteria is known as the co-culture technique. Researchers want to add bioactive substances or functionalize BC matrices with probiotic cells by co-culturing probiotic bacteria with BC producers, which will provide particular therapeutic advantages to BC-based materials (Sabio et al. 2021).

One method that shows promise for creating probiotic BC with specific qualities for use in biological applications is co-culture. In co-culture systems, the main strains that produce BC are *Komagataeibacte*r species, which are well-known for their high BC output. Different probiotic bacteria from *Lactobacillus* and *Bifidobacterium* genera are chosen for their effective health benefits and compatibility with BC synthesizing bacteria for probiotic BC synthesis.

Several studies have been done showing different ways to co-culture probiotic bacteria with the BC-producing bacteria *Komagetaebacter*.

- **Sequential Inoculation:** In this method, *Komagataeibacter* strains are grown in culture media to produce BC, which is followed by the addition of probiotic bacteria to the BC matrix at various timepoints to impart probiotic activity to the cellulose matrix. The sequential inoculation is able to direct where and how the probiotic cells are distributed and incorporated into the BC matrix (Ul-Islam et al. 2013).
- **Simultaneous Inoculation:** During simultaneous inoculation, both *Komagataeibacter* and probiotic bacteria are co-cultured at the onset of BC synthesis. This approach fosters the interactions of BC-forming, and probiotic bacteria, and yields synergism in BC attributes and bioactivity. Despite the observed trends between pH and nutrient content, the media composition needs to be optimized for pH and nutrient content to be able to ensure the least possible disturbance of the microbial balance and the simultaneous maximization of the final BC yield (Rashidian et al. 2021).
- **Immobilisation Strategies:** The bacteria immobilization achieved by entrapping probiotic in BC matrices can provide a localized and sustained delivery system of probiotic cells and thereby maintaining their viability and functionality. Several immobilization methods have been investigated in which probiotic bacteria were incorporated into BC scaffolds, including entrapment, and surface coating of the BC that retained the BC integrity and structure (Sabio et al. 2021).

The co-culture technique for probiotic BC synthesis provides various advantages over traditionl ways of synthesizing Probiotic BC:

1. **Increased Bioactivity:** Probiotic bacteria-loaded BC which release bioactive molecules (i.e., antimicrobial peptides, immunomodulatory factors) provides a

sustained delivery system and increases the therapeutic power of the probiotic BC to be used in biomedical applications (Yao et al. 2022).

2. **Synergistic Effects:** Co-culturing of the probiotic bacteria with *Komagataeibacter* shows synergistic effects. This in turn increases the mechanical strength, porosity, and biocompatibility of the newly synthesised BC. These enhanced properties could help to design new BC-based materials for various pharmaceutical applications, such as tissue engineering or regenerative medicine etc. (Buer Boyetey et al. 2023).
3. **Customised Functionality:** The co-culture technique supports the precise control over the distribution of probiotic bacteria within BC matrices, which allows the modification of BC-based materials with desired functionalities, such as in wound healing and gastrointestinal health remedies and targeted drug delivery etc. (Gregory et al. 2021).

Co-culture method offers a promising process for the synthesis of PBC with improved mechanical properties as well as making it capable to be used in the biomedical field. Through the synergistic interactions of these BC-producing bacteria, *Komagataeibacter*, and various probiotic bacteria, the scientists can craft engineered, microbe-supplemented BC for advanced biomedical applications including wound care, tissue engineering, and drug delivery. Through further investigation in this area, it is inevitable that probiotic bio-BC-based materials will be clinically trialed in the future and provide a direction to solve real-life health problems.

4.6 Healthcare Applications of Probiotic Bacterial Cellulose

BC has emerged as a biomaterial for a vast range of pharmaceutical applications. Normally, *Komagataeibacter xylinus* produce BC as a metabolic byproduct, having unique properties like biocompatibility, biodegradability, and a large surface-to-mass ratio that makes it an ideal polymer for pharmaceutical applications. In combination with probiotic bacteria, BC exhibits greater therapeutic potential, showing a new way for innovative medical solutions. Many instances of clinical uses of probiotic BC are described in this part of the chapter, such as drug delivery, tissue engineering, wound healing, tumour therapy, oral health recovery and others. By going through recent research findings, we aim to demonstrate the broad spectrum of ways in which probiotic BC can revolutionize healthcare practices and improve patient conditions.

1. Wound Healing:

   The combined benefits of the therapeutic effects of probiotic bacteria with the natural properties of BC allows PBC to provide an excellent wound therapeutic agent. Because the cellulose produced by acetic acid bacteria such as *Komagataeibacter xylinus* (BC), is a highly porous material and its properties of biocompatibility, biodegradability and the ability to keep the wound environment moist, are one of the reasons that makes BC a good natural dressing. BC based wound dressings with prepopulated probiotic bacteria such as Bifidobacterium spp., *Lactobacillus* spp., and Pediococcus spp. that effectuate the maximum antimicrobial activity obviously do demostrate immunomodulatory and wound healing properties (Devi and Kumar 2021).

According to the reports, probiotic strain from BC based dressing has antimicrobial activity against various wound pathogens which thereby reduce the infection and promote the wound healing. Moreover, the PBC also modulates the intestinal immune system, modulating the immune responses in the wound site, triggering balanced inflammatory environments and speeding up wound healing processes. Similarly, the ability to immobilize a range of probiotic bacteria provides a way for various bioactive compounds (e.g., antimicrobial peptides and growth factors) to be delivered directly to the wound by BC-based dressings, a process which contributes toward healing (Jabbari and Babaeipour 2024a, b).

In addition, PBC dressings are endowed with certain properties (reduced pain and minimal scarring) and benefits (biodegradable, leading to comfort for the patient and ease of removal upon completion of the wound healing process). By researching on it further an developing it enough PBC can serve as a revolutionary component in wound closure operations and their curing process in the domain of variety of wound management.

2. Tissue Engineering:

Due to its features and the synergistic effects it presents with probiotic bacteria, PBC is destined to have great perspectives for tissue engineering in the near future. These bacteria, which are mainly acetic acid bacteria such as *Komagataeibacter xylinus*, produce bacterial cellulose (BC) a polymer that is almost ideal for the fabrication of scaffolds for tissue regeneration because it is biocompatible, porous, and has astoundingly good mechanical strength (Gorgieva and Trček 2019). It has been reported in different studies that BC yield is increased in presence of probiotic bacteria such as *Lactobacillus* and *Bifidobacterium* species, which in turn showed enhanced bioactivity (increased cell adhesion, proliferation, differentiation, and tissue regeneration).

Additionally, PBC scaffolds are able to support the proliferation of various cells, for instance, osteoblasts, chondrocytes, and neural cells, which characterize PBC as a candidate for bone, cartilage, and nerve tissue engineering. Also, probiotic BC has immunomodulatory properties which create a microenvironment normalising tissue regeneration since it is capable of modulating immune responses and inflammation. The presence of live probiotic bacteria in BC based scaffolds, provides an increased ability to release bioregulators such as cytokines, growth factors and others that potentiate the regeneration phenomenon (Khan et al. 2022). Generally, probiotic BC looks like an excellent, highly efficient biomaterial for tissue engineering solutions and for the production of applicable tissue substitutes in regenerative medicine.

3. Drug Delivery:

The probiotic BC-based drug delivery systems could be a new methodology for the site-specific release under controllable conditions. Due to the high surface area, the porous structure of BC that enables efficient loading and encapsulation of drugs, proteins, and growth factors, as well as probiotic bacteria that help improve bioavailability (Khan et al. 2022). In summary, Probiotic BC carriers offer a promising approach in the field of targeted carriers and have been used to deliver various bioactive molecules viz. antibiotics, anti-cancer drugs and growth factors with desired release kinetics at specific sites (Severino et al. 2019). Through the development of probiotic BC-based DDS, researchers ultimately endeavor to address the limitations observed among conventional drug delivery platforms (e.g. poor solubility, rapid clearance and/or broad systemic toxicity), leading to enhanced patient therapeutic efficacy and adherence.

4. Gastrointestinal Health:
Probiotic bacterial cellulose (PBC) is a novel method with various beneficial properties supported by present research conclusions for better gastrointestinal (GI) health. As PBC acts as an efficient carrier for probiotic bacteria, it supports their survival in the acidic environment of the stomach and promotes their aggregation in the intestines. Whereas the colonisation-induced BC would result in the alteration of gut flora which will help promote epiphytic bacteria such as *Lactobacillus* and *Bifidobacterium* and suppress that of dangerous pathogens (Shi and Walker 2004). Furthermore, a research by Selvamani et al. (2022) demonstrated that probiotics generated from PBC exhibit their anti-inflammatory effects by modulating immune reactions and diminishing the production of pro-inflammatory cytokines to limit inflammation in maladies such as colitis (Selvamani et al. 2022). Studies also suggested that the symbiotic crosstalk of probiotics and cellulose in PBC may play a role in preserving gut barrier function by restricting harmful pathogens and toxic substances through movement across the intestinal epithelium (Selvamani et al. 2022). Furthermore, Vesterhus et al. (2020) have proved that PBC-based probiotics in the GI tract were found to enhance nutrition absorption and metabolism in the gut, and thus increased digestive efficiency (Vesterhus et al. 2020). Also, it is clear that the use of probiotics with PBC on a regular basis has definitely helped boost gut motility and counter symptoms of intestinally-related illnesses such as diarrhoea and irritable bowel syndrome. Altogether, these investigations highlight the promising therapeutical effect of PBC towards digestive health and the management of broad-spectrum GI pathologies, which need to be further validated clinically.
5. Cancer and Tumour Therapy:
Probiotic bacterial cellulose has shown to be an effective solution for cancer and another tumours with the help of its unique properties and advantages of therapeutic benefits of probiotic bacteria. Conclusively, BC offers a desirable platform for the administration of anticancer drugs and also contribute to modification in tumour microenvironment. They are produced by bacteria like *Komagataeibacter xylinus*. Furthermore, BC based formulations exhibit higher anticancer efficacy with reduced systemic toxicity as well as enhanced therapeutic effects when combined along with probiotic bacteria particularly *Lactobacillus* and *Bifidobacterium* species (Singh et al. 2023).

According to the studies conducted so far, it is evident that using probiotic BC carriers may facilitate specific regulation of targeting and triggering drug release by delivering immunomodulators/anticancer peptides/chemotherapeutical agents directly into tumour location. In addition, the probiotic BC also has an immunomodulatory activity that can enhance anticancer immune responses, eventually improving cancer immunotherapy (Nami et al. 2023). Researchers are aiming to devise exciting cancer-therapy strategies that can be implemented with improved safety, efficacy and specificity through the cooperative action of plausible combinations of probiotic bacteria along with BC-matrices.

6. Oral Health:
   The multi-benefit probiotic bacterial cellulose can be a novel biopolymer in oral health. It repopulates the beneficial natural microbiota by introducing good bacteria (like various strains of *Lactobacillus* and *Bifidobacterium*) into the oral environment while at the same time suppressing the harmful bacteria responsible for causing diseases like dental caries (such as *Streptococcus mutans*). Therefore, the restoration of this microbial balance has significant implications for preventive dentistry since it can reduce cavity incidence and enhance overall oral health. The wound-healing abilities of PBC may also be advantageous, helping recover oral tissue and accelerating recovery in postoperative or traumata when implemented (Rowińska et al. 2021).

   Furthermore, together with its probiotic activity in bacterial cellulose, it not only preserves the existing microbiological balance but also treats common oral problems such as halitosis. The result is fresher breath as some strains stabilise the oral pH and prevent odour-causing bacteria from developing by producing organic acids, such as lactic acid. The PBC-based novel composite for oral hygiene not only uses its capacity as an adjuvant to balance the oral microbiota but also eliminates odour (Gregory et al. 2021).

   Although PBC has promising qualities, these findings need to be further explored and confirmed before probiotic bacterial cellulose may become a common material in dental applications. More extensive clinical trials are important to decide whether it is successful in the long term at any dosage or technique of conveyance, even if the underlying research tells us something positive. However, the growing interest in PBC demonstrates its ability to transform oral health care by using the strength of helpful bacteria and their wound-healing capabilities (Swingler et al. 2021).

7. Immunomodulation:
   Because of its immunomodulatory properties, PBC can be proved as an ideal supplement either for immunological responses control or treating diseases related to immune system. BC-based matrices also challenged immune cells to interact with probiotic bacteria, and BC scaffolds group increased immune surveillance against infections and tumor cells while helped immune cell function and even contributed in the production of anti-inflammatory cytokine (Guha et al. 2023). Based on these findings, Probiotic BC represents a natural and biocompatible tool to drive immune homeostasis, thus holding promise for the clinical management of inflammatory disorders, allergies and autoimmune diseases (Han et al. 2024).

8. Skin-Care Applications:
   PBC could be a prospective substitute material for skin care applications as a natural and multilateral bio-material for the skin. Because of these unique

properties, PBC is widely used as a substrate in cosmetics because it has a highly porous structure, biocompatibility and the ability to hold moisture. PBC formulations loaded with probiotics can maintain the balance of the skin microbiome and control inflammation, which will increase wound healing processes. Research by Chilicka et al. (2022) also demonstrated the antimicrobial activity of PBC-based skincare products on acne-related bacteria proving that sustainability can be one of the alternative solutions for skin conditions such as acne (Chilicka et al. 2022). In addition, PBC-derived probiotics improved skin hydration and elasticity by promoting collagen production and increasing the function of the skin barrier, which could further increase overall moisture content in the same. Studies by Jabbari and Babaeipour (2024a, b) highlight the scope of PBC in formulations for skin care to protect against moisture loss and accelerate healing following injury (Jabbari and Babaeipour 2024a, b). These results suggest that PBC has a potential for creating new and advanced skincare formulations contributed to probiotics in skin health and skincare.

Figure 4.2 represents different dirrections healthcare applications of PBC.

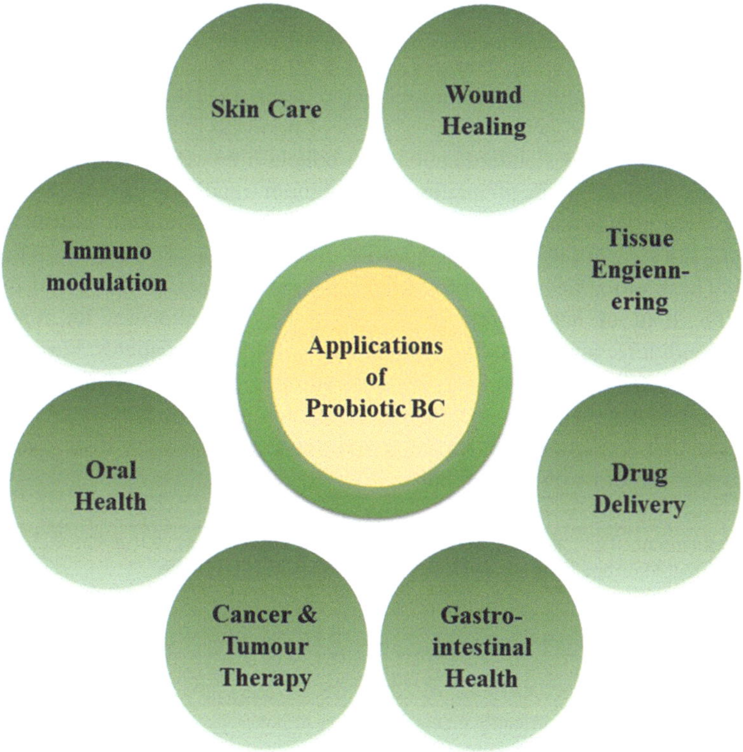

Fig. 4.2 Different applications of probiotic BC

4.7 Conclusion

In conclusion, probiotic bacterial cellulose, or PBC is an incredible bio-mediated nanomaterial that could revolutionize many different medical applications entirely. BC, as a biopolymer with significant material properties and combined benefits of probiotics towards human health, is one of the best candidates from this perspective. Due to the increased interest in green and sustainable biomaterials nowadays, with its outstanding mechanical properties together with a complex nanofibrillar structure it is highly suitable for a range of biological applications. The high degree of biocompatibility and biodegradability make PBC highly desirable in the healthcare industry for reduced adverse effects and environmental consequences. PBC's ability to retain moisture and its porous structure makes it a very good option in cases where wounds need to be kept moist to allow faster healing. The addition of probiotic bacteria into PBC matrices makes the material more therapeutic; in combination with antibacterial properties, it creates conditions favourable for tissue regeneration. Besides wound healing, PBC reveals a promising application in the development of tailored drug delivery systems exploiting probiotic benefits for individual treatment as described. Immobilizing probiotic bacteria in PBC permit researchers to develop new delivery systems with defined release kinetics and enhanced therapeutic effects. In addition, for tissue engineering and regenerative medicine, PBC is an inimitable material owing to its scaffold-like structure. Combining PBC with biomolecules, growth factors, and live cells becomes a feasible procedure. It might provide solutions to significant issues related to organ transplantation and regenerative treatments but also penetrate novel ways of performing tissue regeneration. In any case, there must be standardization of production processes and careful safety evaluations to achieve uniform quality levels without (much) danger, no matter how great the potential. Meanwhile, in order to uncover the therapeutic advantages of PBC and fully functionalize it for specific clinical purposes, further studies are needed. These properties make PBC a promising nanomaterial that could revolutionize the field of healthcare, offering state-of-the-art remedies for numerous medical conditions, and improving the outcomes of different patients. Furthermore, this biopolymer can be used in many sectors more cost-effectively by using different carbon sources as the nutrition media for bacterial fermentation. PBC is such a green biomaterial that it can be considered as an excellent candidate for future sustainable material because of its biodegradability.

Acknowledgement This work is supported by the internal grants IGA/CPS/2023/005 and IGA/CPS/2024/005 from Tomas Bata University in Zlin, Czech Republic.

References

Abdelraof M et al (2019) Ecofriendly green conversion of potato peel wastes to high productivity bacterial cellulose. Carbohydr Polym 211:75–83

Adamopoulou V et al (2024) Optimization of bacterial cellulose production by *Komagataeibacter sucrofermentans* in synthetic media and agrifood side streams supplemented with organic acids and vitamins. Bioresour Technol 398:130511

Andritsou V et al (2018) Synthesis and characterization of bacterial cellulose from citrus-based sustainable resources. ACS Omega 3(8):10365–10373

Arena MP et al (2016) Use of *Lactobacillus plantarum* strains as a bio-control strategy against food-borne pathogenic microorganisms. Front Microbiol 7:464

Ayyash M et al (2020) Physicochemical, bioactive and rheological properties of an exopolysaccharide produced by a probiotic Pediococcus pentosaceus M41. Carbohydr Polym 229:115462

Bi JC et al (2014) Morphology and structure characterization of bacterial celluloses produced by different strains in agitated culture. J Appl Microbiol 117(5):1305–1311

Bilgi E et al (2016) Optimization of bacterial cellulose production by *Gluconacetobacter xylinus* using carob and haricot bean. Int J Biol Macromol 90:2–10

Blanco Parte FG et al (2020) Current progress on the production, modification, and applications of bacterial cellulose. Crit Rev Biotechnol 40(3):397–414

Boltz TP et al (2019) The effect of standard pelleting and more thermally aggressive pelleting utilizing a Hygieniser on feed manufacture and reduction of enterococcus faecium, a salmonella surrogate. J Appl Poult Res 28(4):1226–1233

Bosi A et al (2020) Tryptophan metabolites along the microbiota-gut-brain axis: an interkingdom communication system influencing the gut in health and disease. Int J Tryptophan Res 13: 1178646920928984

Buer Boyetey M-J et al (2023) Bio-scaffold for bone tissue engineering with focus on bacterial cellulose, biological materials for hydroxyapatite synthesis and growth factors. Eur Polym J 194:112168

Cabana MD et al (2017) Early probiotic supplementation for Eczema and Asthma prevention: a randomized controlled trial. Pediatr 140(3):e20163000

Castro-Bravo N et al (2018) Interactions of surface exopolysaccharides from *Bifidobacterium* and *Lactobacillus* within the intestinal environment. Front Microbiol 9:2426

Chilicka K et al (2022) Microbiome and probiotics in acne vulgaris—a narrative review. Life 12. https://doi.org/10.3390/life12030422

Deng L et al (2024) A biological antimicrobial agent functionalized bacterial cellulose-based wound dressing. Cellulose 31(7):4277–4289

Devi S, Kumar P (2021) Use of probiotic bacteria and their bioactive compounds for wound care. In: Kumar P, Kothari V (eds) Wound healing research: current trends and future directions. Springer, Singapore, pp 301–330

dos Santos KMO et al (2015) Artisanal Coalho cheeses as source of beneficial *Lactobacillus plantarum* and *Lactobacillus rhamnosus* strains. Dairy Sci Technol 95(2):209–230

Florea M et al (2016) Engineering control of bacterial cellulose production using a genetic toolkit and a new cellulose-producing strain. Proc Natl Acad Sci 113(24):E3431–E3440

George Kerry R et al (2018) Benefaction of probiotics for human health: a review. J Food Drug Anal 26(3):927–939

Gopu G, Govindan S (2018) Production of bacterial cellulose from *Komagataeibacter saccharivorans* strain BC1 isolated from rotten green grapes. Prep Biochem Biotechnol 48(9): 842–852

Gorgieva S, Trček J (2019) Bacterial cellulose: production, modification and perspectives in biomedical applications. Nanomaterials 9(10):1352. https://doi.org/10.3390/nano9101352

Gregory DA et al (2021) Bacterial cellulose: a smart biomaterial with diverse applications. Mater Sci Eng R Rep 145:100623

Gromovykh TI et al (2017) Bacterial cellulose synthesized by *Gluconacetobacter hansenii* for medical applications. Appl Biochem Microbiol 53(1):60–67

Guha A et al (2023) Cancer stem cell–immune cell crosstalk in breast tumor microenvironment: a determinant of therapeutic facet. Front Immunol 14:1245421

Han Z-Y et al (2024) Microalgal biomass-assisted delivery of probiotics for modulation of gut homeostasis and alleviation of intestinal inflammation. Nano Today 54:102093

He X et al (2020) Novel bacterial cellulose membrane biosynthesized by a new and highly efficient producer *Komagataeibacter rhaeticus* TJPU03. Carbohydr Res 493:108030

Heilbronner S et al (2021) The microbiome-shaping roles of bacteriocins. Nat Rev Microbiol 19(11):726–739

Hill C et al (2014) The international scientific Association for Probiotics and Prebiotics consensus statement on the scope and appropriate use of the term probiotic. Nat Rev Gastroenterol Hepatol 11(8):506–514

Ilavenil S et al (2016) Assessment of probiotic, antifungal and cholesterol lowering properties of Pediococcus pentosaceus KCC-23 isolated from Italian ryegrass. J Sci Food Agric 96(2): 593–601

Jabbari F, Babaeipour V (2024a) Bacterial cellulose as a potential biopolymer for wound care. A review. Int J Polym Mater Polym Biomater 73(6):455–477

Jabbari F, Babaeipour V (2024b) Bacterial cellulose as an ideal potential treatment for burn wounds: a comprehensive review. Wound Repair Regen 32:323–339

Jayani T et al (2020) Bacterial cellulose nano fiber (BCNF) as carrier support for the immobilization of probiotic, *Lactobacillus acidophilus* 016. Carbohydr Polym 250:116965

Jia Y et al (2017) Preparation and characterization of a novel bacterial cellulose/chitosan bio-hydrogel. Nanomat Nanotechnol 7:1847980417707172

Jiang S et al (2021) Pediococcus pentosaceus, a future additive or probiotic candidate. Microb Cell Factories 20(1):45

Juturu V, Wu JC (2018) Microbial production of bacteriocins: latest research development and applications. Biotechnol Adv 36(8):2187–2200

Kamal T et al (2022) Cost-effective synthesis of bacterial cellulose and its applications in the food and environmental sectors. Gels 8. https://doi.org/10.3390/gels8090552

Kaur N et al (2023) Bacterial exopolysaccharides as emerging bioactive macromolecules: from fundamentals to applications. Res Microbiol 174(4):104024

Khalil ES et al (2018) Probiotic properties of exopolysaccharide-producing *Lactobacillus* strains isolated from Tempoyak. Molecules 23. https://doi.org/10.3390/molecules23020398

Khan S et al (2022) Fabrication strategies and biomedical applications of three-dimensional bacterial cellulose-based scaffolds: a review. Int J Biol Macromol 209:9–30

Konuray G, Erginkaya Z (2018) Potential use of Bacillus coagulans in the food industry. Food Secur 7. https://doi.org/10.3390/foods7060092

Kumar V et al (2019) Efficient and economic process for the production of bacterial cellulose from isolated strain of *Acetobacter pasteurianus* of RSV-4 bacterium. Bioresour Technol 275:430–433

Lee Y et al (2017) Effects of *Bifidobacterium animalis* subsp. lactis BB-12® on the lipid/lipoprotein profile and short chain fatty acids in healthy young adults: a randomized controlled trial. Nutr J 16(1):39

Li G et al (2023) The effects of probiotics supplementation on glycaemic control among adults with type 2 diabetes mellitus: a systematic review and meta-analysis of randomised clinical trials. J Transl Med 21(1):442

Luo H et al (2013) Characterization of TEMPO-oxidized bacterial cellulose scaffolds for tissue engineering applications. Mater Chem Phys 143(1):373–379

Ma L et al (2020) Bacterial cellulose: an encouraging eco-friendly nano-candidate for energy storage and energy conversion. J Mater Chem A 8(12):5812–5842

Mahsuli T et al (2019) Mechanical properties of bacterial nanocellulose membrane from pineapple peel waste after homogenization process. AIP Conference Proceedings 2120(1):050019

Marasco G et al (2020) Probiotics, prebiotics and other dietary supplements for gut microbiota modulation in celiac disease patients. Nutrients 12. https://doi.org/10.3390/nu12092674

María Remes-Troche J et al (2020) *Lactobacillus acidophilus* LB: a useful pharmabiotic for the treatment of digestive disorders. Ther Adv Gastroenterol 13:1756284820971201

Martín R et al (2013) Role of commensal and probiotic bacteria in human health: a focus on inflammatory bowel disease. Microb Cell Factories 12(1):71

Mensah A et al (2022) Membrane technological pathways and inherent structure of bacterial cellulose composites for drug delivery. Bioengineering 9. https://doi.org/10.3390/bioengineering9010003

Nami Y et al (2023) Probiotic immunonutrition impacts on colon cancer immunotherapy and prevention. Eur J Cancer Prev 32(1):30–47

Naomi R et al (2020) Plant- vs. bacterial-derived cellulose for wound healing: a review. Int J Environ Res Public Health 17. https://doi.org/10.3390/ijerph17186803

Nayak R et al (2024) Potential of bacterial cellulose for sustainable fashion and textile applications: a review. J Mater Sci 59(16):6685–6710

Oliveira-Alcântara AV et al (2020) Bacterial cellulose/cashew gum films as probiotic carriers. LWT 130:109699

Öz YE, Kalender M (2023) A novel static cultivation of bacterial cellulose production from sugar beet molasses: series static culture (SSC) system. Int J Biol Macromol 225:1306–1314

Pahumunto N et al (2020) Fermented milk containing a potential probiotic *lactobacillus rhamnosus* SD11 with maltitol reduces Streptococcus mutans: a double-blind, randomized, controlled study. J Dent Sci 15(4):403–410

Palem RR et al (2024) In situ fabricated ZnO nanostructures within carboxymethyl cellulose-based ternary hydrogels for wound healing applications. Carbohydr Polym 334:122020

Pang M et al (2020) Application of bacterial cellulose in skin and bone tissue engineering. Eur Polym J 122:109365

Parizzi LP et al (2012) The genome sequence of Propionibacterium acidipropionici provides insights into its biotechnological and industrial potential. BMC Genomics 13(1):562

Park S-K et al (2023) Probiotic properties of Pediococcus pentosaceus JBCC 106 and its lactic acid fermentation on Broccoli juice. Microorganisms 11. https://doi.org/10.3390/microorganisms11081920

Patel A et al (2012) Probiotic properties of exopolysaccharide producing lactic acid bacteria isolated from vegetables and traditional Indian fermented foods. New Delhi Publishers, New Delhi

Prantera C, Scribano ML (2002) Probiotics and Crohn's disease. Dig Liver Dis 34:S66–S67

Rahbar Saadat Y et al (2019) A comprehensive review of anticancer, immunomodulatory and health beneficial effects of the lactic acid bacteria exopolysaccharides. Carbohydr Polym 217:79–89

Rani RP et al (2018) Characterization of a novel exopolysaccharide produced by *lactobacillus gasseri* FR4 and demonstration of its in vitro biological properties. Int J Biol Macromol 109: 772–783

Rashidian E et al (2021) Synthesis and characterization of bacterial cellulose/graphene oxide nano-biocomposites. Polym Compos 42(9):4698–4706

Razmpoosh E et al (2019) The effect of probiotic supplementation on glycemic control and lipid profile in patients with type 2 diabetes: a randomized placebo controlled trial. Diabetes Metab Syndr Clin Res Rev 13(1):175–182

Riaz Rajoka MS et al (2018) Functional characterization and biotechnological potential of exopolysaccharide produced by *Lactobacillus rhamnosus* strains isolated from human breast milk. LWT 89:638–647

Ringel-Kulka T et al (2017) Multi-center, double-blind, randomized, placebo-controlled, parallel-group study to evaluate the benefit of the probiotic *Bifidobacterium infantis* 35624 in non-patients with symptoms of abdominal discomfort and bloating. Off J Am Coll Gastroenterol ACG 112(1):145–151

Rowińska I et al (2021) The influence of diet on oxidative stress and inflammation induced by bacterial biofilms in the human Oral cavity. Materials 14. https://doi.org/10.3390/ma14061444

Sabio L et al (2021) Probiotic cellulose: antibiotic-free biomaterials with enhanced antibacterial activity. Acta Biomater 124:244–253

Savitskaya I et al (2019) Antimicrobial and wound healing properties of a bacterial cellulose based material containing B. subtilis cells. Heliyon 5(10):e02592

Savitskaya I et al (2024) Prebiotic Cellulose–Pullulan Matrix as a "Vehicle" for Probiotic Biofilm Delivery to the Host Large Intestine. Polymers 16. https://doi.org/10.3390/polym16010030

Selvamani S et al (2022) Efficacy of probiotics-based interventions as therapy for inflammatory bowel disease: a recent update. Saudi J Biol Sci 29(5):3546–3567

Severino P et al (2019) Alginate nanoparticles for drug delivery and targeting. Curr Pharm Des 25(11):1312–1334
Shi HN, Walker A (2004) Bacterial colonization and the development of intestinal defences. Can J Gastroenterol 18:690421
Singh P et al (2018) On the viability, cytotoxicity and stability of probiotic bacteria entrapped in cellulose-based particles. Food Hydrocoll 82:457–465
Singh D et al (2023) Probiotics: friend or foe to the human immune system. Bull Natl Res Cent 47(1):126
Swingler S et al (2021) Recent advances and applications of bacterial cellulose in biomedicine. Polymers 13. https://doi.org/10.3390/polym13030412
Tette F-M et al (2022) Therapeutic anti-depressant potential of microbial GABA produced by *Lactobacillus rhamnosus* strains for GABAergic signaling restoration and inhibition of addiction-induced HPA axis hyperactivity. Curr Issues Mol Biol 44:1434–1451. https://doi.org/10.3390/cimb44040096
Torgbo S, Sukyai P (2020) Biodegradation and thermal stability of bacterial cellulose as biomaterial: the relevance in biomedical applications. Polym Degrad Stab 179:109232
Trabelsi I et al (2015) Purification and characterization of a novel exopolysaccharides produced by *Lactobacillus* sp. Ca6. Int J Biol Macromol 74:541–546
Tseng YS et al (2023) Improved production of bacterial cellulose by *Komagataeibacter europaeus* employing fruit extract as carbon source. J Food Sci Technol 60(3):1054–1064
Ul-Islam M et al (2013) Effect of post-synthetic processing conditions on structural variations and applications of bacterial cellulose. Cellulose 20(1):253–263
Ullah MW et al (2016) Structural and physico-mechanical characterization of bio-cellulose produced by a cell-free system. Carbohydr Polym 136:908–916
Ullah MW et al (2017) Recent advancements in bioreactions of cellular and cell-free systems: a study of bacterial cellulose as a model. Korean J Chem Eng 34(6):1591–1599
Vakadaris G et al (2023) The role of probiotics in inducing and maintaining remission in Crohn's disease and ulcerative colitis: a systematic review of the literature. Biomedicines 11. https://doi.org/10.3390/biomedicines11020494
Varela-Trinidad GU et al (2022) Probiotics: protecting our health from the gut. Microorganisms 10. https://doi.org/10.3390/microorganisms10071428
Vesterhus M et al. (2020) Emerging therapies in primary sclerosing cholangitis: pathophysiological basis and clinical opportunities. J Gastroenterol 55(6):588–614
Volova TG et al (2018) Production and properties of bacterial cellulose by the strain *Komagataeibacter xylinus* B-12068. Appl Microbiol Biotechnol 102(17):7417–7428
Wang L et al (2019) Exopolysaccharide, isolated from a novel strain *Bifidobacterium breve* lw01 possess an anticancer effect on head and neck cancer – genetic and biochemical evidences. Front Microbiol 10
Wang G et al (2020) *Lactobacillus acidophilus* JCM 1132 strain and its mutant with different Bacteriocin-producing behaviour have various in situ effects on the gut microbiota of healthy mice. Microorganisms 8. https://doi.org/10.3390/microorganisms8010049
Wang Y et al (2021) Metabolism characteristics of lactic acid bacteria and the expanding applications in food industry. Front Bioeng Biotechnol 9:612285
Wang X et al (2022) Microencapsulating alginate-based polymers for probiotics delivery systems and their application. Pharmaceuticals 15. https://doi.org/10.3390/ph15050644
Wang X et al (2023) Printability of hybridized composite from maleic acid-treated bacterial cellulose with gelatin for bone tissue regeneration. Adv Compos Hybrid Mater 6(4):134
Xie C et al (2015) Probiotics for the prevention of antibiotic-associated diarrhoea in older patients: a systematic review. Travel Med Infect Dis 13(2):128–134
Xu C et al (2014) Antioxidant, antibacterial, and Antibiofilm properties of polyphenols from Muscadine grape (Vitis rotundifolia Michx.) pomace against selected foodborne pathogens. J Agric Food Chem 62(28):6640–6649
Xu Y et al (2019) Purification, characterization and bioactivity of exopolysaccharides produced by *lactobacillus plantarum* KX041. Int J Biol Macromol 128:480–492

Yang Y et al (2019) Bile salt hydrolase can improve *Lactobacillus plantarum* survival in gastrointestinal tract by enhancing their adhesion ability. FEMS Microbiol Lett 366(8):fnz100

Yang Y et al (2024) Delivery of probiotics with cellulose-based films and their food applications. Polymers 16. https://doi.org/10.3390/polym16060794

Yao J et al (2017) Macrofibers with high mechanical performance based on aligned bacterial cellulose nanofibers. ACS Appl Mater Interfaces 9(24):20330–20339

Yao H et al (2022) Design strategies for adhesive hydrogels with natural antibacterial agents as wound dressings: status and trends. Mater Today Bio 16:100429

Ye J et al (2024) Preparation of bacterial cellulose-based antibacterial membranes with prolonged release of drugs: emphasis on the chemical structure of drugs. Carbohydr Polym 323:121379

Yuan F et al (2017) Efficacy of *Bifidobacterium infantis* 35624 in patients with irritable bowel syndrome: a meta-analysis. Curr Med Res Opin 33(7):1191–1197

Zeng Y et al (2020) Evaluation of the antibacterial activity and probiotic potential of *Lactobacillus plantarum* isolated from Chinese homemade pickles. Can J Infect Dis Med Microbiol 2020: 8818989

Zeng Y et al (2022) *Lactobacillus plantarum* disrupts *S. mutans–C. albicans* cross-kingdom biofilms. Front Cell Infect Microbiol 12:872012

Mr. Mainak Chaudhuri, B.Sc. and M.Sc. in Zoology from The University of Burdwan, India. At present, working as a doctoral student at the Centre of Polymer Systems of Tomas Bata University in Zlin, Czech Republic, under the supervision of doc. Nabanita Saha, M.Sc. Ph.D. in the course 'Biomaterials and Biocomposites'. In the meantime, I attended "Training School 2022 Textile Technologies" in July 2022, organized by COSTAction_CONTEXT_CA17107 at Prato, Italy and "Training on Cell Studies" in August 2022, organized by Rudlofs Cimdins Riga Biomaterials Innovation and Development Centre at Riga, Latvia, for INJECT-BIO project "Bioactive injectable hydrogels for soft tissue regeneration after reconstructive maxillofacial surgeries".

Mainak Chaudhuri represented his University in the All-India Science Congress with his team during his master's degree. Besides this, his hobby is Wildlife and nature photography, and he is fond of capturing beautiful photographs of nature.

Dr. Nabanita Saha is a biotechnologist, who received her B.Sc. and M.Sc. degree in "Life Sciences (Botany)" from Visva Bharati University, Santiniketan, India, and was awarded a PhD in "Microbial Biotechnology" from the Indian Institute of Technology, Kharagpur. She received her habilitation 'Associate Professor Degree' in "Technology of Macromolecular Substances" in 2006 from Tomas Bata University in Zlin, Czech Republic. For the last 22 years, she has been working at Tomas Bata University in Zlin, Czech Republic, on biomaterials and biocomposites/bio-based polymeric materials. Dr. Nabanita Saha currently holds an Associate Professor (Faculty of Technology) and Senior Researcher (Centre of Polymer Systems & Footwear Research Centre) position at the same university. Her research group mainly focused their research on the preparation and characterization of bio-based biomaterials (in the form of gel and hydrogels and the production of bacterial cellulose) for health care and commodity applications. She is the author/co-author of more than 89 papers published in international peer-reviewed scientific journals (recorded in Scopus and Web of Science database). In 2012, her research work was nominated for ENI Award-2013 competition on 'Protection of the Environment'. She was associated with e-COST, MP1301-NEW GEN project (2014–2017) as a working group (WG) member, CA15216-ENBA project (2016–2020) as an MC member, CA17107-CONTEXT project (2019–2022) as a WG member. She is/was involved in many, internal/national and international projects running at the Centre of Polymer Systems, TBU in Zlin. She is a member of the Polymer Processing Society (PPS) as well associated with the SPE-European Medical Polymer Division since 2009 as a Board member.

Prof. Petr Saha is the founder of Tomas Bata University in Zlín, Czech Republic, where he served as Rector for four terms of office. He currently holds the post of Director of the University Institute. His expertise includes energy and composite materials. His professional focus is on energy materials, polymer processes and medical polymers. Completed his studies at the Faculty of Technology of the Brno University of Technology, where he also worked. He was employed at Chalmers University of Technology in Sweden for more than 10 years. Moreover, he held the office of Chairman of the Czech Rectors' Conference. Prof. Sáha is also active on the international scene at a number of universities (East China University of Science and Technology, Clemson University v USA, Ton Duc Thang University in Vietnam and the University of Palermo in Italy). Between 2005 and 2007, he held the post of President of the Polymer Processing Society located in the USA, where he focused on the growth of polymer science in the development world. He was instrumental in the establishment of Bata Centre in Vietnam, which supports cooperation between institutions in Vietnam and Czech and European universities. At present, Prof. Sáha is engaged in research into energy materials from renewable resources; his team is mainly focusing on the preparation of components for batteries and supercapacitors from renewable resources, which will replace critical raw materials in the future. Prof. Petr Sáha is the author and co-author of 435 publications (with more than 8000 citations on Web of Science and an H-index of 57 as of May 2024) and several dozen patents, utility models, prototypes and industrial designs.

3D Printing and 4D Printing: Sustainable Manufacturing Techniques for Green Biomaterials

5

Oishani Sarkar, Yukta Mourya, K. L. Kavya, D. Mutthuraj, Pasupuleti Visweswara Rao, and Kanthesh M. Basalingappa

Abstract

The utilization of 3D and 4D printing in biomaterials is reshaping healthcare and biotechnology, enabling the precise fabrication of intricate tissue structures and adaptive materials. These technologies offer vast potential for personalized implants, drug delivery systems, and dynamic structures that respond to environmental changes. Four-dimensional (4D) printing represents a groundbreaking method for creating advanced materials by integrating existing 3D printing techniques. The emergence of 3D and 4D printing has revolutionized industrial processes, with this chapter focusing on sustainable manufacturing methods for eco-friendly biomaterials using these techniques. It showcases their current applications and upcoming trends.

The chapter examines both the advantages and challenges of employing 3D and 4D printing for sustainable manufacturing, addressing technical complexities, financial considerations, and legal implications. It outlines future directions for sustainable manufacturing, identifying emerging trends, potential innovations, and areas for further research. Ultimately, the chapter underscores the transformative potential of 3D and 4D printing as environmentally friendly production processes for green biomaterials. Industries can pave the way for a more sustainable future by harnessing this cutting-edge technology and the

O. Sarkar (✉) · Y. Mourya · K. L. Kavya · D. Mutthuraj · K. M. Basalingappa
Division of Molecular Biology, School of Life Sciences, JSS Academy of Higher Education & Research, Mysuru, Karnataka, India
e-mail: mutthurajd@jssuni.edu.in; kantheshmb@jssuni.edu.in; https://vidwan.inflibnet.ac.in/profile/84555

P. V. Rao
Center for International Relations and Research Collaborations and School of Applied Sciences, REVA University, Bangalore, Karnataka, India

R. Malviya, S. Sundram (eds.), *Sustainable Green Biomaterials As Drug Delivery Systems*, Biomaterials, Bioengineering and Sustainability 1,
https://doi.org/10.1007/978-3-031-79062-1_5

capabilities of eco-friendly biomaterials, thereby minimizing environmental impact while maximizing efficiency and innovation.

In summary, the introduction provides insights into 3D and 4D printing, emphasizing their role in environmentally conscious biomaterial production. An overview of 3D printing explains its principles, processes, and sustainable applications, while an introduction to 4D printing explores its evolution, mechanisms, and potential benefits for eco-friendly manufacturing. Subsequently, the chapter delves into the development and history of these techniques, elucidating why they surpass conventional production methods. Printing technologies utilizing biomaterials are pivotal for sustainable manufacturing, with the chapter exploring various types of green biomaterials, sustainability standards, and eco-friendly alternatives. It also discusses waste reduction strategies, recycling methods, and energy-efficient practices.

Keywords

3D and 4D printing · Sustainable · Green biomaterials · Carrageenan · Polymers · Plant-derived biomaterials

5.1 Introduction

Sustainable manufacturing involves implementing environmentally friendly production methods across various industries. Incorporating eco-friendly manufacturing practices can enhance resource efficiency, decrease waste generation, and optimize energy usage throughout the production process (Jamwal et al. 2021). In recent times, the convergence of sustainable manufacturing and green biomaterials has emerged as a crucial area of focus in efforts to mitigate environmental impact and foster eco-friendly practices. Given the prevailing environmental circumstances, industries globally are compelled to substitute current unsustainable and hazardous products replacing hydrocarbon-based products like plastic and fuel with renewable, biodegradable substitutes as the demand for eco-friendly materials rises. Exploring unused renewable resources becomes essential in this context. Manufacturers are working towards implementing production methods that decrease their reliance on non-renewable resources. Consequently, green manufacturing has become prevalent across various sectors, fueled by an increasing recognition of the adverse environmental and health effects linked to conventional practices (Mosher et al. 2021). This approach is mainly based on utilization of green biomaterials, which are derived from renewable sources and designed to be biodegradable or recyclable. By integrating green biomaterials into sustainable manufacturing processes, industries can significantly lower their carbon footprint and reduce reliance on non-renewable resources.

Biomaterials are substances, whether natural or engineered, that can serve as foundational components and engage to experience intended alterations within biological systems, to repair, reconstruct, restructure, re-establish, regenerate, remodel, redesign structural or functional components (Silva 2019). Green biomaterials are environmentally sustainable and are derived from diverse biological sources or manufactured using environmentally friendly technologies (Biswal et al. 2020). Agricultural waste offers abundant lignocellulosic biomass, serving as a sustainable material for producing a diverse range of bioproducts. Lignocellulose-based bio-composites have been effectively utilized as inks in 3D bio-printing, versatile substrates for environmental cleanup, medical applications, sustainable production, and various other uses (Usmani et al. 2021).

Numerous biomaterials have been created and refined through 3D bio-printing, demonstrating their effectiveness as viable alternatives to traditional materials in various biomedical fields. These environmentally sustainable biomaterials are effectively utilized in medical interventions including cancer treatment, ligament and tendon reconstruction, orthopedic procedures, designing contact lenses for eye care, wound healing, nerve regeneration treatments, breast augmentation, and manufacturing diverse surgical instruments (Biswal et al. 2020). Leveraging intricate geometry manufacturing and cost-effective production, 3D printing with biomaterials simplifies and accelerates detailed prototyping. This enables the precise printing of scaffolds for tissue regeneration, supporting drug delivery, medical interventions, tissue engineering, and various biomedical applications, all while adopting an environmentally sustainable production method (Wang et al. 2018). Although 3D printed constructs lack dynamic properties and fail to replicate the originality of tissues, the emergence of 4D bio-printing allows for controlled conformational changes in printed structures using stimuli-responsive biomaterials. This capability enables the fabrication of tissue structure undergoing morphological changes (Nureddin Ashammakhi et al. 2018). The utilization of intelligent biomaterials, like stimuli-responsive nanomaterials designed to sense the surrounding conditions and react accordingly, will significantly influence bio-printing procedures. In upcoming years, both 3D bio-printing and 4D bio-printing hold promise as advanced scaffold fabrication methods. Presently, 4D bio-printing methodologies are pioneering the development of initial stimuli-responsive, hierarchical, self-morphing and dynamic shape-altering bio-fabricated structures, employing smart biomaterials such as hydrogels, nanoparticles, and other organic polymers (Silva 2019).

5.2 Fundamentals of 3D and 4D Bioprinting

Klebe first introduced the concept of bioprinting through cell labeling technology in 1908; This technology allows micro positioning of cells to form two-dimensional (2D) tissues (Ozbola 2016). Several important events simultaneously at the turn of the century made 3D bioprinting a new field of biomedical domain. In 1986, Charles Hull pioneered the stereolithography technique with liquid photopolymers. That same year, Carl Deckard, a graduate student from the University of Austin in Texas, created the selective laser sintering (SLS) process. Since their inception, 3D printing has become a prevalent biotechnology tool in tissue engineering and regenerative medicine. These techniques are celebrated as groundbreaking advancements in the field of 3D printing.

In 1999, a team led by Anthony Atala at the Wake Forest Institute for Regenerative Medicine used molding technology to create a synthetic human bladder, onto which patient cells were applied. Although the scaffold wasn't made using additive manufacturing technology, these "lab-grown organs" paved the way for bio printed organs. Tom Boland, who was at Clemson at the time, obtained his first patent in 2003 for a bioprinting method utilizing inkjet technology. Meanwhile, Douglas Cheesey's team at the Naval Research Laboratory used laser technology to print bioink and mammalian cells into three-dimensional structures.

In 2009, the industry took a significant leap forward when Organovo and Invetech developed the first commercial bio-printer. Additionally, Wake Forest University researchers led by Anthony Atala produced printed skin structures, which were the closest thing to bio-printing functional textiles. Heart valve models and ear-shaped patterns are examples of recent bioprinting advancements. Finally, in 2014, Organovo developed the first commercially available liver tissue model using bioprinting technology (Jose et al. 2016).

4DP technology is a new form of 3DP that integrates time, allowing objects to be personalized, created or transformed into new forms. MIT's Skylar Tibbits first introduced the 4DP concept in 2013. Tibbits illustrated how prints alter shape over time via a process known as self-assembly (Arif et al. 2022). Despite its short history, 4D bio-printing has grown significantly in recent years because of the creation of novel materials appropriate for 4D printing, the research of new technologies for precise process control, and the exploration of biomedical applications. Materials used in 4D bio-printing allow spatiotemporal control of the structure and/or function of structures, whether they contain cells. Using this technology, scientists have created biostructures that can be transformed into incredibly complex structures that are difficult to create using 3D bio-printing or other methods (Yang et al. 2020).

5.3 Biomaterials in 3D Bioprinting

The term 3D bio-printing refers to the combination of 3D printing and biology. This represents the latest advance in additive manufacturing (AM), where biomaterials are incorporated into biomaterials and then deposited layer by layer (Vanaei et al. 2021; Matai et al. 2020; De Pieri et al. 2020). Printing (3DBP) technology enables various bioactive materials to form bionic structures layer by layer or step by step. The cells resemble normal tissue. The bio-printing method can provide effective tissue change by controlling the position of hormones, cells, and drugs. In 3D bio-printing technology, cell graft, cell type, growth, and material selection are more complex than in 3D printing. The field of 3D bio-printing integrates concepts of tissue engineering, regenerative medicine, and developmental biology into 3D printing. It precisely controls configuration diversity, geographic distribution, architectural precision, and complexity. 3D bio-printing can create the microstructure, structure, material, and biological properties of target tissues and organs. 3D bio-printing for self-healing allows for the controlled delivery of medicines, growth factors, proteins, DNA, cells, and other bioactive materials to the appropriate tissue (Arif et al. 2022; Mudavath and Arvapalli 2023).

The primary printing technologies utilized in 3D bio-printing are inkjet, laser-assisted, and extrusion. Inkjet bio-printing works by depositing a low-viscosity solution, such as a cell culture, at a high shear rate, forming droplets approximately 50 μm in size. Laser-assisted bio-printing, on the other hand, focuses the laser on a layer of biomaterial that absorbs the light, resulting in local pressure and ink deposition. Extrusion-based bio-printing is the simplest and involves the deposition of cells and bio-inks utilizing pneumatically, pistonally, or screwforce-operated nozzles. Despite being slower compared to laser and inkjet printing, extrusion bio-printing offers the capability to deposit patterns and solidify each layer, while incorporating viable cells. Technology plays a crucial role in achieving successful deposition in bio-printing. Droplet and inkjet printers operate optimally with viscosities ranging from 1 to 300 mPa·s, while extrusion-based printers require a minimum viscosity of 30 × 10^7 mPa·s. Therefore, bioinks with adjustable viscosity offer the versatility needed to print on various surfaces (Jovic et al. 2019).

Its increasing acceptance is due to its many advantages, including the ability to create high-performance complex geometries, maximum material savings, and easy design and customization. Metals, polymers, ceramics, and stones are among the most utilized materials in 3D printing. Modern metals and alloys are widely employed in the aerospace sector because traditional methods are laborious, complicated, and costly (Ngo et al. 2018).

The biological materials used in bio-3D printing must be biocompatible, evenly distributed, and printable. Biomaterials are substances, either natural or synthetic, employed for repairing or replacing organs within the human body. Metals, polymers, ceramics, and composites are the four classes based on their chemical composition. Metals, composites, and ceramics are all very strong materials. However, ceramics and composites are more resistant to corrosion than other materials. Polymers are more biocompatible and biodegradable than other materials. Synthetic polymers offer the best materials and designs with fewer limitations. The high molecular weight of natural polymers limits their solubility and increases their viscosity. High strength, specific microstructure, and controlled degradation make synthetic polymers excellent for 3D bio-printing. Hydrolyzed aliphatic polyester is a popular polymer bio-ink due to its biocompatibility, degradability, and printability (Vanaei et al. 2021; Matai et al. 2020; De Pieri et al. 2020; Kim et al. 2012; Zhang et al. 2016; Munaz et al. 2016).

5.3.1 Types of Polymers Used in 3D Bioprinting (Fig. 5.1)

5.3.1.1 Synthetic Polymers

Polylactic Acid (PLA)

PLA is a hydrolysable aliphatic polyester that is widely used as a polymer bioink due to its biocompatibility, degradability, and printability.

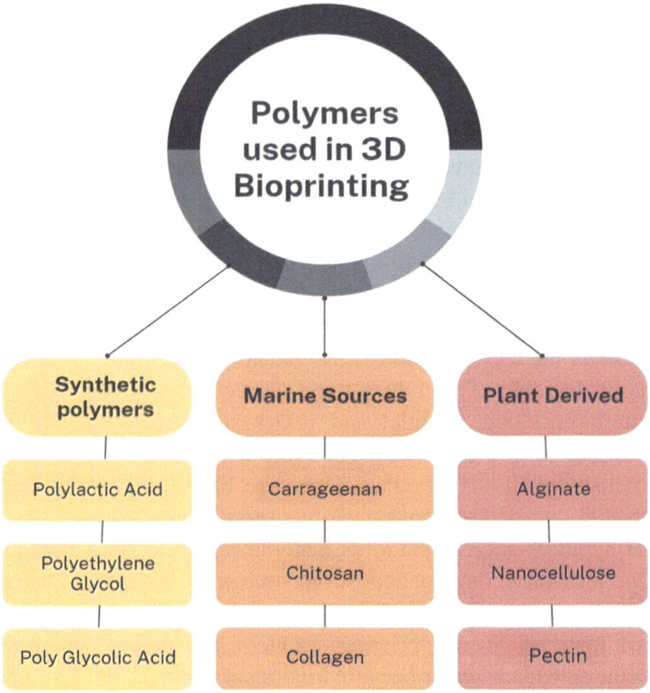

Fig. 5.1 schematic chart representing polymers used in 3D bioprinting

PLA is the primary polymer utilized as a precursor in the FDM process. In musculoskeletal tissue engineering, PLA filaments can replace ligaments and non-biodegradable fibers. When PLA breaks down, it forms acidic by-products that can cause tissue inflammation and cell death, compromising its long-term biocompatibility. The brittleness of PLA reduces its strength compared to bone, limiting its use. These constraints can be circumvented by mixing them with cheap ceramic materials like calcium phosphate. It strengthens bones and lowers acid formation (Vanaei et al., 2021; Garlotta 2001; Guvendiren et al. 2016).

Polyethylene Glycol (PEG)

Radical polymerization yields polyethylene glycol (PEG), a hydrophilic polymer. The tail group of the structure can be an asymmetric or asymmetric hydroxyl ion, and it can be straight or branched. Because of its biocompatibility, PEG is frequently utilized in scaffolds for tissue engineering, drug delivery systems, and surface modifications that produce ionomers and copolymers that are amphiphilic. PEG is suitable for hydrogel formation because of its inherent resistance to adherence of cells and protein adsorption. It is non-biodegradable and has limited mechanical strength. The C-C polymer backbone contributes to the product's non-biodegradability. PEG degradation can occur through hydrolytic and enzymatic processes (Vanaei et al. 2021; Zhu 2010; Ulbricht et al. 2014).

5.4 Polyglycolic Acid (PGA)

Polyglycolic acid (PGA) is widely considered as a key synthetic polymer 3D architecture due to its chemical diversity, ease of fabrication, biocompatibility, and biological properties. The biodegradation of PGA produces glycolic acid monomers, which are removed from the body as carbon dioxide and water through catabolic pathways. This copolymer can control the corporeal and instinctual properties of PGA. PGA is utilized to manufacture internal bone repair devices and absorbable sutures. PDLLA is more hazardous than PGA degradation products. Surface functionalization of PGA by hydrolysis of ester linkages enhances seed density and cell spreading. Hydrolysis approaches offer beneficial properties but may be limited by the reorganization of surface morphology (Vanaei et al. 2021; Asti and Gioglio 2014; Liu and Ma 2004).

Biomaterials Based on Marine Resources

1. **Carrageenan**

Carrageenan is an anionic sulfated polysaccharide that dissolves in water, obtained by alkaline extraction from red sea algae. The carrageenan is a galactan composed of alternating chains of b-d-galactose and a-d-galactose, each having a distinct sulfate group. Seaweed-derived carrageenan is classified into six elements based on the amount and position of sulfate groups: δ-(kappa), δ-(iota), δ-(lambda), μ-(Mu), δ-(Nu), and β-(Theta)-carrageenan. The tissue composition of carrageenan is identical to the glycosaminoglycans found in mammalian ECM. Thermogelation and biocompatibility of κ-carrageenan hydrogel were used to enhance its rheological properties and create accurate cell scaffolds using 3D bioprinting. κ-carrageenan's biocompatibility makes it suitable for use as a biological binder in tissue scaffold printing. Additionally, methacrylated carrageenan (MA-κ-carrageenan) was gelated by a sequential cross-linking process (UV and KCl) and showed excellent suitability for printing cell encapsulation scaffolds. MA-κ-carrageenan hydrogels supported high survival rates (~75%) of human mesenchymal stem cells (hMSCs) for up to 21 days. These chemical modifications to carrageenan result in new functions and features, such as improved physicochemical properties and photo-crosslinking capabilities for SLA-based 3D bioprinting in the field of tissue engineering (Cunha and Grenha 2016; Zhang et al. 2019; Li and Tan 2018; Guibet et al. 2007).

2. **Chitosan**

The second most common natural polymer after cellulose is chitosan, a cationic polysaccharide that is a deacetylated derivative of chitin. One of the most prevalent polysaccharides in the marine environment is chitosan. Numerous marine invertebrates' exoskeletons depend on it for structural support. Chitosan is mostly made up of units of d-glucosamine (70–90%) and N-acetyl-d-glucosamine (10–30%) joined by α-1,4-glycosidic bonds. It is a high toxicity bioactive polymer.

It is used in many different applications, such as nerve tissue engineering. Additionally, chitosan has antibacterial properties due to the interaction of its cargo with the microbial membrane. The utility and printing potential of chitosan as a bioprinting technique are amply demonstrated using chitosan-based hydrogel as bioink in 3D bioprinting of bone tissue engineering. Following 21 days of culture, the liquid adhesive is mineralized and encapsulated in chitosan-based hydrogel cells in bone formation. Chitosan's amino groups can be chemically altered to create further derivatives that have a wide range of biological activities, adequate solubility, and applications. However, since chitosan is not thermoplastic and breaks without melting, it is generally not suitable for thermomechanical processes. The operating temperature must be controlled to prevent the degradation of chitosan (Zhang et al. 2019; Peniche et al. 2008; Lizardi Mendoza et al. 2016; Ravi Kumar 2000).

3. **Collagen**

Three thin, intertwined protein chains make up the triple helix form of collagen, a polymer rich in proteins. The coordination of RGD sites is necessary for the cell attachment domains on these protein chains. In addition, collagen contains hydroxyproline-hip in addition to being abundant in amino acids that are non-polar such as glycine (30%), alanine (10%), and proline (10%). It has a graceful skin tone. Because collagen has these qualities low immunogenicity, biodegradability, and biocompatibility it is frequently employed in tissue engineering 3D bioprinting. Since pure collagen hydrogels have poor properties, collagen can be modified through esterification or used as bioinks to create tissue models or structures by mixing with other materials such as alginate and gelatin. Owing to 3D bioprinting technology, collagen is currently used in thyroid, corneal, and skin restoration procedures. Furthermore, its use in other domains is evident. Due to its enormous volume, low point of melting, low fluidity, good hydrophilicity, and asepsis, marine collagen possesses some advantages not found in collagen from terrestrial species. Collagen has so far been discovered in a wide variety of marine creatures, such as salmon, coral algae, sponges, sea urchins, jellyfish, and corals (Zhang et al. 2019; Kim et al. 2016; Shoulders and Raines 2009; Abraham et al. 2008).

Plant-Derived Biomaterials

1. **Alginate**

One of the materials most utilized in 3D bioprinting is alginate. Brown algae, such as macrocystis, seaweed, and ascophyllum nodosum, can create alginate when given the right conditions. The structural component comprises a copolymer made up of polysaccharides D-mannuronic acid (M) and L-glucuronic acid (G). It was discovered that the copolymer's M and MG sections enhanced flexibility, while the G part increased the copolymer's hardness and gelling qualities. The natural M:G ratio, which also represents seasonal and environmental variations, is influenced by the kind of algae that produces alginate (Jovic et al. 2019; Lee and Mooney 2012; Axpe and Oyen 2016).

Various combinations, therefore, produce anionic molecules with various biomedical applications. Alginate itself does not form adhesion ligands for mammalian cells. On the other hand, the inclusion of gelatin encourages cell adhesion and differentiation and makes it possible to adjust the hydrogel's viscosity in accordance with extrusion and imprinting patterns. Improving the model is achieved by sewing in just a few minutes. Due to the material connection between components, gaps are easily formed after extrusion, and deformation patterns occur even in quick connections. Another way to solve this problem is to use calcium chloride and barium chloride nebulizers, which have fewer side effects and make the process more complicated. The primary technique for crosslinking alginate hydrogel involves the use of Ca2+ ions for ionic crosslinking gel, which is regulated by changing the crosslinking agent's temperature These ions serve as a connection between the chains of the G-block polymer. CaCl2 is among the most often utilized. The fact that high concentrations of Ca 2+ are cytotoxic and that gels dissolve slowly due to the exchange and discharge of Ca 2+ ions into the fluid around it are two drawbacks of binding alginates. Additionally, the build-up period in the body is restricted. If the building blocks break down over time and are replaced by ECM released by cells, then these features might be advantageous (Giuseppe et al. 2018).

2. **Nanocellulose**

Nanocellulose, a versatile and ubiquitous biopolymer, has enormous potential because of its strong physical attributes, superior surface chemistry, favorable biological qualities (high biological compatibility, low biodegradability, and non-toxic), and economic efficiency (Jovic et al. 2019). The primary structural polymer in plants is cellulose, which can be found in the form of paracrystalline tiny filaments or nanofibrils. These polymers are connected to other macromolecules such as proteins, lignin, and heteropolysaccharides. The resultant composite provides plant structures with increased strength and flexibility (Wang et al. 2020).

The architecture is hierarchical, ranging from a complex arrangement of cellulose polymer strands (~1 nm) to a massive fibrillar structure with a diameter of ~5 to 20 micrometers (Jovic et al. 2019). Nanocellulose includes three distinct kinds of nanomaterials: Cellulose Nanocrystals (CNC), Cellulose Nanofibers (CNF), and Bacterial Nanocellulose (BNC). Nanocellulose in various forms is gaining popularity as a nanoscale component for sustainable bioinks in studies on the use of biocompatible polymers. Their potential uses include creating 3D cell culture platforms for drug screening. These include synthetic hydrogel scaffolds and skin tissue mimics that can be printed using scaffold or cell 3D bioprinting to repair damaged cartilage or bone tissue, as well as cancer research. However, widespread acceptance of nano fibrillated cellulose bioinks among the intended users (mostly physicians and cell biologists) has been impeded by concerns about the material's in vivo biodegradability and a lack of proof for its usefulness. Nano safety of cellulose nanoparticles: Because humans lack enzymes (e.g., cellulases) capable of degrading nanocellulose in vivo, nanocellulose is rarely used alone or in combination with composite materials including nanocellulose as the major component to manufacture implants. Treatment, also known as long-term restoration, is employed.

3. **Pectin**

Pectin is a typical component of plant cell walls with a high molecular weight and hydrophilicity, making it a perfect option for hydrogel creation and, ultimately, 3D bioprinting. Ionic crosslinking with CaCl2 considerably raises yield stress, improving extrusion capability and shape accuracy of the printed component. Pectin is produced commercially by using acidic and heat extraction on leftover materials from the apple, cider, and juice industries. Gelation can occur when divalent or trivalent cations and acidic conditions are present, as well as when densely concentrated pectin solutions are used, which increases polymeric connections.

Pectin, like alginate, is easily crosslinked when divalent calcium ions are present. These ions engage with the carboxyl groups in pectin to create bridges between homogalacturonan chains. One significant drawback is that it has a restricted ability to cling to cells, just as alginate and other polysaccharides. However, studies on the use of pectin as a scaffold for bone tissue engineering have shown promising outcomes. Cell adhesion can be increased by chemical modification or by combining pectin with other materials, such as polyvinyl alcohol (Klemm et al. 2011; Dufresne 2013; Du et al. 2017; Masaoka et al. 1993).

5.5 Sustainability in 3D Printing

The sustainable biomaterial represents a synergy between renewable natural resources and biomaterials, offering promising avenues for pioneering sustainable development strategies soon. Currently, there is notable emphasis and progress directed towards crafting materials from sustainable sources, gradually supplanting conventional materials. (Biswal et al. 2020). The arrival of 3D printing technologies has enabled the integration of diverse materials, including biopolymers and green biomaterials into the bio-printing process. These biomaterials are selected for their printability, compatibility with biological systems, and natural decomposability. By choosing sustainable green materials, the resulting 3D-printed constructs not only avoid contributing to pollution but also aid in environmental remediation efforts (Finny 2024). For instance, researchers have investigated the 3D printability of materials derived from algae, particularly polyhydroxyalkanoates (PHAs), which exhibit promising potential as a sustainable substitute. These materials not only uphold superior mechanical properties but also boast eco-friendliness (Grira et al. 2023). Algae derived-PHAs present a viable alternative to fossil-based PHAs, due to their comparable mechanical characteristics in addition to being eco-friendly. Algae exhibit the capacity to generate diverse hydrogel-forming polymers, some of which are presently utilized in 3D printing applications such as bioinks, while others are still being refined to meet specific printing requirements. Exploiting other algae derived products like proteins and polysaccharides not only enhances economic returns but also addresses the growing need for sustainable biomaterials (Grira et al. 2023). Yet another instance lies in the inherent difficulty of merging functional bacteria with 3D bioprinting. This involves striking a delicate balance among material manufacturability, mitigating damage throughout the bioprinting procedure, and maintaining the activity and viability of the bacteria (Zhang et al. 2023).

Cellulose, a crucial polysaccharide in plant structures, stands as one of Earth's most abundant biopolymers. Due to their excellent mechanical properties, cellulose-based bio composites are utilized in material extrusion for 3D printing (Wang et al. 2018). The main benefits of using cellulose-based materials in 3D printing stems from their accessibility, ease of processing, high aspect ratio, sustainability, and suitability for chemical modification (Kuhnt and Camarero-Espinosa 2021). Essential steps in bio-fabrication involve optimizing the selection of bio-inks, considering factors like 3D printability, compatibility with microbes and chemicals, and the degradation of contaminants. Incorporating nanomaterials, such as catalytic nanoparticles, into 3D-printed structures can enhance pollutant degradation efficiency. However, it is crucial to evaluate the sustainability of using these particles. Furthermore, biodegradable and bio-based printing materials research aligns with sustainability objectives in bioremediation, reducing the environmental effect of the remediation tools themselves (Wei et al. 2017) (Fig. 5.2).

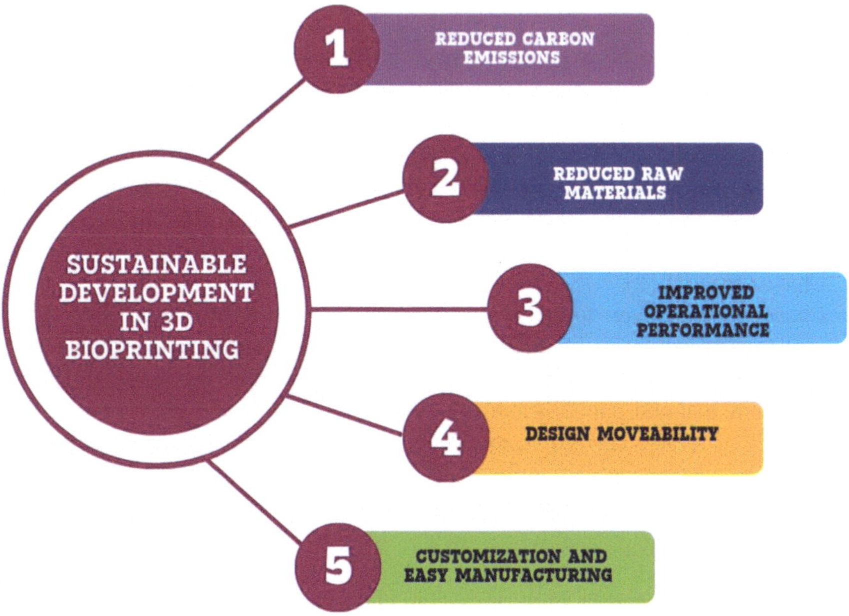

Fig. 5.2 Diagrammatic representation of sustainable development in 3D bioprinting

5.5.1 What Makes your Biomaterial more Sustainable?

- Reducing the amount of raw materials and packaging materials (using plant-based bioplastics for packaging bio-printing materials) (Zimmermann et al. 2020) used, with a greater emphasis on the quality of the manufactured product. In 3D bioprinting, using optimized design software can minimize the amount of bio-polymer needed for scaffolds, reducing waste. For instance, printing a porous scaffold instead of a solid one decreases the material required without compromising structural integrity.
- It's necessary to steer clear of single-use items that can't be recycled or composted. Choosing products made from renewable resources is preferred over those derived from fossil fuels.
- Utilizing alginate, derived from seaweed, as a bio-ink in bio-printing, reduces reliance on petroleum-based products (Grira et al. 2023). Alginate is not only renewable but also biocompatible, making it ideal for medical applications.
- Sustainability should be upheld throughout the lifecycle of manufactured products, including the materials used, potential applications, and the recovery of materials, such as recycling unused bio-ink, from waste and during feedstock growth.
- Materials like collagen and gelatin are both biodegradable and derived from renewable sources (Osidak et al. 2020).
- It's essential to integrate sustainability considerations into social, health, economic, and environmental aspects. Products should be manufactured and designed with recyclability, reusability, or composability in mind.
- Designing modular bio printed structures that can be disassembled, and their components reused in new prints helps with resource conservation.
- Both the state and central governments should encourage sustainable agricultural facilities for farmers and other communities.
- Ensuring that bio-inks and feedstocks are sourced from non-GMO plants protects biodiversity and aligns with organic farming principles. For instance, using non-GMO cornstarch as a raw material for biopolymer production (Álvarez-Chávez et al. 2012).
- Chemicals that adhere to the twelve principles of green chemistry are favored for use in cultivated farms (Ardila Fierro and Hernández 2021).
- It's important to refrain from using nanomaterials and chemicals that pose environmental and public health risks throughout their life cycle. For example, utilizing naturally derived nanoscale cellulose instead of silver nanoparticles (Ee and Li 2021).
- Each stage of the production process should be localized to respective regions.

By adopting these sustainable practices, 3D bio-printing can significantly reduce their environmental impact while advancing medical and industrial applications.

5.6 Advancements in 4D Bioprinting

Even though 3D bio-printing technologies are being continuously utilized to create complex structures with accurate control in an automated way, the current 3D structures cannot accurately replicate the dynamic nature of tissues. This is because of changes that occur in tissue structure over time due to its regeneration and repair mechanism (Ashammakhi et al. 2018). Thus, it is necessary to incorporate time-dependent features into 3D printed tissue structures to replicate these natural changes. To address this limitation, 4D bio-printing has been developed and used. This technique uses stimuli-responsive biomaterials and cell traction forces to create tissue constructs that are structurally dynamic. 4D bio-printing consists of the 3D printing of cell-laden materials that can respond to internal cell forces and external stimulations (Ashammakhi et al. 2018). The advent of 4D bio-printing technology has brought numerous innovations and revolutions to the biomedical field. The method utilizes biomaterials and diverse cell types to fabricate structures for regenerating tissues, such as bones and cartilage, neural and skin tissues etc. Additionally, 4D bio-printing aids in the creation of bioactuators, biorobotics, and biosensors, drug delivery and wound healing systems (Noroozi et al. 2023).

5.6.1 Smart Polymers

Stimuli-responsive polymers, or smart polymers, alter their physical or chemical properties in response to specific stimuli such as heat, light, electric and magnetic fields, pH levels, redox potential, and mechanical force (Wei et al. 2017). Their capacity to translate minor environmental variations into noticeable reactions makes them valuable in fields such as tissue engineering, controlled drug delivery, soft robotics, electronics, and biomaterials with enhanced mechanical strength (Mukhopadhyay et al. 2014).

The benefit of consuming smart polymers in bio-printing lies in their capacity to create intricate yet functional materials that automatically react to environmental stimuli. This eliminates the need for external control. When the sophisticated structures enabled by 4D manufacturing are combined with the responsive nature of smart polymers, the potential applications become endless (Ryan T Shafranek et al. 2019). For instance, cellulose hydrophilic nature allows it to achieve steady bloat in water. This property can be used in 4D printing and drug delivery methods (Khalid and Arif 2022). Numerous hydrogels exhibit responsiveness to stimuli, rendering them suitable for 4D printing applications. However further research is required in this field to overcome the mechanical variability and slow stimulus responsiveness of natural hydrogels (Grira et al. 2023). To produce hydrogels, biopolymers are the most largely used raw material obtained from agriculture, forests and livestock and their byproducts as they are cost effective, renewable, biocompatible, biodegradable, abundant in nature and more suitable from structural and mechanical means.

In recent years, nanomaterials have been greatly in focus for bio-fabrication processes due to their exceptional ability to meet new scientific demands that arise from nanoscale which generates new possibilities. A recent development in this field is 4D bio-fabrication, which seeks to build and refine 3D biological structures through dynamic self-assembly processes. These structures can change their form and functions over time, especially when exposed to specific chemical, biological, or physical stimuli in smart biomaterials (Silva 2019).

5.6.2 Applications of 4D Bio-Printing in Sustainable Manufacturing

In 4D bio-printing, materials that respond to stimuli and then degrade into harmless by-products are preferred. Using biodegradable polymers like polylactic acid (PLA) ensures that post-use, the materials do not contribute to landfill waste (Costa et al. 2021). Using bio-based materials/green biomaterials in 4D printing minimizes waste production and environmental impact. Due to their unique ability of changing dynamics under the effect of external stimuli like heat, temperature, light etc. 4D bio-printing can play a huge role in the medical field such as offering rapid development in customizing scaffolds for implants which makes it time effective over other manufacturing techniques (Ameta et al. 2022).

Bio-printing organs through 4D bio-fabrication technologies can also address health issues by reducing the need for donor organs, thus having a profound societal benefit. 4D bio-printing allows for exact management of the spatial arrangement of different technologies that can autonomously fold or unfold to encapsulate and release drugs or cells in a controlled, programmable way. This ability to have control and optimize spatial and temporal administration of therapeutic agents makes 4D bio-printing a reliable manufacturing technique in drug development industries (Lukin et al. 2019). 4D bio-fabrication has opened opportunities in development of green bio-fabrication related to the growth of sustainable alternatives aiming at the production of bio constructs (Silva 2019). However, the 4D bio-printing approach still needs to be further enhanced to adapt well with the cell's biological behavior and functionalities in a safer and foreseeable manner (Ashammakhi et al. 2018) (Table 5.1).

Table 5.1 Comparison between 3D and 4D bioprinting

3D Bioprinting	4D Bioprinting
Involves creating a three-dimensional structure using biological material like cells and biomaterials.	Adds an additional dimension of time to the printing process.
Primary used for tissue engineering and regenerative medicine applications.	Allows printed structures to change shape or function over time in response to external stimuli.
Static structures with predetermined shape and functions	Enables the creation of self-assembling or self-regulating tissues and organs.
Limited ability to mimic dynamic biological processes.	Holds potential for creating more complex and adaptive biological structures.

5.7 Case Studies on 3D and 4D Bioprinting

5.7.1 Revolutionizing Personalized Medicine: A Case Study of 3D Bioplotter Implementation at Hanover Medical School

Hanover Medical School's Department of Otolaryngology (ENT) embarked on a groundbreaking project in 2019 aimed at developing individualized implants for patients. Led by Dr. Vina Shaper, the Pharmacology of the Inner Ear research group recognized the potential of 3D bioprinting technology to revolutionize personalized medicine. This case study explores their journey in implementing the 3D bio-plotter from Desktop Health, highlighting its applications, challenges, and prospects.

The field of personalized medicine is rapidly evolving, with a growing emphasis on tailored treatments to meet individual patient needs. Traditional manufacturing methods for implants often fall short in delivering precise, patient-specific solutions. Recognizing this gap, Hanover Medical School invested in advanced technology to pioneer a new era of personalized healthcare. In 2020, Hanover Medical School acquired the 3D bio-plotter, a cutting-edge printing system designed for medical applications. Dr. Shaper's team identified key requirements for successful implementation: ease of use, precision, reliability, and ease of cleaning. These criteria were essential to ensure seamless integration into the clinical workflow.

The bio-plotter's versatility opened doors to a myriad of clinical applications. Using x-ray imaging, the team created implantations tailored to each patient's unique anatomy, particularly focusing on cases like congenital atresia of the outer wear. By precisely printing patient-specific stents, the team successfully prevented scar formation and mitigated postoperative inflammation and trauma reactions. While the journey towards personalized medicine presented numerous challenges, Hanover Medical School's research group navigated them with innovation and perseverance. Ensuring reproducibility in printing, maintaining sterility, and optimizing drug delivery were among the hurdles addressed. Through meticulous quality management and continuous refinement, the team overcomes these obstacles, paving the way for broader adoption.

The success of the bio-plotter implementation laid a solid foundation for future advancements. With plans to set up a second printer directly in the clinic, Hanover Medical School aims to streamline the process further, enabling clinicians to print implants onsite for immediate patient care. Additionally, the research group is exploring the development of biodegradable implants, eliminating the need for explanation and offering long-term benefits to patients. Hanover Medical School's pioneering efforts in personalized medicine have far-reaching implications for healthcare worldwide. By leveraging the capabilities of the 3D bio-plotter, the research group has demonstrated the potential to improve patient outcomes, reduce healthcare costs, and revolutionize the standard of care for a wide range of clinical conditions. As they continue to innovate and collaborate, their vision of individualized, drug-eluting implants holds promise for transforming the landscape of medical treatment globally.

The case study of Hanover Medical School's journey with the 3D bio-plotter underscores the transformative power of technology in healthcare. Through visionary leadership, interdisciplinary collaboration, and a commitment to excellence, Dr. Vina Shaper and her team have set a precedent for personalized medicine, inspiring future generations to push the boundaries of innovation in the pursuit of better health for all (Vina 2023).

5.7.2 Advancing Tissue Engineering with Next-Generation Bioprinting: A Case Study of the NGbR System

Recent years have seen notable progress in tissue engineering, driven by advancements in bio-printing technology. This case study explores the application of the Next-Generation Bio-printing Research (NGbR) system, a disruptive platform developed by Poitus in partnership with Scintica Instrumentation for 4D bio-printing. The NGbR system represents a paradigm shift in bio-printing, offering precise control over cell patterning and tissue fabrication. Through a cytocentric approach and laser-assisted bio-printing capabilities, the NGbR system enables high-resolution printing and enhanced cell viability, essential for the development of complex tissues and organs. Tissue engineering holds immense promise for regenerative medicine, offering solutions to tissue loss and organ failure. Bio-printing technology plays a crucial role in this endeavor, allowing researchers to precisely deposit cells and biomaterials to create functional tissues. The NGbR system, introduced by Poitus and distributed by Scintica Instrumentation, represents the next frontier in bio-printing technology. By leveraging laser-assisted bio-printing and advanced robotics, the NGbR system offers unprecedented control and versatility in tissue fabrication.

In this case study, researchers utilized the NGbR system to fabricate complex tissue constructs for various applications, including wound healing, organ regeneration, and disease modeling. The workflow involved designing tissue scaffolds using computer-aided design (CAD) software, optimizing cell and biomaterial formulations, and implementing precise printing parameters. The NGbR system's capabilities in single-cell resolution and 4D bio-printing were harnessed to mimic native tissue architecture and functionality. The application of the NGbR system yielded promising results in tissue engineering research. High-resolution printing enabled the creation of intricate tissue architectures, while laser-assisted bio-printing maintained high cell viability and spatial control. Researchers successfully fabricated tissue constructs with tailored mechanical properties and biological functionalities, demonstrating the system's potential for personalized medicine and drug discovery. Furthermore, the NGbR system facilitated the study of tissue morphogenesis and disease mechanisms, providing valuable insights into complex biological processes.

The NGbR system represents a significant advancement in bio-printing technology, offering researchers unprecedented capabilities in tissue engineering. By combining precision, versatility, and automation, the system accelerates the development of regenerative therapies and biomedical research. Future enhancements, such as integrated imaging and real-time monitoring, hold promise for further advancing tissue engineering applications. Effective collaboration among academia, industry, and healthcare stakeholders will play a pivotal role in maximizing the benefits of the Next-Generation Biomedical Research (NGbR) system to enhance patient outcomes and meet medical needs that are currently unmet. The NGbR system exemplifies the convergence of engineering, biology, and medicine in the pursuit of innovative healthcare solutions. Through collaborative efforts and continued research, bio-printing technology holds the promise of revolutionizing regenerative medicine and personalized healthcare. The case study highlights the transformative potential of the NGbR system and underscores the importance of interdisciplinary collaboration in advancing tissue engineering and biomedical research (Jekyll and Gracia 2022; Coulthard et al. 2022).

5.8 Challenges and Future Directions in 3D and 4D Bioprinting

Challenges

Both 3D and 4D bio-printing encounter obstacles and present future pathways for advancement in biomedical engineering. 3D bio-printing encounters several obstacles in contrast to non-biological printing, spanning material selection, cell viability, and ethical considerations. These challenges are broadly classified into two categories: technical hurdles and ethical or legal dilemmas.

5.8.1 The Technical Challenges in 3D Bio-Printing Include

1. **Printability**: Attaining exact and precise placement of bio-ink requires enhanced resolution to optimize interaction and regulation within the 3D microenvironment. Expediting the printing process and expanding its scale are essential for ensuring commercial feasibility.
2. **Mechanical Properties**: Ensuring the mechanical strength of printed tissues comparable to native tissue remains a challenge. Strategies such as using sacrificial materials during printing or incorporating them into the construct are explored to enhance mechanical properties.
3. **Vascularization**: Overcoming hurdles in vascularization is critical. Approaches include building vasculature during bioprinting using biodegradable or synthetic polymers or mixing angiogenic factors in the bio-ink to attract cells and induce vasculature formation later.
4. **Tissue Complexity**: Creating blueprints for tissues or organs poses challenges due to the intricate architecture and cellular diversity.
5. **Equalizing Mechanical Properties**: Achieving mechanical properties in 3D-printed organs, like bone or cartilage, that mirror those of natural tissues while preserving biological functionality remains a challenge. One potential

strategy involves replicating the elastic moduli of human tissue using scaffolds or components of the extracellular matrix (ECM).

6. **Functional Complexity**: Fully functionalizing complex organs like hair follicles, sweat glands, or the liver, brain, spinal cord, which have heterogeneous cell populations, is a significant challenge.
7. **Immune Reaction**: Managing potential immune responses from recipients to scaffolding material or cells within the printed tissue structure is crucial. Potential remedies include utilizing bioink derived from native decellularized ECM or incorporating immunosuppressive molecules during the pre-implantation incubation phase.
8. **Personalization**: Developing personalized patient-specific tissues or organs requires substantial resources and time due to the procurement and isolation of stem cells and decellularized extracellular matrix.

To address these challenges, continuous updates and advancements in bio-printing hardware, bioink composition, and printing techniques are necessary to move closer to printing biomimetic structures and functional organs. Ethical issues surrounding 3D bioprinting include concerns about what should and should not be printed, and to balance potential risks against benefits. Regulatory frameworks for clinical assessment might require tailored development for 3D bioprinting to tackle concerns regarding standardization and the expansion of personalized medical interventions. Global regulatory bodies, such as the FDA, have encountered difficulties with 3D bioprinting, and there is ambiguity regarding how to manage potential risks linked to the technology. Only a few countries, such as South Korea and Japan, have provided some regulatory guidance, which is still quite broad (Chameettachal et al. 2019).

5.8.2 Challenges in 4D Bio-Printing

While there have been notable advancements in utilizing 4D printing for biomedical purposes, obstacles persist for its broad clinical adoption. The performance relies heavily on materials like shape-memory polymers and hydrogels, which must exhibit desired responses to external stimuli. However, finding materials with suitable properties is challenging, as their mechanical properties may not always meet application requirements, and combining several materials within a single structure can be difficult. Moreover, printing procedures might impose limitations, necessitating materials that can endure high temperatures, mechanical strain, or exposure to light or chemicals, thus narrowing down material choices even further.

Biocompatibility is crucial in biomedical engineering to ensure patient safety, limiting material choices for 4D printing. Developing biocompatible smart materials that meet shape-changing requirements is challenging. Additionally, processing techniques can affect biocompatibility, with some methods leaving chemicals or particles that may trigger negative reactions. Maintaining biocompatibility throughout the 4D printing process is essential for successful biomedical applications.

The speed of deterioration in 4D-printed biomedical devices presents a significant obstacle in fields like drug delivery, tissue engineering, and implants. It's essential for materials to degrade at appropriate rates to avoid harm to patients. Achieving the desired degradation is intricate due to interactions among materials, manufacturing processes, and the biological context. Several factors, including material makeup, structure, and individual patient traits, influence degradation. Resolutions involve creating biomaterials specifically tailored for 4D printing, employing composite materials, and refining printing parameters. Collaboration among scientists, engineers, and biomedical specialists is vital for tackling material challenges in 4D printing for biomedical applications.

One major advantage of 4D printing in biomedical settings is its ability to create structures with a wide range of mechanical properties and responsiveness to stimuli by combining different materials. However, integrating multiple materials poses challenges due to current printing limitations. Achieving precise control over material distribution and interaction is crucial for obtaining the intended functionality, requiring advanced printing techniques and materials that can work together seamlessly. Overcoming these challenges entails the development of innovative printing methods and formulations specifically designed for multi-material 4D printing in the biomedical sector.

In 4D printing for biomedical engineering, attaining exceptional resolution and precision is vital to guarantee the optimal functionality of printed structures. For instance, tissue scaffolds demand exact pore geometry to enable cell attachment, whereas implantable devices require precise dimensions for correct placement. Current printing techniques face challenges in achieving the required resolution and precision, especially with advanced materials. Factors like material viscosity, printing speed, and hardware setup can impact resolution and precision. Therefore, the creation of innovative printing technologies and fine-tuning of parameters are necessary to fulfill the demands of various biomedical applications.

Scalability presents a notable obstacle in 4D printing for biomedical engineering, as printed structures must accommodate diverse patient requirements while remaining cost-efficient to manufacture. Existing methods are frequently slow and lack scalability. Overcoming this challenge involves creating faster printing methods with reduced material wastage and investigating automation and integration with alternative manufacturing technologies to enhance production capacity and lower expenses.

The implementation of innovative medical technologies like 4D printing demands thorough evaluation and approval from regulatory bodies such as the FDA or EMA to guarantee both safety and effectiveness. Navigating the regulatory framework can be intricate, involving preclinical assessments, clinical trials, and post-market monitoring. Due to the innovation of 4D printing, existing regulations may have gaps, underscoring the requirement for customized guidelines to address its distinctive characteristics. Collaboration among researchers, manufacturers, and regulatory authorities is indispensable to ensure adherence to standards of biocompatibility, quality, and safety throughout the validation procedure or 4D-printed biomedical devices.

The incorporation of 4D printing in biomedical engineering raises ethical considerations, including safeguarding patient privacy, securing informed consent, and ensuring equitable access. Preserving patient confidentiality is crucial throughout design and production phases. Obtaining informed consent is vital for patients participating in clinical trials involving 4D-printed devices. Ensuring fair access to these devices is essential given their high costs. Furthermore, each specific application within 4D printing may present distinct challenges, requiring tailored approaches and problem-solving strategies. Factors such as functionality, suitability for biological systems, intricacy, and scalability require thorough consideration in every project. Extensive research and analysis are necessary for overcoming the limitations of 4D printing and maximizing its potential in biomedical engineering.

5.8.3 Future Directions

The emerging trends and research areas in biomedical engineering suggest promising prospects for the future of 4D printing:

1. **Integration of Sensors and Electronics**: Integrating sensors and electronic elements into 4D-printed devices allows for real-time monitoring of device functionality, assisting in the customization of treatment strategies.
2. **Biohybrid Systems**: Combining living cells with 4D-printed frameworks leads to biohybrid systems, offering novel prospects in tissue engineering and regenerative medicine, potentially reducing the need for organ transplants.
3. **Self-Healing Materials**: Integrating self-repair capabilities into 4D-printed devices prolongs their durability and dependability, particularly advantageous for implants and prosthetics, thereby diminishing the necessity for replacements or invasive procedures.
4. **Personalized Medicine**: 4D printing facilitates the creation of tailor-made devices and structures tailored to individual patient requirements, enhancing the effectiveness and results of treatments and interventions.
5. **Environmental and Biodegradable Materials**: Research is dedicated to creating eco-friendly and biodegradable materials suitable for 4D printing, thereby promoting sustainable healthcare approaches.
6. **Multi-Stimuli Responsive Materials**: Utilizing materials responsive to various stimuli enables the advancement of 4D-printed devices with improved functionality and flexibility, including tailored drug delivery systems.
7. **Biomimetic Materials**: Biomimetic materials imitate natural tissues, offering advantages in biocompatibility and functionality, promoting effective healing and recovery in tissue engineering and prosthetics.
8. **Nanomedicine**: The evolution of 4D printing allows to produce intricate nanostructures and drug delivery systems, enhancing treatment effectiveness and facilitating personalized medicine, transforming healthcare and improving patient outcomes.

Progress in production methods is advancing in 4D printing and related cutting-edge advances, resulting in increasingly sophisticated, streamlined, and accurate processes. Significant developments include:

1. **Multi-Material and Multi-Process Printing**: Facilitating the development of complex formations through the simultaneous integration of diverse materials and processes, influencing advancements in biomedical engineering, soft robotics, and intelligent substances.
2. **High-Resolution Printing**: Enabling the generation of complex frameworks with precise configuration details on a micro or nano level, essential for tissue engineering and drug delivery platforms.
3. **Hybrid Fabrication Techniques**: Integrating various manufacturing techniques to produce structures with improved characteristics, such as combining 3D printing and electrospinning for the fabrication of tissue engineering scaffolds.
4. **Self-Assembly and Self-Folding Techniques**: Facilitating the development of dynamic structures without requiring external intervention, crucial for generating adaptable constructs that react to environmental signals.
5. **Embedded Sensors and Actuators**: They provide continuous monitoring and regulation of 4D-printed constructs, improving their versatility across different applications.
6. **Biofabrication**: It encompasses the advancement of techniques like bioprinting, and hydrogel printing infused with cells, allowing the incorporation of living cells and biomolecules into printed constructs, with applications in tissue engineering and regenerative medicine.
7. **Revolutionizing Design**: Leveraging machine learning and AI-powered techniques for design innovation and to anticipate material performance and adjust fabrication parameters for achieving specific functionalities, thereby enhancing productivity and efficacy.
8. **Scalability and Automation**: Enhancing scalability and automation involves speeding up the mass production of 4D-printed structures via efficient manufacturing techniques and automated systems for quality assurance (Silva 2019).

Bioprinting offers hope in alleviating the worldwide scarcity of organs by allowing the creation of transplantable tissues with reduced risks of immune rejection. Innovations in hydrogel technology, such as dynamic switchable and oxygen-generating hydrogels, present novel approaches for regulating cellular environments. However, unlocking the complete capabilities of 3D bio-printing necessitates enhancements in speed, scalability, material availability, tissue vascularization, nerve supply, and scaffold production (Chameettachal et al. 2019).

These advancements have the potential to revolutionize healthcare and provide new therapeutic options across various applications and industries (Silva 2019). Overall, collaboration across disciplines is crucial for advancing this technology towards clinical applications (Ramezani and Mohd 2023).

5.9 Conclusion

The article discusses sustainable manufacturing practices in the context of bio-printing, focusing on the utilization of environmentally friendly biomaterials in 3D and 4D bio-printing for medical applications. It highlights the evolution of bio-printing technology from 3D to 4D, tracing its origins and significant milestones. The role of biomaterials in bio-printing, including synthetic polymers and sustainable options sourced from marine and plant-based origins, is emphasized, along with their contribution to environmental remediation efforts. Strategies for achieving biomaterial sustainability, such as reducing raw material usage and favoring renewable resources, are outlined. It revolves around the transformative capabilities of 4D bio-printing, which is facilitated by intelligent polymers, in pushing the boundaries of tissue engineering and drug delivery. Additionally, it emphasizes the significance of collaborative efforts and ethical contemplations in the realm of bio-printing research. Overall, the article underscores the significance of sustainable manufacturing practices in bio-printing for revolutionizing healthcare while minimizing environmental impact.

References

Abraham LC, Zuena E, Perez-Ramirez B, Kaplan DL (2008) Guide to collagen characterization for biomaterial studies. J Biomed 87B:264–285

Álvarez-Chávez CR, Edwards S, Moure-Eraso R, Geiser K (2012) Sustainability of bio-based plastics: general comparative analysis and recommendations for improvement. J Clean Prod 23(1):47–56

Ameta KL, Solanki VS, Singh V, Devi AP, Chundawat RS, Haque S (2022) Critical appraisal and systematic review of 3D & 4D printing in sustainable and environment-friendly smart manufacturing technologies. Sustain Mater Technol 34:e00481

Ardila Fierro KJ, Hernández JG (2021) Sustainability assessment of mechanochemistry by using the twelve principles of green chemistry. ChemSusChem 14(10):2145–2162

Arif UZ, Khalid YM, Ahmed W, Arshad H (2022) A review on four-dimensional (4D) bioprinting in pursuit of advanced tissue engineering applications. Bioprinting 27

Ashammakhi N, Ahadian S, Zengjie F, Suthiwanich K, Lorestani F, Orive G, Ostrovidov S, Khademhosseini A (2018) Advances and future perspective in 4D bioprinting. Biotechnol J 13(12):e1800148

Asti A, Gioglio L (2014) Natural and synthetic biodegradable polymers: different scaffolds for cell expansion and tissue formation. Int J Artif Organs 37(3):187–205

Axpe E, Oyen ML (1976, 2016) Applications of alginate-based bioinks in 3D bioprinting. Int J Mol Sci 17(12)

Biswal T, BadJena SK, Pradhan D (2020) Sustainable biomaterials and their applications: a short review. Mater Today Proc 30:274–282

Chameettachal S, Yeleswarapu S, Sasikumar S, Shukla P, Hibare P, Bera AK, Bojedla SSR, Pati F (2019) 3D bioprinting: recent trends and challenges. J Ind Inst Sci 99:375

Costa PD, Costa DC, Correia TR, Gaspar VM, Mano JF (2021) Natural origin biomaterials for 4D bioprinting tissue-like constructs. Adv Mater Technol 6(10):2100168

Coulthard T, Jekyll A, Gracia M (2022) A detailed review of 4D bioprinting applications using the NGB-R system. Scintica instrumentation. https://youtu.be/4L-l9D76pzg?si=MJcVcD3rbanobM6s

Cunha L, Grenha A (2016) Sulfated seaweed polysaccharides as multifunctional materials in drug delivery applications. Mar Drugs 14(3):42

De Pieri A, Byerley AM, Musumeci CR, Salemizadehparizi F, Vanderhorst MA, Wuertz Kozak K (2020) Electrospinning and 3D bioprinting for intervertebral disc tissue engineering. JOR Spine: e1117

Du X, Zhang Z, Liu W, Deng Y (2017) Nanocellulose-based conductive materials and their emerging applications in energy devices-a review. Nano Energy 35:299

Dufresne A (2013) Nanocellulose: a new ageless bionanomaterial. Mater Today 16:220

Ee LY, Li SF (2021) Recent advances in 3D printing of nanocellulose: structure, preparation, and application prospects. Nanoscale Adv 3(5):1167–1208

Finny AS (2024) 3D bioprinting in bioremediation: a comprehensive review of principles, applications, and future directions. PeerJ 12:e16897

Garlotta D (2001) A literature review of poly lactic acid. J Polym Environ 9:63–84

Giuseppe MD, Law N, Webb B, Macrae RA, Liew LJ, Sercombe TB, Dilley RJ, Doyle BJ (2018) Mechanical behaviour of alginate-gelatin hydrogels for 3D bioprinting. J Mech Behav Biomed Mater 79:150–157

Grira S, Khalifeh HA, Alkhedher M, Ramadan M (2023) 3D printing algae-based materials: pathway towards 4D bioprinting. Bioprinting 33:e00291

Guibet M, Colin S, Barbeyron T, Genicot S, Kloareg B, Michel G, Helbert W (2007) Degradation of A-carrageenan by Pseudoalteromonas Carrageenan Vora A-carrageenase a new family of glycoside hydrolases unrelated to k and carrageenases. Biochem J 404(1):105–114

Guvendiren M, Molde J, Soares RMD, Kohn J (2016) Designing biomaterials for 3D printing. ACS Biomater Sci Eng 2(10):1679–1693

Jamwal A, Agrawal R, Sharma M, Kumar V (2021) Review on multi-criteria decision analysis in sustainable manufacturing decision making. Int J Sustain Eng 14(3):202–225

Jekyll A, Gracia M (2022) Introduction to the NGB-R, a disruptive 4D bioprinting system. Scintica instrumentation. https://youtu.be/jb9Xh25pBLM?si=4DdtluvgHPwNy_bD

Jose RR, Rodriguez JM, Dixon AT, Omenetto F, Kaplan LD (2016) Evolution of bioinks and additive manufacturing technologies for 3D bioprinting. ACS Biomater Sci Eng 2(10): 1662–1678

Jovic TH, Kungwengwe G, Mills AC, Whitaker IS (2019) Plant-derived biomaterials: a review of 3D bioprinting and biomedical applications. Front Mech Eng 5

Khalid MY, Arif ZU (2022) Novel biopolymer-based sustainable composites for food packaging applications: a narrative review. Food Packag Shelf Life 33:100892

Kim J, McBride S, Tellis B, Alvarez-Urena P, Song Y-H, Dean DD, Sylvia VL, Elgendy H, Ong J, Hollinger JO (2012) Rapid-prototyped PLGA/β-TCP/hydroxyapatite nanocomposite scaffolds in a rabbit femoral defect model. Biofabrication 4(2):025003

Kim JE, Kim SH, Jung Y (2016) Status of three-dimensional printing inks for soft tissue regeneration. Tissue Eng Regen:636–646

Klemm D, Kramer F, Moritz S, Lindström T, Ankerfors M, Gray D, Dorris A (2011) Nanocelluloses: a new family of nature based materials. Chernie Int. Ed, Angew

Kuhnt T, Camarero-Espinosa S (2021) Additive manufacturing of nanocellulose based scaffolds for tissue engineering: beyond a reinforcement filler. Carbohydr Polym 252:117159

Lee KY, Mooney DJ (2012) Alginate: properties and biomedical applications. Prog Polym Sci 37(1):106–126

Li H, Tan YJLI (2018) LA strategy for strong interface bonding by 3D bioprinting of oppositely charged x-carrageenan and gelatin hydrogels. Carbohydr Polym 198:261–269

Liu X, Ma PX (2004) Polymeric scaffolds for bone tissue engineering. Ann Biomed Eng 32:477–486

Lizardi Mendoza J, Arguelles Monal WM, Goycoclea Valencia FM (2016) Chapter 1: Chemical characteristics and functional properties of Chitosan. In: Bautista Baños S, Romanazzi G, Jiménez-Aparicio A (eds) Chitosan in the preservation of agricultural commodities. Academic Press, Cambridge, MA, pp 3–31

Lukin I, Musquiz S, Erezuma I, Al-Tel TH, Golafshan N, Dolatshahi-Pirouz A, Orive G (2019) Can 4D bioprinting revolutionize drug development? Expert Opin Drug Discov 14(10):953–956. https://doi.org/10.1080/17460441.2019.1636781

Masaoka S, Ohe T, Sakota N (1993) Production of cellulose from glucose by Acetobacter Xylinum. J Ferment Bioeng 75:18–22

Matai G, Kaur A, Seyedsalehi A, McClinton CTL (2020) Progress in 3D bioprinting technology for tissue/organ regenerative engineering. Biomaterials 226:Article 119536

Mosher CZ, Brudnicki PA, Gong Z, Childs HR, Lee SW, Antrobus RM, Fang EC, Schiros TN, Lu HH (2021) Green electrospinning for biomaterials and biofabrication. Biofabrication 13(3): 035049

Mudavath S, Arvapalli S (2023) A review on recent advances in 3D bioprinting. Int J Pharm Bio-Med Sci 2, ISSN (print): 2767-827X, ISSN(online): 2767-830X

Mukhopadhyay P, Sarkar K, Bhattacharya S, Bhattacharyya A, Mishra R, Kundu PP (2014) pH sensitive N-succinyl chitosan grafted polyacrylamide hydrogel for oral insulin delivery. Carbohydr Polym 4(112):627–637

Munaz A, Vadivelu RK, John JS, Barton M, Kamble H, Nguyen N-T (2016) Three-dimensional printing of biological matters. J Sci Adv Mater Device 1(1):1–17

Ngo TD, Kashani A, Imbalzano G, Nguyen KTQ, Hui D (2018) Additive manufacturing (3D printing): a review of materials, methods, applications and challenges. Comp Part B Eng 123:172–196

Noroozi R, Arif ZU, Taghvaei H, Khalid MY, Sahbafar H, Hadi A, Sadeghianmaryan A, Chen X (2023) 3D and 4D bioprinting technologies: a game changer for the biomedical sector? Ann Biomed Eng 51(8):1683–1712

Osidak EO, Kozhukhov VI, Osidak MS, Domogatsky SP (2020) Collagen as bioink for bioprinting: a comprehensive review. Int J Bioprinting 6(3):270

Ozbola IT (2016) 3D bioprinting: fundamentals, principles and applications. Academic, London

Peniche C, Arguelles-Monal W, Goycoolca FM (2008) Chapter 25: Chitin and chitosan major sources, properties and applications. In: Monomers, polymers and composites from renewable resources. Elsevier Science, Amsterdam, pp 517–542

Ramezani M, Mohd RZ (2023) 4D printing in biomedical engineering: advancements, challenges, and future directions. J Funct Biomater 14:347

Ravi Kumar MNV (2000) A review of chitin and chitosan applications. React Funct Polym 46:1–27

Shafranek RT, Cem MIlik S, Smith PT, Lee C-U, Boydston AJ, Nelson A (2019) Stimuli-responsive materials in additive manufacturing. Prog Polym Sci 93:36–67

Shoulders MD, Raines RT (2009) Collagen structure and stability. Annu Rev Biochem 78:929–958

Silva LP (2019) Current trends and challenges in biofabrication using biomaterials and nanomaterials: future perspectives for 3D/4D bioprinting. In: 3D and 4D printing in biomedical applications: process engineering and additive manufacturing. Wiley-VCH, Weinheim, pp 373–421

Ulbricht J, Jordan R, Luxenhofer R (2014) On the biodegradability of polyethylene glycol, polypeptoids and poly (2-oxazoline). Biomaterials 35(17):4848–4861

Usmani Z, Sharma M, Awasthi AK, Lukk T, Tuohy MG, Gong L, Nguyen-Tri P, Goddard AD, Bill RM, Nayak SC, Gupta VK (2021) Lignocellulosic biorefineries: the current state of challenges and strategies for efficient commercialization. Renew Sust Energ Rev 148:111258

Vanaei S, Parizi MS, Vanaei S, Salemizadehparizi F, Vanaei HR (2021) An overview on materials and techniques in 3D bioprinting toward biomedical application, vol 2. Eng Regen, pp 1–18

Vina S (2023) Case study: Hannover university hospital bioprints custom ear implants with drug delivery using 3d-bioplotter. Desktop Health. https://youtu.be/rErBmfpnwEY?si=sUbFMu_1PzhiCrkr

Wang Q, Sun J, Yao Q, Ji C, Liu J, Zhu Q (2018) 3D printing with cellulose materials. Cellulose 25: 4275–4301

Wang X, Wang Q, Xu C (2020) Nanocellulose-based inks for 3D bioprinting: key aspects in research development and challenging perspectives in applications. Bioengineering 7:40

Wei M, Gao Y, Li X, Serpe MJ (2017) Stimuli-responsive polymers and their applications. Polym Chem 8(1):127–143

Yang Q, Gap B, Xu F (2020) Recent advances in 4D bioprinting. Biotechnol J 15(1):1900086

Zhang H, Mao X, Du Z, Jiang W, Han X, Zhao D, Han D, Li Q (2016) Three dimensional printed macroporous polylactic acid/hydroxyapatite composite scaffolds for promoting bone formation in a critical-size rat calvarial defect model. Sci Technol Adv Mater 17(1):136–148

Zhang Y, Zhou D, Chen J, Zhang X, Li X, Zhao W, Xu T (2019) Biomaterials based on marine resources for 3D bioprinting applications. Mar Drugs 17(10):555

Zhang S, Chen X, Shan M, Hao Z, Zhang X, Meng L, Zhai Z, Zhang L, Liu X, Wang X (2023) Convergence of 3D bioprinting and nanotechnology in tissue engineering scaffolds. Biomimetics 8(1):94. https://doi.org/10.3390/biomimetics8010094

Zhu J (2010) Bioactive modification of poly (ethylene glycol) hydrogels for tissue engineering. Biomaterials 31(17):4639–4656

Zimmermann L, Dombrowski A, Völker C, Wagner M (2020) Are bioplastics and plant-based materials safer than conventional plastics? In vitro toxicity and chemical composition. Environ Int 145:106066

Ms. Oishani Sarkar is a postgraduate student under Dr. Kanthesh BM an Associate Professor and Coordinator Division of Molecular Biology, in JSS Academy of Higher Education & Research, Mysuru. She is currently pursuing her master's in molecular biology. Before joining JSS, she was in St. Ann's college for women, Osmania University, Hyderabad for her undergraduate where she got a triple major in Genetics, Microbiology and Chemistry. She has also worked as a research trainee at Institute for Genetics and Hospital for Genetic Diseases, Hyderabad. After completing her postgraduate studies, Oishani wants to pursue Ph.D. and continue with research in the field of Human Genetics and Cancer Biology. She is a research-oriented individual who is always looking to expand her reach of science.

Ms. Yukta Mourya, is a postgraduate student under Dr. Kanthesh BM, an Associate Professor and Coordinator of Division of Molecular Biology, in JSS Academy of Higher Education and Research, Mysuru. She is currently pursuing her master's in molecular biology. She has completed her undergraduate in Life Sciences (Hons.) from Amity University, Noida. After completion of her postgraduate studies, Yukta wants to pursue a career in the field of conservation genetics and molecular ecology in wildlife.

Ms. K. L. Kavya is a postgraduate student under Dr. Kanthesh BM an Associate Professor and Coordinator Division of Molecular Biology, in JSS Academy of Higher Education & Research, Mysuru. She is currently pursuing her master's in molecular biology. Before joining JSS, she was in Vijaya College, Bangalore City University, Bangalore for her undergraduate where she got a triple major in Microbiology, Biotechnology and Chemistry. She is a research-oriented individual who is always looking to expand her reach of science.

Mr. D. Mutthuraj, currently a PhD research scholar under the guidance of Dr. Kanthesh BM, Associate Professor, Division of Molecular Biology, JSS Academy of Higher Education & Research, Mysuru, India. He has been actively involved in research since post-graduation with specialization in Biochemistry & Molecular Biology. Identifying various biomarkers are studied involved in autoimmune disorders and their pathways.

He received his undergraduate degree from the prestigious University of Mysore with First class with distinction, later he joined JSS AHER where he worked on auto-immune disorders like Rheumatoid arthritis titled "inhibition of pro-inflammatory molecules by ginger (Zingiber officinale Roscoe) and its anti-inflammatory effects on arthritis patients'' – where he checked for the inhibitory molecules and their effects on arthritis patients and also published 5 review paper and one original paper in peer-reviewed journals and also worked as guest lecturer in various colleges in Mandya before joining as PhD Research Scholar at JSS AHER. The author graduated from the JSS AHER (Deemed to be university) with first class with distinction "A" Grade. Currently focused on ovarian cancer and its chemopreventive studies.

Prof. Pasupuleti Visweswara Rao is currently a Professor in the Department of Biotechnology at REVA University. He holds a position as Associate Dean, School of Applied & Allied Health Sciences and Director of International Relations and Research Collaborations. Prof Rao was the Head for Non-Communicable Diseases Research niche areas and, the faculty research coordinator.

He was heading the internationalization centre at Faculty of Medicine and Health Sciences, Universiti Malaysia Sabah. Prior to this Prof Rao was holding a position as a Founding Deputy Director and Director for the Centre of Excellence—Institute of Food Security and Sustainable Agriculture (Centre of Excellence) and the founding Chairman for Animal Ethics Committee at University Malaysia Kelantan (UMK), Malaysia and as a visiting researcher to Japan, Macao, Indonesia, India and Thailand.

Prof. Pasupuleti previously worked with different universities in India, South Korea, and Malaysia in the field of Biotechnology, Natural Products, Nanoparticles research towards various applications related to microorganisms, environment, engineering, food and agriculture which are finally connected to the human wellness. He is a recipient of several awards including young scientist award, young researcher award, Best Faculty in Biotechnology Award, Excellence Service Award at national and international levels.

Prof Pasupuleti has been elected as an executive council member at International Governing bodies in Malaysia such as Young Scientists Network – Akademi Science Malaysia (YSN- ASM). Prof Pasupuleti has delivered several Keynote and Plenary talks in various international conferences in Biotechnology, Biomedical sciences, Biomedical engineering, and Nanotechnology across the globe. Prof Pasupuleti has also edited seven books two self-learning materials (books) for Biotechnology students at master level and and for the broader audience in related to food biotechnology on stingless bee honey and fruit vinegar, COVID-19 and medicinal plants. He has also published more than 100 research publications with a cumulative impact factor of around 500 in peer reviewed international journals and 6 book chapters in highly reputed publishers such as springer and Elsevier. He is also serving to various journals as editorial member, and as reviewer for several high impact factor journals with international reputation including Nature Publishing Group. He has been awarded with a certificate from Elsevier as Outstanding contributing Reviewer for three high impact factor journals such as Critical reviews in oncology (Impact factor 5.27), Material Science and Engineering C (Impact factor 5.88) and Magnetic science and magnetic materials (2.717).

Dr. Kanthesh M. Basalingappa, is an Associate Professor of Molecular Biology at the School of Life Sciences, JSS Academy of Higher Education and Research, Mysuru, India. His goal of research is to determine the role of RNA Binding proteins in tumor progression and metastasis. Post-transcriptional regulation of gene expression by RNA binding protein is a crucial mechanism in regulating the timing and the amount of expression of genes. Growing evidence indicate that the alteration of the expression and function of RNA binding proteins could potentially play a role in inflammation and cancer. Dr. Kanthesh BM did is Ph.D. from University of Madras (2005), He also did postdoctoral research at the University Malaya, Kuala Lumpur, Malaysia (2007–2009); West Virginia University, Morgantown, USA (2090–2011) and University of Oklahoma Health Sciences, Oklahoma, USA (2011–2014). He received Malaysia Prestigious Bio-Malaysia Gold medal Award (2008). For his research area is Arbovirus infections, in that they done patented work on "Early detection of BK virus using molecular methods". He also Received Dr. Wilson Aruni "Best Research Mentor and Teacher Gold medal Award" from the Indian Association of Applied Microbiology (IAAM) (2018). He has been engaged in teaching and research in Microbiology and Molecular Biology for the past 22 years. He has published over 95 original research papers, 15 book chapters, and 15 review articles. He is also Professional and Scientific Memberships in, American Association for Cancer Research (AACR), Life Member of Indian Association of Applied Microbiology (IAAM), Life Member of Indian Association of Biomedical Scientists (IABMS), Indian Association of Medical Microbiologist (IAMM). He Received Fellowship Award from Indian Association of Applied Micro- biology (FIAAM). At present he is a having collaboration with Royal Research Foundation, a research institute in India.

Proteins as Biocompatible Material for Biomedical Applications

6

Phool Chandra, Rashmi Pathak, Neetu Sachan, and Anurag Verma

Abstract

Natural biocompatible materials have evolved over billions of years into incredibly exact structure-activity interactions that scientists aspire to mimic. Thanks to developments in genetic engineering, it is now possible to conduct in-depth research into how modifications to even a single peptide within a protein sequence might result in biocompatible materials with exceptional mechanical, biological, and thermal properties. With a wide range of biomedical applications, proteins are among the most versatile and extensively researched macromolecules. These materials have several advantages over synthetic counterparts due to their biological and natural origins, including inherent bioactivity, cell recognition, and less immunogenic potential. Moreover, proteins can be readily functionalized by changing the primary amino acid sequence of the protein. Furthermore, proteins can frequently self-assemble into higher-order structures independently or in response to particular environmental stimuli. Protein-based materials' high flexibility, biocompatibility, and biodegradability have led to their use in various biological domains. Nanoscale protein-based polymers are

P. Chandra (✉)
Department of Pharmacology, Teerthanker Mahaveer College of Pharmacy, Teerthanker Mahaveer University, Moradabad, Uttar Pradesh, India

R. Pathak
Department of Pharmacy, Invertis University, Bareilly, Uttar Pradesh, India

N. Sachan
Department of Pharmaceutical Chemistry, Maharana Pratap College of Pharmacy, Mandhana, Kanpur, Uttar Pradesh, India

A. Verma
Department of Pharmaceutics, Teerthanker Mahaveer College of Pharmacy, Teerthanker Mahaveer University, Moradabad, Uttar Pradesh, India

R. Malviya, S. Sundram (eds.), *Sustainable Green Biomaterials As Drug Delivery Systems*, Biomaterials, Bioengineering and Sustainability 1,
https://doi.org/10.1007/978-3-031-79062-1_6

progressively being utilized to deliver drugs into circulation to their molecular sites of effect. Protein-based biomaterials with drug-delivery properties have been shown to have several medical applications, such as the treatment of diabetes, neuroinflammation, cancer, wound healing, and corneal regeneration. This chapter has given an overview of the features and applications of modern protein-based materials in biomedicine.

Keywords

Biocompatible materials · Proteins · Protein sequence · Nanoscale · Vaccine · Biomedicine

Abbreviations

PEG	Poly-Ethylene Glycol
NCAAs	Non-Canonical Amino Acids
CRISPR	Clustered Regularly Interspaced Short Palindromic Repeats
API	Active Pharmaceutical Ingredients
AMP	Antimicrobial Peptides
FDA	Food and Drug Administration
ELP	Elastin-Like Polypeptides
rhTE	Recombinant Human Tropoelastin
DNA	Deoxyribonucleic Acid
PVA	Polyvinyl Alcohol
HIV	Human Immunodeficiency Virus
BSA	Bovine Serum Albumin
PEE	Polyethylene Ether
PLGA	Poly Lactic-co-Glycolic Acid
DOX	Doxorubicin
HSP90	Heat Shock Protein 90
HEMA	2-hydroxyethyl Methacrylate
C.N.	Chitosan
C.P.	Collagen Peptide
GLP-1	Glucagon-like Peptide 1

6.1 Introduction

More advanced drug delivery techniques are required in light of medical advancements and clinical best practices. Historically, the renal clearance process and short half-lives of proteins and oligonucleotides have hampered the therapeutic effectiveness of these molecules. Drugs may be quickly filtered in the kidney and eliminated from circulation by the reticuloendothelial system before reaching their intended destination. When injecting medications, the vasculature may either guide the drug directly to the illness site or take a diversion where the drug can be removed. These drugs must be taken often and in large doses to distribute through the vasculature and ultimately diffuse into the tissue interstitial, which is necessary to have the desired therapeutic effect (Hoffman and Falvo 2004; Layman et al. 2018). Although taking medications orally is the most practical, secure, and extensively used way, these bioactive substances must be insoluble in the stomach to prevent significant losses from the stomach's acid, pepsin, and the small intestine's pancreatic enzymes. Drugs must be soluble and absorbable via the intestinal mucosa once they reach the gut. For chronic and persistent medical disorders, which call for long-term, non-toxic drug delivery systems that are regulated to provide the correct dose at the right time, these conventional techniques mentioned above are not appropriate.

Consequently, there is an excellent demand for mechanically steady polymer materials to be used as drug-delivery vehicles that protect therapies from clearance and degradation while avoiding over-toxicity to the rest of the body. Treating polymer materials into various structures and forms, such as films, gels, microspheres, microcapsules, nanoparticles, and micelles, is simple. Therapeutics and biomolecular proteins may effectively transport drugs to particular sites while preventing immunological reactions and serious adverse effects by using polymer carrier systems to shield them from the harsh gastrointestinal tract environment (Layman et al. 2018).

Polymers are a desirable class of materials for use in biomedical applications because they provide a great degree of control over their physicochemical characteristics. Polymeric biomaterials may have mechanical, biodegradable, biocompatible, and processable qualities customized for a particular tissue and application. Among the most helpful and well-researched biologically generated polymeric materials are proteins. When compared to artificial polymers, proteins have many benefits. Researchers list four significant drawbacks with synthetic polymers (Austin and Rosales 2019; Banskota et al. 2019). First, they frequently rely on poly-ethylene glycol (PEG) to preserve bio-compatibility because it is widely used. Long-term use of this could result in the development of PEG allergies and anti-PEG antibodies, a phenomenon that has already been observed in some people. Second, vinyl addition polymerization produces artificial polymers such as PEG, which are difficult to break down and may build up inside the body with potentially harmful consequences (Garay et al. 2012; Delplace and Nicolas 2015; Qi et al. 2016). Third, the level of polydispersity that results from the polymerization process in the molecular weights of the polymers will always be present, hence restricting the accuracy with which physical-chemical parameters may be adjusted. Finally, chemical conjugation procedures frequently exhibit inadequate site specificity and poorly controlled stoichiometry during polymer synthesis. However, scalable bacterial, insect, or mammalian expression systems can make larger recombinant proteins.

In contrast, chemical synthesis platforms such as solid-phase peptide synthesis can produce shorter peptides. Since they don't need metals, they are also completely biodegradable and biocompatible *in vivo*, unlike synthetic polymers. Because of this, research on protein biomaterials has the potential to benefit human health in the future and has many benefits over other materials for *in vivo* applications (Boyle 2018).

Improved genetic manipulation methods have significantly enhanced control over recombinant expression, leading to the development of protein biomaterials. Since amino acids are the building blocks of proteins, little alterations to their sequence may be used to alter significant changes in a macromolecule's characteristics. With the inclusion of pyrrolysine and seleniocysteine, the number of naturally occurring amino acids has increased to 22. Artificial non-canonical amino acids (NCAAs) have been developed for functional versatility, including fluorination and hydroxylation of peptide backbones. These NCAAs can be included site- or residue-specific, depending on the needed level of information. The Tirrell group was the first to introduce vital chemical functions, including azide groups for cycloaddition and genetically altered microbes that could be added to the NCAA (Voloshchuk and Montclare 2010).

Furthermore, it is now feasible to carefully and precisely create higher-order protein structures with specific functional features thanks to more recent advancements like CRISPR technology. For example, Hernando-Garcia et al. employed Cas12 as nucleation sites to construct viral capsids that self-assemble to carry nucleic acids. Protein-based biomaterials have a promising future ahead of them (Calcines-Cruz et al. 2021).

One of the primary goals of biomaterials has been to improve medication delivery because so many various types of molecules have been investigated for their possible medicinal uses. These substances, also known as active pharmaceutical ingredients, or APIs, fall into many significant categories, including proteins, peptides, genetic material, tiny compounds, cells, and cell fragments. Every one of these loads has benefits of its own and may present different delivery-related difficulties. Small molecules are considered therapeutic when their molecular weight is less than 900 Da. Thirteen small molecule medications include predictable pharmacodynamics and pharmacokinetics and enhanced diffusion and permeability across tissue and cell membranes due to their small size. Small molecule medications can interact unintentionally with untargeted tissues and cells due to their basic chemical structures, and some drugs may become resistant in cells (Yin et al. 2017). Gene therapy involves introducing functional copies of genes or manipulating gene expression in cells, with therapeutic genetic material teaching cells to produce specific proteins and delivering negatively charged nucleic acids with cationic materials (Janarthanan et al. 2019).

Additionally, they frequently require carrier mechanisms to pass across the cell membrane. Furthermore, proteins are frequently utilized as therapeutic drugs; most proteins with FDA approval today are antibody therapies. Protein medications like proapoptotic and antimicrobial peptides (AMPs) disrupt the mitochondrial membrane, forming connections between small molecules and protein antibodies. Also, these peptides are briefly less than 40 amino acids (Fu et al. 2015; Janarthanan et al. 2019). Protein supplements frequently work in a concentrated, targeted manner to reduce the likelihood of unfavourable side effects.

Last, scaffolding is required to lengthen the stability and retention period of cells and cell derivatives, such as exosomes, employed in regenerative medicine and wound healing. Various cell types have been used, depending on the specific needs. These involve primary cells for musculoskeletal tissue engineering and stem cells that can develop into numerous lineages (Zhang et al. 2018).

Protein materials are appealing because their organizational structure can be carefully controlled to provide various forms with unique physical characteristics. It is possible to alter a protein's original amino acid sequence to create specific secondary structural sequences, such as beta sheets, helical coils, and even naturally disorganized structures. Helical coils experience increased mechanical stress due to hydrogen bonds, while fusion proteins utilize the α-helical peptide linker EAAAK for increased protein structure rigidity (Chen et al. 2013; Punia et al. 2019). Certain α-helices with hydrophobic first and fourth residues can form stable coiled coils, with hydrophilic residues in solvent-exposed areas and hydrophobic residues facing the interior. Triple helical configurations are standard in collagen and collagen-like peptides, formed when three left-handed helices combine.

Additionally, proteins can assume the β-strand conformation, which allows them to form pleated sheets or twists. β-sheets may be connected to other sheets in a parallel or antiparallel configuration (Cheng et al. 2013; Li and Yu 2013; Wang et al. 2019). Planar hydrogen bonds are created when the N and C termini of two strands are positioned adjacent to one other in an antiparallel orientation. This arrangement is incredibly stable. Every strand is oriented in a parallel pattern with all of its N-termini next to each other and pointing in the same direction. The conformation is a little less stable because of the out-of-plane hydrogen bonding that emerges from this configuration. Some materials, like silk, exhibit exceptional mechanical properties because of highly ordered secondary structures in their crystalline regions. Last but not least, as demonstrated by the remarkably elastic and rubbery characteristics of elastin and elastin-like polypeptides (ELPs), random coils or disordered secondary structures can also form in proteins (Koh et al. 2015).

6.2 Protein Materials

Drug delivery research has extensively used various fibrous protein materials, including silk, keratin, collagen, and elastin (Fig. 6.1). Protein materials may be handled similarly since they have comparable characteristics (Table 6.1) (van Herwaarden et al. 2011; Brennan et al. 2016).

6.2.1 Keratin

Keratin is found in human and animal integumentary systems, originating from the outermost layer of skin (Ferraro et al. 2016). Regarding biomedical applications, keratin offers a practical and affordable substitute for petroleum-based polymers. About 90% of keratin can be found in wool; the remaining keratins comprise other

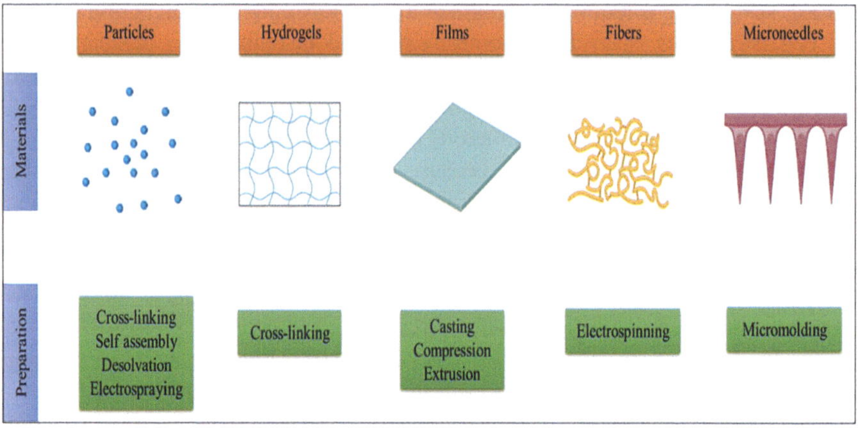

Fig. 6.1 Protein-based biomaterials

Table 6.1 Protein biomaterials in biomedical applications

Material	Applications	Structural design
Collagen	To improve the beneficial drug-delivery capabilities, cartilage, ocular, corneal, skin, nerve, and tendon/ligament Collagen tissues are engineered and combined into a range of composite materials.	Hydrogels, fibers, films (Sell et al. 2009; Sionkowska and Kozłowska 2013; Ghezzi et al. 2017)
Keratin	Antimicrobial, medication administration, tissue engineering, medical devices and trauma, and wound healing	Hydrogels, fibers, films, (Apel et al. 2008; Zoccola et al. 2008; Aboushwareb et al. 2009)
Silk	Adhesive fillers, wound dressing, enzyme immobilization, engineering cartilage or load-bearing tissues, and medication administration	Hydrogels, microcapsules, films, microparticles (Gellynck et al. 2008; Lammel et al. 2011)
Zein	Enhanced mechanical strength, microbiological resistance, positive cell adhesion, biomineralization, controlled drug release, and osteoblast growth	Films, nanofibers, microspheres, nanoparticles (Liu et al. 2005; Li and Yao 2012; Yao et al. 2013)
Elastin	Wound healing applications, highly adjustable thermo-responsive intracellular functionalized peptide medicines, engineering of liver, vascular graft, cartilage, and ocular tissue	Hydrogels, fibers, films (Nowatzki and Tirrell 2004; Srokowski and Woodhouse 2008)

hydrocarbons (Cilurzo et al. 2013). Keratin can be categorized into α-, β-, and γ-keratin, with α-keratin having an α-helix structure containing α-helices forming a protofibril, low sulphur amino acid concentration, and potential disulfide bridge formation. It has a molecular weight between forty and seventy kDa. β-keratin shares a β-sheet shape with other intermediate filament proteins, with a molecular weight extending from 11 to 22 kDa. Tryptophan, lysine, histidine, serine, glycine, and alanine comprise a significant portion of its amino acid composition. Due to its high sulphur amino acid concentration, it can form a high degree of disulfide linkages inside and between molecules (4–6 w/w). Soft and hard keratins may be produced based on the cross-linking cysteine residues for mechanical, chemical, and thermal stability. Wool is an inexpensive source of keratin, a textile industry byproduct, in contrast to collagen and elastin. Wool and human hair are the sources of biocompatible, biodegradable, non-toxic, and highly adaptable keratin biomaterials. This versatile material, shaped into various forms like gels, films, fibres, and sponges, is utilized for neural tissue regeneration, wound dressings, and medicine administration (Srinivasan et al. 2010; Cilurzo et al. 2013).

6.2.2 Collagen

Vertebrates and other species have collagen, an essential protein that promotes cell development and maintains the structural integrity of tissues. It is a three-chain polypeptide protein of animal origin present in various connective tissues (Zhang et al. 2014). Collagen is a very versatile and biocompatible protein substance. It may be transformed into several drug-delivery-suitable forms, including films, hydrogels, and microparticles. The heat presents one difficulty when handling collagen proteins. Collagen must be processed at a somewhat high temperature to become water-soluble. Drugs and other chemicals should not be kept using this heat in microparticles, collagen films, or other materials. Therefore, alternative solvents, including organic acids, may be used. Collagen-based polymers can be used in bone mending and cancer therapy (Zhang et al. 2014).

Chemical cross-linking of collagen can improve its mechanical strength and reduce its hydrothermal and enzymatic degradation rate. Two types of collagen molecules can be cross-linked chemically: amide-type and bi-functional. Some examples of bifunctional reagents that bridge the amine groups of the chain of collagen polypeptide and other naturally occurring protein polymers include genipin, glutaraldehyde, hexamethylene diisocyanate, and polyethylene glycol diacrylate (Ghezzi et al. 2017). Chemical cross-linking reagents can have cytotoxic or chromogenic effects, with genipin and carbodiimide potentially released into the environment. Safe and well-chosen collagen cross-linking alteration methods can prolong drug stability and retention despite potential environmental release (Zhao et al. 2016).

6.2.3 Elastin

Like collagen, elastin is a protein in the extracellular matrix of several flexible tissues. One of the body's most stable proteins, it remains insoluble even after being stretched and relaxed over a billion times. A significant portion of elastic fibres comprises elastin, a highly cross-linked structure with a secondary beta-spiral structure. Although soluble elastin derivatives, including human tropoelastin, are being studied, native elastin is insoluble (Almine et al. 2010; Barenghi et al. 2014). Elastin-like polypeptides (ELP) are biopolymers with amino acids like valine, glycine, proline, and unknown (Xaa) useful for vascular stents due to their low critical solution temperature and phase transition characteristics. It is also feasible to exercise genetic control on the size and sequencing of macromolecular carriers, microparticles, and functionalized elastin nanoparticles. Among the potential conditions of interest for elastin materials are type II diabetes, cancer, and osteoarthritis (Chilkoti et al. 2006; MacEwan and Chilkoti 2014).

Elasticin, another thermoresponsive protein, provides shape and elasticity to the human body's connective tissues, blood vessels, skin, and lungs. A range of thermoresponsive elastin forms, including as elastin-like polypeptides (ELPs), recombinant human tropoelastin (rhTE), and animal-derived soluble elastin, have been used to create promising synthetic tissue scaffolds. The hydrophobic portions of the polypeptide are responsible for producing elastin-like polymers (Zhou et al. 2017).

6.2.4 Silk

Elastin-derived structures exhibit a vital phase change at specific transition temperatures and are commonly used as thermoresponsive units in biomaterials. These structures can take the form of a star or linear structure. The lower critical solution temperature can be adjusted by changing the sequence of amino acids. Elastin-derived structures can also change flow and retention in porous hydrogels, elastomeric films, and electrospun scaffolds, promoting cell division and impacting mesenchymal stem cell destiny. The insoluble beta-sheet crystal structures found in silkworm cocoons and spider threads are the primary sources of silk fibroin protein. Drug delivery applications have been implemented with several silkworm species, including Bombyx mori, Tussah, and Eri silks. Various techniques may be used to create silk proteins, including films, micro and nanoparticles, and three-dimensional porous scaffolds, all of which have controllable breakdown rates (Pritchard et al. 2013). Due to its moderate processing conditions, silk lowers costs and is a perfect material for drug delivery applications. Silk-based materials have controlled biodegradability and good biocompatibility, making them ideal for long-term drug-eluting depots. The solubilization/degradation of the silk material and the drug payload's dispersion are the two primary methods used to describe the features of silk-controlled release. For instance, when a thin silk film degrades in the designated area, it may release the necessary material according to a predetermined solubility rate (Seib and Kaplan 2013).

Spider dragline silks offer numerous desirable physical characteristics, such as far better tensile strength and flexibility than natively spun silkworm fibres. There is great interest in producing silk fibres with spider silk qualities. However, since spiders are naturally very territorial, developing a method for producing spider silk via spider farming presents significant challenges. For this reason, standard recombinant silk protein manufacturing systems were used. Due to the inability to properly assembly native-sized recombinant silk protein into spider silk fibres, only limited amounts of artificial spider silk have been produced (Teule et al. 2012). Although spider silk may be preferentially produced in *E. coli* that has undergone metabolic engineering, the recombinant proteins may also produce copies with lower molecular weights and worse fibre qualities. Researchers used recombinant DNA to create the novel genetic engineering vector known as "PiggyBac" to get around these restrictions (Xia et al. 2010). These vectors are transposon-derived bits of DNA that may enter the silkworm's genetic machinery. They are derived from spiders. These transgenic silkworms generated silk fibres that combined the proteins seen in spider and chimaera silkworms. The resulting genetically modified silk proteins are a stable composite material stronger and more elastic than natural dragline spider silk fibres. Gene modifications into the present commercial silkworm production process by recombinant DNA may provide a feasible way to produce modified protein fibres on a wide scale that surpass natural caterpillar silk's strength and flexibility. This biotechnological approach offers solid, lightweight textiles with optimized mechanical qualities for drug-delivery applications, including bandages, suture materials, and tendon and ligament repair scaffolds (Xia et al. 2010).

6.2.5 Resilin

The cuticles of many jumping insects were shown to contain the elastomeric protein resilin. The cross-linking of di- and tri-tyrosine connections results in an elastic and flexible network structure due to the coiled amino acid chains. High-frequency contraction, extension, and deformation do not impair the resilience and long-range reversible elasticity of dityrosine cross-linked protein structures. Resililine, with its exceptional flexibility, low stiffness, durability, and efficient energy storage, is ideal for biomedical applications like drug delivery and tissue scaffolds due to its rubber-like properties and exceptional mechanical and biocompatibility (Qin et al. 2012; Li et al. 2013; McGann et al. 2013).

6.2.6 Zein

Numerous plant-based protein varieties have been integrated into medication delivery systems. Maize kernel endosperm contains zein, a significant plant-based storage protein with a high prolamine concentration. It is the end product of grinding maize corn into ethanol using wet and dry milling procedures. Ethanol is then used to make corn gluten meal. Zein, a 40 kDa molecular weight protein, is commonly found in glutamine, leucine, alanine, and proline residues (Nonthanum et al. 2012; Sousa et al. 2012; Bouman et al. 2016). Zein, a hydrophobic protein, is soluble in ethanol, acetone, and acetylacetone but insoluble in water. Its non-polar helical cores contain glutamine-rich twists and loops, allowing self-assemblement into layers and particles (Sousa et al. 2012). Zein can fold helical segments consecutively in an antiparallel form to exhibit thermoplastic and chemical properties that are pH and heat-stable. Zein is a desirable material for application in medication administration and food and pharmaceutical coatings due to its other remarkable properties, including biodegradability, mechanical robustness, and water barrier capacity. Zein is an ideal matrix material for delayed release since it may also trap and agglomerate solutes like amino acids or medicines (Sousa et al. 2012).

6.3 Proteins as Adaptable and Biocompatible Building Blocks for Biomedical Applications in Biomaterials

Proteins may come from either plants or animals, are widely distributed in nature, biodegrade, and can often be modified into materials using gentle methods and environmental circumstances (Zhu et al. 2018; Grigsby et al. 2020). Because they are quickly produced in vast quantities, emit little greenhouse gases, and originate from renewable resources, plant-based proteins offer a sustainable and ecologically benign source of polymers. Plant-based proteins are often free from animal-related diseases, making them an appealing option for those who choose not to consume animal products due to ethical, religious, or personal reasons (Nijdam et al. 2012; Jones et al. 2013; Eshel et al. 2019). Animal-derived proteins, while expensive, can be obtained from other sources or byproducts of food or agricultural industries, making them an affordable and environmentally friendly alternative for biomedical applications. As an alternative to mammalian collagen, collagen derived from marine waste materials has been utilized to create osteoinductive bio-composite scaffolds for bone tissue engineering.

Furthermore, keratin, the building block of biomaterials, may be recovered from human hair or biological wastes produced by the chicken industry (Barros et al. 2015; Arslan et al. 2017; He et al. 2017; Rapa et al. 2020). The need for more sustainable, eco-friendly, safe, and green technology and a circular bioeconomy is now being recognized. Proteins are utilized in biomaterial production, promoting green chemistry principles such as renewable and biodegradable building blocks, waste reduction, and finite raw material use (Park et al. 2015).

Regarding mechanical strength, biomaterials based on animal proteins often exhibit more strength than those based on plant proteins. Moreover, the spectrum of suitable solvents for most plant proteins is restricted, which might complicate producing biomaterials derived from plants. Selecting the suitable solvent, cross-linking, and combining with natural or synthetic bio-polymers like polysaccharides can enhance the adaptability and pertinency of plant and animal proteins (Pinheiro et al. 2016; Samadian et al. 2020; Fiorentini et al. 2021; Kamada et al. 2021).

6.3.1 Building Blocks for Biomaterials: Biocompatible and Biodegradable Proteins

Proteins are biocompatible and biodegradable, making them ideal for medicinal applications. Their degradation rate impacts biomaterial performance, with specific proteases affecting protein susceptibility to degradation (Guo et al. 2020). Protein folding, including secondary and tertiary structures, can obscure recognition motifs and limit protein accessibility to proteolytic cleavage, making it resistant to protease digestion. The protein controls degradation speed, but material type also influences this process. Müller-Herrmann and Scheibel studied the degradation processes of various materials, including spider silk protein films, nanoparticles, and electrospun nanofibers (Muller-Herrmann and Scheibel 2015).

Interestingly, many companies are already selling goods based on proteins as alternatives to current materials utilized in a range of applications. Xampla, a spinoff company, has successfully detangled and reassembled pea protein into a spider silk protein structure, producing solid and flexible materials like films, microcapsules, and coatings for various applications (Chandregowda et al. 2009). GelaCellTM, a Gelatex product, creates nanofibrous scaffolds for tissue engineering and wound care using biobased, eco-friendly gelatin and corn-based zein (Kishore et al. 2008). Sofregen Medical Inc. owns SERI®, a silk-based scaffold used in U.S. surgical procedures (Hikal et al. 2017). Furthermore, WounDres collagen hydrogel from Coloplast facilitates wound healing (Boussahel et al. 2015). Lastly, Matriderm® Dermal MatriX is a cross-linked collagen-elastin template that doesn't require chemicals. It is used as a scaffold for dermal replacement to treat ailments related to the skin, burns, and poorly healing wounds (Pistritto et al. 2016).

6.4 Protein Biomaterials as Minimally Invasive and Non-invasive Approaches for Biomedical Uses

Various preparation techniques may create materials with distinctive physical shapes starting from particular protein-building components. Protein inherent qualities, such as bioactivity, may be retained in the substance. The protein's chemical and physical characteristics are used in various applications to create materials with specific qualities. To produce materials with optimized and specified qualities, choosing the suitable protein and having total control over the processing conditions is essential. Table 6.2 showcases protein-based compounds, showcasing their remarkable adaptability in various applications, which will be further explored in subsequent sections.

6.4.1 Protein-Based Particle Systems

One of the crucial functions of proteins *in vivo* is to function as natural transporters of various compounds, including ions, small molecules, and other macromolecules.

Table 6.2 The table explores the specific applications of protein biomaterials

Materials	Proteins	Applications
Nanoparticles	Gliadin	The possible therapy for skin conditions is by encapsulating all trans retinoic acid (Freitag et al. 2020).
	Serum albumin	To decontaminate nerve poisons that are employed as murder weapons. Aspirin is administered intraocularly to treat diabetic retinopathy. For bevacizumab administration in the treatment of proliferative (neovascular) ocular disorders (Das et al. 2012; Wang et al. 2020)
	Zein	Enables the topical encapsulation of various medications, including proteins and small compounds (Liu et al. 2023).
Films	Silk	For ciprofloxacin distribution using gelatin/silk sheets coated in polyethylene glycol (PEG) to promote wound healing (Alam and Shubhra 2015a).
	Collagen	For the long-term release of human growth hormone to repair wounds (Maeda et al. 2001).
Hydrogels	Collagen	As chitosan-infused composite for use in corneal implants (Rafat et al. 2008).
	Silk fibroin	For the administration of nanoparticles laden with curcumin to treat psoriasis (Rafat et al. 2008).
	Keratin	Like poly (ethylene imine) and polyvinyl alcohol (PVA) wound dressings may hasten the healing process (Park et al. 2015).
Nanofibers	Alpha-lactalbumin	For ampicillin topical encapsulation (Stie et al. 2020). For nicotine replacement treatment by oromucosal administration (Kalouta et al. 2020).
Microneedles	Gelatin	Developed for administering insulin using calcium sulphate, sodium carboxymethyl cellulose, or genipin cross-linking (Lee et al. 2017; Yu et al. 2017; Chen et al. 2018a).
	Silk	For a long-term, steady release of levonorgestrel spanning many months. To administer insulin (Lau et al. 2017; Yavuz et al. 2020; Zhu et al. 2020).

Protein materials have thus been investigated as possible delivery systems for biopharmaceuticals and small molecules. A range of techniques can generate particles from proteins, including electro-spraying, desolvation, physical aggregation/self-assembly, and/or cross-linking to stabilize the colloidal system (Lopez-Sagaseta et al. 2016; Mendes et al. 2020). Gliadin, a gluten-extracted protein, produces nanoparticles containing all-trans-retinoic acid, which can treat skin conditions like acne, ichthyosis, hyperkeratosis, psoriasis, and malignant skin tumours. These nanoparticles are produced through desolvation, containing up to 97% drug loading. Interestingly, these protein-based nanoparticles showed retinoic acid's zero-order sustained release for at least 3 h after the drug's first, brief burst release (20%) within 15 min. Furthermore, gliadin-based nanoparticles are attractive therapeutic carriers for mucosal applications because these particles' mucoadhesive qualities have been established *in vitro* and ex vivo (Arangoa et al. 2000).

Lactoferrin-based nanoparticles ranging in size from 40 to 70 nm were loaded with either anti-HIV Efavirenz or antimicrobial-spermicidal curcumin. When compared to soluble curcumin and Efavirenz, rats administered lactoferrin-based nanoparticles *in vivo* demonstrated decreased drug absorption and lower inflammation in vaginal tissue and plasma. Histopathological analysis of vaginal tissue compared the protein-based nanoparticles to soluble curcumin and Efavirenz, revealing improved pharmacokinetic profiles and decreased toxicity. The scientists hypothesize that the protein's inherent abundance in vaginal discharge may account for the safety and lack of an inflammatory reaction to lactoferrin-based nanoparticles (Lakshmi et al. 2016).

Bovine serum albumin (BSA) is bonded to a polymer like DMAEMA, resulting in particles ranging from 100 to 400 nm. These particles are then used to clean nerve poisons that are employed as assassination instruments. When incorporated into a nanocomposite, the BSA protein demonstrated comparable or even more significant decontamination impact than conventional agents. Nevertheless, neither BSA nor the polymers alone or in physical combination show appreciable decontamination action. In a poisoned rat model, BSA and polymer-based particles outperformed $NaCO_3$ in *in vivo* decontamination of nerve agent Soman on skin. Rats given conventional cleaning chemicals, such as $NaCO_3$ and NaOH, exhibited skin irritation. However, decontaminating nanocomposite particles based on bovine serum albumin did not cause skin damage in rats treated *in vivo*, suggesting that this method might be less risky.

In conclusion, the authors show that the globular proteins in the nanocomposite, such as BSA, have better decontaminating effectiveness than fibrous proteins, suggesting that the effect of decontamination is independent of the specific protein (Wang et al. 2020). Patents and licences have also been issued for protein particle compositions. Specifically, (U.S. 2012/0195947 A1) was used to patent zein-based protein nanoparticles, which were then licenced to the research company Transderm Solutions for topical use. Since the U.S. Food and Drug Administration (FDA) has acknowledged zein, a protein derived from maize, as generally considered safe (GRAS), it is desirable for medicinal applications. Zein-based particles are effective in encapsulating and absorbing a wide range of medications for topical distribution. Pharmaceuticals like retinol, 5-fluorouracil, BSA, platelet-rich plasma, FITC, and Rhodamine 123 are used in anticancer treatments (Alqahtani 2012).

6.4.2 Protein-Based Hydrogels

Hydrogels are insoluble in aqueous phases because of their three-dimensional, water-retaining polymeric networks, which are stabilized by chemical and physical cross-linking. Hydrogels may absorb water at 10% to several thousand times their weight. The high-water content of hydrogels is thought to be responsible for their biomimetic properties, which mimic genuine tissue. Hydrophilic amino acids, including amide groups or amino acids, enhance hydrogels' hydrophilicity and swelling potential (Hoffman 2012). Notably, the air has recently been proposed as a substitute for the water component in hydrogels to generate lightweight protein-based solid materials. These aerogels might represent a new type of very stable three-dimensional structures since they were created using amyloid fibril gel as a template. Their improved retention qualities might expand the range of applications that may be accessed (Betz et al. 2012; Nystrom et al. 2017; Kleemann et al. 2018).

Protein hydrogels can be formed through chemical processes like covalent bonds or non-covalent physical cross-linkings like hydrogen bonding, hydrophobic interactions, or electrostatic interactions (Panahi and Baghban-Salehi 2019). Protein-based hydrogels are formed through physical cross-linking induced by destabilizing conditions, with new intermolecular interactions like hydrophobic areas or free SH-groups attaching to specific molecular locations. Through non-covalent and disulfide bonding, supramolecular structures may be formed as a result. Intermolecular interaction can create stable α-helical coiled-coil domains regulated by temperature, pH, ionic strength, and supramolecular interactions and can be influenced by physicochemical parameters (Vaiana et al. 2001). Compared to hydrogels produced by chemical cross-linking, those produced by physical cross-linking often have lower mechanical strength and stability. For example, thermally induced hydrogels frequently fail to swell to their original volume during dehydration.

Furthermore, proteins' combined properties can be exploited to produce synergistic effects with other materials. Graft copolymerization or chemical modification of acrylic acid can enhance water absorption by adding a higher negative charge density. However, due to the potential dangers of organic solvents and cross-linking agents, their use in biomedical applications is limited (Cuadri et al. 2017; Wattie et al. 2018).

The study evaluated the use of in situ establishing hydrogels made from gelatin and oxidized alginate as a wound-healing preparation. Gelatin was cross-linked in the wound because, in the presence of minuscule amounts of borax, periodate-oxidized alginate quickly cross-links the hydrogel's proteins (Balakrishnan et al. 2005; Park et al. 2015). Figure 6.2 shows protein-based hydrogel applications.

6.4.3 Protein-Based Films

Because thin films may come into close touch with the skin's or mucosa's surface, they are particularly well suited for topical delivery. The casting, compression, and extrusion methods employed to produce films based on proteins are comparable to those employed to produce films based on conventional polymers. Neat protein films are brittle and inflexible, making them difficult to work with and modify to fit the curved surfaces of the body. Adding plasticizers like glycerol and PEG can enhance protein films' mechanical properties, which enhance free volume and chain mobility (Sothornvit and Krochta 2001; Rouilly et al. 2006; Ullsten et al. 2016). Additionally, it has been shown that integrating proteins with various physicochemical properties into a single material might improve the mechanical properties of protein-based films. For instance, films created from a silk and gelatin composite demonstrated

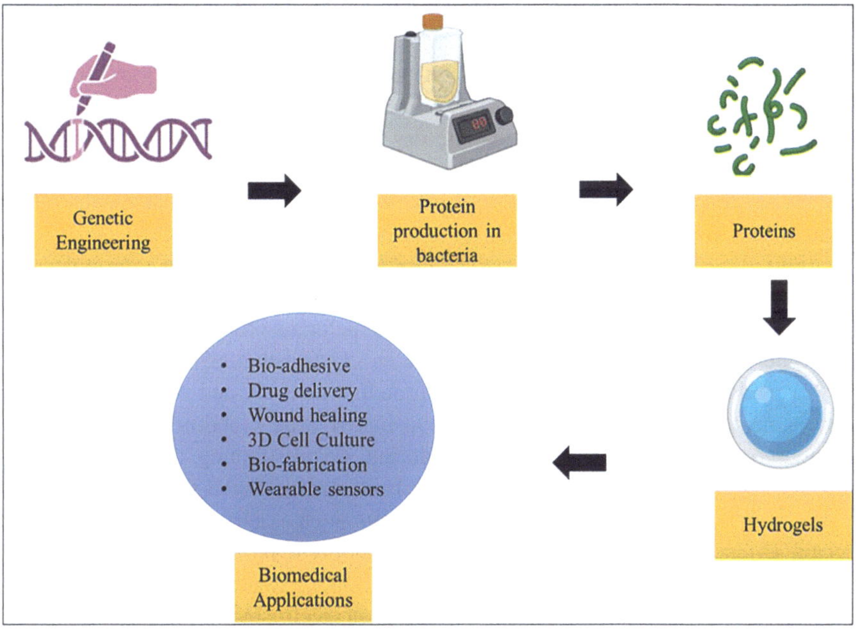

Fig. 6.2 Protein-based hydrogel applications

both gelatin's beneficial gelation qualities and silk's robust mechanical qualities. Casting was used to manufacture PEG-coated silk/gelatin films carrying the antibiotic ciprofloxacin, and their ability to treat wounds was examined (Alam and Shubhra 2015b). Because of the healing qualities of gelatin and the sericin contained in silk, PEGylated gelatin/silk films have been shown to speed up wound healing *in vivo* within 7 days. Silk/gelatin wound dressings fared better in a clinical trial than paraffin gauze bandages, proving the effectiveness and security of this protein-based wound dressing (Kanokpanont et al. 2012; Hasatsri et al. 2015). Materials intended for use as wound dressings must fulfil several criteria to promote natural healing and reduce scarring. The process of wound healing is intricate. For the materials to protect against mechanical stress and conform to the curved surfaces of the skin, they must exhibit appropriate mechanical characteristics. To permit the exchange of moisture and oxygen, the material must be sufficiently porous and capable of absorbing wound exudates. As was previously indicated, proteins are biocompatible, and their influence on biological pathways pertaining to cell attachment and proliferation may facilitate tissue repair. Protein-based films have the potential to function as drug delivery systems for bioactive materials such as antibiotics, hence preventing infections. Mucoadhesive gelatin-based films with thymol/β-cyclodextrin-encapsulated antibacterial patches have been proposed recently as a therapy for oral infections. The process of solvent evaporation produces these films (Zhao et al. 2018; Santos et al. 2021).

6.4.4 Protein Electrospun Fibers

Electrospinning creates materials based on nanofibers, such as patches and scaffolds for topical application. A strong electric field may be used to create the entwined network of tiny fibres that make up the dry formulation by electrospinning. The electrospinning method needs the right circumstances to promote intermolecular entanglement, solvent evaporation, and smooth fibre creation. It works under moderate settings, such as room temperature and air pressure, with short processing durations. Because of their restricted ability to form fibres due to intermolecular entanglement, proteins may be challenging to electrospin in their original state. Elevated temperatures or excessive pH may denaturize proteins, which enhances their intermolecular entanglement and makes electrospinning into fibres easier. However, denaturation might reduce the protein's activity, and exposure to harsh environments could also reduce an encapsulated drug's action. A well-known method for promoting fibre production during electrospinning is to use organic solvents with intense volatility. It has also been demonstrated that this technique works well for electrospinning a range of proteins, such as silk, collagen, elastin, and gelatin (Mendes et al. 2017). Organic solvents harm protein-based composites' environmental impact and industrial production, weakening biocompatibility. Water-borne electrospinning and aqueous solution electrospinning are gaining interest in therapeutic applications. Hydrophilic polymers like PEE and PVA can enhance solution viscosity, making it suitable for electrospun proteins like alpha-lactalbumin, silk, and BSA (Tang et al. 2012).

6.4.5 Protein-Based Microneedles

Overcoming significant biological barriers like the skin and mucosal membranes is one of the biggest problems in medication delivery today. This poses a particular difficulty for absorbing hydrophilic medications with a large molecular weight, such as therapeutic proteins and peptides. The most popular way to give medications with low bioavailability and restricted permeability is by injections; nevertheless, this can cause pain, discomfort, and poor patient compliance. Smaller than a few millimetres in height, microneedles are becoming increasingly popular as a solution to some of the main problems that normal injections face. Microneedles are less intrusive than hypodermic needles because they only penetrate the topmost and most important layer of mucosae and skin (Fonseca et al. 2019). Because microneedles are smaller than hypodermic needles, applying patches with a variety of them results in less tissue damage and generally less discomfort (Kaushik et al. 2001; Gill et al. 2008). Microneedles come in various types: solid, dissolving, hollow, coated, and hydrogel-forming. The most common method for producing microneedles from protein is micromolding, which involves filling and drying the needles (Fonseca et al. 2019). Because proteins are very biocompatible and biodegradable, they are ideal for creating microneedles. To reduce inflammatory reactions *in vivo*, the microneedles' biocompatibility and their breakdown products' biocompatibility are crucial.

Proteins, primarily silk and gelatin, have been used to create microneedles. Other protein-based microneedle developments include kanamycin-loaded squid suckerin microneedles for antibacterial therapy and zein-based microneedles for anticancer treatment and vaccination administration (Tsioris et al. 2012; Cai et al. 2015; Ding et al. 2017).

The skin's stratum corneum is made up of resistant layers of keratinized epithelium, which must be penetrated with sufficient mechanical strength. Silk has mechanical qualities that make it ideal for creating robust microneedles. It has been demonstrated that, under-regulated circumstances *in vitro*, microneedles made from silk fibroin may release rhodamin-B, horse radish peroxidase, FITC-Dextran, levonorgestrel, and BSA. The drug release can be controlled by adjusting needle shape, surfactants, number, silk fibroin concentration, molecular weight, secondary structure, and cross-linking degree (Lee et al. 2015; Yavuz et al. 2020). Silk fibroin microneedles offer a promising alternative to synthetic polymers like PLGA for sustained drug release. Insulin and immunizations have also been delivered *in vivo* using silk fibroin-based microneedles. After administering silk-based microneedles loaded with insulin, the microneedles released the insulin in less than 2 h, resulting in a relative bioavailability of 98.812.6%.

Furthermore, the insulin delivered by the protein microneedles had a longer-lasting effect *in vivo* than intraperitoneal injection. The *in vivo* effectiveness of antigens from influenza, *Clostridium difficile*, and *Shigella* as possible vaccinations administered transdermally using microneedles made of silk fibroin was assessed in a different investigation. Unlike antigen-loaded needles, silk fibroin microneedles did not exhibit any discernible inflammatory reaction, indicating that silk-based microneedles offer promise as a biocompatible transdermal immunization delivery technique (Lau et al. 2017; Stinson et al. 2017; Zhu et al. 2020).

Gelatin can generate microneedles of superior quality since it possesses less mechanical strength and form stability than silk fibroin. Blending gelatin with starch and gold nanoclusters improves microneedle quality and mechanical properties. The glucose-sensitive AuNC acts as an insulin release mechanism, healing type 1 diabetic mice and maintaining blood glucose levels for 1–2 days (Yu et al. 2017; Chen et al. 2018b; Zhang et al. 2021).

6.4.6 Keratin Composites

Utilizing freeze-drying processes, scaffolds for medicated wound dressings have been created. Based on fibrin (F), keratin (K), and gelatin (G), SEM research reveals that these solid sponge scaffolds contain densely packed pores with a macroporous shape. The extracellular matrix that may be created by simulating cell-binding motifs in fibrin and keratin can promote the proliferation of NIH 3T3 fibroblasts and keratinocytes. Fluorescence pictures captured *in vitro* may demonstrate adhesion. The study evaluated the effectiveness of KFG-composite scaffolds, using mupirocin as a model medication, and found that the antibiotic was released gradually and consistently, suggesting potential use in skin tissue rejuvenation (Singaravelu et al. 2016).

Researchers have also investigated the bulk interaction and inter-molecular surface characteristics of decreased keratin proteins (kerateines) in tissue regeneration using BMP-2 and gold as possible drug carriers (Singaravelu et al. 2016). Studying interactions between gold and kerateines, they found intermolecular solid covalent disulfide bonds. They investigated the binding and release of BMP-2 from the kerateine network, providing insights for potential drug delivery in bone regeneration and repair (Singaravelu et al. 2016).

Nanoparticles of Keratin-G PEG can be produced through thiolene click chemistry and self-assemble in aquatic conditions due to their thiol groups and amphiphilicity. Disulfide bonds can cross-link the nanoparticle cores on keratin backbones. A model drug, Doxorubicin hydrochloride salt (DOX·HCl), was evaluated for its drug-delivery system. The study found that the amount of GSH in cells affects DOX release. The nanoparticles showed an 18.1% DOX-loading capacity and faster release into the cell nucleus (Li et al. 2012).

Keratin is a non-toxic, biodegradable, porous scaffold for regenerative medicine that encourages positive cell interactions, much like other naturally occurring polymers. The researcher developed alkylated kerateine hydrogels conjugated with rhBMP-2, ciprofloxacin and rhIGF-1 to encourage stem cells to grow into osteogenic lineages. The process of iodoacetamide alkylation, sometimes known as "capping," of cysteine residues on keratin that had undergone reductive removal produced the scaffold or kerateines. Kerateines may form a more stable cross-linked hydrogel network than oxidatively produced keratin because they cannot form disulfide cross-links (keratose). The first can. The degree of disulfide cross-linking in keratin hydrogels may be controlled by alkylation, which can result in controlled gel erosion and drug release rates. Changing kerateines did not result in enhanced cellular toxicity in pre-osteoblasts, and it preserved the cells' capacity to adhere to the biomaterial at concentrations higher than collagen. The pace at which keratin hydrogel breaks down without disulfide cross-linking seems to be related to the release of medications. Controlled dosing in combination with alkylated kerateine hydrogels resulted in excellent cell adhesion and proliferation and drug concentrations suitable for bone and tissue-engineering applications (Han et al. 2015).

6.4.7 Elastin Composites

The work proposes thermosensitive ELP-based diblock biopolymers to present functional aspartic acid/poly-glutamic blocks for drug conjugation. By targeting and inhibiting heat shock protein 90 (HSP90), a critical molecular chaperone in several pro-oncogenic pathways, these blocks may be utilized to treat hyperthermic carcinoma edges. The study found adjustable thermoresponsiveness and high drug loadings between the polymer and G.A. conjugates. The release of the pH-sensitive medication G.A. into the acidic tumour microenvironment is enabled by thermal precipitation at hyperthermic isotherms, reducing off-target effects and systemic harm and offering a potential therapeutic platform for cancer treatment (Chen et al. 2011).

Dash and colleagues created moveable hollow ELP spheres for gene transmission using polystyrene beads. They used plasmid DNA (pDNA) as a model drug in permeability studies. The spheres were cross-linked using microbial transglutaminase for stability. When pDNA was added to the polyplex, loading efficiency increased and controlled release occurred, enhancing the efficiency of gene transmission. Furthermore, polyplex-loaded hollow spheres prevented endosomal degradation, which increased luciferase output. Surface functional groups were updated throughout this procedure. ELP hollow spheres may be made in enormous numbers because to the process's efficiency and ease of usage. Combining targeted ligands may produce Complex ELP hollow spheres, resulting in fascinating new gene delivery applications (Dash et al. 2011).

Kim et al. created thermosensitive ELP-grafted liposomes for controlled drug release and tumour-specific administration using a lipid film hydration process. They used an $\alpha v\beta 3$ integrin targeting moiety to create an ELP with a heat-activated trigger. The study found that doxorubicin (DOX) was liberated from the thermosensitive ELP at 45 °C and 42 °C. The researchers found that targeted RELs, triggered by external heat-producing techniques like ultrasound or radiofrequency, showed eight to tenfold expression in HUVEC and U87MG cells (Kim et al. 2014).

6.4.8 Collagen Composites

Alginate polymer-coated porous collagen-based scaffold that is highly biocompatible has been created for use as a drug-release mechanism or in drug-eluting implants. The collagen scaffold was constructed layer by layer using a cryogenic plotting technique. The model pharmaceutical used to assess the efficacy of the drug-delivery device was rhodamine B. While the amount of alginate covered on the collagen scaffold allowed for a prolonged drug release, the permeability of the CAC scaffolds resulted in an early burst release in the system. The Young's modulus of the pure collagen scaffolds is nine times lower than that of the CAC scaffolds, which is 30 MPa. An 88% porosity may nevertheless allow for the preservation of biological function. Following implantation, it has been shown that osteoblast-like cells (MG63) multiply and move into CAC scaffolds. It is possible to alter CAC scaffolds in tissue engineering for long-term medicinal uses (van Herwaarden et al. 2011).

Using a freeze-dried method, researchers created type-I collagen gels cross-linked with glutaraldehyde for use in dentistry. The model drug used in the drug penetration investigations on the 3D collagen matrices was nifummic acid. The collagen sponges were mixed with nifumic acid, an anti-inflammatory medication, using glutaraldehyde as a cross-linking agent. Collagen matrices with a consistent rate of niflumic acid release were formed when cross-linking agents were added, and the cross-linked collagen gels showed less enzymatic degradation and swelling. It was suggested that the non-Fickian kinetic mechanism of drug release was caused by niflumic acid's enhanced capacity for water absorption. The percentage of released niflumic acid may be adjusted by varying the quantity of cross-linking agent, allowing for the most effective release of the required dosage based on the application area and therapeutic recommendations (Constantin Barbaresso et al. 2013).

Hydrogel films with antibiotic-loaded collagen for topical drug delivery have been developed by combining collagen with artificial monomers like acrylamide and HEMA. The effectiveness of antibiotics was tested with naproxen and gallic acid *in vitro*. The p(coll-co-HEMA) loaded by N.P. and p(coll-co-AAm) loaded by G.A. composite films exhibited linear drug releases that persisted for a maximum of 32 and 36 h, respectively. Prior research has shown that collagen cross-linking might increase levels of active pharmaceutical ingredients. Hydrogel films with improved mechanical strength and absorbency can be enhanced by adding metal nanoparticles like Ag and Cu to enhance their antibacterial properties against common bacteria. The research suggests that antibiotic-coated collagen hydrogel films might be made rapidly and used as a viable material for burn and/or wound dressings to help distribute medicine and promote healing (Sahiner et al. 2014).

Creating controlled and targeted-release nanoparticles using peptides or biopolymers is also possible. Researchers synthesized collagen peptides to create chitosan (C.N.) nanoparticles by precipitation of chitosan and collagen peptide (C.P.) in the ionic gelation process. Electrostatic interactions and hydrogen bonding work together to form the CPCN nanoparticles. They are stable under physiological circumstances and sensitive to pH. Doxorubicin (DOX) hydrochloride was the model pharmaceutical used to assess the efficacy of the drug-delivery system. CPCN nanoparticles showed biphasic release, releasing antitumor medications in two stages. They showed improved antiproliferative effects, suggesting protein-based nanoparticles have potential for cancer drug delivery (Wu et al. 2016).

6.5 Biomedical Applications of Protein-Based Drug-Delivery Materials

6.5.1 Bone Healing

Any injury to the bone is referred to as a bone injury because the bone is an area of concentrated cellular and molecular organization. Reconstruction of bone tissue is possible after an injury. However, diseases like osteoporosis cause the bones to become weak and fragile. Moreover, bone is a challenging location to treat with medications requiring high blood flow. Many different protein materials mentioned above have been employed to promote bone repair. The future of bone strengthening and regeneration will include protein-based drug delivery, which makes medications, stimulants, and other chemicals accessible (Athanasou 2009; MacEwan and Chilkoti 2014).

For instance, ELP depots may distribute medications difficult for the blood to reach. The depots may be used for a long time because of their lengthy half-life. These depots serve as storage facilities for medications and other therapeutic materials. Drug-loaded nanoparticles would make up the depots. Because of its extended half-life, the ELP depots may stay in place until enough medication has been released to promote appropriate healing (MacEwan and Chilkoti 2014).

It is also possible to rebuild bone tissue using collagen-based microparticles. The application of scaffolds made of collagen composite microparticles was assessed by comparing them to scaffolds made of collagen without any collagen. This made it possible to statistically compare the differences in porosity, cytocompatibility, degradation, and Young's modulus between the non-collagen and collagen models. Furthermore, new studies are being conducted to improve the possibility of bone regeneration by the combination of collagen-based microparticles with growth factors, antimicrobial agents, and other substances. Silk may also be used to heal bone tissue. Like collagen, silk scaffolds are mechanically stable enough to be inserted into high-stress regions like bone. Unlike collagen, silk materials may provide bone with mechanical strength without requiring artificial protein cross-linking, reducing the risk of cellular toxicity, an inflammatory response, and other undesirable outcomes. Mice were used in research where aspirin was loaded into silk scaffolds, and the pace of skull tissue regeneration was monitored by controlled drug release. Tested scaffolds made of silk and silk-composite materials demonstrated efficient drug release and *in vivo* tissue regeneration (Wenk et al. 2011; Zugravu et al. 2013).

6.5.2 Antibiotic Release

The use of antibiotics began in the early 1940s. During this period, penicillin was developed, which completely changed the medical field. Creating antibiotics with reduced bacterial resistance is a critical field of research. However, the administration is another important issue that shouldn't be disregarded. Certain body parts, such as the eyes and bones, are more challenging to treat with antibiotics. Protein materials laden with drugs may help deliver medication to rather challenging locations. Drug release may be regulated in a material that matches tissue's mechanical characteristics and strain in certain situations because of their adjustable qualities (Friess 1998).

Penicillin and ampicillin studies have shown the *in vivo* effectiveness of drug-loaded silk films, hydrogels, and microspheres in a mouse-infected wound model. For example, it has been shown that using silk nanoparticles in combination with metallic implants may be very successful. When metallic and metal-based implants are initially implanted, several problems may occur. These consist of inflammation, implant loosening, and bacterial infection. One common metal seen in metallic and metal-infused implants is titanium. A recent work added Gentamicin, an antibiotic, to silk nanoparticles. Next, in an *in vitro* investigation, the particles were layered with a titanium surface. Compared to pure titanium, the silk nanoparticles loaded with the drug demonstrated more effective bacterial suppression over an 18-day (Chen and Thouas 2015; Sharma et al. 2016).

Antibiotics may also be delivered continuously using collagen hydrogels. For instance, daily eye drop administration has been replaced with unique collagen hydrogel compositions. Initially, medications such as antibiotics were made into gels for eye drops. The gel may be kept in place until the required doses have been given by regulating the collagen network's biodegradation rate (Friess 1998).

6.5.3 Diabetes

A disorder called diabetes alters how the body responds to insulin. Type 1 and type 2 diabetes are the most prevalent types. Diabetes type 1 is a result of insufficient insulin production by the pancreas. Patients need to give exogenous insulin routinely. 85–90% of people with diabetes have type 2 diabetes, which is the most prevalent kind of the disease. Insulin resistance or decreased insulin production and insulin resistance indicate this kind of diabetes. Globally, 280 million people are expected to have diabetes by 2025 (Donath and Shoelson 2011; Pal et al. 2020). As a result, much research is being done on diabetes treatment advancements. It is possible to enhance medication release and decrease the number of administrations by using, for instance, protein materials. The secret to helping people with diabetes lower their blood glucose levels is insulin. Insulin administration with hypodermic injections has been the norm for almost eight decades. However, insulin action in the circulation is eliminated within 2–3 h. The half-life was almost 2.5 times longer when insulin bound with silk nanoparticles was prepared. Insulin and crystalline silk nanoparticles were covalently linked, and the insulin-silk systems demonstrated a high recovery rate (>90%) (Yan et al. 2009).

Diabetes and its side effects may also be treated using ELP depots. Even if the disease is not precisely at the depot, treatment may still be closely monitored. Diabetes medications, especially glucagon-like peptide 1 (GLP-1), may be therapeutically effective even if they must be taken often. Owing to GLP-1's relatively rapid degradation, an ELP-glucagon-like peptide one fusion depot was created to extend the drug's duration of administration. This medication may be genetically attached to elastin-like peptide molecules with arginine before the proteases degrade the arginine residues and release the glucagon-like peptide one peptide medicines (Chow et al. 2013; MacEwan and Chilkoti 2014).

6.5.4 Cancer Treatment

One illness that has an impact on cell division is cancer. It leads to unchecked cell proliferation, which accumulates into tumours. Since the turn of the century, cancer has emerged as one of the leading causes of mortality. Drug administration for cancer treatment is a fascinating field of research. Because of its nature, cancer usually requires several medication delivery processes, frequent administration, and a large number of clinical visits. Thus, research has shifted to proteins in search of alternatives to the routine visits and administrations. Materials based on silk and elastin can potentially address these problems (Chow et al. 2013).

ELPs are perfect for local and long-term medication administration because of their adjustable phase transition (temperature control). For instance, treating cancerous tumours requires continuous medication administration. As a result, radiation treatment for malignant tumours was administered using ELP depots. The depot's effectiveness in treating cancer was further enhanced by modifying ELP characteristics, including injection formulation and concentration (MacEwan and Chilkoti 2014).

Medications like curcumin may also be added to silk nanoparticles and injected locally into cancerous regions. For instance, *in vitro* experiments utilizing drug-loaded silk nanoparticles demonstrated the eight-day-long death of breast cancer cells. Desolvation was used in another work to create silk nanoparticles. Doxorubicin, a medication often used in chemotherapy, was more effectively delivered to the malignant region thanks to these particles (Seib and Kaplan 2013).

6.5.5 Neuroinflammation

Inflammation of the neurological system brought on by trauma, strokes, intracerebral haemorrhages, and other brain injuries is known as neuroinflammation. It is recommended that anti-inflammatory medications be delivered locally. By optimizing the medication's interaction with the site of inflammation, temperature-triggered ELP accumulation of nanoparticle depots may dramatically reduce drug loss to circulation (Chow et al. 2013; Pathak et al. 2022).

6.5.6 Wound Healing

Usually, surface wounds need to be cleaned before a bandage is applied. For a complete course of therapy, specific injuries call for additional medications. Drug delivery and skin surface coating have been explored using silk films. A combination of silk and gelatin was tested *in vitro* on rats in one research. The mix was supplemented with polyethylene glycol to slow down the breakdown rate. The outcome was encouraging; the wound healed 1 week after the medication was fully released from the protein-composite film (Alam and Shubhra 2015b; Chandra et al. 2022).

6.5.7 Corneal Regeneration

Blindness is most often caused by corneal disease. Keratoplasty, or a transplant from a human donor in good health, is the only generally recognized and successful course of therapy. The need for new corneal grafts has increased because of the scarcity of donor tissue, making allogenic and synthetic materials the preferred option (Salehi et al. 2014). Despite allogenic materials' outstanding mechanical and physiological qualities, their use is restricted by the danger of viral transmission, immunological rejection, and the availability of quality donor corneal grafts. Polymers based on synthetics were created to alleviate the shortages. Protein polymers like silk, elastin, and collagen proteins could replace corneal regeneration in tissue banking, reducing costs and preventing complications like inflammation, necrosis, and endophthalmitis. Ghezzi and colleagues developed a 3D silk culture method that mimics natural corneal stroma tissue, demonstrating smooth cell organization, tissue regeneration and host-graft integration in stacked silk protein films. These materials offer the physical and biofunctional characteristics needed for tissue healing, cell adhesion, and proliferation, reducing the need for donor corneal tissue for synthetic keratoprostheses (Zhao et al. 2016; Ghezzi et al. 2017).

6.6 Conclusion

Proteins are adaptable and versatile, making them useful in various applications like food packing, textiles, wound bandages, and drug delivery systems. They are biocompatible, biodegradable, and a renewable resource, making them ideal building blocks for health and medical sciences. This is because many synthetic polymers lack their superior characteristics. Furthermore, proteins possess distinct and often intricate physicochemical properties that may be used as specific molecular structural components in developing biomaterials. Furthermore, proteins are the building blocks for very potent and specific drugs. One approach that shows promise is using protein-based active pharmaceutical ingredients (APIs) as a crucial aspect of drug delivery systems, such as protein microneedles or self-assembling protein particles. Much effort has thoroughly evaluated the mechanical, chemical, and physical properties of novel protein materials in the interdisciplinary field of protein-based materials. Utilizing the vast information generated to tailor protein materials for specific therapeutic uses is crucial for maximizing their potential. More preclinical and clinical research is needed to advance the field and develop solutions that improve patient lives, but protein-based materials have already shown usefulness in various clinical scenarios.

References

Aboushwareb T, Eberli D, Ward C et al (2009) A keratin biomaterial gel hemostat derived from human hair: evaluation in a rabbit model of lethal liver injury. J Biomed Mater Res B Appl Biomater 90(1):45–54. https://doi.org/10.1002/jbm.b.31251

Alam A, Shubhra QTH (2015a) Surface modified thin film from silk and gelatin for sustained drug release to heal wound. J Mater Chem B 3(31):6473–6479. https://doi.org/10.1039/c5tb00920k

Alam AKMM, Shubhra QTH (2015b) Surface modified thin film from silk and gelatin for sustained drug release to heal wound. J Mater Chem B 3(31):6473–6479. https://doi.org/10.1039/C5TB00920K

Almine JF, Bax DV, Mithieux SM et al (2010) Elastin-based materials. Chem Soc Rev 39(9): 3371–3379. https://doi.org/10.1039/b919452p

Alqahtani OPPKAKP (2012) Protein nanocarriers for topical delivery. South Dakota State University, Brookings

Apel PJ, Garrett JP, Sierpinski P et al (2008) Peripheral nerve regeneration using a keratin-based scaffold: long-term functional and histological outcomes in a mouse model. J Hand Surg Am 33(9):1541–1547. https://doi.org/10.1016/j.jhsa.2008.05.034

Arangoa MA, Ponchel G, Orecchioni AM et al (2000) Bioadhesive potential of gliadin nanoparticulate systems. Eur J Pharm Sci 11(4):333–341. https://doi.org/10.1016/S0928-0987(00)00121-4

Arslan YE, Sezgin Arslan T, Derkus B et al (2017) Fabrication of human hair keratin/jellyfish collagen/eggshell-derived hydroxyapatite osteoinductive biocomposite scaffolds for bone tissue engineering: from waste to regenerative medicine products. Colloids Surf B: Biointerfaces 154: 160–170. https://doi.org/10.1016/j.colsurfb.2017.03.034

Athanasou NA (2009) Pathology of bone injury. Diagn Histopathol 15(9):437–443. https://doi.org/10.1016/j.mpdhp.2009.06.001

Austin MJ, Rosales AM (2019) Tunable biomaterials from synthetic, sequence-controlled polymers. Biomater Sci 7(2):490–505. https://doi.org/10.1039/c8bm01215f

Balakrishnan B, Mohanty M, Umashankar PR et al (2005) Evaluation of an in situ forming hydrogel wound dressing based on oxidized alginate and gelatin. Biomaterials 26(32):6335–6342. https://doi.org/10.1016/j.biomaterials.2005.04.012

Banskota S, Yousefpour P, Kirmani N et al (2019) Long circulating genetically encoded intrinsically disordered zwitterionic polypeptides for drug delivery. Biomaterials 192:475–485. https://doi.org/10.1016/j.biomaterials.2018.11.012

Barenghi R, Beke S, Romano I et al (2014) Elastin-coated biodegradable photopolymer scaffolds for tissue engineering applications. Biomed Res Int 2014:624645. https://doi.org/10.1155/2014/624645

Barros AA, Aroso IM, Silva TH et al (2015) Water and carbon dioxide: green solvents for the extraction of collagen/Gelatin from marine sponges. ACS Sustain Chem Eng 3(2):254–260. https://doi.org/10.1021/sc500621z

Betz M, García-González CA, Subrahmanyam RP et al (2012) Preparation of novel whey protein-based aerogels as drug carriers for life science applications. J Supercrit Fluids 72:111–119. https://doi.org/10.1016/j.supflu.2012.08.019

Bouman J, Belton P, Venema P et al (2016) Controlled release from Zein matrices: interplay of drug hydrophobicity and pH. Pharm Res 33(3):673–685. https://doi.org/10.1007/s11095-015-1818-8

Boussahel S, Speciale A, Dahamna S et al (2015) Flavonoid profile, antioxidant and cytotoxic activity of different extracts from Algerian Rhamnus alaternus L. bark. Pharmacogn Mag 11 (Suppl 1):S102–S109. https://doi.org/10.4103/0973-1296.157707

Boyle A (2018) Applications of de novo designed peptides. In: Peptide Applications in Biomedicine, Biotechnology and Bioengineering. Woodhead Publishing, Cambridge

Brennan DA, Jao D, Siracusa MC et al (2016) Concurrent collection and post-drawing of individual electrospun polymer nanofibers to enhance macromolecular alignment and mechanical properties. Polymer 103:243–250. https://doi.org/10.1016/j.polymer.2016.09.061

Cai B, Xia W, Bredenberg S et al (2015) Bioceramic microneedles with flexible and self-swelling substrate. Eur J Pharm Biopharm 94:404–410. https://doi.org/10.1016/j.ejpb.2015.06.016

Calcines-Cruz C, Finkelstein IJ, Hernandez-Garcia A (2021) CRISPR-guided programmable self-assembly of artificial virus-like nucleocapsids. Nano Lett 21(7):2752–2757. https://doi.org/10.1021/acs.nanolett.0c04640

Chandra P, RoomiKhan SN et al (2022) Protective effect of Zingiber officinale rhizomes against experimental induced ulcers in diabetic rats. Pharm Chem J 56(8):1107–1115. https://doi.org/10.1007/s11094-022-02760-6

Chandregowda V, Kush A, Reddy GC (2009) Synthesis of benzamide derivatives of anacardic acid and their cytotoxic activity. Eur J Med Chem 44(6):2711–2719. https://doi.org/10.1016/j.ejmech.2009.01.033

Chen Q, Thouas GA (2015) Metallic implant biomaterials. Mater Sci Eng R Rep 87:1–57. https://doi.org/10.1016/j.mser.2014.10.001

Chen Y, Youn P, Furgeson DY (2011) Thermo-targeted drug delivery of geldanamycin to hyperthermic tumor margins with diblock elastin-based biopolymers. J Control Release 155(2): 175–183. https://doi.org/10.1016/j.jconrel.2011.07.040

Chen X, Zaro JL, Shen WC (2013) Fusion protein linkers: property, design and functionality. Adv Drug Deliv Rev 65(10):1357–1369. https://doi.org/10.1016/j.addr.2012.09.039

Chen BZ, Ashfaq M, Zhu DD et al (2018a) Controlled delivery of insulin using rapidly separating microneedles fabricated from genipin-crosslinked gelatin. Macromol Rapid Commun 39(20): 1800075. https://doi.org/10.1002/marc.201800075

Chen C-H, Shyu VB-H, Chen C-T (2018b) Dissolving microneedle patches for transdermal insulin delivery in diabetic mice. Potential Clin Appl 11(9):1625

Cheng PN, Pham JD, Nowick JS (2013) The supramolecular chemistry of β-sheets. J Am Chem Soc 135(15):5477–5492. https://doi.org/10.1021/ja3088407

Chilkoti A, Christensen T, MacKay JA (2006) Stimulus responsive elastin biopolymers: applications in medicine and biotechnology. Curr Opin Chem Biol 10(6):652–657. https://doi.org/10.1016/j.cbpa.2006.10.010

Chow JP, Simionescu DT, Warner H et al (2013) Mitigation of diabetes-related complications in implanted collagen and elastin scaffolds using matrix-binding polyphenol. Biomaterials 34(3): 685–695. https://doi.org/10.1016/j.biomaterials.2012.09.081

Cilurzo F, Selmin F, Aluigi A et al (2013) Regenerated keratin proteins as potential biomaterial for drug delivery. Polym Adv Technol 24(11):1025–1028. https://doi.org/10.1002/pat.3168

Constantin Barbaresso R, Rau I, Gabriela Zgarian R et al (2013) Niflumic acid-collagen delivery systems used as anti-inflammatory drugs and analgesics in dentistry. C R Chim 17(1):12–17. https://doi.org/10.1016/j.crci.2013.07.007

Cuadri AA, Romero A, Bengoechea C et al (2017) Natural superabsorbent plastic materials based on a functionalized soy protein. Polym Test 58:126–134. https://doi.org/10.1016/j.polymertesting.2016.12.024

Das S, Bellare JR, Banerjee R (2012) Protein based nanoparticles as platforms for aspirin delivery for ophthalmologic applications. Colloids Surf B: Biointerfaces 93:161–168. https://doi.org/10.1016/j.colsurfb.2011.12.033

Dash BC, Mahor S, Carroll O et al (2011) Tunable elastin-like polypeptide hollow sphere as a high payload and controlled delivery gene depot. J Control Release 152(3):382–392. https://doi.org/10.1016/j.jconrel.2011.03.006

Delplace V, Nicolas J (2015) Degradable vinyl polymers for biomedical applications. Nat Chem 7(10):771–784. https://doi.org/10.1038/nchem.2343

Ding D, Pan J, Lim SH et al (2017) Squid suckerin microneedle arrays for tunable drug release. J Mater Chem B 5(43):8467–8478. https://doi.org/10.1039/C7TB01507K

Donath MY, Shoelson SE (2011) Type 2 diabetes as an inflammatory disease. Nat Rev Immunol 11(2):98–107. https://doi.org/10.1038/nri2925

Eshel G, Stainier P, Shepon A et al (2019) Environmentally optimal, nutritionally sound, protein and energy conserving plant based alternatives to U.S. Meat Sci Rep 9(1):10345. https://doi.org/10.1038/s41598-019-46590-1

Ferraro V, Anton M, Santé-Lhoutellier V (2016) The "sisters" α-helices of collagen, elastin and keratin recovered from animal by-products: functionality, bioactivity and trends of application. Trends Food Sci Technol 51:65–75. https://doi.org/10.1016/j.tifs.2016.03.006

Fiorentini F, Suarato G, Grisoli P et al (2021) Plant-based biocomposite films as potential antibacterial patches for skin wound healing. Eur Polym J 150:110414. https://doi.org/10.1016/j.eurpolymj.2021.110414

Fonseca DFS, Vilela C, Silvestre AJD et al (2019) A compendium of current developments on polysaccharide and protein-based microneedles. Int J Biol Macromol 136:704–728. https://doi.org/10.1016/j.ijbiomac.2019.04.163

Freitag TL, Podojil JR, Pearson RM et al (2020) Gliadin nanoparticles induce immune tolerance to gliadin in mouse models of celiac disease. Gastroenterology 158(6):1667–1681.e12. https://doi.org/10.1053/j.gastro.2020.01.045

Friess W (1998) Collagen – biomaterial for drug delivery. Eur J Pharm Biopharm 45(2):113–136. https://doi.org/10.1016/S0939-6411(98)00017-4

Fu B, Long W, Zhang Y et al (2015) Enhanced antitumor effects of the BRBP1 compound peptide BRBP1-TAT-KLA on human brain metastatic breast cancer. Sci Rep 5:8029. https://doi.org/10.1038/srep08029

Garay RP, El-Gewely R, Armstrong JK et al (2012) Antibodies against polyethylene glycol in healthy subjects and in patients treated with PEG-conjugated agents. Expert Opin Drug Deliv 9(11):1319–1323. https://doi.org/10.1517/17425247.2012.720969

Gellynck K, Verdonk PC, Van Nimmen E et al (2008) Silkworm and spider silk scaffolds for chondrocyte support. J Mater Sci Mater Med 19(11):3399–3409. https://doi.org/10.1007/s10856-008-3474-6

Ghezzi CE, Marelli B, Omenetto FG et al (2017) 3D functional corneal stromal tissue equivalent based on corneal stromal stem cells and multi-layered silk film architecture. PLoS One 12(1): e0169504. https://doi.org/10.1371/journal.pone.0169504

Gill HS, Denson DD, Burris BA et al (2008) Effect of microneedle design on pain in human volunteers. Clin J Pain 24(7):585–594. https://doi.org/10.1097/AJP.0b013e31816778f9

Grigsby WJ, Scott SM, Plowman-Holmes MI et al (2020) Combination and processing keratin with lignin as biocomposite materials for additive manufacturing technology. Acta Biomater 104:95–103. https://doi.org/10.1016/j.actbio.2019.12.026

Guo C, Li C, Kaplan DL (2020) Enzymatic degradation of Bombyx mori silk materials: a review. Biomacromolecules 21(5):1678–1686. https://doi.org/10.1021/acs.biomac.0c00090

Han S, Ham TR, Haque S et al (2015) Alkylation of human hair keratin for tunable hydrogel erosion and drug delivery in tissue engineering applications. Acta Biomater 23:201–213. https://doi.org/10.1016/j.actbio.2015.05.013

Hasatsri S, Angspatt A, Aramwit P (2015) Randomized clinical trial of the innovative Bilayered wound dressing made of silk and Gelatin: safety and efficacy tests using a split-thickness skin graft model. Evid Based Complement Alternat Med 2015:206871. https://doi.org/10.1155/2015/206871

He M, Zhang B, Dou Y et al (2017) Fabrication and characterization of electrospun feather keratin/poly(vinyl alcohol) composite nanofibers. RSC Adv 7(16):9854–9861. https://doi.org/10.1039/c6ra25009b

Hikal DM, Awad NS, Abdein MA (2017) The anticancer activity of cashew (Anacardium occidentale) and almond (Prunus dulcis) kernels. Adv Environ Biol 11:31+

Hoffman AS (2012) Hydrogels for biomedical applications. Adv Drug Deliv Rev 64:18–23. https://doi.org/10.1016/j.addr.2012.09.010

Hoffman JR, Falvo MJ (2004) Protein – Which is Best? J Sports Sci Med 3(3):118–130

Janarthanan G, Kim IG, Chung EJ et al (2019) Comparative studies on thin polycaprolactone-tricalcium phosphate composite scaffolds and its interaction with mesenchymal stem cells. Biomater Res 23:1. https://doi.org/10.1186/s40824-018-0153-7

Jones BA, Grace D, Kock R et al (2013) Zoonosis emergence linked to agricultural intensification and environmental change. Proc Natl Acad Sci USA 110(21):8399–8404. https://doi.org/10.1073/pnas.1208059110

Kalouta K, Stie MB, Janfelt C et al (2020) Electrospun α-Lactalbumin nanofibers for site-specific and fast-onset delivery of nicotine in the Oral cavity: an in vitro, ex vivo, and tissue spatial distribution study. Mol Pharm 17(11):4189–4200. https://doi.org/10.1021/acs.molpharmaceut.0c00642

Kamada A, Rodriguez-Garcia M, Ruggeri FS et al (2021) Controlled self-assembly of plant proteins into high-performance multifunctional nanostructured films. Nat Commun 12(1):3529. https://doi.org/10.1038/s41467-021-23813-6

Kanokpanont S, Damrongsakkul S, Ratanavaraporn J et al (2012) An innovative bi-layered wound dressing made of silk and gelatin for accelerated wound healing. Int J Pharm 436(1):141–153. https://doi.org/10.1016/j.ijpharm.2012.06.046

Kaushik S, Hord AH, Denson DD et al (2001) Lack of pain associated with microfabricated microneedles. Anesth Analg 92(2):502–504. https://doi.org/10.1213/00000539-200102000-00041

Kim MS, Lee DW, Park K et al (2014) Temperature-triggered tumor-specific delivery of anticancer agents by cRGD-conjugated thermosensitive liposomes. Colloids Surf B Biointerfaces 116:17–25. https://doi.org/10.1016/j.colsurfb.2013.12.045

Kishore AH, Vedamurthy BM, Mantelingu K et al (2008) Specific small-molecule activator of Aurora kinase A induces autophosphorylation in a cell-free system. J Med Chem 51(4): 792–797. https://doi.org/10.1021/jm700954w

Kleemann C, Selmer I, Smirnova I et al (2018) Tailor made protein based aerogel particles from egg white protein, whey protein isolate and sodium caseinate: influence of the preceding hydrogel characteristics. Food Hydrocoll 83:365–374. https://doi.org/10.1016/j.foodhyd.2018.05.021

Koh L-D, Cheng Y, Teng C-P et al (2015) Structures, mechanical properties and applications of silk fibroin materials. Prog Polym Sci 46:86–110. https://doi.org/10.1016/j.progpolymsci.2015.02.001

Lakshmi YS, Kumar P, Kishore G et al (2016) Triple combination MPT vaginal microbicide using curcumin and efavirenz loaded lactoferrin nanoparticles. Sci Rep 6(1):25479. https://doi.org/10.1038/srep25479

Lammel A, Schwab M, Hofer M et al (2011) Recombinant spider silk particles as drug delivery vehicles. Biomaterials 32(8):2233–2240. https://doi.org/10.1016/j.biomaterials.2010.11.060

Lau S, Fei J, Liu H et al (2017) Multilayered pyramidal dissolving microneedle patches with flexible pedestals for improving effective drug delivery. J Control Release 265:113–119. https://doi.org/10.1016/j.jconrel.2016.08.031

Layman DK, Lönnerdal B, Fernstrom JD (2018) Applications for α-lactalbumin in human nutrition. Nutr Rev 76(6):444–460. https://doi.org/10.1093/nutrit/nuy004

Lee J, Park SH, Seo IH et al (2015) Rapid and repeatable fabrication of high A/R silk fibroin microneedles using thermally-drawn micromolds. Eur J Pharm Biopharm 94:11–19. https://doi.org/10.1016/j.ejpb.2015.04.024

Lee I-C, Lin W-M, Shu J-C et al (2017) Formulation of two-layer dissolving polymeric microneedle patches for insulin transdermal delivery in diabetic mice. J Biomed Mater Res A 105(1):84–93. https://doi.org/10.1002/jbm.a.35869

Li Y, Yao C (2012) Mineralization of hydroxyapatite crystallites on zein microspheres. Poly Comp 33(6):961–966. https://doi.org/10.1002/pc.22202

Li Y, Yu SM (2013) Targeting and mimicking collagens via triple helical peptide assembly. Curr Opin Chem Biol 17(6):968–975. https://doi.org/10.1016/j.cbpa.2013.10.018

Li Q, Zhu L, Liu R et al (2012) Biological stimuli responsive drug carriers based on keratin for triggerable drug delivery. J Mater Chem 22(37):19964–19973. https://doi.org/10.1039/C2JM34136K

Li L, Tong Z, Jia X et al (2013) Resilin-like polypeptide hydrogels engineered for versatile biological functions. Soft Matter 9(3):665–673. https://doi.org/10.1039/c2sm26812d

Liu X, Sun Q, Wang H et al (2005) Microspheres of corn protein, zein, for an ivermectin drug delivery system. Biomaterials 26(1):109–115. https://doi.org/10.1016/j.biomaterials.2004.02.013

Liu G, An D, Li J et al (2023) Zein-based nanoparticles: preparation, characterization, and pharmaceutical application. Front Pharmacol 14:1120251. https://doi.org/10.3389/fphar.2023.1120251

Lopez-Sagaseta J, Malito E, Rappuoli R et al (2016) Self-assembling protein nanoparticles in the design of vaccines. Comput Struct Biotechnol J 14:58–68. https://doi.org/10.1016/j.csbj.2015.11.001

MacEwan SR, Chilkoti A (2014) Applications of elastin-like polypeptides in drug delivery. J Control Release 190:314–330. https://doi.org/10.1016/j.jconrel.2014.06.028

Maeda M, Kadota K, Kajihara M et al (2001) Sustained release of human growth hormone (hGH) from collagen film and evaluation of effect on wound healing in db/db mice. J Control Release 77(3):261–272. https://doi.org/10.1016/S0168-3659(01)00512-0

McGann CL, Levenson EA, Kiick KL (2013) Resilin-based hybrid hydrogels for cardiovascular tissue engineering. Macromolecules 214(2):203–213. https://doi.org/10.1002/macp.201200412

Mendes AC, Stephansen K, Chronakis IS (2017) Electrospinning of food proteins and polysaccharides. Food Hydrocoll 68:53–68. https://doi.org/10.1016/j.foodhyd.2016.10.022

Mendes AC, Saldarini E, Chronakis IS (2020) Electrohydrodynamic processing of potato protein into particles and Fibers. Molecules 25(24). https://doi.org/10.3390/molecules25245968

Muller-Herrmann S, Scheibel T (2015) Enzymatic degradation of films, particles, and nonwoven meshes made of a recombinant spider silk protein. ACS Biomater Sci Eng 1(4):247–259. https://doi.org/10.1021/ab500147u

Nijdam D, Rood T, Westhoek H (2012) The price of protein: review of land use and carbon footprints from life cycle assessments of animal food products and their substitutes. Food Policy 37(6):760–770. https://doi.org/10.1016/j.foodpol.2012.08.002

Nonthanum P, Lee Y, Padua GW (2012) Effect of γ-zein on the rheological behavior of concentrated zein solutions. J Agric Food Chem 60(7):1742–1747. https://doi.org/10.1021/jf2035302

Nowatzki PJ, Tirrell DA (2004) Physical properties of artificial extracellular matrix protein films prepared by isocyanate crosslinking. Biomaterials 25(7):1261–1267. https://doi.org/10.1016/S0142-9612(03)00635-5

Nystrom G, Fong W-K, Mezzenga R (2017) Ice-Templated and cross-linked amyloid fibril aerogel scaffolds for cell growth. Biomacromolecules 18(9):2858–2865. https://doi.org/10.1021/acs.biomac.7b00792

Pal D, Chandra P, Sachan N (2020) Chapter 15 – Sesame seed in controlling human health and nutrition. In: Preedy VR, Watson RR (eds) Nuts and seeds in health and disease prevention, 2nd edn. Academic, London

Panahi R, Baghban-Salehi M (2019) Protein-based hydrogels. In: Mondal MIH (ed) Cellulose-based superabsorbent hydrogels. Springer International Publishing, Cham

Park M, Shin HK, Kim B-S et al (2015) Effect of discarded keratin-based biocomposite hydrogels on the wound healing process in vivo. Mater Sci Eng C 55:88–94. https://doi.org/10.1016/j.msec.2015.03.033

Pathak R, Sachan N, Chandra P (2022) Mechanistic approach towards diabetic neuropathy screening techniques and future challenges: a review. Biomed Pharmacother 150:113025. https://doi.org/10.1016/j.biopha.2022.113025

Pinheiro A, Cooley A, Liao J et al (2016) Comparison of natural crosslinking agents for the stabilization of xenogenic articular cartilage. J Orthop Res 34(6):1037–1046. https://doi.org/10.1002/jor.23121

Pistritto G, Trisciuoglio D, Ceci C et al (2016) Apoptosis as anticancer mechanism: function and dysfunction of its modulators and targeted therapeutic strategies. Aging (Albany NY) 8(4):603–619. https://doi.org/10.18632/aging.100934

Pritchard EM, Hu X, Finley V et al (2013) Effect of silk protein processing on drug delivery from silk films. Macromol Biosci 13(3):311–320. https://doi.org/10.1002/mabi.201200323

Punia K, Kronenberg JB, Montclare JK (2019) Protein biomaterials for theranostic applications. Mol Syst Design Eng 4(6):1074–1094. https://doi.org/10.1039/c9me00143c

Qi Y, Simakova A, Ganson NJ et al (2016) A brush-polymer conjugate of exendin-4 reduces blood glucose for up to five days and eliminates poly(ethylene glycol) antigenicity. Nat Biomed Eng 1. https://doi.org/10.1038/s41551-016-0002

Qin G, Hu X, Cebe P et al (2012) Mechanism of resilin elasticity. Nat Commun 3:1003. https://doi.org/10.1038/ncomms2004

Rafat M, Li F, Fagerholm P et al (2008) PEG-stabilized carbodiimide crosslinked collagen–chitosan hydrogels for corneal tissue engineering. Biomaterials 29(29):3960–3972. https://doi.org/10.1016/j.biomaterials.2008.06.017

Rapa M, Gaidau C, Stefan LM et al (2020) New nanofibers based on protein by-products with bioactive potential for tissue engineering. Materials (Basel) 13(14). https://doi.org/10.3390/ma13143149

Rouilly A, Meriaux A, Geneau C et al (2006) Film extrusion of sunflower protein isolate. Poly Eng Sci 46(11):1635–1640. https://doi.org/10.1002/pen.20634

Sahiner M, Alpaslan D, Bitlisli BO (2014) Collagen-based hydrogel films as drug-delivery devices with antimicrobial properties. Polym Bull 71(11):3017–3033. https://doi.org/10.1007/s00289-014-1235-x

Salehi S, Fathi M, Javanmard SH et al (2014) Generation of PGS/PCL blend nanofibrous scaffolds mimicking corneal stroma structure. Macromol Mater Eng 299(4):455–469. https://doi.org/10.1002/mame.201300187

Samadian H, Zamiri S, Ehterami A et al (2020) Electrospun cellulose acetate/gelatin nanofibrous wound dressing containing berberine for diabetic foot ulcer healing: in vitro and in vivo studies. Sci Rep 10(1):8312. https://doi.org/10.1038/s41598-020-65268-7

Santos LS, Andrade TA, Barbosa Gomes de Carvalho YM et al (2021) Gelatin-based mucoadhesive membranes containing inclusion complex of thymol/β-cyclodextrin for treatment of oral infections. Int J Polym Mater Polym Biomater 70(3):184–194. https://doi.org/10.1080/00914037.2019.1706509

Seib FP, Kaplan DL (2013) Silk for drug delivery applications: opportunities and challenges. Isr J Chem 53(9–10):756–766. https://doi.org/10.1002/ijch.201300083

Sell SA, McClure MJ, Garg K et al (2009) Electrospinning of collagen/biopolymers for regenerative medicine and cardiovascular tissue engineering. Adv Drug Deliv Rev 61(12):1007–1019. https://doi.org/10.1016/j.addr.2009.07.012

Sharma S, Bano S, Ghosh AS et al (2016) Silk fibroin nanoparticles support in vitro sustained antibiotic release and osteogenesis on titanium surface. Nanomedicine 12(5):1193–1204. https://doi.org/10.1016/j.nano.2015.12.385

Singaravelu S, Ramanathan G, Raja MD et al (2016) Biomimetic interconnected porous keratin-fibrin-gelatin 3D sponge for tissue engineering application. Int J Biol Macromol 86:810–819. https://doi.org/10.1016/j.ijbiomac.2016.02.021

Sionkowska A, Kozłowska J (2013) Properties and modification of porous 3-D collagen/hydroxyapatite composites. Int J Biol Macromol 52:250–259. https://doi.org/10.1016/j.ijbiomac.2012.10.002

Sothornvit R, Krochta JM (2001) Plasticizer effect on mechanical properties of β-lactoglobulin films. J Food Eng 50(3):149–155. https://doi.org/10.1016/S0260-8774(00)00237-5

Sousa FF, Luzardo-Alvarez A, Blanco-Mendez J et al (2012) NMR techniques in drug delivery: application to zein protein complexes. Int J Pharm 439(1–2):41–48. https://doi.org/10.1016/j.ijpharm.2012.09.046

Srinivasan B, Kumar R, Shanmugam K et al (2010) Porous keratin scaffold-promising biomaterial for tissue engineering and drug delivery. J Biomed Mater Res B Appl Biomater 92(1):5–12. https://doi.org/10.1002/jbm.b.31483

Srokowski EM, Woodhouse KA (2008) Development and characterisation of novel cross-linked bio-elastomeric materials. J Biomater Sci Polym Ed 19(6):785–799. https://doi.org/10.1163/156856208784522038

Stie MB, Corezzi M, Juncos Bombin AD et al (2020) Waterborne electrospinning of α-Lactalbumin generates Tunable and biocompatible nanofibers for drug delivery. ACS Appl Nano Mater 3(2):1910–1921. https://doi.org/10.1021/acsanm.9b02557

Stinson JA, Raja WK, Lee S et al (2017) Silk fibroin microneedles for transdermal vaccine delivery. ACS Biomater Sci Eng 3(3):360–369. https://doi.org/10.1021/acsbiomaterials.6b00515

Tang C, Ozcam AE, Stout B et al (2012) Effect of pH on protein distribution in electrospun PVA/BSA composite nanofibers. Biomacromolecules 13(5):1269–1278. https://doi.org/10.1021/bm2017146

Teule F, Miao YG, Sohn BH et al (2012) Silkworms transformed with chimeric silkworm/spider silk genes spin composite silk fibers with improved mechanical properties. Proc Natl Acad Sci USA 109(3):923–928. https://doi.org/10.1073/pnas.1109420109

Tsioris K, Raja WK, Pritchard EM et al (2012) Fabrication of silk microneedles for controlled-release drug delivery. Adv Funct Mater 22(2):330–335. https://doi.org/10.1002/adfm.201102012

Ullsten NH, Mikael G, Mikael SH (2016) Plasticizers for protein-based materials. In: Mohamed Fathy E-A (ed) Viscoelastic and viscoplastic materials. IntechOpen, Rijeka

Vaiana SM, Manno M, Emanuele A et al (2001) The role of solvent in protein folding and in aggregation. J Biol Phys 27(2):133–145. https://doi.org/10.1023/A:1013146530021

van Herwaarden S, Iervolino E, van Herwaarden F et al (2011) Design, performance and analysis of thermal lag of the UFS1 twin-calorimeter chip for fast scanning calorimetry using the Mettler-Toledo Flash DSC 1. Thermochim Acta 522(1):46–52. https://doi.org/10.1016/j.tca.2011.05.025

Voloshchuk N, Montclare JK (2010) Incorporation of unnatural amino acids for synthetic biology. Mol BioSyst 6(1):65–80. https://doi.org/10.1039/b909200p

Wang Y, Katyal P, Montclare JK (2019) Protein-Engineered Functional Materials. Adv Healthc Mater 8(11):e1801374. https://doi.org/10.1002/adhm.201801374

Wang J, Li Y, Huang J et al (2020) A protein nanocomposite for ultra-fast, efficient and non-irritating skin decontamination of nerve agents. Nanoscale 12(7):4400–4409. https://doi.org/10.1039/C9NR09015K

Wattie B, Dumont M-J, Lefsrud M (2018) Synthesis and properties of feather keratin-based superabsorbent hydrogels. Waste Biomass Valoriz 9(3):391–400. https://doi.org/10.1007/s12649-016-9773-0

Wenk E, Merkle HP, Meinel L (2011) Silk fibroin as a vehicle for drug delivery applications. J Control Release 150(2):128–141. https://doi.org/10.1016/j.jconrel.2010.11.007

Wu H, Liu S, Xiao L et al (2016) Injectable and pH-responsive silk nanofiber hydrogels for sustained anticancer drug delivery. ACS Appl Mater Interfaces 8(27):17118–17126. https://doi.org/10.1021/acsami.6b04424

Xia XX, Qian ZG, Ki CS et al (2010) Native-sized recombinant spider silk protein produced in metabolically engineered Escherichia coli results in a strong fiber. Proc Natl Acad Sci USA 107(32):14059–14063. https://doi.org/10.1073/pnas.1003366107

Yan H-B, Zhang Y-Q, Ma Y-L et al (2009) Biosynthesis of insulin-silk fibroin nanoparticles conjugates and in vitro evaluation of a drug delivery system. J Nanopart Res 11(8):1937–1946. https://doi.org/10.1007/s11051-008-9549-y

Yao C, Li Y, Wu F (2013) Zein nanofibrous membranes as templates for biomineralization of hydroxyapatite crystallites. Poly Comp 34(7):1163–1171. https://doi.org/10.1002/pc.22525

Yavuz B, Chambre L, Harrington K et al (2020) Silk fibroin microneedle patches for the sustained release of Levonorgestrel. ACS Appl Bio Mater 3(8):5375–5382. https://doi.org/10.1021/acsabm.0c00671

Yin L, Yuvienco C, Montclare JK (2017) Protein based therapeutic delivery agents: contemporary developments and challenges. Biomaterials 134:91–116. https://doi.org/10.1016/j.biomaterials.2017.04.036

Yu W, Jiang G, Liu D et al (2017) Fabrication of biodegradable composite microneedles based on calcium sulfate and gelatin for transdermal delivery of insulin. Mater Sci Eng C 71:725–734. https://doi.org/10.1016/j.msec.2016.10.063

Zhang X, Zhang J, Shi B (2014) Mesoporous bioglass/silk fibroin scaffolds as a drug delivery system: fabrication, drug loading and release in vitro and repair calvarial defects in vivo. J Wuhan Univ Technol Mater Sci Ed 29(2):401–406. https://doi.org/10.1007/s11595-014-0929-0

Zhang K, Zhao X, Chen X et al (2018) Enhanced therapeutic effects of mesenchymal stem cell-derived exosomes with an injectable hydrogel for Hindlimb ischemia treatment. ACS Appl Mater Interfaces 10(36):30081–30091. https://doi.org/10.1021/acsami.8b08449

Zhang Y, Wu M, Tan D et al (2021) A dissolving and glucose-responsive insulin-releasing microneedle patch for type 1 diabetes therapy. J Mater Chem B 9(3):648–657. https://doi.org/10.1039/D0TB02133D

Zhao X, Song W, Liu S et al (2016) Corneal regeneration by utilizing collagen based materials. SCIENCE CHINA Chem 59(12):1548–1553. https://doi.org/10.1007/s11426-016-0248-6

Zhao Y, Wang Z, Zhang Q et al (2018) Accelerated skin wound healing by soy protein isolate–modified hydroxypropyl chitosan composite films. Int J Biol Macromol 118:1293–1302. https://doi.org/10.1016/j.ijbiomac.2018.06.195

Zhou M, Shmidov Y, Matson JB et al (2017) Multi-scale characterization of thermoresponsive dendritic elastin-like peptides. Colloids Surf B Biointerfaces 153:141–151. https://doi.org/10.1016/j.colsurfb.2017.02.014

Zhu C, Lei H, Fan D et al (2018) Novel enzymatic crosslinked hydrogels that mimic extracellular matrix for skin wound healing. J Mater Sci 53(8):5909–5928. https://doi.org/10.1007/s10853-017-1956-y

Zhu M, Liu Y, Jiang F et al (2020) Combined silk fibroin microneedles for insulin delivery. ACS Biomater Sci Eng 6(6):3422–3429. https://doi.org/10.1021/acsbiomaterials.0c00273

Zoccola M, Aluigi A, Vineis C et al (2008) Study on cast membranes and electrospun nanofibers made from keratin/fibroin blends. Biomacromolecules 9(10):2819–2825. https://doi.org/10.1021/bm800579a

Zugravu MV, Smith RA, Reves BT et al (2013) Physical properties and in vitro evaluation of collagen-chitosan-calcium phosphate microparticle-based scaffolds for bone tissue regeneration. J Biomater Appl 28(4):566–579. https://doi.org/10.1177/0885328212465662

Dr. Phool Chandra, M. Pharm. (BU, Jhansi), Ph.D. (AKTU, Lucknow) is presently working as Professor and Head at the Teerthanker Mahaveer College of Pharmacy, Faculty of Pharmacy, Teerthanker Mahaveer University, Moradabad, India. He has more than eighteen years of teaching and research experience. He is the author of seven books and 21 book chapters with different publishers of international repute. He has published 9 Indian Patents and more than 100 research and review articles in National and International Journals. He is also, the editorial member of different Journals of National and International repute. He has supervised 36 M. Pharm. and seven Ph.D. research scholars and is supervising three M. Pharm. and seven Ph.D. research scholars. Dr. Chandra received the Distinguished Scientist Award (2022) for outstanding and remarkable contribution to Pharmaceutical Research during the International Conference on Drug Engineering, SERB, DST, Govt. of India, New Delhi, and Shobhit University. He is a member of BOS of TMU and other Universities. Also, He is an alumnus of Jawahar Navodaya Vidyalaya Unnao and Amethi (UP), India.

Ms. Rashmi Pathak is currently working as an Assistant Professor in the Department of Pharmacy at Invertis University Bareilly. She secured a silver medal in B. Pharm. at Invertis University, Bareilly, and a gold medal in M.Pharm. (Pharmacology) at IFTM University, Moradabad. She possesses 2 years of experience in academics. She has published 4 research papers, 11 review articles, 7 book chapters with national and international publishers, and 1 Patent. She had organized one FDP and One workshop as an Organizing secretory and Co-convenor, respectively.

Dr. Neetu Sachan, M. Pharm. (UPTU, Lucknow), Ph.D. (IFTMU, Moradabad) is presently working as a Professor at Maharana Pratap College of Pharmacy, Kanpur, Uttar Pradesh, India. She has more than seventeen years of teaching and research experience. She is the author of four books and 17 book chapters with different publishers of international repute. She has published 6 Indian Patents and more than 80 research and review articles in national and international journals. Dr. Sachan has supervised 23 M. Pharm. and eight Ph.D. research scholars. Dr. Sachan has organized two weeks of FDP and four three-day Entrepreneurship awareness camps sponsored by the National Science & Technology Entrepreneurship Development Board (NSTEDB), Department of Science & Technology (DST), Govt. of India. For her role in Entrepreneurship awareness, she received an Award of Excellence (Entrepreneurship) during a national conference sponsored by the National Commission for Women, Govt of India on "Expanding Women's Role in Developing Technology: Increasing Productivity, Improving Lives," New Delhi organized by Shobhit University, Gangoh in 2020.

Dr Anurag Verma is currently working as a Professor and Principal, at the College of Pharmacy, Teerthankar Mahaveer University, Moradabad, Uttar Pradesh. He completed his M Pharm (Pharmaceutics) from KLE's College of Pharmacy in 1999 and PhD in Pharmaceutical Sciences from Uttar Pradesh Technical University in 2012. His research supervisors were Prof Jayanta K Pandit and Prof A K Wahi (Ex-Professor BHU-IT). He has 25 years of teaching experience, 2.5 years of Industrial experience, and 14 years of research experience. He has published more than 150 research papers in National and International journals of repute (Elsevier/Wiley/Taylor & Francis/Nature) and supervised PhD research work of 14 scholars. He has produced 25 Postgraduates and is presently an advisor to 8 PhD scholars. Ongoing research programs in his lab focus on Cosmetics, Material science (evaluation of polymers/polymer composites for oral drug delivery applications), and target delivery through nanoparticles.

Graphene-Based Carbonaceous Materials: A Sustainable Biomaterial for Biomedical Application

7

Adam Aberra Challa, Nabanita Saha, and Petr Saha

Abstract

Biomaterials interact with biological systems, revolutionizing fields like tissue engineering, drug delivery, and implant development due to their biocompatibility and structural adaptability. However, concerns about the sustainability of their synthesis and sourcing drive the search for eco-friendly alternatives. Fortunately, carbonaceous materials, particularly graphene and its derivatives, have emerged as promising candidates. Graphene's one-atom-thick, sp^2-bonded carbon structure provides high tensile strength, large surface area, stability, and elasticity. Its derivates share similar extended characteristics. These properties lead to exceptional conductivity, rapid electron transfer, and high adsorption potential, making graphene valuable in catalysis, energy storage, and material synthesis. In biomedicine, their targeted surface immobilization, efficient drug loading, and high biocompatibility enhance applications in drug delivery, biosensing, and antimicrobial coatings. Despite the promise of graphene-based materials, challenges remain regarding their biodegradability and environmental impact. To this effect, increasing application of sustainable synthesis from organic sources is taking place, enhancing their value as viable alternatives. They can be synthesized from plant extracts, agro-residues, and bio/food waste, offering renewable solutions with high porosity, structural configurability, and surface functionality. This chapter explores carbonaceous biomaterials in-depth, examining their structure, functionality, and diverse biomedical applications. Through a comprehensive analysis of recent advancements and prospects, the chapter aims to highlight

A. A. Challa · N. Saha (✉) · P. Saha
Centre of Polymer Systems, University Institute, Tomas Bata University in Zlin, Zlin, Czech Republic
e-mail: nabanita@utb.cz

R. Malviya, S. Sundram (eds.), *Sustainable Green Biomaterials As Drug Delivery Systems*, Biomaterials, Bioengineering and Sustainability 1,
https://doi.org/10.1007/978-3-031-79062-1_7

the pivotal role of graphene-based biomaterials in advancing biomedical research and catalyzing healthcare innovations.

Keywords

Carbonaceous nanomaterials · Graphene-based biomaterials · Sustainability · Drug delivery · Tissue engineering · Biodegradability

Abbreviations

LEDs	Light-emitting diodes
rGO	Reduced graphene oxide
GO	Graphene oxide
GQDs	Graphene quantum dots
PLA	Polylactic acid
PEG	Polyethylene glycol
RB5	Reactive Black 5

7.1 Introduction

Biomaterials are functional materials produced to interact with the biological systems of the human body. Their interactions could be on the surface (such as wound dressing materials), inside the body (such as drug carriers), as aids to the function of organs (such as heart pacemakers), or as supports to the system of the body that guide cell growth and tissue regeneration (such as dental applications and scaffolds in tissue engineering). Advanced fabrication techniques have enabled the precise tailoring of biomaterials at the molecular level, unlocking more opportunities for customizing their properties to meet specific biomedical needs.

From biocompatible polymers to nanomaterials and bioceramics, the diversity of biomaterials available today allows for multiple applications, from enhancing the biocompatibility of implants to promoting tissue regeneration and engineering complex organoids for dictating disease characteristics and testing drugs. Furthermore, linked with emerging technological advances such as 3D printing, artificial intelligence, and microfluidics, the world of biomaterials is currently revolutionizing personalized medicine by enabling the fabrication of patient-specific implants, organs-on-chips, and wearable biosensors. As biomaterials continue to evolve, their impact is increasingly advancing medical treatments, diagnostic tools, and therapeutic interventions.

The need for biomaterials is well described by the current market that it holds. From 120 billion USD in 2020 to 140 billion USD in 2023 and an expected market of 432 billion USD in 2032, the global need for biomaterials is high and rising steadily (*Biomaterials Market Size, Share, & Growth Analysis Report* 2023).

Recently, the metallic biomaterials sector (such as titanium, chromium, nickel, stainless steel, cobalt, and their alloys) has held the highest value of market share within the worldwide biomaterials market and the highest usage is for orthopedics (Biomaterials Market by Type, Application, and Region – Global Forecast to *2029* 2024). This list of application areas where the use of biomaterials is rising includes dental appliances, wound healing, and tissue engineering fields as well.

Since these statistics indicate the usage of a large amount of materials, it is clear that it would create an impact on the environment and a disbalance in their development route. Aspects of material sourcing, processing techniques, and testing methods need innovative solutions. Thus, researchers have long been prioritizing such phenomena and the viable options are ever-evolving.

One of the notable advances toward such solutions is the utilization of carbon-based materials. Wide availability, alternative sourcing options, an array of techniques of production, and excellent desired properties made such materials highly favored. Graphene, carbon nanotubes, carbon dots, fullerenes, silicon carbide, and others show the level at which carbon technology has reached. Their potential in biomedicine such as wound dressing, targeted drug delivery, bioimaging, antimicrobial coatings, and tissue engineering is promising, prompting continuous research efforts to fully obtain their potential.

Graphene is a planar carbon allotrope that is built with carbon atoms in a crystal hexagonal honeycomb structure. It exhibits one of the highest mechanical strengths (a tensile strength of nearly 130 GPa, and a Young's modulus of up to 1 TPa) (Zhu et al. 2010) when it comes to building blocks. It can be shaped into distinct geometrical features. It can be wrapped into a 0-dimension (fullerene), rolled into a 1-dimension (carbon nanotubes), or stacked into 3-dimensions (graphite). Graphene-based materials have a lot to offer to the biomedical field given their properties. For instance, their increased surface area facilitates efficient drug delivery, while their superior mechanical properties bolster developments in tissue engineering. Additionally, these materials demonstrate notable capabilities for surface immobilization, efficient drug loading, proven antimicrobial properties, and heightened biocompatibility.

One of the most important advantages of using carbon-based biomaterials, including graphene, is the fact that they can be obtained from cheap, alternative, natural origins. This shines not only on their potential for mass production but also on the development of improved properties. Using straightforward processes such as graphitization and pyrolysis for converting these sources to nanomaterials, combined with innovative graphene synthesis techniques, including epitaxial growth and chemical vapor deposition, these carbonaceous materials show exceptional characteristics. Their high porosity, structural configurability, and surface functionality make them highly versatile for multiple purposes.

Within the context of biomedical engineering, however, the persistent challenges related to the biodegradability of carbonaceous nanomaterials require deeper investigation into their possible environmental impact. Nonetheless, extracting these materials from organic sources like plant residues and agricultural waste as mentioned above, offers a potential solution, as they are prone to enzymatic degradation.

This not only addresses concerns regarding their ecological impact but also enhances their value as sustainable and viable alternatives. This chapter dives deep into the world of carbonaceous biomaterials where graphene-based materials are centered. It iterates their exceptional properties and applications in diverse areas by focusing on biomedicine. The sustainability aspect of biomaterial usage is also well explained in the sub-sections of this piece.

7.2 Graphene and Its Family

Dubbed "wonder material" graphene is considered one of the most revolutionary nanomaterials in material science. This is mainly because it boosts superb physical, mechanical, electrical, and optical characteristics among others. These phenomena have granted it a preference in the material science world. For instance, its high mechanical strength is a recipe for a good bone tissue scaffold, whereas its excellent electrical conductivity makes it capable of being used in neural and heart tissue engineering (Ghosal et al. 2021). Its surface topography and mechanism make it suitable for diversified execution. Having an elaborate surface area and the ability to be functionalized, Graphene is suitable for producing composite materials. The following sub-sections deal with its structure, properties, and functionalities.

7.2.1 Structure of Graphene

The one-atom-thick hexagonal lattice structure gives graphene its uniqueness. It has a stable configuration because of the tight packing through the in-plane σ bond formed by its sp^2 orbital hybridization. Carbon has four valence electrons. Hence it has a $1s^22s^22p^2$ electron configuration. In graphene, from the four orbital electrons, three orbitals (s, p_x, and p_y) hybridize to form covalent bonds with one carbon atom each. Whereas the unhybridized p_z orbitals interlock with other atoms on a parallel plane forming π bonds in either π state (valence band) or an empty π^* state (conduction band) (Samantara et al. 2019). The sp^2 covalent bonds give it its exceptional mechanical properties even higher than that of diamond which has sp^3 hybridized carbon bonds. Whereas, the π bond which is located perpendicular to the surface of the lattice, is associated with the electrical conductivity of graphene (Deretzis and La Magna 2011; Yang et al. 2018).

As can be seen in the schematic in Fig. 7.1, the cut section of the graphene structure shows either zigzag or armchair edges. The zigzag edges form what are known as edge states, which show superior quality in terms of electronic and magnetic activities (Enoki et al. 2012). Hence, structural configurations are important in defining the usage of graphene.

There are two carbon atoms in every unit cell of a one-layered structure of graphene. The distance between the two atoms is 1.42 Å (0.142 nm), with a lattice constant of 2.46 Å (0.246 nm). Such a layer is 0.34 nm thick. Evidently, graphene's lattice structure is interpreted as two intertwined triangular sub-lattices, termed A

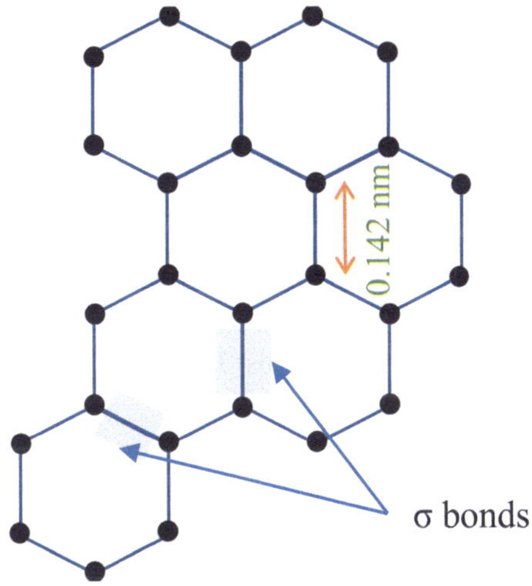

Fig. 7.1 The planar lattice description of a graphene layer

and B, originating from the two atoms found in its unit cell (Avouris and Dimitrakopoulos 2012).

7.2.2 Properties of Graphene-Based Biomaterials

The structural configuration and unique architecture of graphene are responsible for the range of capabilities that it has, linking it to the description, "a wonder material". Graphene has a thermal conductivity that reaches 3000–5000 W/m.K which puts it above its close contenders such as copper and diamond. Its electron mobility is measured as high as 200,000 cm^2/V.s, indicating the ease with which electrons can be carried through. In addition, graphene's optical transparency as a single layer can be about 97.3% which can be deemed as completely transparent (Moosa and Abed 2021). Graphene is inert, and relatively stable due to its σ bonds, and because it lacks hydroxide groups on its surface, it has poor water solubility. On top of that, it is embodied with a large surface area measurement of 2630 m^2/g.

Other materials in the graphene family also encompass exceptional properties. Graphene oxide, as the name implies, is an oxygen-containing form of graphene formed through the intercalation and oxidation of graphite. Through its synthesis, it is equipped with various functional groups such as hydroxyl (C-OH), carboxyl (C-COOH), carbonyl (C=O), and epoxy (C-O-C) attached to the surface. These functional groups are found on the basal planes (hydroxyl and epoxy) and edges (carboxylic) of the GO sheets, where aliphatic regions are formed (Aliyev et al. 2019). Because of these functional groups, GO forms stable suspensions in aqueous or many other solutions.

Reduced graphene oxide (rGO) is considered to have a graphene-like behavior even though it does not match the properties of that of pristine graphene. However, it is a viable material because of its manageable functionality, and superb electric and thermal conductance (Gürsel 2020). In addition, the ease of attaining its precursor and the scalable synthesis routes make it more attractive among the graphene family. Valid to applicable areas, rGO shows hydrophobic behavior much like graphene, because of the lack of oxygen-containing compounds.

Graphene quantum dots (GQDs) are essentially sheets of graphene that are less than 100 nm in their lateral dimension. They have a monoatomic layered planar structure, broad surface area, and oxygen-containing functional groups on their surface which can provide significant active sites for molecules to attach (Zhao et al. 2020). GQDs are mostly soluble or dispersible in water and are biocompatible with human neural cells. They also possess stable photoluminescence and adjustable band gap (Smith et al. 2019).

7.2.3 Synthesis of Graphene Compounds

Graphene is synthesized using either top-down or bottom-up processes (Fig. 7.2). The former includes mechanical exfoliation, electrochemical exfoliation, or reduction of graphene oxide. These processes, although easy and cost-effective, pose difficulty in managing the quality of the output such as the quantity of the individual layers of graphene sheets. The latter include organic synthesis, epitaxial growth, or chemical vapor deposition. These methods result in better quality products but are expensive to maintain in comparison.

- Mechanical exfoliation was the primary method used for the discovery of graphene. This is done by peeling off a highly oriented pyrolytic graphite using scotch tape. The challenge in such exfoliation is the lack of control over having a desired output. Hence, researchers have been working on optimizing this route to synthesize single-layer graphene with a significant structural quality (Moosa and Abed 2021).
- Chemical vapor deposition is a process that utilizes a substrate upon which an element in a gaseous form is deposited. For a long time, it has stood as a scalable and economical method for generating high-quality graphene films, making it the foremost choice for large-scale graphene production (Saeed et al. 2020). This method is preferred when it comes to graphene's electronic applications. However, achieving precise control over the thickness is challenging.
- Electrochemical exfoliation involves applying a certain voltage to electrolytes to chemically release their ions (such as sulfates) and integrate them into the interlayer of graphite rods to break them into graphene sheets (Achee et al. 2018).
- Epitaxial growth of graphene mainly requires a silicon carbide substrate. This substrate is subjected to a high temperature during which thermal decomposition of the silicon atoms takes place. This leaves the carbon atoms to diffuse and grow into and graphene honeycomb lattice (Zebardastan et al. 2023). It is mainly beneficial to the semiconductor industry as the graphene is directly grown on the intended substrate.

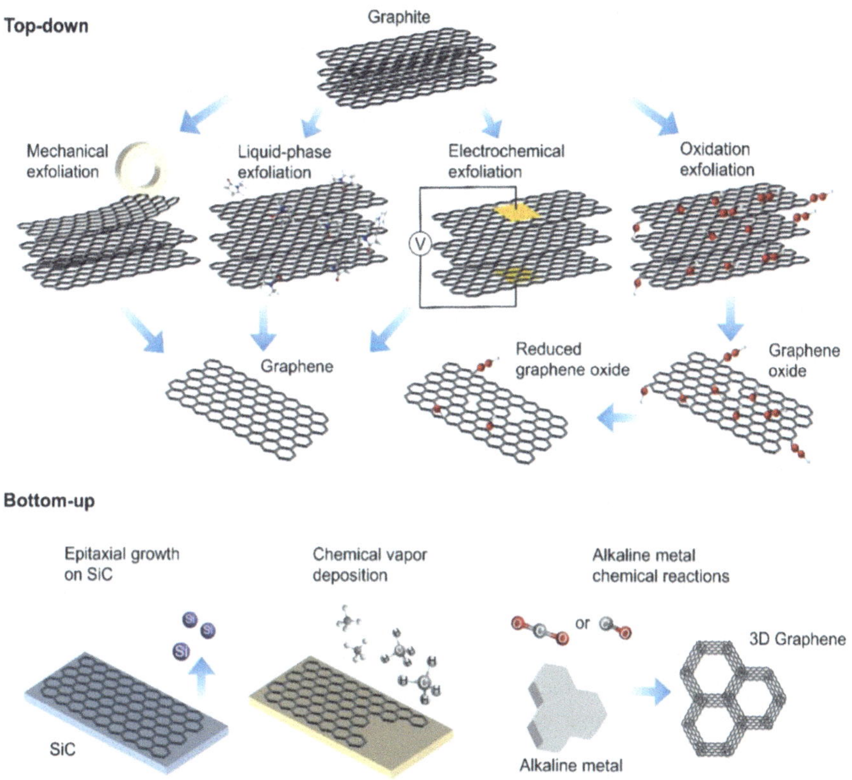

Fig. 7.2 Process routes of the two synthesis methods of graphene. (Reproduced under the terms of the CC-BY license (Su and Hu 2021) © 2020 The Authors. Energy Science & Engineering published by the Society of Chemical Industry and John Wiley & Sons Ltd)

Although graphene's derivates (GO, rGO, GQDs) could share similar synthesis routes as graphene itself, they have their dedicated methods that relate to their properties.

GO is an intercalated and oxidized derivative of graphite sheets. The conventional method to produce GO is the modified Hummer's method. A carbon source (usually graphite flakes or powders) is mixed with a protonated solvent (such as sulfuric acid, phosphoric acid, or a combination of these) and a strong oxidizing agent (potassium permanganate) is introduced. The solvent will be the intercalating agent and the latter will induce the formation of the functional groups. After diluting with deionized water, hydrogen peroxide is added to the resulting mixture to remove any metal ions that remain from the oxidizer. The final product is obtained by repeated centrifugation and sonication processes.

To synthesize rGO, chemical, thermal, or electrochemical means can be used as reducing mechanisms for GO. The following specific methods can be utilized to obtain rGO (Smith et al. 2019):

(a) chemical reduction of GO using hydrazine, metal hydrides, or hydrohalic acids
(b) photocatalytic reaction of GO under UV light and the presence of a catalyst
(c) exposure to high-intensity pulsed light produced from xenon flash tubes
(d) heat treatment of GO using urea as an expansion-reduction agent
(e) thermal treatment of GO in a furnace at high temperatures

The chemical reduction of GO involves the use of harmful chemicals and this has been a cause for concern. Fortunately, there are greener alternatives to solve this issue and will be discussed in a later sub-section.

GQDs, as quantum derivatives of graphene, are formed through top-down or bottom-up procedures as well. The top-down method includes, for instance, the exfoliation of carbon into GQDs that are nanosized using methods like electron beam lithography and liquid exfoliation. Conversely, the bottom-up techniques include growing molecular components, such as polymers or graphene-like aromatic hydrocarbons, into GQDs by hydrothermal, microwave-assisted hydrothermal, and metal-catalyzed methods (Tian et al. 2018).

Graphene is capable of creating composites, as are its derivatives, for various applications. Graphene and a chosen component can be bonded by covalent or noncovalent connections through a variety of chemical processes. Covalent connections are attained by nucleophilic, cycloaddition, condensation, and electrophilic reactions. Whereas noncovalent connections are made possible through π–π bonding, electrostatic interactions, hydrogen bonding, and ionic interactions (Karthik et al. 2021; Kumar et al. 2023). Through these means, graphene can be functionalized with many polymers, such as chitosan, gelatin, and polyethylene glycol (PEG), to mention a few.

7.2.4 Applications of Graphene Compounds

The areas of application of graphene and its compounds are vast. They exhibit exceptional conductivity, fast electron transfer capabilities, efficient heat transfer, and adsorption capabilities. As a result, they find widespread application in catalysis improvement, energy generation and storage enhancement, and material synthesis.

For instance, in fuel cells, the many anchoring sites found on the surfaces of graphene-containing materials attract active metal particles ensuring uniform dispersion. This phenomenon coupled with the large surface area and superb electrical conductivity, enhances electrochemical reactions such as fuel oxidation and oxygen reduction reaction (Su and Hu 2021). Such higher activity prompts the immobilization of the typical platinum catalyst used in the anode of the fuel cell and improves its stability.

The atomic-level architecture of graphene and its composites cultivated their use as active materials in electrical components including supercapacitors, electrochemical batteries, and Lithium-ion capacitors (Wang et al. 2019). Their potential in this field is immense, signaled by their affordability, abundance, diverse allotropes and transformations, as well as exceptional physical and chemical stability.

Graphene and its family have come as versatile materials with wide-ranging applications, particularly in the realm of optical devices. Their utility extends to, various sensors, photodetectors, LEDs, and an array of other functionalities. Beyond optics, graphene-based materials find use in temperature and gas sensors, transducers, and even carbon storage and conversion technologies (Abbas et al. 2022). These diverse applications underscore the multifaceted potential of graphene and its derivatives across various scientific and technological domains. Graphene nanocomposites find extensive application across various environmental sectors as well, showcasing their versatility and potential impact. Membranes incorporating graphene nanocomposites demonstrate remarkable efficiency in detecting and purifying contaminants from wastewater, offering a sustainable solution to water treatment challenges. They are considered one of the most vital adsorbent materials for water pollutants among carbonaceous materials since they are endowed with unique characteristics. Organic pollutants such as phenolic compounds, dyes, and the like are considered to pose significant dangers to the human body. These compounds contain hydroxyl functional groups connected to their aromatic rings (Mohamad Said et al. 2021). This allows them to easily dissolve in water and hence are considered as major water contaminants and their presence in drinking water is very much restricted by law. Nonetheless, these compounds find their way to water supplies. Runoffs from chemical industries, domestic and municipal wastewater, and day-to-day consumables such as detergents and disinfectants contribute to the existence of phenols in water (Ladeia Ramos et al. 2024). Whereas, paper, textile, leather treatment, and paint industries release significant amounts of detrimental dyes into the water environment (Al-Tohamy et al. 2022). These heavily affect microorganisms and general life in water, in addition to that of humans. Moreover, the situation is exacerbated due to the fact that conventional water treatment processes fail to remove these compounds. Here, the many advantages of graphene-based materials, including the numerous active sites, proven chemical stability, extended surface area, and large delocalized π-electron system, come into effect. One of the researches made in this regard is that of Shen et al. In their work, a magnetic diazonium functionalized-reduced graphene oxide was synthesized with adsorptive properties to eliminate 4-chlorophenol and 2,4-dichlorophenol in an aqueous mixture (Shen et al. 2018). The intermolecular π—π interactions generated from the reduced graphene oxide structure were instrumental in the adsorption process of these phenolic pollutants. The external magnetic field helped to easily separate the compound from the solution.

A 3D network of graphene oxide/chitosan composite was synthesized by Lai et al. for adsorption purposes (Lai et al. 2020). They reported that the composite proved to be an effective adsorbent for the synthetic RB5 dye. The adsorption capacity was seen to be 638.93 mg/g. This high adsorption was possible through the $\pi - \pi$ arrangement of graphene oxide and the electrostatic interaction between the parts.

As can be gathered from the above studies, graphene and its family of materials engage in a multitude of applications. This extends to another vital area—the field of biomedicine, and the following section is dedicated to it.

7.3 Carbonaceous Materials in Biomedical Applications

Biomedicine is a broad, multifaceted sector that spans various disciplines such as:

(a) Material chemistry that deals with the compounds that make up the individual materials, their synthesis methods, and behaviors,
(b) Material sciences that relate to the composition, properties, and requirements of the biomaterials,
(c) Medical engineering that is concerned with medical devices.

Such fields and more must converge to bring the intended application into fruition.

There are different materials used for biomedical applications. Generally, they could be classified as metals, ceramics, polymers, or composites. As their applications are numerous, so are their elemental forms. This section discusses the usage of carbonaceous biomaterials.

Carbonaceous materials have emerged as versatile platforms in the realm of biomedical applications given their unique physicochemical properties and biocompatibility. They offer various opportunities for addressing challenges in healthcare through their qualities in terms of thermal, electrical, mechanical, and morphological properties. Their tunable surface chemistry, exceptional mechanical properties, and ability to resemble the extracellular matrix are the reasons why they are good choices for several biomedical applications.

One of the most notable uses of carbonaceous biomaterials is in tissue engineering and regenerative medicine. Scaffolds composed of carbon nanomaterials offer a biomimetic environment that supports cell adhesion, proliferation, and differentiation. These scaffolds can be designed to handle the mechanical exertions imposed on native tissues while acting as a structural support for the process of tissue regeneration. Furthermore, the electrical conductivity of certain carbonaceous materials can modulate cellular behavior and facilitate the integration of electronic devices for real-time monitoring and control of tissue growth.

Apart from the above, carbon-containing materials are also important in the field of drug delivery. For instance, carbon nanotubes and graphene oxide are regarded as promising for their ability to encapsulate therapeutic agents and deliver them to specific targets with high precision. Carbon Quantum Dots, nano-diamonds, and fullerenes have a large surface area-to-volume ratio. This facilitates efficient drug loading and makes it possible for their surfaces to be coated with different components such as peptides, antibodies, and aptamers for detecting cancer cells (Singh et al. 2021). Their ability to be functionalized prompts the attachment of targeting ligands, allowing selective drug delivery to diseased tissues while not affecting healthy cells. Along with this, their biocompatibility ensures minimal adverse effects on biological systems.

These materials also exhibit remarkable optical properties, making them valuable tools for various biomedical imaging modalities. Carbon dots, for instance, are excellent contrast agents for fluorescence imaging and photothermal therapy. Their small size, low toxicity, and high photostability make them ideal candidates for tracking biological processes at the cellular and molecular levels. Additionally, carbon-based nanoprobes can be functionalized with targeting ligands to enable specific imaging of diseased tissues, offering clinicians valuable insights for diagnosis and treatment planning. Through these unique optical properties, researchers are advancing the frontier of biomedical imaging and paving the way for more accurate and personalized healthcare interventions.

The following subtopics will be dedicated to highlighting the importance of carbon-based materials in the biomedicine field; specifically tissue engineering, drug delivery, biosensing, and smart materials.

7.3.1 Tissue Engineering

Tissue engineering encompasses a multifaceted realm of scientific and engineering endeavors, connecting the disciplines of biomedicine, materials science, and engineering. This interdisciplinary approach involves understanding biological processes, developing suitable biomaterials, and employing engineering principles to fabricate tissues as well as organs. It involves the use of scaffolds, made from biomaterials, to provide an environment that mimics the extracellular matrix of the body and provides support for cells to adhere and grow. The process is summarized in Fig. 7.3 (Karkan et al. 2019).

Scaffolds are porous three-dimensional structures that are designed with hollow pathways and interconnected substructures that allow cell adhesion, growth, and differentiation. Since they are foreign, they are expected to degrade after a suitable length of time, biologically, giving way to the growth of new tissues. Tissue engineering is thus concerned with the appropriation of the best-suited biomaterials for the development of scaffolds. The selection of biomaterials is highly dependent on the mechanical properties of tissues, as mentioned previously. It is expected that the mechanical strength of the scaffold and that of the tissue ready for modeling should be close enough, whether in healthy or pathological conditions. The

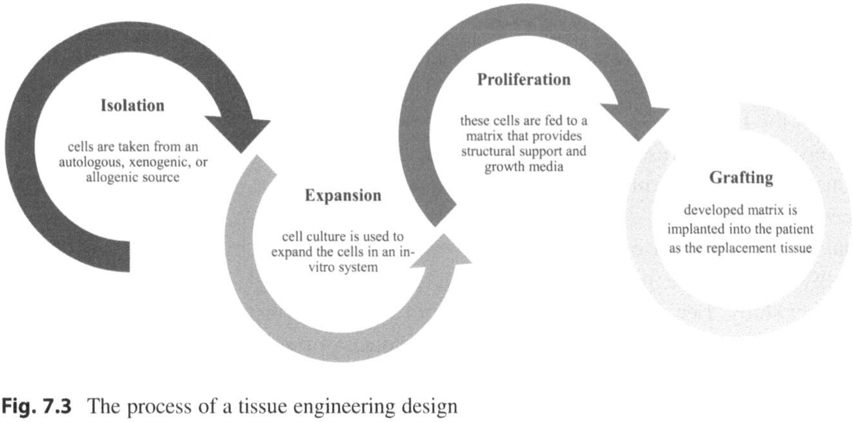

Fig. 7.3 The process of a tissue engineering design

material's surface is the primary gateway for contact with cells and hence guides the cells' behavior and fate (Caddeo et al. 2017).

Carbonaceous materials offer multiple benefits for tissue engineering. The proven excellent electrical conductivity of these materials can enhance electrical stimulation to culturing cells. This phenomenon positively impacts cell proliferation and osteogenic activity in tissues. In addition, their enhanced electrochemical properties induce self-sensing activities in the cell-culture medium (Islam et al. 2022). The intended use of carbon-based materials in three of the most notable tissue engineering fields is discussed here in detail.

A. **Bone tissue engineering**

Bone is naturally capable of healing itself after a certain injury. Most of the damage incurred to bones can be healed. However, this self-healing proves to be insufficient when the level of injury is above the threshold. This could be due to severe trauma, chronic diseases, or pathological fractures. Bone's intrinsic ability to heal also degrades due to aging. In such cases, bone tissue regeneration methods come into effect. The most preferred and practiced standard so far is bone grafting (mostly autologous) (Orciani et al. 2017; Putra et al. 2024). However, this comes with its consequences as it is a complex operation. In the case of autologous grafting, patients who had such treatments could experience morbidity at the graft harvest site and there would be a general limited availability of tissue grafts. In the case of allografts and xenografts, there is the risk of rejection by the receiver and lack of osteogeneses and vascularization even if the procedure is successful.

Thus, bone tissue engineering introduced substitutes to bone grafts in the form of scaffolds. These are designed to resemble the physical attributes of the natural bone and facilitate the growth of new tissues. A scaffold designed for such a purpose must fulfill several requirements such as (Qu et al. 2019):

(a) biocompatibility that allows cell attachment and that proves non-toxicity,
(b) biodegradability or bioresorbability within the intended safe period,
(c) interconnected porosity for proper nutrient and waste movement,
(d) surface chemistry that mimics that of the applied area,
(e) adequate mechanical strength to be able to handle the exerted weight.

One of the biomaterials that are highly preferred for such properties is carbon-based materials. Several researchers have attained ways of using carbonaceous materials in bone tissue engineering outputs. For instance, single-layer graphene was found to accelerate osteogenic differentiation, both in vitro and in vivo, of human mesenchymal stem cells (Liu et al. 2016). This was mainly done by upregulating H3K4 methylation at the promoter regions of genes associated with osteogenesis. In another study, Lee et al. described that their reduced graphene oxide/hydroxyapatite nanocomposites showed an enhanced osteoinductive potential in an in-vitro model (Lee et al. 2015). The model used cells from the murine preosteoblastic cell line which showed successful osteogenic proliferation. Their in-vivo tests also exhibited the formation of new bone in full thickness from a defect of calvarium specimens.

Multi-walled carbon nanotubes offer structural reinforcement and controlled release capabilities, aiding in tasks like regulating cell proliferation in biomedicine. Whereas single-walled carbon nanotubes demonstrate superior performance in precision targeting and cellular imaging applications (Kim et al. 2024). Taale et al. produced 3D porous scaffolds from bioactive glass and hydroxyapatite nanoparticles integrated with carbon nanotubes (Taale et al. 2018). The tunable hollow microtubule structure of the scaffolds resulted in an enhanced ion release and electrical conductivity, whereas the added surface area provided a platform for osteoblasts and fibroblasts adhesion.

Mxenes are another class of carbon-based materials with tunable optical and electrical properties. Several studies are being conducted on their applications in biomedicine. Diedkova et al. showed that a few layers of MXene deposited on polycaprolactone electrospun nanofibers resulted in an optimum environment that promoted cellular attachment and proliferation with a mild antibacterial effect. This composite displayed good conductivity across a range of temperatures, which is a testimonial to its potentiality as a conductive biomaterial (Diedkova et al. 2023).

B. **Cardiac tissue engineering**

In this tissue engineering field, there is an essential link between cells and scaffolds, biomimicking the native myocardial environment to facilitate cell adhesion, proliferation, and maturation. Biomaterials serve as the cornerstone, providing structural support while fostering cellular interactions vital for tissue development. From natural polymers like collagen, gelatin, and fibrin to synthetic counterparts such as polylactide, polyethylene glycol, and polycaprolactone, the selection of biomaterials is tailored to meet the specific mechanical, biological, and electrical properties requisite for functional cardiac tissue regeneration (Mohammadi Nasr et al. 2020).

Biomaterials containing carbon are effectively used in this field as well. Małgorzata et al. used scaffolds of graphene oxide and reduced graphene oxide for cardiac tissue regeneration on mesenchymal stem cells (Sekuła-Stryjewska et al. 2021). The graphene oxide sheets with large flake surfaces and low reduced graphene oxide presented high biocompatibility and were found not to affect the stem cells proliferation. This was possible without the introduction of cytokines and growth factors to stimulate the cell culture. Their work also showed that both materials promoted myocardial and angiogenic differentiation. This ascertained the value of graphene-based materials to heart tissue regeneration by advancing the regenerative potential of mesenchymal stem cells. Smith et al. showed the successful use of hybrid scaffolds of graphene-PEG in repairing cardiac tissues (Smith et al. 2017). The topography of the scaffold surfaces together with their anisotropic electrical conductivity promoted the structural properties of myofibrils and sarcomeres. On top of that, the existence of graphene nanomaterials improved the protein expression, calcium ion transfer, cell morphology, and duration of action potential. In another study, silk nanofibrous biomaterials modified with reduced graphene oxide particles were produced to regenerate cardiac tissues (Zhao et al. 2018). The method used was effective in controlling the electrical conductivity of the biomaterial and its fibrous morphology for the intended outcome. The reduced graphene oxide was applied as an ultrathin coating over the surface of the silk which resulted in a high conductivity. This in turn enhanced the cardiac-specific spreading of cells, protein expression, sarcomere structures formation, and junctions in the cell-cell gap. Moreover, the new morphology did not affect the flexibility and topological integrity of the original matrices.

C. **Neural Tissue engineering**

In this discipline, the main aspect is to treat injuries to the nervous system, mainly to the axon bundles. After a traumatic brain injury, the axons in the central nervous system are extremely difficult to mend, while those in the peripheral nervous system have a comparatively better regeneration recovery, based on the extent of the damage (Doblado et al. 2021). This aspect is even more severe since it is hard to treat the injuries using conventional medications. This is where the tissue regeneration technology comes into effect. In neural tissue engineering, the prevailing approach involves applying biocompatible 3D biomaterials, along with cells and bioactive molecules, to rebuild damaged tissues. This methodology prioritizes the restoration of tissue integrity while endeavoring to preserve the original anatomical structure and functionality as closely as feasible. This endeavor thus focuses on the design and engineering of biomimetic scaffolds that can handle the physical and functional complexity of native neural tissue, providing a conducive environment for cell adhesion, growth, and differentiation. Moreover, studies in the biology of stem cells and advanced tissue culture techniques have opened up new avenues for generating patient-specific neural cells for transplantation, advancing the potential of neural tissue engineering to enhance clinical treatments for ailments such as spinal cord injury, stroke, and neurodegenerative diseases (Boni et al. 2018).

Carbonaceous biomaterials have emerged as promising candidates in neural tissue engineering because of their excellent biocompatibility, tunable mechanical properties, and conductance. Graphene, carbon nanotubes, and conductive polymers offer opportunities to mimic the biochemical and biomechanical cues of native neural tissue. Moreover, their ability to conduct electrical signals enables seamless integration with the electrical activity of the nervous system, prompting their potential use in neural interfaces and bioelectronic devices. This electrical conductivity/stimulation makes them capable of promoting neuron differentiation and the development of longer neurites. As a testimony, neurons cultured on graphene and carbon nanotube substrates exhibited the capability to establish close connections between their proximal regions and distal regions (Cellot et al. 2022; Villanueva-Flores et al. 2023). They also demonstrated heightened synaptic frequencies in the process. In another study, graphene oxide nano-scaffolds combined with polycaprolactone demonstrated excellent functional and morphological recovery of damaged nerve tissue (Qian et al. 2018). The conductive property of GO coupled with the stiff nature of the polycaprolactone helped in utilizing essential cues of the design which contributed to a successful nerve guidance aspect, much needed for peripheral nerve tissue engineering.

7.3.2 Biosensing

Biosensing and biosensors represent pivotal elements in the landscape of biomedical technology, offering innovative solutions across a spectrum of applications. Biosensing involves the detection and analysis of biological elements or processes, providing crucial insights into health, disease, and environmental factors. Biosensors, sophisticated devices designed to detect biological analytes, play a central role in this field. They operate by converting biological cues into readable output signals in the form of electrical, optical, or others, enabling real-time monitoring and diagnostics (Haleem et al. 2021). In a typical setting, as shown in Fig. 7.4, they include Analyte, Bioreceptor, Transducer, Electronics, and Reader Display (Tripathy et al. 2021).

In biomedical contexts, biosensors facilitate early disease detection, personalized medicine, and precise drug delivery, revolutionizing healthcare practices. This has been particularly evident in the areas of cardiovascular diseases, diabetes, cancer as well as wound healing procedures (Shafiee et al. 2018). With advancements in nanotechnology, microfluidics, and bioinformatics, biosensors continue to evolve, promising enhanced sensitivity, specificity, and portability for a wide array of biomedical applications (Chuang 2023).

Carbonaceous materials have shown promising applications as biosensors. Graphene, for example, can serve as a transducer by forming a discernible signal by transforming the link between a receptor and target molecules (Joshi et al. 2021). Its high surface area and ability to be surface functionalized make it ideal for such a purpose. The functional groups on a modified graphene's surface allow the attachment of various molecules. These essentials could be proteins, carbohydrates, DNA, aptamers, and polymers. Whereas enzymes can be immobilized onto graphene surfaces through electrostatic interactions (between the enzyme's negative charge

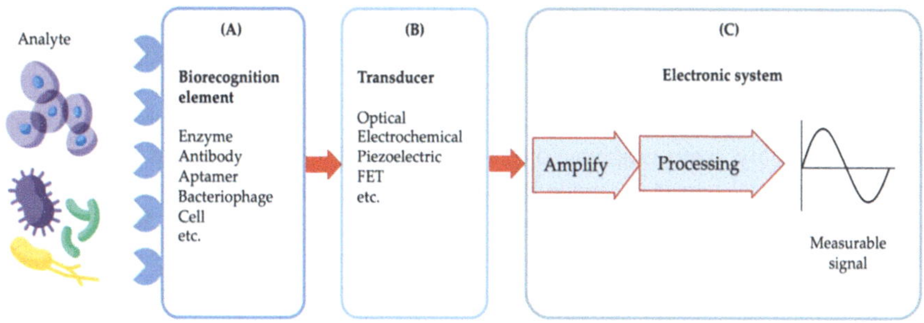

Fig. 7.4 Typical setting of a biosensor; the analyte and the three main components. (Reproduced under the terms of the open access CC-BY license (Awang et al. 2021))

and the opposite on the surface of graphene), covalent bonds (between graphene and the enzymatic molecules), or through entrapping the enzyme (using another polymer such as a polysaccharide) (Du et al. 2012). Liu et al. were able to produce a fluorescent biosensor by employing a colloidal interaction of graphene and RNA-cleaving molecule termed DNAzyme (a DNA oligonucleotide with an ability to specifically detect bacteria) through non-covalent interactions (Liu et al. 2018). The result was a suitable selective detection system for *E. coli* bacterial pathogens. Esmail et al. synthesized a nanocomposite sensor consisting of nitrogen-doped carbon nano-onions mixed with gold nanoparticles (Sohouli et al. 2020). They used this composite to detect acetaminophen both in aqueous and biological samples. They reported that the sensor showed a sufficient result in the electrochemical determination of the drug.

7.3.3 Drug Delivery

Currently, several researchers are working on methodologies for controlled methods of drug delivery to carry and deploy disease-fighting elements right at target sites in the body. Carbon-based materials are also chosen here, mainly, as composites.

For example, graphene oxide or reduced graphene oxide monolayers are considered here since their surface area is sufficient for drug encapsulation. The electrons on the surface of nanographene enable the formation of π–π bonds with diverse aromatic drug molecules. Following this interaction, the aforementioned oxidized graphene surfaces can be functionalized to attach chosen molecules, facilitating the selective delivery of drugs to specific cellular targets. This approach holds promise for enhancing drug efficacy while minimizing the possibilities of applying off-target.

Xie et al. reported on a magnetic graphene oxide nanocomposite modified by chitosan/sodium alginate to load the drug doxorubicin hydrochloride for killing cancer cells (Xie et al. 2019). It was found that the loaded nanocomposites had improved dispersion and drug release ability with pH-sensitive components. In addition, the functionalized graphene oxide nanosheets had an enhanced loading capacity through the magnetization feature. Similarly, Pan et al. synthesized a heat-sensitive drug carrier attached to graphene sheets. Poly (N-isopropyl acrylamide), was grafted to the graphene oxide surface chemically through π–π and hydrophobic connections, ensuring a very stable composite (Pan et al. 2011). This composite was used to load camptothecin, a water-insoluble anticancer drug, and it showed a successful tumor-cell-killing ability under physiological conditions without harming cells. Comparably, nano-graphene oxide combined with PLA-PEG was used to load the drug paclitaxel in a study made by Angelopoulou et al. (2015). The copolymers played the role of stabilizing the graphene oxide aqueous dispersion. Whereas, the effective loading and the control on the release of the drug was possible due to graphene oxide's π electrons on the sp^2 carbon configuration and those in the aromatic rings of paclitaxel. Their in-vitro test resulted in the intended cytotoxicity of the composite against the cancer cells in culture.

Lin et al. used another carbon group. They utilized fullerenes to delay the release of drugs through self-assembly (Lin et al. 2017). Amphiphilic functionalized fullerenes ($C_{60}R_5Cl$) were assembled into spherical hollow vesicles to control anticancer drugs release for; fluorouracil, cyclophosphamide, and cisplatin. By combining the structural properties and functionalization capability of fullerenes, it was able to control the rate of drug release.

Carbon dots are also a different class of carbon that is revolutionizing the drug delivery sector. By utilizing their ability for surface modifications, several agents can be loaded onto their functional groups. Through this combination of carbon dots with antimicrobial agents, some of the drawbacks of the free agents such as poor water solubility can be solved. Antiviral, antifungal, and antibacterial drugs such as Metronidazole, Penicillin, Amphotericin B, and locked nucleic acid have been loaded onto the carbon dots and presented good results (Wang et al. 2022).

7.3.4 Smart Biomaterials

With the growing trend of materials science and engineering, technological might has presented the world of living biomaterials. Researchers have been able to synthesize materials that respond in a certain way upon introduction to physical, chemical, or biological triggers (Amukarimi et al. 2021). These external triggers or stimuli could be light/photo, electrical, ultrasound, magnetic field stimulation, or biological. Each of these parameters can be modified and fine-tuned to control their level or strength, ultimately achieving the desired cellular reaction (Gelmi and Schutt 2021). This is a highly sought-after behavior especially in the biomedicine sector, since the need for control over the impact of applying biomaterials in the body is crucial. Carbon nanomaterials can be tailored to deploy intended drugs in response to specific light wavelengths, ensuring targeted therapy with minimal side effects. Similarly, pH-sensitive carbon-based materials enable drug release in acidic environments typical of certain disease sites, enhancing therapeutic efficacy while minimizing systemic toxicity. Moreover, temperature-responsive carbon-based scaffolds offer opportunities for regulating cellular behavior and promoting tissue regeneration in a controlled manner.

Nano-graphene oxide composed of a redox-responsive, detachable polyethylene glycol shell was formed by Wen et al. for tumor-selective drug delivery (Wen et al. 2012). The cross-linkage was made via a disulfide linkage which reacts to the specific stimulus of a redox environment. This PEG-nano graphene oxide was developed for a fast release of an encapsulated drug, doxorubicin hydrochloride, at certain levels of tumor-relevant glutathione. The study revealed that the controlled delivery significantly enhanced the effectiveness of treatments on cancerous cells.

Self-assembly and self-healing is also another avenue of smart materials. Self-assembly is an elaborate technique in which disorganized units are arranged to assume an organized structure. The organized fragments pertain to distinct molecular attributes, including electrostatic, geometric, and photonic properties. These molecular or atomic interactions encompass various weak noncovalent forces (Rajapaksha

2020). Since healing incorrect bonds and growing ordered domains would require reversible interactions between building units, the aforementioned noncovalent forces play a vital role (Talapin et al. 2020).

Here again, carbon-based materials are considered as part of the emerging technologies. Self-healing carbon-based hydrogels have shown promise to be used as injectable carriers for controlled delivery of drugs, allowing for sustained release of therapeutic agents and improved treatment outcomes. In biosensing applications, self-healing carbon-based materials can be integrated into wearable or implantable sensors for real-time monitoring of physiological parameters, such as glucose levels or biomarker concentrations. Furthermore, self-healing coatings based on carbon materials can increase the biocompatibility and longevity of implantable devices, reducing the feared risk of infection and implant failure.

The self-healing properties of graphene-containing materials stem from the unique atomic structure and bonding configurations that they possess. The sp^2-hybridized carbon atoms in graphene sheets form robust covalent bonds, endowing the material with exceptional mechanical strength and stability. Upon encountering structural defects or fractures, these covalent bonds facilitate the migration of carbon atoms, enabling the material to autonomously repair damage at the atomic scale (Özçelik et al. 2013). Diffusion-driven mechanisms, such as vacancy migration and Stone-Wales transformations, play pivotal roles in facilitating the reformation of bonds and the restoration of material integrity. Additionally, functional groups and dopants in compounds containing graphene derivatives can further enhance self-healing capabilities by promoting cross-linking and facilitating dynamic bonding interactions. Self-healing graphene composites offer exciting prospects for developing implantable devices and tissue scaffolds with prolonged lifespans and improved biocompatibility.

7.4 Biomaterials and Sustainability

Biomaterials have a very extended utilization in today's world. The main reason is the simple fact of the global population rise. With an increased number of humans in today's technological and capitalist age, come a multitude of health issues that need material intervention. It is almost impossible to think of a functioning society without the involvement of biomaterials in one way or the other.

Biomaterials can be classified as natural, synthetic, or composite as shown in Fig. 7.5 below. Biologically driven ones include chitosan, collagen, gelatin, cellulose, carboxymethyl cellulose, and alginate. These can also be classified as natural/biopolymers. Metallic biomaterials are considered to be inert and could be degradable (magnesium, iron, and their alloys) or non-biodegradable (stainless steel, titanium, and cobalt-chromium alloys) (Morsiya 2022). Ceramic biomaterials are mainly produced from metallic elements linked with non-metallic elements through covalent or ionic bonds (Chong et al. 2023). Whereas polymeric biomaterials are divided into natural and synthetic.

The following are the desired properties of biomaterials:

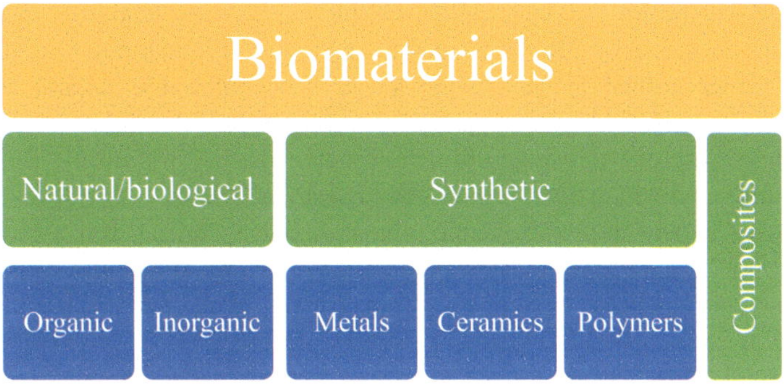

Fig. 7.5 Classifications of biomaterials

- Stable mechanical properties: for biomaterials needed to support a certain load in the human body, they will need to perform in various conditions and for a prolonged time.
- Bioactivity: the materials should be geared to induce a response from the biological system such as tissue uptake, metabolism, or physiological response
- Biocompatibility: must be non-toxic, non-inflammatory, and corrosion resistant. In addition, they need to facilitate the adhesion and propagation of cells in the desired part of the system of the body.
- Biodegradability: biomaterials are expected to degrade on time, to nontoxic and easily metabolized units that can be removed from the body. The biomaterial's degradation time should match the regeneration or healing process of the target element in the body to ensure the required result (Song et al. 2018).

To achieve these characteristics, there are many production sources and routes. However, most of these techniques come with significant challenges. Either in terms of the capacity it requires to produce or the effect that it will have on the environment, the increasing need for biomaterials is met with a growing concern. For instance, once they are used up to their intended life span, biomaterials will either be replaced or discarded. These consist of implants, scaffolds, dental appliances, catheters, prostheses, and stents. Most of these end up in landfills which leads to their degradation that will result in water and soil pollution. Otherwise, they are incinerated as medical waste, another means of atmospheric pollution (Joseph et al. 2021; Wiśniewska et al. 2023). On the other hand, there is the issue of synthetic polymers that make up the majority of today's global biomaterials usage. Synthetic polymers derived from petro-materials. The petro-polymer/plastics industry is encountering mounting challenges for its future trajectory. These include the depletion of fossil fuels, escalating costs of petroleum, and the enduringly calamitous effect on the environment (Zhang et al. 2023).

Thus, the concerns regarding the in-time degradability and sustainability of their synthesis and sourcing prompted a pressing need for eco-friendly alternatives. Responding to this need, researchers have tuned in to renewable, natural biomaterials as promising candidates for biomedical applications. Studies are increasingly focusing on the sustainability aspect of biomaterial synthesis. The world of sustainability in this aspect deals with renewable natural resources maintained through biodegradation and recycling approaches (Arif et al. 2023). Among these resources, residues of crops/plants, marine organisms, and animal waste are the most widely due to their abundance (Li 2019).

One of the widely used sustainable biomaterials is cellulose. It is a biopolymer that contains D-glucose units that are connected by β-glycosidic bonds. Its individual components are repetitive units of cellobiose (Zhang et al. 2023). It is synthesized from several sources including plants, bacteria, or animals. Plant cellulose has a structure with a high crystallinity and can be a source of nanocellulose using chemical or mechanical decomposition. This can be taken as one of the top-down approaches to obtaining cellulose biomaterials. Whereas, an example of a viable bottom-up approach would be the growth of bacterial cellulose by a Gram-negative bacterial strain such as the Acetobacter (Komagataeibacter) xylinum.

7.4.1 Sustainability in Graphene-Based Materials

Sustainable carbon materials are generally categorized into the following (Kumar Gupta et al. 2016).

(a) graphitic nanostructures where graphene and its derivatives as well as other carbon nanostructures are categorized;
(b) carbon materials synthesized from deep eutectic solvents (such as choline chloride), usually used to produce porous and hierarchical carbon;
(c) 'Starbons' derived from polysaccharides such as starch and chitosan, which are used to form xero- or aerogels which in turn will be used to synthesize 'starbonised' porous carbon;
(d) biopolymer-derived hydrothermal and chiral carbon materials including chitin and oligosaccharides

This section focuses on the aspects and progress of the category: graphitic carbon structures. It is beneficial to deal with such materials because of their comparatively easy means of being synthesized from sustainable sources. A wide variety of plant extracts, agro-residues, and bio/food waste can be used as carbon precursors, which makes the process renewable as well.

Liu et al. reported a scalable production of highly porous nitrogen-doped carbon nanosheets by pyrolyzing chitosan and urea precursors eliminating the need for a metal catalyst (Liu et al. 2014). Ammonia that is released as the urea decomposes triggers nitrogen doping to the amorphous carbon gained from chitosan, which, in turn, crystallizes as the temperature is raised. The large surface area combined with the interconnected nanostructure of the synthesized nanosized sheets indicated the possible usage for electronic capacitors or hydrogen storage.

Wood waste, an abundantly found source, was converted to graphene materials using the processes of ball milling and a vacuum method. It was then further treated by direct heating and catalytic pyrolysis at a low temperature (Liou and Huang 2013). Whereas, Chen et al. described the formation of few-layer graphene using the common agricultural waste, wheat straw (Chen et al. 2016). Potassium hydroxide was used as an activation agent and pristine graphene sheets were produced through hydrothermal treatment that followed. These sheets were found to have very good electrochemical properties which makes the product efficient for lithium ion batteries. In another study, Wang et al. used the plant *Plumeria rubra* to synthesize high-quality porous graphene nanosheets (Wang et al. 2018). Similar to the aforementioned research, potassium hydroxide was used for activation. The activation reaction brought the exfoliation of the precursor as gases were released. The sample was then calcined for a desired degree of graphitization. The result exhibited a heightened photocatalysis activity which could be used to split water. Tade et al. produced graphene quantum dots using bamboo fibers (Tade and Patil 2020). These quantum dots were formed in a single, eco-friendly hydrothermal method and were used in the process of selective quantitation for curcumin.

Kraft lignin was employed as a precursor by Li et al. (2019). To perform catalytic graphitization, Iron (III) nitrate was used. Then, the formed lignin graphite was transformed into lignin graphene oxide by using the modified Hummer's method. It was found that the nanoscale lateral size and the oxidation reaction time exhibited by this conversion were evidently lower than that of graphene oxide synthesized from the conventional natural graphite. On the other hand, a research by Ahmad et al. was able to synthesize reduced graphene oxide using different algal strains as reducers (Ahmad et al. 2019). This eliminated the use of hazardous chemicals that are conventionally used to reduce graphene oxide to reduced graphene oxide, highlighting the applicability of green chemistry that was mentioned previously. The product was utilized as an efficient adsorbent and decontaminating agent for the elimination of the heavy metals copper and lead from aqueous media. In our team's previous work, we used spent coffee grounds as a natural biomass source to synthesize graphene oxide (Challa et al. 2023). The washed coffee grounds were soaked in an iron chloride salt for graphitization purposes. The iron component was removed by washing it with a concentrated acid and the sample was then subjected to carbonization in a tube furnace. The graphitized carbon char was then treated with the modified Hummer's method to obtain graphene oxide. This was used to produce electrospun cellulose acetate/graphene oxide scaffolds used for bone tissue regeneration purposes. The results showed integrated nano-fibers with embedded GO particles as shown in Fig. 7.6. Subsequent tests proved that the scaffold was mechanically sound and biocompatible.

By utilizing the ubiquitous nature of the abundance of carbon, the possibilities of producing carbon-based materials are numerous. However, sustainability goes beyond sustainable sourcing. It also relates to the synthesis routes. Green chemistry principles can be applied to develop eco-friendly synthesis routes that minimize hazardous byproducts and solvent usage. For instance, aqueous-based methods using non-toxic reagents and mild reaction conditions offer a greener alternative to

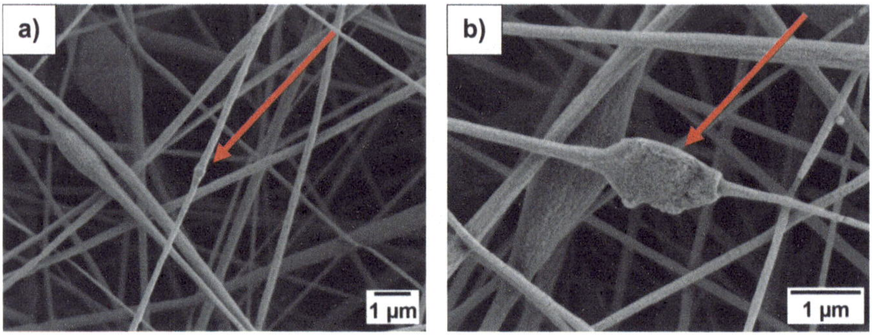

Fig. 7.6 SEM images of cellulose acetate/GO nanofibers where GO nanoparticles are shown with the arrows ((**a**) and (**b**) are images taken at two localities)

traditional organic solvents and harsh chemical treatments. One such method used the exfoliation of the flower petals of lotus and hibiscus plants to produce graphene and nickel-decorated graphene (Ray et al. 2012). This took place by soaking the precursors in nickel catalyst and subsequent heating at 1600 °C. The method avoided the use of hazardous chemicals and, hence proved to be eco-friendly and efficient for large-scale synthesis. Similarly, Babhuiya et al. obtained 'Flash graphene' via Flash Joule heating (Barbhuiya et al. 2021). Municipal waste was converted to Flash graphene using a high-voltage discharge with an instantaneous result. This method did not require any solvents, furnaces, or reactive gases, eliminating the need for harsh chemical treatments.

7.5 Conclusions and Futuristic Aspects of Carbon-Based Biomaterials

Graphene-based carbonaceous materials are endowed with superb characteristics as has been mentioned in the above sections. While progress has been made in nanotechnology, there are still hurdles in precisely tailoring the physical and chemical properties, breakdown in the body (biodegradation), potential harm to living cells (cytotoxicity) during animal testing, and responsiveness to external cues (stimuli-responsiveness) of carbon-based biomaterials designed for medical use. The problems associated with size-, shape-, environment-, cell-, and performance-dependent toxicity remain pressing issues that require comprehensive investigation and resolution. The complexity of navigating these challenges is compounded by issues related to large-scale production and standardization, further emphasizing the need for meticulous attention to detail throughout the development process.

Additionally, there are concerns regarding whether nano-sized graphene-based materials might harm cells, damage genetic material, or cause cancer when placed inside the body (physiological environments), particularly within the heart muscle. Understanding the prolonged effects of these materials, both in their pristine form and after degradation, is crucial for assessing their safety and efficacy in biomedical applications.

Incorporating insights from various sources, it becomes evident that addressing these challenges requires interdisciplinary collaboration and rigorous scientific investigation. Unleashing the true potential of graphene-based biomaterials in medicine requires a collaborative effort, prioritizing both patient safety and the effectiveness of these materials.

Acknowledgments This work is supported by the Internal Grant Agency (IGA) project registered as IGA/CPS/2024/005 provided by the Center of Polymer Systems, Tomas Bata University in Zlin, Czech Republic. The first author would also like to acknowledge the support provided by the Ministry of Education, Youth and Sports, Czech Republic, for foreign nationals, registered under the reference number MSMT-44726/2013.

References

Abbas Q, Shinde PA, Abdelkareem MA, Alami AH, Mirzaeian M, Yadav A, Olabi AG (2022) Graphene synthesis techniques and environmental applications. Materials (Basel) 15(21). https://doi.org/10.3390/ma15217804

Achee TC, Sun W, Hope JT, Quitzau SG, Sweeney CB, Shah SA, Habib T, Green MJ (2018) High-yield scalable graphene nanosheet production from compressed graphite using electrochemical exfoliation. Sci Rep 8(1):14525. https://doi.org/10.1038/s41598-018-32741-3

Ahmad S, Ahmad A, Khan S, Ahmad S, Khan I, Zada S, Fu P (2019) Algal extracts based biogenic synthesis of reduced graphene oxides (rGO) with enhanced heavy metals adsorption capability. J Ind Eng Chem 72:117–124. https://doi.org/10.1016/j.jiec.2018.12.009

Aliyev E, Filiz V, Khan MM, Lee YJ, Abetz C, Abetz V (2019) Structural characterization of graphene oxide: surface functional groups and fractionated oxidative debris. Nanomaterials (Basel) 9(8). https://doi.org/10.3390/nano9081180

Al-Tohamy R, Ali SS, Li F, Okasha KM, Mahmoud YAG, Elsamahy T, Jiao H, Fu Y, Sun J (2022) A critical review on the treatment of dye-containing wastewater: ecotoxicological and health concerns of textile dyes and possible remediation approaches for environmental safety. Ecotoxicol Environ Saf 231:113160. https://doi.org/10.1016/j.ecoenv.2021.113160

Amukarimi S, Ramakrishna S, Mozafari M (2021) Smart biomaterials—a proposed definition and overview of the field. Curr Opin Biomed Eng 19:100311. https://doi.org/10.1016/j.cobme.2021.100311

Angelopoulou A, Voulgari E, Diamanti EK, Gournis D, Avgoustakis K (2015) Graphene oxide stabilized by PLA–PEG copolymers for the controlled delivery of paclitaxel. Eur J Pharm Biopharm 93:18–26. https://doi.org/10.1016/j.ejpb.2015.03.022

Arif ZU, Khalid MY, Noroozi R, Hossain M, Shi HH, Tariq A, Ramakrishna S, Umer R (2023) Additive manufacturing of sustainable biomaterials for biomedical applications. Asian J Pharm Sci 18(3):100812. https://doi.org/10.1016/j.ajps.2023.100812

Avouris P, Dimitrakopoulos C (2012) Graphene: synthesis and applications. Mater Today 15(3): 86–97. https://doi.org/10.1016/S1369-7021(12)70044-5

Awang MS, Bustami Y, Hamzah HH, Zambry NS, Najib MA, Khalid MF, Aziah I, Abd Manaf A (2021) Advancement in salmonella detection methods: from conventional to electrochemical-based sensing detection. Biosensors 11(9):346. https://www.mdpi.com/2079-6374/11/9/346

Barbhuiya NH, Kumar A, Singh A, Chandel MK, Arnusch CJ, Tour JM, Singh SP (2021) The future of flash graphene for the sustainable management of solid waste. ACS Nano 15(10): 15461–15470. https://doi.org/10.1021/acsnano.1c07571

Biomaterials Market by Type, Application, and Region – Global Forecast to 2029 (2024) Markets and Markets. https://www.marketsandmarkets.com/Market-Reports/biomaterials-393.html
Biomaterials Market Size, Share, & Growth Analysis Report (2023) Precedence Research Pvt. Ltd. https://www.precedenceresearch.com/biomaterials-market
Boni R, Ali A, Shavandi A, Clarkson AN (2018) Current and novel polymeric biomaterials for neural tissue engineering. J Biomed Sci 25(1):90. https://doi.org/10.1186/s12929-018-0491-8
Caddeo S, Boffito M, Sartori S (2017) Tissue engineering approaches in the design of healthy and pathological in vitro tissue models [review]. Front Bioeng Biotechnol 5. https://doi.org/10.3389/fbioe.2017.00040
Cellot G, Franceschi Biagioni A, Ballerini L (2022) Nanomedicine and graphene-based materials: advanced technologies for potential treatments of diseases in the developing nervous system. Pediatr Res 92(1):71–79. https://doi.org/10.1038/s41390-021-01681-6
Challa AA, Saha N, Szewczyk PK, Karbowniczek JE, Stachewicz U, Ngwabebhoh FA, Saha P (2023) Graphene oxide produced from spent coffee grounds in electrospun cellulose acetate scaffolds for tissue engineering applications. Mater Today Commun 35:105974. https://doi.org/10.1016/j.mtcomm.2023.105974
Chen F, Yang J, Bai T, Long B, Zhou X (2016) Facile synthesis of few-layer graphene from biomass waste and its application in lithium-ion batteries. J Electroanal Chem 768:18–26. https://doi.org/10.1016/j.jelechem.2016.02.035
Chong ETJ, Ng JW, Lee P-C (2023) Classification and medical applications of biomaterials–a mini review. BIOI 4(2):54–61. https://doi.org/10.15212/bioi-2022-0009
Chuang H-S (2023) Editorial: advanced biosensing Technologies in Medical Applications. Biosensors 13(1):11. https://www.mdpi.com/2079-6374/13/1/11
Deretzis I, La Magna A (2011) Coherent electron transport in quasi-one-dimensional carbon-based systems. Eur Phys J B 81(1):15–36. https://doi.org/10.1140/epjb/e2011-20134-x
Diedkova K, Pogrebnjak AD, Kyrylenko S, Smyrnova K, Buranich VV, Horodek P, Zukowski P, Koltunowicz TN, Galaszkiewicz P, Makashina K, Bondariev V, Sahul M, Čaplovičová M, Husak Y, Simka W, Korniienko V, Stolarczyk A, Blacha-Grzechnik A, Balitskyi V et al (2023) Polycaprolactone–MXene Nanofibrous scaffolds for tissue engineering. ACS Appl Mater Interfaces 15(11):14033–14047. https://doi.org/10.1021/acsami.2c22780
Doblado LR, Martínez-Ramos C, Pradas MM (2021) Biomaterials for neural tissue engineering [review]. Front Nanotechnol 3. https://doi.org/10.3389/fnano.2021.643507
Du D, Yang Y, Lin Y (2012) Graphene-based materials for biosensing and bioimaging. MRS Bull 37(12):1290–1296. https://doi.org/10.1557/mrs.2012.209
Enoki T, Fujii S, Takai K (2012) Zigzag and armchair edges in graphene. Carbon 50(9):3141–3145. https://doi.org/10.1016/j.carbon.2011.10.004
Gelmi A, Schutt CE (2021) Stimuli-responsive biomaterials: scaffolds for stem cell control. Adv Healthc Mater 10(1):2001125. https://doi.org/10.1002/adhm.202001125
Ghosal K, Mondal P, Bera S, Ghosh S (2021) Graphene family nanomaterials- opportunities and challenges in tissue engineering applications. FlatChem 30:100315. https://doi.org/10.1016/j.flatc.2021.100315
Gürsel H (2020) What is the difference between graphene oxide and reduced graphene oxide? Blografi. https://nanografi.com/blog/what-is-the-difference-between-graphene-oxide-and-reduced-graphene-oxide/
Haleem A, Javaid M, Singh RP, Suman R, Rab S (2021) Biosensors applications in medical field: a brief review. Sens Int 2:100100. https://doi.org/10.1016/j.sintl.2021.100100
Islam M, Lantada AD, Mager D, Korvink JG (2022) Carbon-based materials for articular tissue engineering: from innovative scaffolding materials toward engineered living carbon. Adv Healthc Mater 11(1):2101834. https://doi.org/10.1002/adhm.202101834
Joseph B, James J, Kalarikkal N, Thomas S (2021) Recycling of medical plastics. Adv Industrial Eng Polym Res 4(3):199–208. https://doi.org/10.1016/j.aiepr.2021.06.003
Joshi P, Mishra R, Narayan RJ (2021) Biosensing applications of carbon-based materials. Curr Opin Biomed Eng 18:100274. https://doi.org/10.1016/j.cobme.2021.100274

Karkan SF, Davaran S, Rahbarghazi R, Salehi R, Akbarzadeh A (2019) Electrospun nanofibers for the fabrication of engineered vascular grafts. J Biol Eng 13(1):83. https://doi.org/10.1186/s13036-019-0199-7

Karthik V, Selvakumar P, Senthil Kumar P, Vo D-VN, Gokulakrishnan M, Keerthana P, Tamil Elakkiya V, Rajeswari R (2021) Graphene-based materials for environmental applications: a review. Environ Chem Lett 19(5):3631–3644. https://doi.org/10.1007/s10311-021-01262-3

Kim C-H, Lee S-Y, Rhee KY, Park S-J (2024) Carbon-based composites in biomedical applications: a comprehensive review of properties, applications, and future directions. Adv Compos Hybrid Mater 7(2):55. https://doi.org/10.1007/s42114-024-00846-1

Kumar Gupta G, De S, Franco A, Balu AM, Luque R (2016) Sustainable biomaterials: current trends, challenges, and applications. Molecules 21(1):48. https://www.mdpi.com/1420-3049/21/1/48

Kumar R, Singh DP, Muñoz R, Amami M, Singh RK, Singh S, Kumar V (2023) Graphene-based materials for biotechnological and biomedical applications: drug delivery, bioimaging and biosensing. Mater Today Chem 33:101750. https://doi.org/10.1016/j.mtchem.2023.101750

Ladeia Ramos R, Rezende Moreira V, Santos Amaral MC (2024) Phenolic compounds in water: review of occurrence, risk, and retention by membrane technology. J Environ Manag 351: 119772. https://doi.org/10.1016/j.jenvman.2023.119772

Lai KC, Lee LY, Hiew BYZ, Yang TC-K, Pan G-T, Thangalazhy-Gopakumar S, Gan S (2020) Utilization of eco-friendly and low cost 3D graphene-based composite for treatment of aqueous reactive black 5 dye: characterisation, adsorption mechanism, and recyclability studies. J Taiwan Inst Chem Eng 114:57–66. https://doi.org/10.1016/j.jtice.2020.09.024

Lee JH, Shin YC, Lee S-M, Jin OS, Kang SH, Hong SW, Jeong C-M, Huh JB, Han D-W (2015) Enhanced Osteogenesis by reduced graphene oxide/hydroxyapatite nanocomposites. Sci Rep 5(1):18833. https://doi.org/10.1038/srep18833

Li Y-CE (2019) Sustainable biomass materials for biomedical applications. ACS Biomater Sci Eng 5(5):2079–2092. https://doi.org/10.1021/acsbiomaterials.8b01634

Li J, Yan Q, Zhang X, Zhang J, Cai Z (2019) Efficient conversion of lignin waste to high-value bio-graphene oxide nanomaterials. Polymers 11(4):623. https://www.mdpi.com/2073-4360/11/4/623

Lin M-S, Chen R-T, Yu N-Y, Sun L-C, Liu Y, Cui C-H, Xie S-Y, Huang R-B, Zheng L-S (2017) Fullerene-based amino acid ester chlorides self-assembled as spherical nano-vesicles for drug delayed release. Colloids Surf B: Biointerfaces 159:613–619. https://doi.org/10.1016/j.colsurfb.2017.08.007

Liou Y-J, Huang W-J (2013) Quantitative analysis of graphene sheet content in wood char powders during catalytic pyrolysis. J Mater Sci Technol 29(5):406–410. https://doi.org/10.1016/j.jmst.2013.03.008

Liu Q, Duan Y, Zhao Q, Pan F, Zhang B, Zhang J (2014) Direct synthesis of nitrogen-doped carbon Nanosheets with high surface area and excellent oxygen reduction performance. Langmuir 30(27):8238–8245. https://doi.org/10.1021/la404995y

Liu Y, Chen T, Du F, Gu M, Zhang P, Zhang X, Liu J, Longwei L, Xiong C, Zhou Y (2016) Single-layer graphene enhances the osteogenic differentiation of human mesenchymal stem cells in vitro and in vivo. J Biomed Nanotechnol 12(6):1270–1284. https://doi.org/10.1166/jbn.2016.2254

Liu M, Zhang Q, Brennan JD, Li Y (2018) Graphene-DNAzyme-based fluorescent biosensor for Escherichia coli detection. MRS Communications 8(3):687–694. https://doi.org/10.1557/mrc.2018.97

Mohamad Said KA, Ismail AF, Abdul Karim Z, Abdullah MS, Hafeez A (2021) A review of technologies for the phenolic compounds recovery and phenol removal from wastewater. Process Saf Environ Prot 151:257–289. https://doi.org/10.1016/j.psep.2021.05.015

Mohammadi Nasr S, Rabiee N, Hajebi S, Ahmadi S, Fatahi Y, Hosseini M, Bagherzadeh M, Ghadiri AM, Rabiee M, Jajarmi V, Webster TJ (2020) Biodegradable nanopolymers in cardiac tissue engineering: from concept towards Nanomedicine. Int J Nanomedicine 15:4205–4224. https://doi.org/10.2147/ijn.S245936

Moosa AA, Abed MS (2021) Graphene preparation and graphite exfoliation. Turk J Chem 45(3): 493–519. https://doi.org/10.3906/kim-2101-19

Morsiya C (2022) A review on parameters affecting properties of biomaterial SS 316L. Aust J Mech Eng 20(3):803–813. https://doi.org/10.1080/14484846.2020.1752975

Orciani M, Fini M, Di Primio R, Mattioli-Belmonte M (2017) Biofabrication and bone tissue regeneration: cell source, approaches, and challenges [review]. Front Bioeng Biotechnol 5. https://doi.org/10.3389/fbioe.2017.00017

Özçelik VO, Gurel HH, Ciraci S (2013) Self-healing of vacancy defects in single-layer graphene and silicene. Phys Rev B 88(4):045440. https://doi.org/10.1103/PhysRevB.88.045440

Pan Y, Bao H, Sahoo NG, Wu T, Li L (2011) Water-soluble poly(N-isopropylacrylamide)–graphene sheets synthesized via click chemistry for drug delivery. Adv Funct Mater 21(14): 2754–2763. https://doi.org/10.1002/adfm.201100078

Putra NE, Zhou J, Zadpoor AA (2024) Sustainable sources of raw materials for additive manufacturing of bone-substituting biomaterials. Adv Healthc Mater 13(1):2301837. https://doi.org/10.1002/adhm.202301837

Qian Y, Song J, Zhao X, Chen W, Ouyang Y, Yuan W, Fan C (2018) 3D fabrication with integration molding of a graphene oxide/polycaprolactone nanoscaffold for neurite regeneration and angiogenesis. Adv Sci 5(4):1700499. https://doi.org/10.1002/advs.201700499

Qu H, Fu H, Han Z, Sun Y (2019) Biomaterials for bone tissue engineering scaffolds: a review. RSC Advances 9(45):26252–26262. https://doi.org/10.1039/C9RA05214C

Rajapaksha RDAA (2020) 7 – Self-assembling smart materials for biomaterials applications. In: Bouhfid R, Qaiss AEK, Jawaid M (eds) Polymer nanocomposite-based smart materials. Woodhead Publishing, Duxford, pp 121–147. https://doi.org/10.1016/B978-0-08-103013-4.00007-8

Ray AK, Sahu RK, Rajinikanth V, Bapari H, Ghosh M, Paul P (2012) Preparation and characterization of graphene and Ni-decorated graphene using flower petals as the precursor material. Carbon 50(11):4123–4129. https://doi.org/10.1016/j.carbon.2012.04.060

Saeed M, Alshammari Y, Majeed SA, Al-Nasrallah E (2020) Chemical vapour deposition of graphene-synthesis, characterisation, and applications: a review. Molecules 25(17). https://doi.org/10.3390/molecules25173856

Samantara AK, Ratha S, Raj S (2019) Chapter 4 – Functionalized graphene nanocomposites in air filtration applications. In: Jawaid M, Bouhfid R, Kacem Qaiss AE (eds) Functionalized graphene nanocomposites and their derivatives. Elsevier, pp 65–89. https://doi.org/10.1016/B978-0-12-814548-7.00004-0

Sekuła-Stryjewska M, Noga S, Dźwigońska M, Adamczyk E, Karnas E, Jagiełło J, Szkaradek A, Chytrosz P, Boruczkowski D, Madeja Z, Kotarba A, Lipińska L, Zuba-Surma EK (2021) Graphene-based materials enhance cardiomyogenic and angiogenic differentiation capacity of human mesenchymal stem cells in vitro – focus on cardiac tissue regeneration. Mater Sci Eng C 119:111614. https://doi.org/10.1016/j.msec.2020.111614

Shafiee A, Ghadiri E, Kassis J, Pourhabibi Zarandi N, Atala A (2018) Biosensing technologies for medical applications, manufacturing, and regenerative medicine. Curr Stem Cell Rep 4(2): 105–115. https://doi.org/10.1007/s40778-018-0123-y

Shen X, Chen X, Sun D, Wu T, Li Y (2018) Fabrication of a magnetite/diazonium functionalized-reduced graphene oxide hybrid as an easily regenerated adsorbent for efficient removal of chlorophenols from aqueous solution. RSC Advances 8(14):7351–7360. https://doi.org/10.1039/C8RA00503F

Singh G, Kaur H, Sharma A, Singh J, Alajangi HK, Kumar S, Singla N, Kaur IP, Barnwal RP (2021) Carbon based nanodots in early diagnosis of cancer [mini review]. Front Chem 9. https://doi.org/10.3389/fchem.2021.669169

Smith AST, Yoo H, Yi H, Ahn EH, Lee JH, Shao G, Nagornyak E, Laflamme MA, Murry CE, Kim D-H (2017) Micro- and nano-patterned conductive graphene–PEG hybrid scaffolds for cardiac tissue engineering. Chem Commun 53(53):7412–7415. https://doi.org/10.1039/C7CC01988B

Smith AT, LaChance AM, Zeng S, Liu B, Sun L (2019) Synthesis, properties, and applications of graphene oxide/reduced graphene oxide and their nanocomposites. Nano Mater Sci 1(1):31–47. https://doi.org/10.1016/j.nanoms.2019.02.004

Sohouli E, Shahdost-Fard F, Rahimi-Nasrabadi M, Plonska-Brzezinska ME, Ahmadi F (2020) Introducing a novel nanocomposite consisting of nitrogen-doped carbon nano-onions and gold nanoparticles for the electrochemical sensor to measure acetaminophen. J Electroanal Chem 871:114309. https://doi.org/10.1016/j.jelechem.2020.114309

Song R, Murphy M, Li C, Ting K, Soo C, Zheng Z (2018) Current development of biodegradable polymeric materials for biomedical applications. Drug Des Devel Ther 12:3117–3145. https://doi.org/10.2147/dddt.S165440

Su H, Hu YH (2021) Recent advances in graphene-based materials for fuel cell applications. Energy Sci Eng 9(7):958–983. https://doi.org/10.1002/ese3.833

Taale M, Schütt F, Zheng K, Mishra YK, Boccaccini AR, Adelung R, Selhuber-Unkel C (2018) Bioactive carbon-based hybrid 3D scaffolds for osteoblast growth. ACS Appl Mater Interfaces 10(50):43874–43886. https://doi.org/10.1021/acsami.8b13631

Tade RS, Patil PO (2020) Green synthesis of fluorescent graphene quantum dots and its application in selective curcumin detection. Curr Appl Phys 20(11):1226–1236. https://doi.org/10.1016/j.cap.2020.08.006

Talapin DV, Engel M, Braun PV (2020) Functional materials and devices by self-assembly. MRS Bull 45(10):799–806. https://doi.org/10.1557/mrs.2020.252

Tian P, Tang L, Teng KS, Lau SP (2018) Graphene quantum dots from chemistry to applications. Mater Today Chem 10:221–258. https://doi.org/10.1016/j.mtchem.2018.09.007

Tripathy A, Nine MJ, Silva FS (2021) Biosensing platform on ferrite magnetic nanoparticles: synthesis, functionalization, mechanism and applications. Adv Colloid Interf Sci 290:102380. https://doi.org/10.1016/j.cis.2021.102380

Villanueva-Flores F, Garcia-Atutxa I, Santos A, Armendariz-Borunda J (2023) Toward a new generation of bio-scaffolds for neural tissue engineering: challenges and perspectives. Pharmaceutics 15(6). https://doi.org/10.3390/pharmaceutics15061750

Wang Y, Song T, Zhang P, Huang T, Wang T, Wang T, Zeng H (2018) Gas-exfoliation assisted fabrication of porous graphene nanosheets derived from Plumeria rubra for highly efficient photocatalytic hydrogen evolution. ACS Sustain Chem Eng 6(9):11536–11546. https://doi.org/10.1021/acssuschemeng.8b01723

Wang X, Liu L, Niu Z (2019) Carbon-based materials for lithium-ion capacitors. Mater Chem Front 3(7):1265–1279. https://doi.org/10.1039/C9QM00062C

Wang Z-X, Wang Z, Wu F-G (2022) Carbon dots as drug delivery vehicles for antimicrobial applications: a minireview. ChemMedChem 17(13):e202200003. https://doi.org/10.1002/cmdc.202200003

Wen H, Dong C, Dong H, Shen A, Xia W, Cai X, Song Y, Li X, Li Y, Shi D (2012) Engineered redox-responsive PEG detachment mechanism in PEGylated nano-graphene oxide for intracellular drug delivery. Small 8(5):760–769. https://doi.org/10.1002/smll.201101613

Wiśniewska P, Saeb MR, Bencherif SA (2023) Biomaterials recycling: a promising pathway to sustainability [mini review]. Front Biomater Sci 2. https://doi.org/10.3389/fbiom.2023.1260402

Xie M, Zhang F, Peng H, Zhang Y, Li Y, Xu Y, Xie J (2019) Layer-by-layer modification of magnetic graphene oxide by chitosan and sodium alginate with enhanced dispersibility for targeted drug delivery and photothermal therapy. Colloids Surf B: Biointerfaces 176:462–470. https://doi.org/10.1016/j.colsurfb.2019.01.028

Yang G, Li L, Lee WB, Ng MC (2018) Structure of graphene and its disorders: a review. Sci Technol Adv Mater 19(1):613–648. https://doi.org/10.1080/14686996.2018.1494493

Zebardastan N, Bradford J, Lipton-Duffin J, MacLeod J, Ostrikov K, Tomellini M, Motta N (2023) High quality epitaxial graphene on 4H-SiC by face-to-face growth in ultra-high vacuum. Nanotechnology 34(10):105601. https://doi.org/10.1088/1361-6528/aca8b2

Zhang Y, Poon K, Masonsong GSP, Ramaswamy Y, Singh G (2023) Sustainable nanomaterials for biomedical applications. Pharmaceutics 15(3):922. https://www.mdpi.com/1999-4923/15/3/922

Zhao G, Qing H, Huang G, Genin GM, Lu TJ, Luo Z, Xu F, Zhang X (2018) Reduced graphene oxide functionalized nanofibrous silk fibroin matrices for engineering excitable tissues. NPG Asia Mater 10(10):982–994. https://doi.org/10.1038/s41427-018-0092-8

Zhao C, Song X, Liu Y, Fu Y, Ye L, Wang N, Wang F, Li L, Mohammadniaei M, Zhang M, Zhang Q, Liu J (2020) Synthesis of graphene quantum dots and their applications in drug delivery. J Nanobiotechnol 18(1):142. https://doi.org/10.1186/s12951-020-00698-z

Zhu Y, Murali S, Cai W, Li X, Suk JW, Potts JR, Ruoff RS (2010) Graphene and graphene oxide: synthesis, properties, and applications. Adv Mater 22(35):3906–3924. https://doi.org/10.1002/adma.201001068

Adam Aberra Challa is a doctoral student of Biomaterials and Biocomposites at the Center of Polymer Systems, Tomas Bata University (TBU) in Zlin, Czech Republic. He received his B.-Sc. degree in Civil Engineering and his M.SC. in Structural Engineering, both from the Addis Ababa Institute of Technology, Addis Ababa University, Addis Ababa, Ethiopia. He is currently working on developing biocomposites for bone tissue engineering applications. His research is based on graphene-based biomaterial composites synthesized from natural sources such as agricultural biomass. Adam studied structural engineering for his master's degree. As a structural engineer, he worked in a design office in his home city, Addis Ababa, Ethiopia where he was involved in building design projects, review of structural designs, and investigation of building materials for quality applications. In addition, before his Ph.D. studies in the Czech Republic, he worked as a lecturer at the Addis Ababa Institute of Technology. He lectured on courses related to structural analysis, design, building systems, and materials in construction. Adam is a proud Rotarian and a member of Toastmasters International.

Dr. Nabanita Saha is a biotechnologist, who received her B.Sc. and M.Sc. degree in "Life Sciences (Botany)" from Visva Bharati University, Santiniketan, India, and was awarded a PhD in "Microbial Biotechnology" from the Indian Institute of Technology, Kharagpur. She received her habilitation 'Associate Professor Degree' in "Technology of Macromolecular Substances" in 2006 from Tomas Bata University in Zlin, Czech Republic. For the last 22 years, she has been working at Tomas Bata University in Zlin, Czech Republic, on biomaterials and biocomposites/bio-based polymeric materials. Dr. Nabanita Saha currently holds an Associate Professor (Faculty of Technology) and Senior Researcher (Centre of Polymer Systems & Footwear Research Centre) position at the same university. Her research group mainly focused their research on the preparation and characterization of bio-based biomaterials (in the form of gel and hydrogels and the production of bacterial cellulose) for health care and commodity applications. She is the author/co-author of more than 89 papers published in international peer-reviewed scientific journals (recorded in Scopus and Web of Science database). In 2012, her research work was nominated for ENI Award-2013 competition on 'Protection of the Environment'. She was associated with e-COST, MP1301-NEW GEN project (2014–2017) as a working group (WG) member, CA15216-ENBA project (2016–2020) as an MC member, CA17107-CONTEXT project (2019–2022) as a WG member. She is/was involved in many, internal/national and international projects running at the Centre of Polymer Systems, TBU in Zlin. She is a member of the Polymer Processing Society (PPS) as well associated with the SPE-European Medical Polymer Division since 2009 as a Board member.

Prof. Petr Saha is the founder of Tomas Bata University in Zlín, Czech Republic, where he served as Rector for four terms of office. He currently holds the post of Director of the University Institute. His expertise includes energy and composite materials. His professional focus is on energy materials, polymer processes and medical polymers. Completed his studies at the Faculty of Technology of the Brno University of Technology, where he also worked. He was employed at Chalmers University of Technology in Sweden for more than 10 years. Moreover, he held the office of Chairman of the Czech Rectors' Conference. Prof. Sáha is also active on the international scene at a number of universities (East China University of Science and Technology, Clemson University v USA, Ton Duc Thang University in Vietnam and the University of Palermo in Italy). Between 2005 and 2007, he held the post of President of the Polymer Processing Society located in the USA, where he focused on the growth of polymer science in the development world. He was instrumental in the establishment of Bata Centre in Vietnam, which supports cooperation between institutions in Vietnam and Czech and European universities. At present, Prof. Sáha is engaged in research into energy materials from renewable resources; his team is mainly focusing on the preparation of components for batteries and supercapacitors from renewable resources, which will replace critical raw materials in the future. Prof. Petr Sáha is the author and co-author of 435 publications (with more than 8000 citations on Web of Science and an H-index of 57 as of May 2024) and several dozen patents, utility models, prototypes and industrial designs.

Green Approach for Synthesizing Silk Fibroin Biomaterial Scaffolds

8

Shristy Verma, Rishabha Malviya, Lavanya Gupta, and Sathvik Belagodu Sridhar

Abstract

The protein called silk fibroin, which comes from the Bombyx mori silkworm, has exceptional mechanical, biodegradable, and biocompatibility properties that make it a great biomaterial for use in medical applications. Chemical-intensive methods that endanger human health and the environment are used to create silk. Green synthesis, a novel and environmentally friendly method, had to be developed for such a reason. Applications for silk fibroin include wound healing, tissue engineering, bone engineering, and many more. The advantages of green synthesis over traditional methods for creating silk fibroin biomaterial scaffolds for biomedical applications are explored in this chapter. Many techniques and alternatives to conventional techniques are covered, including the use of DES, enzymatic degumming, supercritical CO_2 fluid, and many more. In conclusion, the mentioned observations indicate that green synthesized silk fiber scaffolds, which are also more environmentally friendly, can be used to produce enhanced silk fiber (SF) for biomedical applications.

S. Verma
Department of Pharmacy, School of Medical and Allied Sciences, Galgotias University, Greater Noida, Uttar Pradesh, India

R. Malviya (✉)
Department of Pharmacy, Galgotias University, Greater Noida, Uttar Pradesh, India
e-mail: rishabha.malviya@galgotiasuniversity.edu.in

L. Gupta
Faculty of Biosciences, Heidelberg University, Heidelberg, Germany
e-mail: Lavanya.Gupta@medma.uni-heidelberg.de

S. B. Sridhar
RAK College of Pharmacy, RAK Medical & Health Sciences University, Ras Al Khaimah, United Arab Emirates
e-mail: sathvik@rakmhsu.ac.ae

R. Malviya, S. Sundram (eds.), *Sustainable Green Biomaterials As Drug Delivery Systems*, Biomaterials, Bioengineering and Sustainability 1,
https://doi.org/10.1007/978-3-031-79062-1_8

Keywords

Silk fibroin · Green · Degumming · Water annealing · Deep eutectic solvent

8.1 Introduction

Silk fibroin is a type of protein that comes from the silk of the Bombyx mori silkworm. It is an excellent biomaterial for medical applications because of its impressive mechanical properties, biocompatibility, and controlled biodegradability. These characteristics make it perfect for tissue engineering, especially for bone regeneration and skin repair. However, the traditional methods for processing and modifying silk fibroin often involve harsh chemicals and conditions that can be harmful to the environment and health. Therefore, there is a growing interest in developing more eco-friendly alternatives that are just as effective (Lu et al. 2010). It is necessary to advance the field in alignment with ecological and health standards. Three-dimensional porous matrices with robust mechanical characteristics support cell adhesion and expansion in tissue engineering (Grabska-Zielińska et al. 2021). The rate at which these scaffolds break down should also be similar to the growth of new tissue. Porous scaffolds promote cell adhesion and expansion and often require interconnected pores larger than 100 μm in diameter. Moreover, the scaffolds should be structurally sound to hold and release growth factors at a regulated pace to optimize cell functions and the development of particular tissues. Consequently, developing porous scaffolds in an environmentally friendly manner by avoiding harsh processing procedures and organic solvents would be a helpful strategy for maximizing the functionalization of these scaffolds for cell interactions and possible delivery of bioactive growth factors (Heseltine et al. 2022).

Biomedical sutures made of silk fibroin have been in use for many years. Because of silk fibroin's remarkable mechanical, biodegradable, and biocompatibility qualities, numerous other biomedical uses have been investigated in recent years. As biomaterial scaffolds, silks offer support matrices for various cell types, such as fibroblasts, osteoblasts, hepatocytes, stem cells, and ligaments. They also serve as scaffolding for tissue engineering applications, including bone, cartilage, blood vessels, ligaments, and cartilage. Additionally, silks can be used to create controlled release systems for drugs and growth factors. Scaffolds made of silk have been prepared using various techniques, including electrospinning, gas foaming, freeze-drying, and salt leaching (Huang et al. 2017; Liu et al. 2017). Nevertheless, none of the aforementioned techniques can eliminate the need for severe processing conditions, high salt concentrations, or organic solvents like methanol or hexafluoroisopropanol (HFIP). In some situations, scaffolds are needed to hold labile medications or bioactive substances that would otherwise break down or become denatured owing to any or all of the aforementioned factors. In these situations, it would be helpful to prevent these problems (Tahamtan et al. 2015).

Moreover, biocompatibility, tunable degradability, minimal toxicity, and mechanical characteristics make silk fibroin a promising biomaterial. To employ silk fibroin's natural qualities, proteins and detritus from native silkworm cocoons are removed to prevent inflammation. In purification and manufacturing, chemical solvents, energy-intensive machinery, and lots of water are utilized to reverse engineer silk fibroin into an aqueous solution and then the final substance. Greener extraction and manufacturing processes replace toxic chemical additives and energy-inefficient equipment while producing mechanically robust biomaterials (Nguyen et al. 2019; Goldberg and Kim 2023). Silkworms and spiders are the most studied silk sources. They have excellent mechanical qualities, biostability, and biodegradability and can be processed greenly. Spider silk is mechanically superior to silkworm silk, although silkworm cocoons have been harvested for ages due to their convenience, availability, and affordability (Smith and Doe 2021). Silkworm cocoons comprise mostly fibroin and sericin. Sericin is a globular gluelike protein that keeps fibroin fibers together. Due to their inflammatory effects, the two proteins are usually separated in biological applications (Rahimpour et al. 2023). Purified and sterilized silk fibroin is cytocompatible with poly(lactic acid) and collagen.

8.2 Properties of Silk Fibroin Scaffolds as a Biomedical Application

Silk fibroin is immunologically inert and low-inflammatory after sericin silk protein elimination (Diez-Echave et al. 2021). Silk fibroin is a disulfide-bonded copolymer with light and heavy chains, Mw = 26 and 390 kDa, respectively. The heavy chain has 12 big repeating hydrophobic amino acid domains—glycine, alanine, serine, valine, and tyrosine—separated by 11 short hydrophilic areas. Protein secondary structure depends on hydrophobic domains (Asakura 2021). Silk fibroin has two process-dependent secondary structures: silk I, an amorphous, water-soluble structure with random coils and α-helices, and silk II, characterized by at least 55% crystalline antiparallel β-sheet structures. Processing processes cause protein rearrangement from α-helix to β-sheet structures. Hydrogen bond creation allows recurring hydrophobic domains in the heavy chain segment to produce antiparallel β-sheet crystallites. Controlling the β-sheet composition can adjust the mechanical characteristics of silk fibroin materials. Increased β-sheet content leads to stronger material. The crystalline domains absorb impact pressure and distribute it along the fibroin network, giving silk fibroin elasticity at significant deformations. At minor deformations, the semi-amorphous matrix controls silk fibroin's toughness and extensibility due to its modulus, breaking strength, and elongation (Suzuki 2016; Hashimoto et al. 2020). Silk biodegradability is another benefit as shown in Fig. 8.1. A proteolytic enzyme breaks down silk fibroin into amino acids the body can absorb. Adjusting β-sheet content and processing factors can influence the degradation rate, similar to silk's mechanical qualities. Material degradation slows with higher β-sheet concentrations. Outside of material processing, the insertion site affects the degradation rate (Wang et al. 2021b).

The production of three-dimensional porous silk fibroin matrices with control over structural and morphological aspects is explained using a novel,

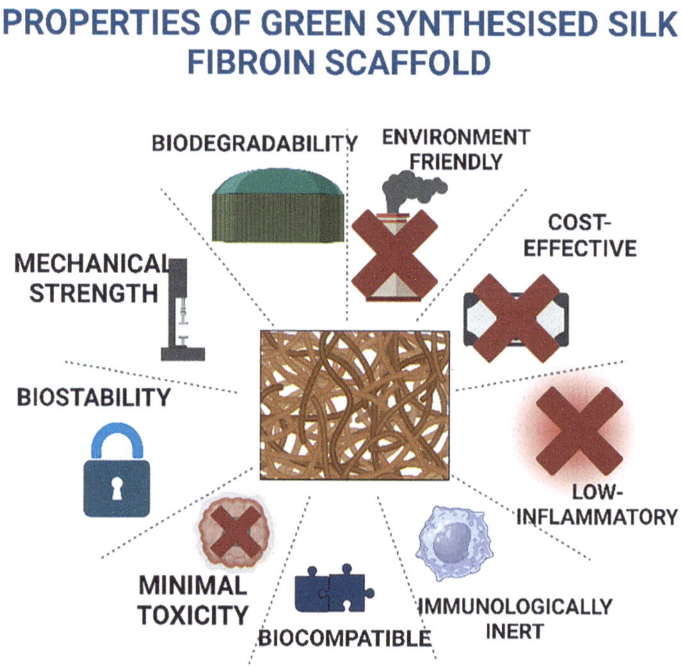

Fig. 8.1 Various properties of silk fibroin scaffold

environmentally friendly, all-aqueous method. Lyophilization is a method used to create scaffolds made of silk. To modify the conformation of silk fibroin and the interactions between silk fibroin and water, gelatin is added to the aqueous silk fibroin solution (Liu et al. 2021). This is done by modifying the hydrophilic interactions in silk fibroin-gelatin-water systems, which inhibits the formation of distinct sheet-like structures in the material and produces a more homogenous structure. By creating insolubility in the silk fibroin-gelatin scaffold system through water annealing, the need for organic solvents like methanol to lock in the beta-sheet structure is avoided (Wang et al. 2020; Rajput et al. 2020). The morphological and functional characteristics of the scaffolds can be controlled by varying the concentrations of silk fibroin and gelatin. Depending on the amount of silk fibroin utilized in the procedure, the scaffolds' interconnecting pores ranged in size from 100 to 300 micrometers (Liu et al. 2020). Due to the rise in gelatin content, the scaffolds' morphology altered at the same time from lamellar sheets to porous structures. When compared to scaffolds obtained from aqueous salt leaching and scaffolds derived from hexafluoroisopropanol (HFIP), these freeze-dried scaffolds exhibited a greater degree of hydrophilicity due to their reduced beta-sheet concentration. Without the burst release in the phosphate buffer saline solution, the majority of the gelatin was trapped in the silk fibroin-gelatin scaffolds. These silk fibroin-gelatin scaffolds were more cell-compatible during in vitro cell culture than the salt-leaching silk fibroin scaffolds. The following new method offers scaffold structures based on silk fibroin that are beneficial for tissue engineering applications. Moreover, the entire procedure is environmentally friendly, utilizing just water, ambient temperature, pressure, and no hazardous chemicals or solvents. It opens up new avenues for including growth hormones or bioactive medications in the process (Farokhi et al. 2020).

The two main crystalline structures of SF are called silk I and silk II. Silk I is a metastable crystalline structure that contains bound water molecules, while silk II is the most stable state because of strong hydrogen bonds between neighboring peptide blocks, which increase mechanical properties like tensile strength and stiffness (Qi et al. 2017). The crystalline and amorphous structures found in the secondary structure derived from regenerated silk fibroin (RSF) solutions will be covered in the sections that follow. Silk comprises β-turns (silk I) and insoluble structures made of folded β-sheets (silk II) when it has a crystalline structure, and α-helices, turns, and random coil structures when it is in an amorphous state. Silk I can be easily converted to silk II using methanol or potassium chloride; this process is widely used in biomaterial engineering applications. At the air-water interface of RSF solutions, Silk III is the unstable crystal structure of SF (Cho et al. 2015; Cebe et al. 2017; Jaramillo-Quiceno et al. 2017). To guarantee that RSF has good mechanical properties, numerous approaches have been tested. For instance, the breaking stress of RSF fibers produced by a dry-spinning method contrasts with the breaking stress of composite silk fibers made of graphene oxide and RSF. The mechanical qualities of silk scaffolds made from RSF can be enhanced, among other things, by crosslinking, porogens, and 3D bioprinting technologies. As a result, the resulting SF-based scaffolds are strong enough to be handled during the surgical procedures required for implantation (Valluzzi et al. 1999).

Furthermore, the ability of cells to adhere to scaffold surfaces and migrate into the scaffold while undergoing proliferation and differentiation within the scaffold is a critical component of successful scaffold implementation. This is made possible by biocompatibility. Furthermore, the scaffold must induce either no immune response at all or a very mild one following implantation (Zhang et al. 2016; Floren et al. 2016; Nazarov et al. 2004). Since SF is a naturally occurring polymer, it is known to be biocompatible and biologically inert. In in vivo experiments, SF has demonstrated blood compatibility since 1989. Successful scaffold materials should also have the ability to degrade and resorb biologically since during recovery, the patient's cells and extracellular matrix should gradually replace the scaffolds. It is crucial that biodegradation byproducts are non-toxic and do not affect other tissues, organs, or bodily functions during bodily metabolism (Gregory et al. 2016; Cao and Wang 2009). It has been demonstrated that SF, an enzymatically degradable polymer, does not elicit an immunogenic response. Enzymes are adsorbed onto the surface of the SF scaffold through surface-bonding domains, initiating the degradation process (Arai et al. 2004).

8.3 Various Approaches for Green Synthesis of Silk Fibroin Biomaterial Scaffolds

The silk fibroin synthesis process, traditionally reliant on chemical-intensive methods, has been increasingly adapted towards greener approaches due to environmental concerns and sustainability goals. Silk fibroin, a natural polymer obtained from the Bombyx mori silkworm, is highly valued in biomedical applications for its remarkable biocompatibility, strength, and biodegradability (DeBari et al. 2021). Traditional methods, however, often utilize harsh chemicals and processes not aligned with green chemistry principles. There are various green techniques for silk fibroin synthesis that not only reduce environmental impact but also enhance the material's applicability in medical fields. Degumming is the initial critical step in the processing of silk fibroin, where sericin, the sticky protein surrounding fibroin fibers, is removed. Traditional methods like sodium carbonate or soap solutions are effective but environmentally demanding, involving high water and energy consumption (Lujerdean et al. 2022). Using a freeze-drying approach, researchers created scaffolds with good porosity architectures by observing that collagen interacted with silk fibroin in an aqueous solution in some research, although the exact process remains unknown. Drugs might be included in these scaffolds and released at predetermined times at the same time. But to get the scaffolds ready to produce insolubility in water, methanol was needed. Researchers have demonstrated that a water-annealing technique can convert soluble silk films into water-stable silk films with a modified β-sheet composition. This method was not applied to the manufacture of porous scaffolds (Ansari and Sheikh 2023). The advantages and disadvantages of green synthesized silk fibroin biomaterial scaffold are summarised in Table 8.1.

Some researchers provided a novel approach for the production of 3D silk fibroin scaffolds that would be stable in aqueous systems and not require the use of harsh

Table 8.1 Various advantages and disadvantages of green synthesized silk fibroin biomaterial scaffold

S.no.	Green approach	Advantages	Disadvantages
1	Green degumming	Non-immunogenic	Expensive
2	Water annealing	Environment friendly	Low viscosity
3	Green lyophilisation	Biocompatible	Low production
4	Use of supercritical CO_2 fluid	Biodegradable	No data regarding clinical efficacy
5	Inducing secondary structures	Excellent mechanical properties	Cannot incorporate cells within
6	Enzymatic degumming	Good transparency	Rheology needs to be optimized
7	Use of alternative dissolution medium	Well characterised	Allergic response to silk
8	Use of DES	Higher cellular viability	

processing conditions or organic solvents for cell and tissue culture requirements. Therefore, by combining lyophilization and water annealing, a revolutionary green technology was devised. During the procedure, silk-based scaffolds were made without using salt or organic solvents at room temperature utilizing an all-aqueous processing technique. Notably, as compared to earlier silk fibroin scaffold systems, these new scaffolds based on silk fibroin have a lower β-sheet content, which gives them advantageous characteristics for tissue engineering (Guo et al. 2021).

8.3.1 Green Alternatives for Degumming

Supercritical CO_2 Fluid is the alternative which utilizes high-pressure carbon dioxide to remove sericin without water. This method is praised for its minimal environmental impact and ability to recycle the CO2 used in the process. It reduces water usage and energy compared to traditional methods. It also enables the extraction of clean sericin, which can be used in other industries, thus promoting zero-waste principles (Lo 2021). Enzymatic Degumming is another alternative as shown in Fig. 8.2 that employs natural enzymes from sources such as papaya skin, pineapple skin, and microbial cultures to selectively break down sericin without damaging fibroin fibers. It is highly efficient and specific, this method lowers the environmental burden by avoiding harsh chemicals and reducing energy usage. It is particularly effective in preserving the mechanical integrity of fibroin fibers (Zhu et al. 2022). Citric Acid is another one that uses a mild, biodegradable organic acid that is less harsh on fibers and the environment than synthetic chemicals. It is comparable to traditional methods in efficacy, citric acid is less resource-intensive and more

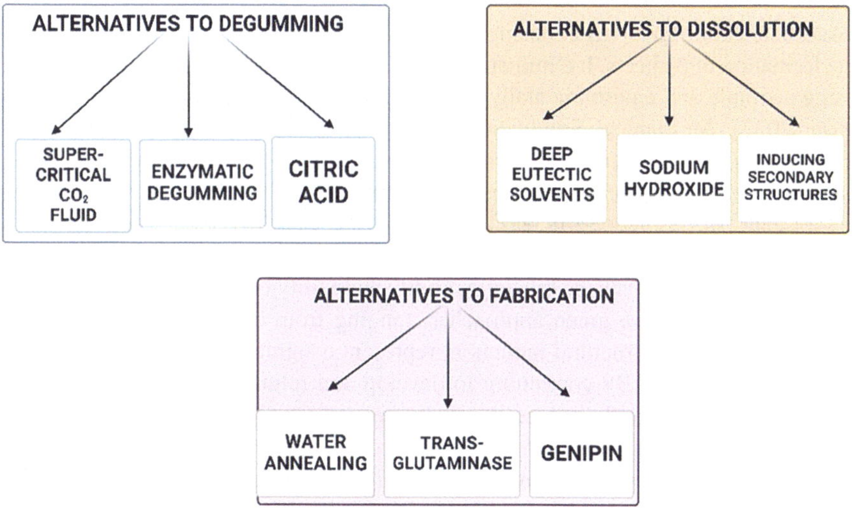

Fig. 8.2 Schematic representation of various alternative approaches to synthesizing silk fibroin scaffold

environmentally friendly, enhancing the sustainability of the silk processing (Wöltje et al. 2021). Sodium Hydroxide and Urea is another one that offers an alternative alkali treatment that is more energy-efficient than sodium carbonate, though still demanding in terms of chemical use. While not the greenest, modifications in process design, such as reduced temperature and optimized reagent concentrations, can enhance its eco-friendliness (DeBari et al. 2021; Medronho et al. 2019). Once degummed, silk fibroin must be dissolved into a processable solution. Traditional solvents like lithium bromide are effective but environmentally harmful.

8.3.2 Green Alternative to Dissolution Techniques

Deep Eutectic Solvents (DES) involve a mixture of biodegradable compounds that dissolve silk under less harsh conditions than typical chemical solvents. DESs can be tailored for specific applications and are reusable, significantly reducing solvent waste and energy consumption (Cao and Wang 2009). Also, sodium hydroxide acts as both a degumming agent and a dissolution medium, simplifying the silk processing pipeline. It uses less water and energy than other methods, though it requires careful management to avoid excessive fiber damage (Hu et al. 2020). Inducing secondary structures is another technique where its secondary structure largely determines the final material properties of silk fibroin, typically the β-sheet configuration, which provides the material with its needed stability and strength for medical applications (Wang et al. 2021a).

8.3.3 Green Alternative to Fabrication Techniques

Water Annealing involves exposing silk films or fibers to moisture, which induces the formation of β-sheets. It eliminates the need for harmful chemicals like methanol, using a simple and environmentally benign process that can be easily scaled (Egan et al. 2022). Mechanical Stimulation uses physical forces, such as stretching or ultrasonication, to induce alignment and crystallization of fibroin molecules. Water annealing reduces reliance on chemical treatments and emphasizes physical processes that can be energy-efficient and minimize waste. Moreover, adopting green synthesis methods for silk fibroin not only aligns with global sustainability goals but also potentially improves the biocompatibility and utility of silk in various biomedical applications. These green approaches, ranging from enzymatic degumming to water annealing for structural induction, represent a significant advancement in the field of biomaterials. By continuing to develop and refine these methods, the silk industry can significantly reduce its environmental footprint while enhancing the functionality and application scope of silk fibroin products (Hu et al. 2011).

The synthesis of silk fibroin scaffolds through green chemistry involves the application of principles that reduce or eliminate the use or generation of hazardous substances. Key principles include the use of safer solvents and renewable materials, energy efficiency, and designing for degradation. These principles aim to minimize the ecological footprint of biomaterial production while ensuring the functionality and safety of the scaffolds for biomedical use. Traditionally, silk fibroin is extracted from silk using sodium carbonate or other harsh chemicals that can harm the environment (Li et al. 2024). A greener approach is a milder, aqueous-based processes that reduce toxic waste and energy consumption. For instance, the degumming of silk, necessary to remove sericin (another protein in silk), can be achieved using enzymatic treatments that are less polluting and preserve the integrity of the fibroin fibers better than chemical methods (Wang et al. 2023). In green synthesis, cross-linking of silk fibroin—necessary to improve its stability and mechanical properties—is achieved using non-toxic agents such as genipin, a natural crosslinker derived from the gardenia fruit. This contrasts with commonly used chemical cross-linkers like glutaraldehyde, which are effective but potentially harmful. Additionally, enzyme-based methods for cross-linking, such as using transglutaminase, offer specificity and can be conducted under mild conditions, reducing energy use and avoiding toxic by-products (Dorishetty et al. 2020).

8.4 Comparing the Traditional Approach to the Green Approach

Silk fibroin, a biomaterial used in many medical applications, is synthesized using chemical-intensive techniques that pose environmental and health hazards. As sustainability becomes a top priority in industries, green methods are being embraced. These green methods reduce ecological footprints while maintaining or improving silk fibroin functioning for tissue engineering, wound healing, and medication delivery (Kostag et al. 2021).

Silk fibroin is traditionally produced by degumming with sodium carbonate, dissolving with lithium bromide, and chemically forming secondary structures. Each stage has serious environmental impacts. Conventional degumming uses a lot of water and strong chemicals that pollute the environment and kill aquatic life. Due to the use of hazardous, volatile solvents that emit significant VOCs, the dissolution process poses serious occupational and environmental health concerns (Khan et al. 2013). While, green silk fibroin synthesis uses safer chemicals, less waste, and energy-efficient techniques to reduce environmental consequences. Greener degumming methods like enzymatic degumming have replaced sodium carbonate. Natural enzymes target sericin proteins, conserving fibroin integrity and decreasing chemical inputs and energy use. Studies show that papaya enzymes can degum similarly, if not better while being biodegradable and safer for workers and the environment (Zhou et al. 2019b). Green synthesis also uses deep eutectic solvents (DESs), a blend of biodegradable chemicals that break down silk fibroin at lower temperatures without the toxicity of standard solvents. Reusing and recycling these DESs reduces waste and exposure to dangerous compounds (Zhou et al. 2001).

The traditional synthesis of silk fibroin scaffolds often involves solvents and conditions that are not environmentally friendly. For example, the use of organic solvents and high temperatures leads to significant energy consumption and environmental pollution. Green synthesis methods, on the other hand, utilize water as a solvent and operate at lower temperatures, significantly reducing the environmental burden (Wenk et al. 2011). Lifecycle assessments (LCA) of silk fibroin scaffold production reveal that green synthesis methods have a lower environmental impact compared to traditional methods. Green approaches contribute less to carbon emissions, use less hazardous substances, and generate less waste, making them preferable from an ecological standpoint (Lu and Kaplan 2010). Studies comparing the biological performance of silk fibroin scaffolds synthesized via green and traditional methods have shown that green scaffolds are equally effective, if not superior, in supporting cell attachment, proliferation, and differentiation. For instance, scaffolds cross-linked using genipin exhibit excellent mechanical properties and enhanced biocompatibility, crucial for successful tissue engineering applications (Teule et al. 2012).

Environmentally friendly technologies use less energy and produce less carbon dioxide due to lower temperatures and chemical conditions. Supercritical CO2 degumming reduces water and energy use and recovers and reuses CO2, an advanced recycling capacity not seen in previous approaches. This technique produces sericin, a clean and possibly lucrative byproduct that can be used in other industries, fostering a circular economy (Numata and Kaplan 2010). Due to softer processing conditions, green silk fibroin has improved biocompatibility and mechanical integrity. This is important for medical applications since natural protein structures improve performance and minimize immune reactivity when implanted (Kundu et al. 2013a). Transitioning to green synthesis processes is difficult despite the benefits. Due to specialized equipment and pricier materials, setup and operation expenses may be greater. Some green procedures are difficult to scale, especially for medical industry output. However, the long-term benefits—reduced environmental impact, increased worker safety, and possibly lower regulatory and cleaning costs—make further research and adoption worthwhile. Green chemistry will likely improve efficiency and cost-effectiveness to increase industrial applicability. Traditional methods are optimized for yield and cost-efficiency, but they lack sustainability. Green silk fibroin synthesis addresses environmental and health concerns and improves product quality and safety. As the world prioritizes green production, the silk sector can lead by example, showing that environmental responsibility can coexist with technological and industrial growth (Unger et al. 2004).

8.5 Applications of Silk Fibroin Biomaterial Scaffolds

Stents and flow-diverting devices made of regenerated vascular tissues based on silk are used in clinical settings. Two of the three patients in a flow device study have positive results, indicating that silk is a desirable treatment option for fragile blood-blister-like aneurysms. An intracranial aneurysm artery is also constructed using silk stents (Causin et al. 2011; Leonardi et al. 2011). Furthermore, vascular constructs can be effectively created using composites of silk fibroin and human-like collagen or double-raschel knitted silk poly (ethylene glycol di-glycol glycidyl ether). Although the latter composition can prevent early thrombosis, the former composition can give the construct good tensile strength (Yagi et al. 2011; Lovett et al. 2010). A tubular structure that resembles a blood vessel is also designed using an aqueous gel spinning process. One of the most thoroughly researched tissue engineering strategies to date is bone tissue engineering based on silk fibroin. When designed inside a bioreactor, porous fibroin scaffold-based bone constructs can promote the advanced development of bone tissues in as little as five weeks (Meinel et al. 2006; Li et al. 2006). In mice, human mesenchymal stem cell-based femoral defect healing is also facilitated by silk fibroin scaffolds. Additionally, bone morphogenic protein-2 was loaded into a silk fibroin polyethylene oxide nano-fibrous composite scaffold by human mesenchymal stem cells, which resulted in the regeneration of bone-like tissue. Animal bone regeneration is enhanced when hydroxyapatite nanoparticles are added to the silk matrix (Wang et al. 2010; Kweon et al. 2011).

RSF (regenerated silk fibroin) scaffolds are a preferred biopolymer for use in ligament and tendon TE because of their special mechanical qualities (such as high toughness values and good elasticity) and structural integrity (Kundu et al. 2013a, b). The first RSF matrix was effectively used in 2002 to engineer an anterior cruciate ligament (ACL) that closely resembled the mechanical characteristics of the ACL in humans (Altman et al. 2002). These encouraging findings led researchers to begin creating knitted scaffolds made of SF for the regeneration of tendons and ligaments. In contrast to RSF hydrogel knitted scaffolds, Liu et al. created web-like SF sponges on knitted scaffolds on which HMSCs were seeded. These scaffolds exhibit higher cellular activity. The findings showed that knitted scaffolds made of SF demonstrated structural strength and that RSF sponges with microporous web-like structures could increase cellular activity (Liu et al. 2008; Cai et al. 2010). Compared to untreated MCL or only SF knitted scaffold, the combination of RSF knitted scaffolds and collagen was demonstrated to enhance scaffold–ligament interface healing when implanted into rabbit medial collateral ligament (MCL) defected regions. Also, Because of its biocompatibility, biodegradability, and antimicrobial properties, chitosan has been used extensively in skin (Farokhi et al. 2018; Jeong et al. 2009). It also has the reputation of encouraging fibroblast cells to form collagen, which raises the tensile strength of the regenerated tissue in the damaged area. After creating RSF–chitosan scaffolds by electrospinning, the researchers observed that the mechanical strength and antimicrobial activity of the scaffolds rose as the RSF concentration increased (Kong et al. 2010; Xu et al. 2012). Furthermore, it was demonstrated that RSF–chitosan scaffolds enhanced cellular growth and had antimicrobial qualities against Escherichia coli. Moreover, when employed as a crosslinker, alginate dialdehyde (ADA) has lower toxicity and improves cell attachment and proliferation. Consequently, ADA exhibits good water absorption, high water transmission, and increased cell activity, making it suitable for crosslinking RSF–chitosan scaffolds in skin defects (Gu et al. 2013). Various applications of silk fibroin scaffolds are summarised in Table 8.2.

In recent years, scientists have attempted to create SF hydrogels that are appropriate for corneal reconstruction using various cross-linking techniques. As a

Table 8.2 Various applications of green synthesized silk fibroin biomaterial scaffolds

S.no.	Silk fibroin scaffold	Application	Reference
1	Silk fibroin and human-like collagen or double-raschel knitted silk poly (ethylene glycol di-glycol glycidyl ether).	Prevent early thrombosis	Yagi et al. (2011)
2	Web-like SF sponges	Increase cellular activity	Liu et al. (2008)
3	RSF–chitosan scaffolds	Anti-microbial activity	Gu et al. (2013)
4	Sf-genipin	Encourage keratinocyte migration	Chen et al. (2021)
5	SF scaffold	Alzheimer treatment	Chei et al. (2020)
6	SF scaffold	Neuroprotective	Chae et al. (2004)
7	SF-HA scaffold	CNS regeneration	Sha et al. (2023)
8	SF-based microneedle	Cancer therapy	Lujerdean et al. (2022)
9	SF-based nanoparticle	Controlled release	De Giorgio et al. (2024)
10	SF-based microgel	Tissue engineering	Karimi et al. (2024)

naturally occurring cross-linking agent, genipin (GP) can interact with the free amino groups in fibroin (Avila et al. 2016). Using GP cross-linking, researchers created porous polyvinyl alcohol/fibroin/nano-hydroxyapatite (PVA/SF/n-HA) composite hydrogels with favorable physical characteristics. The sacrificial portions of PVA/SF/n-HA hybrid hydrogels exhibit improved tensile strength, thermal resistance, and biocompatibility upon chemical cross-linking with GP (Zhou et al. 2019a). A promising technique for repairing damaged corneas is the use of human corneal fibroblasts (HCFs), which grow in the small spaces of composite hydrogels and have good adhesion and proliferation properties. Research has demonstrated that SF interpenetrating network hydrogels possess exceptional strength and toughness, making them highly valuable in the biomedical domain. Created semi-interpenetrating hydrogels using varying ratios of SF to polyacrylamide (PA), showing that these semi-interpenetrating networks could enhance cell adhesion, encourage keratinocyte migration during healing, and boost the effectiveness of in situ corneal tissue regeneration (Chen et al. 2021; Bhattacharjee and Ahearne 2022). Moreover, other researchers used high-molecular-weight RSF (HMWRSF) and carbamide as raw materials to create an extrudable printed RSF-based pre-solidified hydrogel that showed good cytocompatibility for dental pulp mesenchymal stem cells and the capacity to promote their growth and proliferation (Chen et al. 2022). By using visible light to induce riboflavin cross-linking as a photoinitiator, fibroin hydrogels containing rapidly gelatinized cells have also demonstrated good compatibility with human dental pulp stem cells (hDPSCs) (Piluso et al. 2020). Human exfoliated deciduous tooth stem cells can multiply on silk sponges, but more in vivo research is needed to determine whether scaffolds can be utilized for root canal therapy (Collado-González et al. 2018).

The translucent, elastic, grayish film that separates the middle and external ears is called the tympanic membrane (TM). It is composed of keratinocytes, fibroblasts, and collagen (types II and III), and it has three layers: the outermost layer is the epithelium, the middle layer is the fibrous layer, and the innermost layer is the mucosal layer (Levin et al. 2013; Ghassemifar et al. 2010). The tympanic membrane can rupture due to mechanical trauma, infection, or pressure changes, which can result in hyperacusis or even hearing loss. Some researchers studied that SF scaffolds effectively supported the adhesion and growth of human tympanic membrane keratinocytes while preserving their cell lines. The tensile strength and sound transmission capacity of regenerated SF membranes (RSF) are comparable to cartilage (Allardyce et al. 2016; Shen et al. 2013). Using a guinea pig acute TM perforation model, researchers showed that SF scaffolds (SFSs) improved both the structure and the function of TM restorations in comparison to conventional paper patches, leading to complete closure of the perforation site and early restoration of hearing. Simultaneously, SFSs result in therapeutically acceptable tissue responses, such as minimal inflammation as shown in Fig. 8.3, defined neovascularization, and scaffold degradation, and are well tolerated by the host (Shen et al. 2014; Lee et al. 2014). The role of SF-based biomaterials in treating Parkinson's and Alzheimer's disease has been thoroughly examined because of their amino-acidic sequences (Chei et al. 2020). As SF hydrolysis byproducts inhibit inflammation and delay the apoptotic process caused by reactive oxygen species and β-amyloid, they have neuroprotective effects. It has also been discovered that SF peptides function as neuroprotectants against the neurodegenerative effects of 6-hydroxydopamine, a neurotoxin, thereby preserving the viability of dopaminergic neurons (Chae et al.

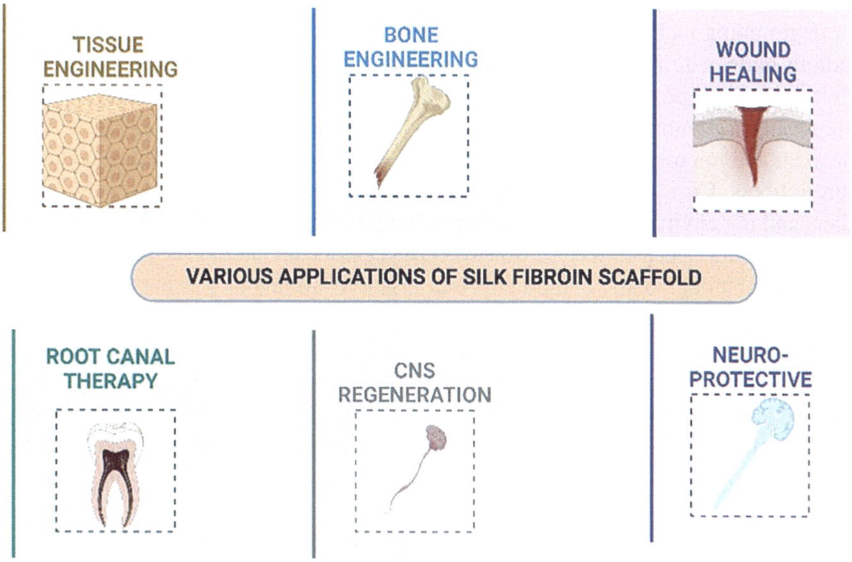

Fig. 8.3 Various applications of silk fibroin scaffold

2004; Kim et al. 2011). Furthermore, a hydrogel coated with hyaluronic acid, silk fibroin, and poly-dopamine has been shown to have the potential to improve spinal cord regeneration through the extension of neurotrophin-3 release rate, a finding that suggests the hydrogel's great potential for CNS regeneration drug delivery. However, due to extensive research on the PNS—a system made up of nerve bundles that span numerous body regions—many practical applications for SF-based biomaterials have been made possible, opening up new avenues for biomedical advancements, particularly in the context of traumatic injuries (Sha et al. 2023; Jiang et al. 2021). Because of their elasticity and tensile strength, silk fibers have been used to promote the growth and reunification of cut nerves. They have also been used to improve the differentiation of stem cells into neurons, and numerous pre-clinical studies are currently being completed to support the clinical use of fibers (Zhang et al. 2010; Guedan-Duran et al. 2020; Bojnordi et al. 2018).

8.6 Conclusion

Silk fibroin, a protein from the Bombyx mori silkworm, is an excellent biomaterial for medical applications due to its mechanical properties, biocompatibility, and controlled biodegradability. It is ideal for tissue engineering, particularly for bone regeneration and skin repair. However, traditional methods often involve harsh chemicals and conditions that can be harmful to the environment and health. A green synthesis approach is necessary to advance the field in alignment with ecological and health standards. A novel, environmentally friendly, all-aqueous method called lyophilization is used to create three-dimensional porous silk fibroin matrices with control over structural and morphological aspects. Green alternatives for degumming include supercritical CO2 Fluid, enzymatic degumming, citric acid, sodium hydroxide and urea, and induced secondary structures. Supercritical CO2 Fluid uses high-pressure carbon dioxide to remove sericin without water, while enzymatic degumming uses natural enzymes from sources like papaya skin, pineapple skin, and microbial cultures to selectively break down sericin without damaging fibroin fibers. Citric Acid is a mild, biodegradable organic acid that is less harsh on fibers and the environment than synthetic chemicals, enhancing the sustainability of silk processing. Deep Eutectic Solvents (DES) can be an alternative that involves a mixture of biodegradable compounds that dissolve silk under less harsh conditions than typical chemical solvents. Green fabrication techniques include water annealing and mechanical stimulation. Enzymatic cross-linking of silk fibroin is achieved using non-toxic agents like genipin, a natural crosslinker derived from the gardenia fruit, and enzyme-based methods for cross-linking, such as transglutaminase, which offer specificity and can be conducted under mild conditions, reducing energy use and avoiding toxic by-products. The aforementioned observations suggest that enhanced SF can be obtained for biomedical applications using green synthesized silk fibroin scaffolds, which are also more environmentally friendly.

References

Allardyce BJ, Rajkhowa R, Dilley RJ, Xie Z, Campbell L, Keating A, Wang X (2016) Comparative acoustic performance and mechanical properties of silk membranes for the repair of chronic tympanic membrane perforations. J Mech Behav Biomed Mater 64:65–74

Altman GH, Horan RL, Lu HH, Moreau J, Martin I, Richmond JC, Kaplan DL (2002) Silk matrix for tissue-engineered anterior cruciate ligaments. Biomaterials 23(20):4131–4141

Ansari AI, Sheikh NA (2023) Biocompatible scaffold based on silk fibroin for tissue engineering applications. J Inst Eng India Ser C 104(1):201–217

Arai T, Freddi G, Innocenti R, Tsukada M (2004) Biodegradation of Bombyx mori silk fibroin fibers and films. J Appl Polym Sci 91(4):2383–2390

Asakura T (2021) Structure of silk I (Bombyx mori silk fibroin before spinning)-type II β-turn, not α-helix. Molecules 26(12):3706

Avila MY, Narvaez M, Castañeda JP (2016) Effects of genipin corneal crosslinking in rabbit corneas. J Cataract Refract Surg 42(7):1073–1077

Bhattacharjee P, Ahearne M (2022) Silk fibroin-based interpenetrating network hydrogel for corneal stromal regeneration. Int J Biol Macromol 223:583–594

Bojnordi MN, Ebrahimi-Barough S, Vojoudi E, Hamidabadi HG (2018) Silk nanofibrous electrospun scaffold enhances differentiation of embryonic stem like cells derived from testis in to mature neuron. J Biomed Mater Res A 106(10):2662–2669

Cai ZX, Mo XM, Zhang KH, Fan LP, Yin AL, He CL, Wang HS (2010) Fabrication of chitosan/silk fibroin composite nanofibers for wound-dressing applications. Int J Mol Sci 11(9):3529–3539

Cao Y, Wang B (2009) Biodegradation of silk biomaterials. Int J Mol Sci 10(4):1514–1524

Causin F, Pascarella R, Pavesi G, Marasco R, Zambon G, Battaglia R, Munari M (2011) Acute endovascular treatment (<48 hours) of uncoilable ruptured aneurysms at non-branching sites using silk flow-diverting devices. Interv Neuroradiol 17(3):357–364

Cebe P, Partlow BP, Kaplan DL, Wurm A, Zhuravlev E, Schick C (2017) Silk I and silk II studied by fast scanning calorimetry. Acta Biomater 55:323–332

Chae HS, Kang YK, Shin YK, Lee HJ, Yu JI, Lee KG, Kim SS (2004) The role of BF-7 on neuroprotection and enhancement of cognitive function. Korean J Physiol Pharmacol 8(4): 173–179

Chei S, Oh HJ, Lee K, Jin H, Lee JY, Lee BY (2020) Dietary silk peptide inhibits LPS-induced inflammatory responses by modulating toll-like receptor 4 (TLR4) signaling. Biomol Ther 10(5):771

Chen WW, Kang K, Lv J, Yue L, Zhang WQ (2021) Galactose-NlGr11 inhibits AMPK phosphorylation by activating the PI3K-AKT-PKF-ATP signaling cascade via insulin receptor and Gβγ. Insect Sci 28(3):735–745

Chen N, Zhang X, Lyu J, Zhao G, Gu K, Xia J, Shao Z (2022) Preparation of a novel regenerated silk fibroin-based hydrogel for extrusion bioprinting. Soft Matter 18(38):7360–7368

Cho SY, Yun YS, Lee S, Jang D, Park KY, Kim JK, Jin HJ (2015) Carbonization of a stable β-sheet-rich silk protein into a pseudographitic pyroprotein. Nat Commun 6(1):7145

Collado-González M, Pecci-Lloret MP, García-Bernal D, Aznar-Cervantes S, Oñate-Sánchez RE, Moraleda JM, Rodríguez-Lozano FJ (2018) Biological effects of silk fibroin 3D scaffolds on stem cells from human exfoliated deciduous teeth (SHEDs). Odontology 106:125–134

De Giorgio G, Matera B, Vurro D, Manfredi E, Galstyan V, Tarabella G, D'Angelo P (2024) Silk fibroin materials: biomedical applications and perspectives. Bioengineering 11(2):167

DeBari MK, King CI III, Altgold TA, Abbott RD (2021) Silk fibroin as a green material. ACS Biomater Sci Eng 7(8):3530–3544

Diez-Echave P, Ruiz-Malagón AJ, Molina-Tijeras JA, Hidalgo-García L, Vezza T, Cenis-Cifuentes L, Lozano-Pérez AA (2021) Silk fibroin nanoparticles enhance quercetin immunomodulatory properties in DSS-induced mouse colitis. Int J Pharm 606:120935

Dorishetty P, Dutta NK, Choudhury NR (2020) Silk fibroins in multiscale dimensions for diverse applications. RSC Adv 10(55):33227–33247

Egan G, Phuagkhaopong S, Matthew SA, Connolly P, Seib FP (2022) Impact of silk hydrogel secondary structure on hydrogel formation, silk leaching and in vitro response. Sci Rep 12(1): 3729

Farokhi M, Mottaghitalab F, Fatahi Y, Khademhosseini A, Kaplan DL (2018) Overview of silk fibroin use in wound dressings. Trends Biotechnol 36(9):907–922

Farokhi M, Mottaghitalab F, Reis RL, Ramakrishna S, Kundu SC (2020) Functionalized silk fibroin nanofibers as drug carriers: advantages and challenges. J Control Release 321:324–347

Floren M, Migliaresi C, Motta A (2016) Processing techniques and applications of silk hydrogels in bioengineering. J Funct Biomater 7(3):26

Ghassemifar R, Redmond S, Zainuddin, Chirila TV (2010) Advancing towards a tissue-engineered tympanic membrane: silk fibroin as a substratum for growing human eardrum keratinocytes. J Biomater Appl 24(7):591–606

Goldberg E, Kim S (2023) Silk fibroin in tissue engineering: a green synthesis approach. Green Chem Lett Rev 16(1):88–99

Grabska-Zielińska S, Sionkowska A, Carvalho Â, Monteiro FJ (2021) Biomaterials with potential use in bone tissue regeneration—collagen/chitosan/silk fibroin scaffolds cross-linked by EDC/NHS. Materials 14(5):1105

Gregory DA, Zhang Y, Smith PJ, Zhao X, Ebbens SJ (2016) Reactive inkjet printing of biocompatible enzyme powered silk micro-rockets. Small 12(30):4048–4055

Gu Z, Xie H, Huang C, Li L, Yu X (2013) Preparation of chitosan/silk fibroin blending membrane fixed with alginate dialdehyde for wound dressing. Int J Biol Macromol 58:121–126

Guedan-Duran A, Jemni-Damer N, Orueta-Zenarruzabeitia I, Guinea GV, Perez-Rigueiro J, Gonzalez-Nieto D, Panetsos F (2020) Biomimetic approaches for separated regeneration of sensory and motor fibers in amputee people: necessary conditions for functional integration of sensory–motor prostheses with the peripheral nerves. Front Bioeng Biotechnol 8:584823

Guo X, Lin N, Lu S, Zhang F, Zuo B (2021) Preparation and biocompatibility characterization of silk fibroin 3D scaffolds. ACS Appl Bio Mater 4(2):1369–1380

Hashimoto T, Nakamura Y, Tamada Y, Kurosu H, Kameda T (2020) The influence of thermal treatments on the secondary structure of silk fibroin scaffolds and their interaction with fibroblasts. PeerJ Mater Sci 2:e8

Heseltine PL, Bayram C, Gultekinoglu M, Homer-Vanniasinkam S, Ulubayram K, Edirisinghe M (2022) Facile one-pot method for all aqueous green formation of biocompatible silk fibroin-poly (ethylene oxide) fibers for use in tissue engineering. ACS Biomater Sci Eng 8(3):1290–1300

Hu X, Shmelev K, Sun L, Gil ES, Park SH, Cebe P, Kaplan DL (2011) Regulation of silk material structure by temperature-controlled water vapor annealing. Biomacromolecules 12(5): 1686–1696

Hu Y, Liu L, Yu J, Wang Z, Fan Y (2020) Preparation of natural multicompatible silk nanofibers by green deep eutectic solvent treatment. ACS Sustain Chem Eng 8(11):4499–4510

Huang Y, Bailey K, Wang S, Feng X (2017) Silk fibroin films for potential applications in controlled release. React Funct Polym 116:57–68

Jaramillo-Quiceno N, Álvarez-López C, Restrepo-Osorio A (2017) Structural and thermal properties of silk fibroin films obtained from cocoon and waste silk fibers as raw materials. Procedia Eng 200:384–388

Jeong L, Yeo IS, Kim HN, Yoon YI, Jung SY, Min BM, Park WH (2009) Plasma-treated silk fibroin nanofibers for skin regeneration. Int J Biol Macromol 44(3):222–228

Jiang J, Dai C, Liu X, Dai L, Li R, Ma K, Zhang S (2021) Implantation of regenerative complexes in traumatic brain injury canine models enhances the reconstruction of neural networks and motor function recovery. Theranostics 11(2):768

Karimi F, Farbehi N, Ziaee F, Lau K, Monfared M, Kordanovski M, Rnjak-Kovacina J (2024) Photocrosslinked silk fibroin microgel scaffolds for biomedical applications. Adv Funct Mater 34:2313354

Khan MMR, Tsukada M, Zhang X, Morikawa H (2013) Preparation and characterization of electrospun nanofibers based on silk sericin powders. J Mater Sci 48:3731–3736

Kim TK, Park D, Yeon S, Lee SH, Choi YJ, Bae DK, Kim YB (2011) Tyrosine-fortified silk amino acids improve physical function of Parkinson's disease rats. Food Sci Biotechnol 20:79–84

Kong M, Chen XG, Xing K, Park HJ (2010) Antimicrobial properties of chitosan and mode of action: a state of the art review. Int J Food Microbiol 144(1):51–63

Kostag M, Jedvert K, El Seoud OA (2021) Engineering of sustainable biomaterial composites from cellulose and silk fibroin: fundamentals and applications. Int J Biol Macromol 167:687–718

Kundu B, Kurland NE, Bano S, Patra C, Engel FB, Yadavalli VK (2013a) Silk proteins for biomedical applications: bioengineering perspectives. Prog Polym Sci 38(3):1392–1414

Kundu B, Rajkhowa R, Kundu SC, Wang X (2013b) Silk fibroin biomaterials for tissue regenerations. Adv Drug Deliv Rev 65(4):457–470

Kweon H, Lee KG, Chae CH, Balázsi C, Min SK, Kim JY, Kim SG (2011) Development of nano-hydroxyapatite graft with silk fibroin scaffold as a new bone substitute. J Oral Maxillofac Surg 69(6):1578–1586

Lee H, Jang CH, Kim GH (2014) A polycaprolactone/silk-fibroin nanofibrous composite combined with human umbilical cord serum for subacute tympanic membrane perforation; an in vitro and in vivo study. J Mater Chem B 2(18):2703–2713

Leonardi M, Cirillo L, Toni F, Dall'Olio M, Princiotta C, Stafa A, Agati R (2011) Treatment of intracranial aneurysms using flow-diverting silk stents (BALT): a single centre experience. Interv Neuroradiol 17(3):306–315

Levin B, Redmond SL, Rajkhowa R, Eikelboom RH, Atlas MD, Marano RJ (2013) Utilising silk fibroin membranes as scaffolds for the growth of tympanic membrane keratinocytes, and application to myringoplasty surgery. J Laryngol Otol 127(S1):S13–S20

Li C, Vepari C, Jin HJ, Kim HJ, Kaplan DL (2006) Electrospun silk-BMP-2 scaffolds for bone tissue engineering. Biomaterials 27(16):3115–3124

Li B, Chai CHT, Koh XQ, Tang KY, Chan CY, Heng JZX, Li Z (2024) Green preparation of silk nanofiber membranes from natural silkworms for effective dye/salt fractionation. J Membr Sci 701:122741

Liu H, Fan H, Wang Y, Toh SL, Goh JC (2008) The interaction between a combined knitted silk scaffold and microporous silk sponge with human mesenchymal stem cells for ligament tissue engineering. Biomaterials 29(6):662–674

Liu Q, Liu H, Fan Y (2017) Preparation of silk fibroin carriers for controlled release. Microsc Res Tech 80(3):312–320

Liu J, Chen H, Wang Y, Li G, Zheng Z, Kaplan DL, Wang X (2020) Flexible water-absorbing silk-fibroin biomaterial sponges with unique pore structures for tissue engineering. ACS Biomater Sci Eng 6(3):1641–1649

Liu Q, Ying G, Jiang N, Yetisen AK, Yao D, Xie X, Liu H (2021) Three-dimensional silk fibroin microsphere-nanofiber scaffolds for vascular tissue engineering. Med Nov Technol Devices 9: 100051

Lo CHH (2021) Degumming silk by CO2 supercritical fluid and their dyeing ability with plant indigo. Int J Cloth Sci Technol 33(3):465–476

Lovett M, Eng G, Kluge J, Cannizzaro C, Vunjak-Novakovic G, Kaplan DL (2010) Tubular silk scaffolds for small diameter vascular grafts. Organogenesis 6(4):217–224

Lu Q, Kaplan DL (2010) In vitro stabilization of a silk fibroin hydrogel with a silk protein. Biomaterials 31(6):896–906

Lu Q, Zhang X, Hu X, Kaplan DL (2010) Green process to prepare silk fibroin/gelatin biomaterial scaffolds. Macromol Biosci 10(3):289–298

Lujerdean C, Baci GM, Cucu AA, Dezmirean DS (2022) The contribution of silk fibroin in biomedical engineering. Insects 13(3):286

Medronho B, Filipe A, Napso S, Khalfin RL, Pereira RF, de Zea Bermudez V, Cohen Y (2019) Silk fibroin dissolution in tetrabutylammonium hydroxide aqueous solution. Biomacromolecules 20(11):4107–4116

Meinel L, Betz O, Fajardo R, Hofmann S, Nazarian A, Cory E, Kirker-Head C (2006) Silk based biomaterials to heal critical sized femur defects. Bone 39(4):922–931

Nazarov R, Jin HJ, Kaplan DL (2004) Porous 3-D scaffolds from regenerated silk fibroin. Biomacromolecules 5(3):718–726

Nguyen TP, Nguyen QV, Nguyen VH, Le TH, Huynh VQN, Vo DVN, Le QV (2019) Silk fibroin-based biomaterials for biomedical applications: a review. Polymers 11(12):1933

Numata K, Kaplan DL (2010) Silk-based delivery systems of bioactive molecules. Adv Drug Deliv Rev 62(15):1497–1508

Piluso S, Gomez DF, Dokter I, Texeira LM, Li Y, Leijten J, Malda J (2020) Rapid and cytocompatible cell-laden silk hydrogel formation via riboflavin-mediated crosslinking. J Mater Chem B 8(41):9566–9575

Qi Y, Wang H, Wei K, Yang Y, Zheng RY, Kim IS, Zhang KQ (2017) A review of structure construction of silk fibroin biomaterials from single structures to multi-level structures. Int J Mol Sci 18(3):237

Rahimpour S, Jabbari H, Yousofi H, Fathi A, Mahmoodi S, Jafarian MJ, Shotorbani SS (2023) Regulatory effect of sericin protein in inflammatory pathways; a comprehensive review. Pathol Res Pract 243:154369

Rajput M, Mandal M, Anura A, Mukhopadhyay A, Subramanian B, Paul RR, Chatterjee J (2020) Honey loaded silk fibroin 3D porous scaffold facilitates homeostatic full-thickness wound healing. Materialia 12:100703

Sha Q, Wang Y, Zhu Z, Wang H, Qiu H, Niu W, Qian J (2023) A hyaluronic acid/silk fibroin/poly-dopamine-coated biomimetic hydrogel scaffold with incorporated neurotrophin-3 for spinal cord injury repair. Acta Biomater 167:219–233

Shen Y, Redmond SL, Teh BM, Yan S, Wang Y, Atlas MD, Marano RJ (2013) Tympanic membrane repair using silk fibroin and acellular collagen scaffolds. Laryngoscope 123(8): 1976–1982

Shen Y, Redmond SL, Papadimitriou JM, Teh BM, Yan S, Wang Y, Dilley RJ (2014) The biocompatibility of silk fibroin and acellular collagen scaffolds for tissue engineering in the ear. Biomed Mater 9(1):015015

Smith J, Doe A (2021) Comparative analysis of chemical and green synthesis of silk fibroin scaffolds. J Biomed Mater Res A 105(4):1124–1134

Suzuki Y (2016) Structures of silk fibroin before and after spinning and biomedical applications. Polym J 48(11):1039–1044

Tahamtan A, Tabarraei A, Moradi A, Dinarvand M, Kelishadi M, Ghaemi A, Atyabi F (2015) Chitosan nanoparticles as a potential nonviral gene delivery for HPV-16 E7 into mammalian cells. Artif Cells Nanomed Biotechnol 43(6):366–372

Teule F, Miao YG, Sohn BH, Kim YS, Hull JJ, Fraser MJ, Zhao H (2012) Silkworms transformed with chimeric silkworm/spider silk genes spin composite silk fibers with improved mechanical properties. Proc Natl Acad Sci 109(3):923–928

Unger RE, Wolf M, Peters K, Motta A, Migliaresi C, Kirkpatrick CJ, Kirkpatrick CJ (2004) Growth of human cells on a non-woven silk fibroin net: a potential for use in tissue engineering. Biomaterials 25(6):1069–1075

Valluzzi R, Gido SP, Muller W, Kaplan DL (1999) Orientation of silk III at the air-water interface. Int J Biol Macromol 24(2–3):237–242

Wang G, Yang H, Li M, Lu S, Chen X, Cai X (2010) The use of silk fibroin/hydroxyapatite composite co-cultured with rabbit bone-marrow stromal cells in the healing of a segmental bone defect. J Bone Joint Surg Br 92(2):320–325

Wang Y, Fan S, Li Y, Niu C, Li X, Guo Y, Wang X (2020) Silk fibroin/sodium alginate composite porous materials with controllable degradation. Int J Biol Macromol 150:1314–1322

Wang HY, Zhang YQ, Wei ZG (2021a) Dissolution and processing of silk fibroin for materials science. Crit Rev Biotechnol 41(3):406–424

Wang Z, Cui Y, Feng Y, Guan L, Dong M, Liu Z, Liu L (2021b) A versatile Silk Fibroin based filtration membrane with enhanced mechanical property, disinfection and biodegradability. Chem Eng J 426:131947

Wang K, Ma Q, Zhou HT, Zhao JM, Cao M, Wang SD (2023) Review on fabrication and application of regenerated Bombyx mori silk fibroin materials. AUTEX Res J 23(2):164–183
Wenk E, Merkle HP, Meinel L (2011) Silk fibroin as a vehicle for drug delivery applications. J Control Release 150(2):128–141
Wöltje M, Kölbel A, Aibibu D, Cherif C (2021) A fast and reliable process to fabricate regenerated silk fibroin solution from degummed silk in 4 hours. Int J Mol Sci 22(19):10565
Xu Y, Li L, Yu X, Gu Z, Zhang X (2012) Feasibility study of a novel crosslinking reagent (alginate dialdehyde) for biological tissue fixation. Carbohydr Polym 87(2):1589–1595
Yagi T, Sato M, Nakazawa Y, Tanaka K, Sata M, Itoh K, Asakura T (2011) Preparation of double-raschel knitted silk vascular grafts and evaluation of short-term function in a rat abdominal aorta. J Artif Organs 14:89–99
Zhang F, Liu R, Zuo BQ, Qin JZ (2010) Electrospun silk fibroin nanofiber tubes for peripheral nerve regeneration. In: In 2010 4th international conference on bioinformatics and biomedical engineering. IEEE, pp 1–4
Zhang C, Zhang Y, Shao H, Hu X (2016) Hybrid silk fibers dry-spun from regenerated silk fibroin/graphene oxide aqueous solutions. ACS Appl Mater Interfaces 8(5):3349–3358
Zhou CZ, Confalonieri F, Jacquet M, Perasso R, Li ZG, Janin J, Tang RK (2001) Silk fibroin: structural implications of a remarkable amino acid sequence. Proteins Struct Funct Bioinf 44(2): 119–122
Zhou H, Wang Z, Cao H, Hu H, Luo Z, Yang X, Zhou L (2019a) Genipin-crosslinked polyvinyl alcohol/silk fibroin/nano-hydroxyapatite hydrogel for fabrication of artificial cornea scaffolds—a novel approach to corneal tissue engineering. J Biomater Sci Polym Ed 30(17):1604–1619
Zhou J, Zhang Y, Yang Y, Chen Z, Jia G, Zhang L (2019b) Silk fibroin-graphene oxide functionalized melamine sponge for efficient oil absorption and oil/water separation. Appl Surf Sci 497:143762
Zhu L, Lin J, Pei L, Luo Y, Li D, Huang Z (2022) Recent advances in environmentally friendly and green degumming processes of silk for textile and non-textile applications. Polymers 14(4):659

Ms. Shristy Verma is an ideal professional in the pharmaceutics field whose contributions have greatly enhanced the scholarly environment. Her extensive body of work in the field, which includes multiple published book chapters and review papers, is evidence of her commitment to and proficiency in the field. By selecting esteemed outlets like Elsevier and Bentham Science to publish her work, Ms. Verma has demonstrated her dedication to quality and her astute awareness of the norms maintained by respectable publishers. Ms. Verma has actively participated in over five national and international conferences in addition to her impressive publications. These conferences gave her a forum to share the results of her research and a great chance to network with other researchers, academics, and industry professionals. Ms. Verma's academic accomplishments demonstrate her dedication to remaining up to date with the most recent advancements in the field of pharmaceutics. Her capacity to make a significant contribution to conference talks as well as scholarly literature demonstrates a unique combination of theoretical understanding and real-world application.

Dr. Rishabha Malviya completed B. Pharmacy from Uttar Pradesh Technical University and M. Pharmacy (Pharmaceutics) from Gautam Buddha Technical University, Lucknow Uttar Pradesh. His PhD (Pharmacy) work was in the area of Novel formulation development techniques. He has 13 years of research experience and presently working as Associate Professor in the Department of Pharmacy, School of Medical and Allied Sciences, Galgotias University since past 10 years. His area of interest includes formulation optimization, nanoformulation, targeted drug delivery, localized drug delivery and characterization of natural polymers as pharmaceutical excipients. He has authored more than 170 research/review papers for national/international journals of repute. He has 58 patents (19 grants, 38 published, 1 filed) and publications in reputed National and International journals with total of 240 cumulative impact factor. He has also received an Outstanding Reviewer award from Elsevier. He has authored/edited/editing 50 books (Wiley, CRC Press/Taylor and Francis, Springer, IOP Publishing, River Publisher, Apple academic Press/Taylor and Francis Groups and OMICS publication) and authored 123 book chapters. His name has included in word's top 2% scientist list for the year 2020, 2021, 2022 and 2023 by Elsevier BV and Stanford University. He is Reviewer/Editor/Editorial board member of more than 50 national and international journals of repute. He has invited as author for "Atlas of Science" and pharma magazine dealing with industry (B2B) "Ingredient south Asia Magazines".

Lavanya Gupta completed her Bachelor's in Biochemistry Hons in 2018 from Sri Venkateswara College, University of Delhi, New Delhi. Afterward, she completed her Master's from All India Institute of Medical Sciences (AIIMS), New Delhi, in 2023. During her master's, she identified the Estrogen-regulated miRNA network in the Human trophoblast invasion process and its correlation with Gestational trophoblast disorders like preeclampsia. She joined Prof Bibli's lab in 2024 exploring novel cysteinolytic regulators. She has worked with large datasets to obtain differentially regulated genes in the pathology of pregnancy-related disorders, i.e. preeclampsia. She worked with various databases, R programming language, and software packages to identify novel genes and their association with disease pathophysiology. She have used many databases such as MsigDB, Target scan, mirDB, mirtarBase, RNA22 version 2.0, Agilent, miRDeep2 (software package), R/Bioconductor package Edge R, R package gplots, Kyoto Encyclopedia of Genes and Genomes (KEGG) database. During her PhD at Heidelberg University at the faculty of Biosciences, she will focus on identifying novel regulators of cysteinolysis within endothelial cells which are regulated by two enzymes named cystathionine beta-synthase (CBS) and cystathionine gamma-lyase (CSE).

Dr. Sathvik Belagodu Sridhar is currently working as an Associate Dean and Professor at RAK College of Pharmacy (RAKCOP), RAK Medical and Health Sciences University located in Ras Al Khaimah, United Arab Emirates. Dr. Sathvik has completed his Ph.D. in Clinical Pharmacy from Rajiv Gandhi University of Health Sciences, Bangalore, India. He has received an FIP fellowship for higher training in renal clinical pharmacy at Royal Adelaide Hospital and Queen Elizabeth Hospital in Adelaide, South Australia. Dr. Sathvik has worked as an Assistant Professor and Clinical Pharmacist at JSS College of Pharmacy and JSS Medical College Hospital, Mysore, India, for eight years and later Joined International Medical University (IMU) Malaysia as a senior lecturer in the Department of Pharmacy Practice from 2008 to 2010. Dr. Sathvik joined as an assistant professor in the Department of Clinical Pharmacy at RAK College of Pharmacy, RAK Medical, and Health Sciences University in 2010 and worked as a professor and chairperson of the Department of Clinical Pharmacy and Pharmacology from 2018 to 2023. Dr. Sathvik has been honored with a prestigious teacher award by the Society of Pharmaceutical Education & Research [SPER]. He has also received training in Pharmacovigilance at the Uppsala Monitoring Centre in Uppsala, Sweden. Dr. Sathvik has 25 years of experience teaching and practicing clinical pharmacy in India, Malaysia, and the UAE. His research interests include Medication Safety, Pharmacovigilance, Pharmacoepidemiology, and Renal and Psychiatry Clinical Pharmacy. Dr. Sathvik has more than 75 research papers published in reputable national and international journals, along with book chapters. He has presented more than 75 research papers at numerous national and international conferences and has been honored with many awards. He also serves as a reviewer for various national and internal pharmacy practice-related journals.

Green Catalysts in the Synthesis of Biomaterials for Biomedical Applications

9

Murugiah Krishani, Nonni Soraya Sambudi, and Hazwani Suhaimi

Abstract

Green chemistry witnessed an enhanced development in making green catalysts by reducing the usage of toxic substances and harnessing renewable resources. Sustainable materials from bio sources, light sources, nano sources, and heteropolyacids act as green catalysts and exhibit potential candidates for synthesizing biomaterials for biomedical applications. This chapter highlights the use of various green catalysts in synthesizing biomaterials for biomedical purposes, including biosensors in clinical diagnosis, drug delivery, tissue engineering, wound healing, polymer coatings and inks for bioprinting. This chapter then discusses the methods involved in synthesizing green catalyst sources of green catalyst and finally concludes with future perspectives.

Keywords

Green catalyst · Biomaterials · Biomedical applications · Biosensors · Drug delivery · Tissue engineering · Wound healing · Polymer coating and bioprinting

M. Krishani (✉) · H. Suhaimi
Faculty of Integrated Technologies, Universiti Brunei Darussalam, Jalan Tungku Link, Gadong, Brunei
e-mail: 20h8400@ubd.edu.bn; hazwani.suhaimi@ubd.edu.bn

N. S. Sambudi
Department of Chemical Engineering, Universitas Pertamina, Simprug, Jakarta, Indonesia
e-mail: nonni.ss@universitaspertamina.ac.id

R. Malviya, S. Sundram (eds.), *Sustainable Green Biomaterials As Drug Delivery Systems*, Biomaterials, Bioengineering and Sustainability 1,
https://doi.org/10.1007/978-3-031-79062-1_9

Abbreviations

AgNO3	Silver nitrate
aNSCs	Adult neural stem cells
AuNPs	Gold nanoparticles
BET	Brunauer-Emmett-Teller
Bio-MOFs	Bio-metal organic frameworks
BMSCs	Bone marrow mesenchymal stem cells
BP	Black phosphorus
CD	Carbon dot
CD11a	α chain of the $\alpha L\beta 2$ integrin
CNT	Carbon nanotubes
CNTpega	Carbon nanotubes poly ethylene glycol acrylate
Co_3O_4	Cobalt oxide
COD	Choline oxidase
COOH	Carboxyl group
COX-1	Cyclooxygenase - 1
COX-2	Cyclooxygenase - 2
DNA	Deoxyribonucleic acid
ECM	Extracellular matrix
EDC	1-ethyl-3-(3-dimethylaminopropyl)-carbodiimide hydrochloride
F	Fluorine
Fe2O3	Ferric oxide
Fe3O4	Black iron oxide
FeNi3	Iron;nickel
FN	Fibronectin
GAG	Glycosaminoglycans
GO	Graphene oxide
GR	Graphene
Gtn-HPA	Gelatin-hydroxyphenyl propionic acid
H_2O	Water
H_2O_2	Hydrogen peroxide
H_2SO_4	Sulphuric acid
HLC	Human-like collagen
HPA	Heteropolyacids
HRP	Horse radish peroxidase
MAS	Microwave-aided synthesis
MCM-41	Mobil Crystalline Materials
MTT	(3-(4,5-Dimethylthiazol-2-yl)-2,5-Diphenyltetrazolium Bromide)
MOF	Metal-organic framework
MoS2	Molybdenum disulfide
Na_2CO_3	Sodium carbonate
NH_2	Amine functional group
NHS	N-hydroxysuccinimide

NiFe	Nickel–iron alloy
$NiFe_2O_4$	Nickel ferrite
NIR	Near-infrared
NO_3	Nitrate ion
NPs	Nanoparticles
O	Oxygen
OH	Hydroxyl group
OPF	Oligo(poly(ethylene glycol) fumarate
PCL	Polycaprolactone
PDLSCs	Proliferation, adherence, and viability of human periodontal ligament mesenchymal stem cells
PDT	Photodynamic therapy
PEG	Polyethylene glycol
PTT	Photothermal therapy
RNA	Ribonucleic acid
ROS	Reactive oxygen species
SBA-15	Santa Barbara Amorphous
SiO2	Silicon dioxide
Ti	Titanium
TiO2	Titanium dioxide
TMDCs	Transition metal dichalcogenides
TPABr	Tetrapropylammonium bromide
WDH	Wound dressing hydrogel
ZSM-5	Zeolite Socony Mobil–5

9.1 Introduction

Green chemistry involves inventing chemicals and protocols or procedures using environmentally friendly methods. The main goals of green chemistry are the minimum exhaustion of natural sources, waste reduction, energy conservation, and the reduction of non-toxicities. The catalysts developed to fulfil the requirements of green chemistry by being selective, efficient and, most importantly, environmentally friendly are termed green catalysts. Thus, green chemistry and green catalysts correlate (Chhangani 2016). Green catalysts are sustainable, produced from non-hazardous renewable starting materials, provide high efficiency, are non-toxic, and yield minimal byproducts (Shoda et al. 2016). Research on using greener materials in chemical synthesis has become a focus in the past few decades. This idea from green chemistry accelerates profitable pharmaceutical or chemical production and reduces waste generation (Rai and Gupta 2021). Many researchers are focused on developing toxic and solvent-free synthetic pathways affordably to utilize green catalysts in the biomedical sector. Green catalysts can also yield biomaterial and biopolymers used in biomedical applications due to their environmentally benign and accuracy (Gildner and Colacot 2015; Tao and Xu 2009).

Biomaterial is a material developed to interact with biological systems to improve, treat, or replace damaged tissue or organs. Biomaterial should be biocompatible and biodegradable. Various biomaterials are being developed from different sources. These biomaterials are used in biomedical applications, including medical treatment like repairing tendons and ligaments, wound healing and cancer therapy, tissue engineering, human body parts manufacturing, preparation of bio-inks for bio printing and biosensors in clinical diagnosis. Biomaterials are produced using green technology or taken from biological substances, so they are environmentally friendly (Krishani et al. 2023a; Biswal et al. 2020). Catalysts are crucial in many industries, from oil refining to pharmaceutical production. The catalyst speeds up the reaction by decreasing the activation energy. Catalysts also marked their footprints in biomaterial synthesis for biomedical applications, including diagnosis, therapeutics and regenerative medicine. As mentioned, the catalysts produced using greener technology and creating less environmental impact are termed green catalysts. Materials from bio sources include enzymes, microorganisms, and biomass-based materials. Materials from nanotechnology include metal and metal oxide nanoparticles; 2D nanomaterials like graphene, black phosphorous, MXene and transition metal dichalcogenides; porous nanomaterials like zeolite, metal-organic framework (MOF) and mesoporous silica nanomaterials and heteropolyacids, are widely used as a green catalyst in the synthesis of biomaterial for biomedical application. High acidity, redox activity, stability, efficiency, selectivity, high surface area, ease of functional group modification, strong electrical conductivity, high tensile strength, and biocompatible and non-toxic properties make the materials mentioned above green catalysts. It also catalyzes some reactions like hydrogenation, oxidation, esterification and dehydration (Zhang et al. 2019b; Mokhtar et al. 2022; Khalaf et al. 2024; Gómez-lópez et al. 2020; Park et al. 2023; Amiryaghoubi et al. 2023).

This chapter highlights how green catalysts are applied in synthesizing biomaterial for biomedical purposes, including biosensors in clinical diagnosis, drug delivery, tissue engineering, wound healing, polymer coatings and inks for bioprinting. With this chapter, we desire to give an overview to help researchers grasp the requirements of sophisticated biomedical applications and expand the implementation of green catalysts to synthesize novel biomaterial for biomedical applications.

9.2 Green Catalyst and Its Classification

Catalysts developed using renewable materials and non-hazardous substances in an environmentally sustainable way are termed green catalysts. Sustainable materials, including photocatalyst, biocatalyst, nanocatalyst and heteropolyacids, act as green catalysts (Fig. 9.1).

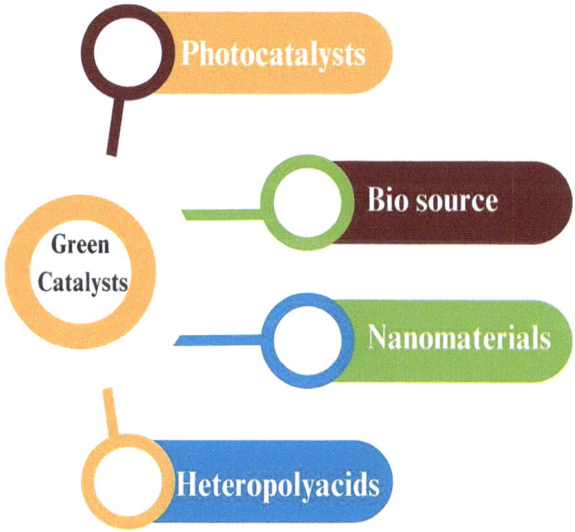

Fig. 9.1 Types of green catalysts

9.2.1 Green Catalyst from the Light Source

Light is the ideal component to create environmentally friendly chemical processes. Photosynthesis is the process by which organisms such as bacteria, algae, and plants convert solar energy into carbohydrates, hydrogen, and oxygen through complex chemical reactions (Dogutan and Nocera 2019). Nature has created effective light garner systems and energy metamorphosis mechanisms that ladle out as researchers' templates. Replicating and applying such complexity in biomedical applications is challenging. Understanding and synthesizing biological processes is therefore crucial (Bigham et al. 2024).

Photocatalysts use light energy to operate chemical processes, resulting in targeted drug release, improved tissue renewal, and accurate visualization of biological structures. Biomimetic and bio-inspired photocatalysts are valuable in biomedicine due to their therapeutic potential and environmental benefits. Biomimetic photocatalytic nanomaterials can serve multiple functions, notably for therapeutic use and regeneration, by releasing inorganic ions. Using biomimetic and bio-inspired photocatalytic substances in tissue renewal yields healing and revival results. In bone or skin cancer treatment, light-responsivity can eliminate malignant cells and release therapeutic ions to promote tissue regeneration (Bigham et al. 2024; Stratakis et al. 2020; Levin et al. 2020).

Photocatalysts often originate from semiconducting precursors like titanium dioxide or zinc oxide, which absorb light energy and stimulate electrons. Excited electrons can combine with other molecules, forming new compounds. Nanoparticles that exhibit semiconducting properties such as light absorbance, charge relocation, and advantageous electronic organization are called nanophotocatalysts. Nanophotocatalysts can produce reactive oxygen species (ROS) when exposed to light. ROS are very reactive chemicals that can cause various biological consequences, including cell death. This feature has been used to produce photodynamic therapy (PDT), a non-invasive therapeutic method that uses light activation of nano photocatalysts to select and kill malignant cells specifically. PDT has demonstrated encouraging outcomes in several forms of cancer, presenting a possible substitute or supplement to established therapy techniques (Fooladi et al. 2023; You et al. 2020). Green photoactive nanoparticles can be produced using various biotic origins, including microbes and plants. This synthesis technique is sustainable, biocompatible, and economical (Hassaan et al. 2023). Nanophotocatalysts from the biogenic method display excellent catalytic activity while eradicating costly and hazardous chemicals (Agarwal et al. 2017; Abdul Salam et al. 2014).

9.2.2 Green Catalyst from Bio Source

Enzymes are one type of bio source catalyst. Enzymes work as natural catalysts in living things to speed up particular chemical reactions. Self-assembled structures are generally formed via enzyme-regulated processes under perpetual circumstances. Enzymatic reactions are one of the primary biotic mechanisms that create macromolecular hydrogel networks for biomedical purposes. These enzymatic mechanisms offer a distinctive opportunity to incorporate hydrogel fabrication. Due to the enzymes' stereo-, regio-, and chemo-selectivity, enzyme catalysis is commonly used to modify the structure and substrates of hydrogels. When processed with diverse enzyme schemes, hydrogels demonstrated outstanding properties as dynamic frames for cells, bioactive compounds, and pharmaceuticals in tissue engineering, drug delivery, and regenerative medicine. Enzyme-mediated crosslinking hydrogels mimic extracellular matrices by exhibiting distinct physicochemical properties and functionalities such as bioactivity, biostability, biocompatibility, biodegradability, water retention capacity, optoelectronic properties, shape memory ability, and self-healing ability (Badali et al. 2021). Enzymes like horse radish peroxidase are catalysts used in bio-printing (Kotani et al. 2023).

Microorganisms, including bacteria and fungi, can produce enzymes or other catalytic particles, such as nanoparticles (NPs). So, microbes can act as a green catalyst in the production process. Bacteria have their systems for living and sustenance uptake that allow them to thrive and multiply. They may reduce metallic compounds and use the energy for themselves. Bacteria have evolved defence mechanisms such as internal sequestration, altering metal ion concentrations, pumping efflux and extracellular precipitation to deal with varied challenges. These bacterial methods can be used in the biogenic synthesis of nanoparticles. Bacterial strains of *Lactobacillus sp.*, *Escherichia coli*, *Acinetobacter calcoaceticus*, *Bacillus licheniformis*, *Bacillus amyloliquefaciens*, *Pseudomonas stutzeri* and *Bacillus megaterium* were used to synthesize silver nanoparticles (Ali et al. 2020).

Ascomycetes (Microscopic filamentous fungi) are tolerant of heavy metals and can internalize and bioaccumulate them. As a result, these organisms were used to reduce and stabilize NPs throughout their production processes (Šebesta et al. 2023).

Microalgae can be used as an effective bionanofactory to produce metallic nanoparticles by reducing metal ions such as gold and silver. Metal nanoparticles are readily synthesized from both live and dried microalgae biomass. Several microalgae, such as Lyngbya majuscule (Duraisamy et al. 2020), Spirulina platensis (Rashad et al. 2019), and Chlorella vulgaris (Mahajan et al. 2019), have been used for silver nanoparticle synthesis.

Biomass-derived materials are either from carbohydrates or proteins; hence, they contain the repeating units of saccharides or amino acids that are biocompatible and biodegradable, making them ideal for biomedical purposes. Proteins and carbohydrate biomass-derived material possess $-NH_2$, $-OH$ and $-COOH$ as functional groups in their configurations. These features of bioactive moieties enable crosslinking of biomass materials with polymers and metal ions, as well as chemical grafting of biomolecules, resulting in versatile functions such as printability, selectivity, and sensitivity for biomedical applications. Chitosan, cellulose, lignin, gelatin and alginate are some biomass-derived materials used as biosource catalysts (Li 2019). Egg white is a catalyst for synthesizing phenothiazinium Schiff base and its nano silver complex. Synthesized compounds displayed excellent antibacterial activity against *Staphylococcus aureus* and *Escherichia coli* strains (El-Kebir et al. 2016).

Nanocellulose, a renewable, nanoscaled cellulose fibril generated from plant cell walls, has emerged as a promising candidate in the search for perfect catalysts. With a large surface area, outstanding mechanical strength, and simplicity of chemical modification, its usage in catalytic processes offers a sustainable and efficient substitute for traditional catalysts (Thomas et al. 2018). Hydroxyapatite is a biomineral in the bone that acts as a green catalyst due to its multifunctionality and high flexibility (Yook et al. 2023). Hydroxyapatite is widely used in bone tissue engineering, and it can be synthesized from calcium-rich materials, counting waste shells (Murugiah et al. 2021; Krishani et al. 2023b).

9.2.3 Green Catalyst from Nanotechnology

Nanotechnology is a potential field for addressing various public health concerns. Nanoparticles are extensively utilized in tissue renewal and medicine because of their distinctive characteristics, including physiochemical rigidity, expanded surface area, and being toxic-free and biocompatible (Bharathala and Sharma 2019). Metal and metal oxide nanoparticles have been the most extensively researched material groups. Countless studies have found that metals and metal oxide nanoparticles have considerable therapeutic advantages. These advantages lead to their use in various sectors, including biosensors, drug delivery systems, diagnostic imaging applications, and scaffolds for tissue engineering (Yaqoob et al. 2020; Nadaf et al. 2022). Using natural plant extracts to synthesize inorganic NPs is straightforward, cheap, and environmentally benign. Specifically, extracts from leaves, seeds, roots, flowers, and fruits were employed. Plant extracts have gained popularity for their capacity to decrease and stabilize metal nanoparticles in a single step, harnessing their natural properties. Plant extracts contain terpenoids, alkaloids, saponins, flavonoids, tannins, and steroids, stabilizing and reducing agents (Ishak et al. 2019). Metals like gold (Au) and silver (Ag) NPs are the most commonly utilized in the catalytic sector due to their catalytic solid reduction, antimicrobial, and optoelectronic capabilities (Ahmed et al. 2017). For example, Folorunso et al. developed AuNPs using Annona muricata leaf extracts (Folorunso et al. 2019). Moreno-Luna synthesized gold nanoparticles using a quick, unique and straightforward approach using an aqueous extract from Agave potatorum (Moreno-Luna et al. 2019). Anandan et al. synthesized silver nanoparticles from *Dodonaea viscosa* aqueous leaf extract and showed effective inhibition against developing A549 NSCLC cancer cells (Anandan et al. 2019). Nabi et al. synthesized TiO_2 nanoparticles from cinnamon powder using an environmentally friendly green method and showed excellent photocatalytic properties (Nabi et al. 2020).

Two-dimensional (2D) biomaterials, as a unique nano platform with planar topology, have sparked interest in sectors such as biomedicine over the last decade due to their distinct arrangement, physicochemical properties, and biological effects. Motivated by graphene's success in biomedicine, hundreds of ultrathin 2D biomaterials have found diverse bio-applications, including biosensing, biomedical imaging, therapeutic agent delivery, cancer therapeutics, and tissue engineering (Huang et al. 2021). Biomaterials comprise several biomolecules, including polysaccharides, amino acids, proteins, peptides, enzymes, lipids, DNA, RNA and lipids. Adsorption of biomolecules onto the surface of nanoparticles is a ubiquitous process in nature (Du et al. 2019. 2022). 2D materials exhibit extraordinary surface sensitivity. Biomolecules bind to the surface and edge of 2D materials, enhancing their properties and applications. Biomolecules vary in shape and size, with more complex designs. Biomolecules interact with 2D materials, leading to numerous possibilities for generating biomaterials with distinct functionality (Ji et al. 2019; Wang et al. 2021b). Graphene, black phosphorus, Mxenes, and transition metal dichalcogenides (TMDCs) are well-known 2D materials used in biomedical applications (Wang et al. 2021a, b).

Graphene (GR) was used to create scaffolds for tissue engineering. GR is a highly effective material for tissue engineering because of its wide surface area, strong electrical conductivity, high tensile strength, and ease of functional group modification. Graphene oxide (GO) and biocomposites have been used to promote chondrogenesis in mesenchymal stem cells. GR/GO-based materials may serve as a planar culture platform for stem cell differentiation, increasing interest in articular tissue engineering and regeneration. GO may stimulate growth factors and proteins. It can transport chondrogenic chemicals, such as fibronectin (FN) and transform growth factor-β3 through electron clouds π-π bonding, facilitating stem cell differentiation and proliferation. Furthermore, GR can stimulate the expression of biomolecules such as aggrecan, collagen II, and glycosaminoglycans (GAG), resulting in improved extracellular matrix (ECM) composition. As a result, GR-based materials are extremely useful in cartilage and articular tissue engineering applications (Amiryaghoubi et al. 2023).

Phosphorene, often known as black phosphorus (BP), is a two-dimensional (2D) substance that has acquired significant attention and is being paid to several contemporary research topics. Its distinctive qualities include an adjustable bandgap width, exceptional surface activity, advantageous on/off current ratios, excellent biocompatibility, rapid biodegradation and infrared-light responsiveness. Distinguish this material from conventional two-dimensional materials. BP's applicability in the biomedical field has been overgrown recently, especially in cancer, bone, skin, nerve, kidney, and biosensing-related treatments (Liu et al. 2021).

MXene, a two-dimensional substance, can be produced in various ways. The synthesis technique determines MXene's electrical conductivity, surface functional groups, thermal stability, and other features. MXene is currently used in tissue engineering and regenerative medicine due to its hydrophilicity, light solid absorbance, photothermal conversion capabilities and vast surface area. MXene can rebuild nerves, skin, bones, and other tissues. MXene-based scaffolds can be employed for tumour ablation via Photothermal therapy (PTT), new bone production, and antimicrobial compounds. MXene also uses electrical stimulation to release drugs efficiently in wound-healing regions. Furthermore, MXenes have been used to develop artificial nerves with implantable neural interfaces and high electrical conductivity. Recently, the tetrapropylammonium bromide (TPABr) high electrical conductivity of MXene has been used in muscle and cardiac tissues (Park et al. 2023).

Transition metal dichalcogenides (TMDCs) contain transition metal atoms like Mo, W, and Re sandwiched covalently between two hexagonal planes of chalcogen atoms such as Se, S, and Te. Solid and powerful chemical bonds exist inside each layer, yet bonds between layers are fragile. As a result, the crystals have a simple primary fissure, lubricity, and determined anisotropy, which explains why the scientific community is so interested in this family of chemicals. Bone regeneration scaffolds composed of TMDC compounds demonstrated increased mechanical strength, positioning them as viable materials for tissue engineering scaffold design and 3D printing (Anju and Mohanan 2021).

MoS_2 is a well-studied material in the TMDC family due to its resilience. MoS_2 has a unique combination of atomic-scale thickness, spin-orbit solid coupling, direct tunable bandgap, and electrical and mechanical properties, making them promising candidates for basic research. These intriguing capabilities can impact optoelectronics, nanomedicine, DNA sequencing, customized medicine, therapies, and other biomedical purposes (Choi et al. 2017).

The inadequate auto-healing and regeneration capacity of human bone necessitates using bone regeneration scaffolds. Various osteoinductive materials have been considered for this purpose, with TMDC-based compounds receiving the most attention recently. TMDC-based materials can create promising biocompatible scaffolds with impressive renewal capacity. One such study reported MoS_2 nanosheets formed on the strut surface of bioceramic material, demonstrating the properties of a potential dual functional scaffold with photothermal therapeutic efficacy. In vivo, the scaffold favourably allowed osteogenic bone marrow mesenchymal stem cells (BMSCs) to proliferate and differentiate (Li et al. 2017). It is a viable clinical technique for effectively treating tumour-induced bone abnormalities. A further theory associated with TMDC family materials is their ability to substitute 3D printing inks. Non-hazardous water-based inks were created using exfoliated 2D materials. Grotta et al. found that by employing MoS_2 nanosheets, they could effectively formulate mechanically strong and chemically stable 3D-printed inks (Grotta et al. 2017).

Ordered micropores, high surface area, stability, tunable acidity and active site access are the attractive properties of the porous nanomaterial. Zeolite, metal-organic framework (MOF) and mesoporous silica nanomaterials like Santa Barbara Amorphous (SBA-15) and Mobil Crystalline Materials (MCM-41) are some examples of porous nanomaterials which are used as a green catalyst in biomaterial synthesis in biomedical applications (Gómez-lópez et al. 2020).

SBA-15 is assembled in hexagonal pores with uniform pore sizes up to 30 nm, and it also acts as a carrier of large biomolecules like proteins. It has more structural stability, which helps in premature protein degradation from body fluids and provides localized and sustained release of therapeutics (Gkiliopoulos et al. 2022). Zhou. P et al. prepared a composite scaffold for bone tissue engineering using SBA-15 nanoparticle, in which rhBMP-2 was incorporated into SBA-15 to treat bone defects. The developed composite scaffold displayed better biocompatibility, bioactivity and antibacterial activity. It also induced in-vitro osteogenic differentiation and in vivo ectopic ossification (Zhou et al. 2014).

A multidisciplinary approach is used by tissue engineering and regenerative medicine to develop and use novel materials to treat various tissue abnormalities. Appropriate tissue regeneration usually requires both positive cellular activity and associated biocompatibility. It could be the outcome of carefully choosing a platform from the startling number of structural options with different porosity features (pore size, connectivity, etc.). Zeolite is a type of porous structure recognized as a microporous tectosilicate and may be used in tissue engineering applications to create a biological microenvironment. Furthermore, zeolite has shown great promise in wound dressing and tooth- and bone-oriented scaffolds. Zeolitic materials have a distinct character due to their wide variety of composition and hierarchical pore structure, mainly used for tissue engineering applications (Zarrintaj et al. 2020).

9.2.4 Green Catalyst from Heteropolyacids

Heteropolyacids (HPAs) are inorganic chemicals with remarkable potential as green catalysts. These metal-oxygen compounds have unique catalytic characteristics due to their redox activity and high acidity. One of the primary benefits of employing HPAs as a catalyst is that they are environmentally friendly. Unlike conventional catalysts, HPAs are safe and can be utilized in mild reaction conditions. Hence, HPA are suitable for use as green catalysts. HPAs operate as catalysts in various processes, including hydrogenation, oxidation, esterification and dehydration. They are also utilized in polymer and pharmaceutical productions. In addition to their environmental credentials, HPAs have several other benefits as catalysts. They are exceedingly firm and used repeatedly without losing catalytic activity. They are highly selective towards specific reactions, leading to less waste and increased efficiency (Zhang et al. 2019b; Mokhtar et al. 2022; Khalaf et al. 2024). HPAs are composed of nanometric transition metal-oxygen octahedral anions. They form a steady and tightly packed backbone of polymeric oxoanions as a fundamental organization and exhibit various configurations. HPAs with a Keggin-type structure are the most investigated and employed, notably as acid catalysts, because of their accessibility and chemical firmness (Escobar et al. 2021).

9.3 Biomedical Applications

Using green catalysts for biomaterial synthesis is one of the significant achievements in biomedical applications. Some of the green catalysts used for a synthesising biomaterial in biomedical applications are shown in Fig. 9.2, including drug

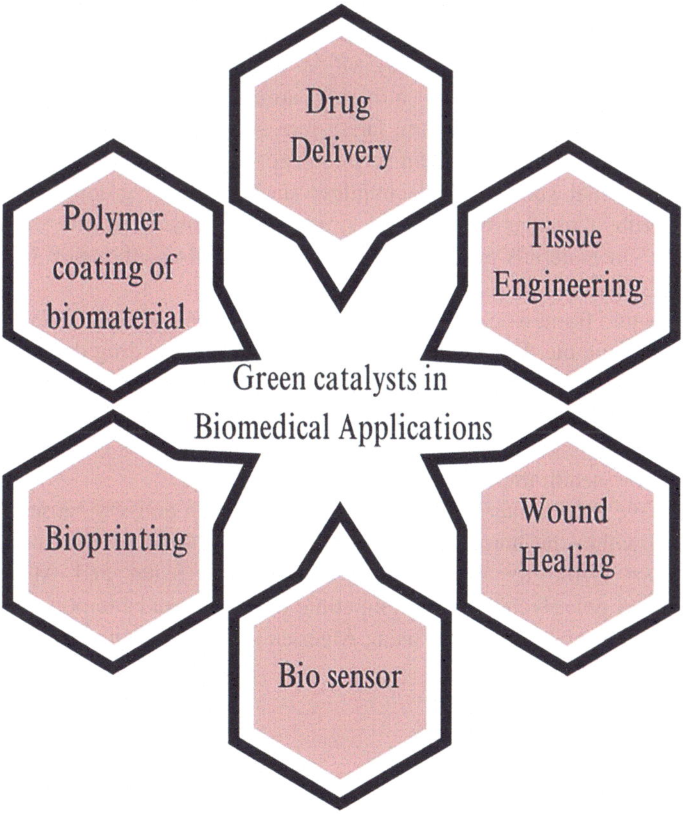

Fig. 9.2 Green catalysts in biomedical applications

delivery, tissue engineering, polymer coating of biomaterials, bio sensor for clinical diagnosis, wound healing and inks for bioprinting.

9.3.1 Drug Delivery

Drug delivery is one of the most extensively explored applications of carbon-based nanomaterials. Considering the possibility of carbon nanomaterials and the countless interactions, such as hydrophobic, pi-pi stacking and electrostatic attraction, which favour loading a wide range of drug molecules. The enormous surface area of these carbon entities allows for great encapsulation and drug-loading efficiency. Carbon dots are used for medicine delivery in addition to their other features. Considering the biocompatibility, low relative cytotoxicity, and good cell uptake, they are excellent materials for drug and gene delivery (Kumar et al. 2020). Azizi A. et al. developed a Fe_3O_4/cellulose nanocomposite using magnetic iron oxide nanoparticles synthesized from the biogenic method using an aqueous extract of spent tea waste. The obtained nanocomposite showed controlled magnetic conveyability, biodegradability, toxic-free, and high absorption capacity, and it can be applied in metronidazole drug carrier systems (Azizi 2020). MCM-41 is a silica derivative with several outstanding properties, including a highly arranged porous system with pore diameters ranging from 2 to 10 nm, large pore volumes, high Brunauer-Emmett-Teller (BET) surface areas, excellent sorption capacity, biocompatibility, toxic free, exceptional thermal stability, and countless surface silanol groups. As a result, MCM-41, with a regular mesoporous structure, is considered one of the most excellent lay-on materials and drug carriers for many drugs/therapies (Shariatinia and Pourzadi 2021).

Metal-organic frameworks (MOFs) have considerable uses, particularly in drug storage. More suitable drug carriers are needed for effective drug delivery in vitro and in vivo—bio-metal organic frameworks (Bio-MOFs) effectively extend MOFs. Biomolecule linkers and biocompatible metal cations can be used to create bio-MOFs. These materials can be effective medication carriers in the human body without adverse health impacts.

The massive MOF materials employed for medication delivery caused compatibility issues within the human body. Over time, nano-sized MOF materials have become more viable for drug storage and release (Sattar and Athar 2018). Hyperbranched polyglycidol, a biocompatible and biodegradable polymer used in drug delivery systems, is obtained using Algerian montmorillonite clay as a catalyst (El-Kebir et al. 2016).

9.3.2 Polymer Coating

Cardiovascular stents are often coated with a medication designed to bear the artery physically and, more recently, to decrease extravagant scar tissue generation attributable to the arterial wall's exposure to the balloon. To help with drug delivery to the stent, a polymer coating has been created to prevent the drug from washing off during device deployment, giving accurate loading and firm release of the drug (Driver 2012; Gobalakrishnan et al. 2021). Hataminia F et al. used Fe_3O_4 nanoparticles as a novel catalyst to create thermoplastic as a stable gel coating on cardiovascular stents. The oxidation of starch by iron oxide nanoparticles as a catalyst was investigated in that research (Hataminia et al. 2022). Metal oxides have substantial advantages in various applications, including catalysts, since changes in surface characteristics impact the material's bandgap energy (Saleh 2020). Iron oxide nanoparticles are familiar due to their low toxicity and production costs. The application of nanoparticles in polymer composites to increase mechanical properties is prevalent. Indeed, at acidic pH, the electrostatic interaction between the positively charged Fe_3O_4 nanoparticles' outer surface and the negatively charged starch's oxygen atoms improves their mechanical characteristics (Dung et al. 2009).

Furthermore, after oxidation, COOH plays an essential role in electrostatic interactions on the surface of NP. COOH is unsuitable because it enhances the structure's hydrophilicity, yet it is pretty effective for electrostatic coupling on the NP surface. The carbonyl groups generated by the oxidation reaction boost the polymer's hydrophobicity, and both functional groups upgrade their characteristics overall (Hataminia et al. 2022).

9.3.3 Biosensor

Preliminary detection of diseases can improve the efficacy of therapies, such as prompt diagnosis and treatment. Biosensors, or devices incorporating biological recognition components to identify disease biomarkers, have been used for superior health management. The most recent advancements in biosensor technology have led to the evolution of exact, error-free, sturdy, and speedy devices that can recognize disease-dependent changes in analyte levels, allowing for quick diagnosis and treatment of numerous diseases (Akolpoglu et al. 2019). Biosensors are analytical instruments that detect biomarkers and produce quantitative readings of a target substance in a sample. Biosensors are primarily based on biomolecular recognition, which creates a quantifiable signal. This data is created when the target analyte interacts with a biological receptor, such as enzymes, antibodies, or DNA (de Assis et al. 2023). Enzymes, metal nanoparticles synthesized using green methods, graphene and graphene oxide and nano cellulose can act as green catalysts mentioned earlier in this chapter. These green catalysts are widely applied in the biosensors. Arginine deiminase is an enzyme that catalyzes the conversion of L-arginine to ammonia and citrulline. Due to its specific activity on arginine, Arginine deiminase is used in developing biosensors for arginine (Kawatra et al. 2022). Arginase, L-arginine decarboxylase and urease are the other enzymes used in biosensors for arginine (Berketa et al. 2022). Nanocellulose, as a natural nanomaterial, has unique properties that make it a suitable choice for use in biosensing. It is a renewable, biodegradable, readily accessible, and adaptable material with excellent mechanical, thermal, and physical-chemical properties.

Nanocellulose's large surface area enables extensive chemical modification and greater immobilization of biological recognition receptors. The plethora of customizable functional groups on the surface of nanocellulose presents potential opportunities for surface functionalization and customization of a biosensor to satisfy specific electro-catalytic demands (Khorsandi et al. 2024). Nanocellulose-based biosensors detect high urea levels in blood, which helps prevent kidney failure (Nwosu et al. 2020). Nanocellulose-based biosensors are also used in disease detection, including dengue fever (Pirich et al. 2017) and S*taphylococcus aureus detection* in blood serum (Ranjbar and Shahrokhian 2018). Bollella et al. synthesized silver and gold nanoparticles using quercetin as an environmentally friendly reducing agent and showed better lactose detection results (Bollella et al. 2017). Single-stranded DNA is quickly immobilized on the surface of graphene layers by π-π interactions between the hexagonal cells of graphene or graphene oxide and the aromatic bases. These hybrids are biosensors for detecting DNA sequences connected to numerous diseases and hereditary disorders (Georgakilas et al. 2016). The biosensors currently developed for diagnosis purposes using green catalysts are listed in Table 9.1.

9.3.4 Tissue Engineering

During regular metabolic processes, unstable, extremely reactive, and partially reduced oxygen atoms are produced. These chemically reactive molecules, reactive oxygen species (ROS), are crucial for controlling nearly every biological process (Kim et al. 2021). ROS is essential for preserving intracellular homeostasis because it regulates transcription factors and phosphorylates proteins at physiological

Table 9.1 List of biosensors developed using green catalysts and its source

Green catalyst used	Source	Analyte	References
Copper oxide nanoparticles	*Caesalpinia bonducella*	Vitamin B_2	Sukumar et al. (2020)
Copper nanoparticles	*Ocimum tenuiflorum*	Glucose	Dayakar et al. (2017)
Gold nanoclusters	Onion membranes	Sucrose	Bagal-Kestwal et al. (2015)
Gold nanoparticles	Eggshell membrane	Glucose	Zheng et al. (2010)
Silver nanoparticles	Quercetin	Lactose	Bollella et al. (2017)
Gold nanoparticles	Papaya juice	L-lysine	Yu et al. (2019)
Silver nanoparticles	*Araucaria angustifolia*	Paracetamol	Zamarchi and Vieira (2021)
Zinc oxide nanoparticles	Peach extract	Glucose	Muthuchamy et al. (2018)
Graphene oxide with gold nanoparticles	Rose water	Glucose	Tabrizi and Varkani (2014)

quantities and signals (Sassetti et al. 2021). Thus, cell development and survival depend on keeping these intracellular ROS homeostasis (Rhee 2006). Accumulation of ROS damages proteins, lipids, DNA, and RNA and is crucial for cytokine and growth factor signalling (Jastrzębska et al. 2021). Maintaining the proper equilibrium of reactive oxygen species (ROS) is vital in tissue engineering due to the physiological properties of free radicals (Castro et al. 2021). MXene causes the production of ROS and modifies the degree of oxidative stress that cells go through (Zheng et al. 2021). The ability to generate ROS has been extensively investigated for several MXenes (Szuplewska et al. 2019; Jastrzębska et al. 2017; Lim et al. 2021). MXene and cells come into direct physiochemical contact, which causes oxidative stress and ROS production, which in turn causes cellular damage and eventual death (Szuplewska et al. 2019). Some MXene often has a negative zeta potential on its surface because of functional groups like Ti-O, Ti-OH, and Ti-F. These groups help MXene have a high affinity for cell surfaces and facilitate simple chemical processes. Ti-F groups, mainly, are unstable at high pH values and rapidly generate reduced iron species, followed by another redox reaction. ROS are produced by this complex redox process, which involves the termination of oxygen, hydroxyl group and fluorine surfaces. Fe-layered titanium carbide MXene may have a noticeably higher capacity for ROS formation since it elevates the quantity of TiO_2 on the MXene surface (Rosales et al. 2020).

Furthermore, MXenes with high conductivity can generate ROS by forming an electron-transfer bridge with the lipid bilayer of cells. ROS are produced when direct contact with MXenes, which transports electrons from the intracellular compartment to the external environment despite the cell membrane's insulation (Rasool et al. 2016). These findings suggest that MXene's ability to affect cell survival and activity can be helpful in tissue engineering (Park et al. 2023).

Graphene-based nanoparticles have also been utilized in tissue engineering to repair injured connective tissue. For example, composites based on graphene oxide (GO) and silk fibroin displayed better proliferation, adherence, and viability of human periodontal ligament mesenchymal stem cells (PDLSCs) than silk fibroin alone (Hasan et al. 2018).

Using the sonochemical method, Kumar et al. developed carbon dot(CD)—based nanomaterials for neural engineering. The investigations revealed excellent biocompatibility, even at low CD concentrations. Sonochemically produced CDs are considered more promising materials for biomedical and tissue engineering applications due to their minimal cytotoxicity compared to chemical-based formulations (Kumar et al. 2019).

Liu et al. developed an injectable BP hydrogel infused with carbon nanotubes (CNT) to enhance mechanical strength, electrical conductivity, and phosphate ion release for bone tissue engineering. The hydrogel was created using biodegradable and cross-linkable oligo(poly(ethylene glycol) fumarate) (OPF) polymer as the primary matrix. CNTs are known to improve hydrogel conductivity and stimulate cell proliferation. Crosslinkable CNT-PEG acrylate (CNTpega) improved gel mechanics and conductivity. BPNSs were introduced to release phosphate ions through in-situ phosphorus oxidation. In vitro experiments demonstrated that this gel improves preosteoblast cell adhesion, proliferation, and osteogenic differentiation. The gel promotes osteogenic pathway gene expressions in preosteoblast cells when stimulated with electricity. X-ray imaging revealed that the BP-CNTpega-gel has outstanding in situ gelation ability and could potentially be used as injectable formulations for bone applications such as posterolateral spinal fusion, femur defects, and vertebral body cavities (Liu et al. 2020).

Chen et al. suggested a unique technique for developing a biomimetic hydrogel scaffold with a dual-differential microenvironment for articular osteochondral defect repair. Kartogenin is an effective chondrogenic drug, whereas atorvastatin has a synergetic impact on cartilage development and bone regeneration. The upper cartilage layer was created by grafting Kartogenin into gelatin using an enzyme crosslinking reaction based on p-gt (HPA). The bone-specific microenvironment was created by grafting atorvastatin into gelatin using a dual-crosslinked network that included both enzyme crosslinking and glycidyl methacrylate-based photo-crosslinking reactions as the lower subchondral bone layer. Including a tidemark-like middle layer promotes the creation of well-defined cartilage-bone integrated architecture. Furthermore, in vitro investigations revealed a considerable mechanical difference between three layers: successful drug grafting, high cytocompatibility, and tissue-specific induced function. Meanwhile, in vivo investigations demonstrated the feasibility of mending articular osteochondral lesions with trilayered biomimetic hydrogel scaffolds by activating endogenous recovery, indicating a possible alternative for future therapeutic therapy (Chen et al. 2024).

Osteoporosis is a condition that results in decreased bone density and bone fragility; it is typically identified in older adults and has a protracted fracture healing period (Pérez-Amodio and Engel n.d.). A biomimetic hydrogel of graphene and chitosan can deliver anti-osteoporotic drugs when triggered with Near-infrared (NIR) irradiation. Teriparatide, a popular osteoporosis drug, has minimal efficacy when delivered systemically. However, local release has been shown to improve outcomes. A study compared the effects of constant and pulsatile Teriparatide release on bone marrow mesenchymal stem cells in vitro, revealing that pulsatile release showed better results. The hydrogel was tested in a bone defect in vivo, and it found that NIR irradiation triggered pulsatile release of the encapsulated drugs. The replicated localized release of Teriparatide promoted osteoporotic bone regeneration by retaining the drug's concentration in the therapeutic window while avoiding undesired systemic adverse effects (Wang et al. 2021a, b).

Osteoarthritis is a joint illness diagnosed in various pathological circumstances, including cartilage loss and synovial inflammation. Anti-inflammatory drugs can protect joint cartilage in the short term, but their effectiveness is limited. Whereas surgery for joint replacement is effective, it comes with high risks of failure and expensive costs (Jafari et al. 2021). Copper sulfide nanoparticles incorporating dexamethasone sodium phosphate were produced using cell membrane coating for osteoarthritis treatment. The coating layer comprises neutrophil and erythrocyte membranes. Nanoparticles exhibited good biocompatibility, regulated drug release, and photothermal conversion in vitro. Compared to free dexamethasone sodium phosphate and drug-loaded nanoparticles without NIR, the ones triggered by external laser light had more anti-inflammatory effects. Nanoparticles may target inflammatory joint locations by recognizing CD11a on neutrophil membranes. In vivo studies on nanoparticle efficacy concluded that combining PTT with drug release lowered cartilage degeneration and synovial inflammation (Xue et al. 2022).

Wang et al. developed an injectable HP-based (Gtn–HPA) hydrogel using an oxidative coupling reaction of phenol moiety catalyzed by horse radish peroxidase (HRP) and hydrogen peroxide (H_2O_2) for neural tissue engineering. This Gtn–HPA hydrogel system offers an easy and effective approach to investigating cell functional responses in a 3D environment. It features outstanding cell adhesion, variable mechanical properties, and in situ forming capacity. The results revealed that Gtn–HPA hydrogel is a viable stem cell-based in vivo therapeutics method. It provides an encouraging platform for regenerative uses of stem cells in tissue engineering because of its biocompatibility and biodegradability. Injectable Gtn-HPA hydrogels with mechanical characteristics that can be adjusted for 3D cell culture and differentiation represent a significant tactical tool for treating future brain injuries and neurological conditions (Wang et al. 2010). Because of their flexible crosslinking, Gtn-HPA hydrogels can control the migration and proliferation of adult neural stem cells (aNSCs). In addition, Gtn-HPA hydrogels can favour neural cells by encouraging neuronal differentiation and upholding a greater degree of cell survival than neuronal differentiation (Lim et al. 2012).

A scaffold for tissue engineering was developed by combining the carboxyl group of HPA and the amino group of human-like collagen (HLC) using N-hydroxysuccinimide (NHS) and 1-ethyl-3-(3-dimethylaminopropyl)-carbodiimide hydrochloride (EDC) and followed by crosslinking with H_2O_2 and HRP. It is an injectable HLC-HPA hydrogel based on HLC. The developed HLC-HPA hydrogel revealed good biodegradability and excellent biocompatibility, as confirmed by in vitro cell viability and in vivo degradation studies. Interestingly, the HLC-HPA hydrogel demonstrated anti-inflammatory activity in cultured macrophages by increasing the secretion of the anti-inflammatory cytokine interleukin-10, which was induced by lipopolysaccharide, and decreasing the amount of pro-inflammatory cytokines tumour necrosis factor-α and interleukin-6. On the other hand, HLC-HPA hydrogels displayed outstanding biocompatibility, hemostasis, and a reduced inflammatory response in animal studies. Thus, there is much encouragement for developing the HLC-HPA hydrogel for use in the biomedical domain (Gao et al. 2020). Some examples of green catalyst-based biomaterial composite and its application in tissue engineering are listed in Table 9.2.

9.3.5 Wound Healing

Enzymes are essential in wound healing by dampening ROS, facilitating cell growth and infiltration, averting microbial inflammation, and allowing in-built infection detection systems (Guebitz and Nyanhongo 2018). Enzymes like laccases, peroxidases, tyrosinases, and transglutaminases are commonly used to produce wound dressing hydrogels. Enzymes' specificity and selectivity and their capability

Table 9.2 List of green catalyst-based biomaterial in tissue engineering applications

Green chemistry based biomaterial composite	Application	Inference	References
Calcium phosphate-recombinant human bone	Bone tissue engineering	Toxic-free and environmentally friendly solvents were used	Wang et al. (2017a)
Polycaprolactone, chitosan; and gelatin	Skin tissue engineering	Atom efficiency—a green chemistry principle is applied in the production method	Gomes et al. (2017)
Chitosan and graphene oxide composite	Skeletal muscle tissue engineering	Showed excellent biodegradability and biocompatibility	Jing et al. (2017)
Healing-inspired collagen targeting glue	Skin tissue engineering	Atom efficiency—a green chemistry principle is applied in the production method	Jeon et al. (2017)
Polyurethane/small intestinal submucosa	Soft tissue engineering	Less cytotoxic product was obtained	Da et al. (2017)
Gellan gum	Skeletal muscle tissue engineering	Toxic-free and environmentally friendly solvents were used	Berti et al. (2017)
Gelatin/sulfated or nonsulfated hyaluronan	Skin tissue engineering	Atom efficiency—a green chemistry principle is applied in the production method	Bhowmick et al. (2017)
Beta vulgaris	Skin tissue engineering	Less toxic chemicals are used in the synthesis method	Ranjbarvan et al. (2018)
Nanofluorohydroxyapatite/ poly(lactic-co-glycolic acid)/	Bone tissue engineering	Atom efficiency—a green chemistry principle is applied in the production method	Tahriri et al. (2018)
Chitosan and gelatin composite	Vascular tissue engineering	Showed enhanced biological and mechanical properties	Badhe et al. (2017)
Bacterial cellulose	Skin tissue engineering	Toxic-free and environmentally friendly solvents were used	Keskin et al. (2017)
Hydroxylapatite nanorods	Bone tissue engineering	Less toxic chemicals are used in the synthesis method	Moeini et al. (2017)
Alginate, gelatin and heparin composition	Skeletal muscle tissue engineering	Cheap to manufacture	Yi et al. (2017)
Poly(l-lactic acid)/ polyaniline	Cardiac tissue engineering	Toxic-free and environmentally friendly solvents were used	Wang et al. (2017b)
Hydroxyethyl cellulose / silver nanoparticles	Skin tissue engineering	Toxic-free and environmentally friendly solvents were used	Zulkifli et al. (2017)
Polycaprolactone (PCL)-functionalized gelatin nanofiber	Vascular tissue engineering	Showed excellent biodegradability and biocompatibility	Coimbra et al. (2017)
Gold nanoparticle-coated Cholecyst-derived extracellular matrix.	Skeletal muscle tissue engineering	Toxic-free and environmentally friendly solvents were used	Nair et al. (2017)

to imitate in vivo biosynthetic processes make it simple and secure to incorporate wound healing-boosting molecules and cells during wound dressing hydrogel (WDH) synthesis. Hydrogels are hydrophilic polymeric networks that keep hold of large amounts of water while maintaining shape. Their three-dimensional (3D) structure effectively mimics cell and tissue culture conditions, making them popular for encapsulating cells in a microenvironment. Hydrogels effectively deliver physiologically active substances like growth factors and organize cells and tissues due to their multilayered structure (Moreira Teixeira et al. 2012). This feature is consistent with the increasing acceptance of in situ synthesis for complicated sprayable, injectable, and bioprintable wound dressing hydrogels. Transglutaminases catalyze the coupling of a free lysine amine group from a protein or peptide to a deamidated glutamine protein or peptide-bound g-carboxamide group, synthesizing protein/peptide-based WDHs. Oxidative enzymes like peroxidases, laccases, and tyrosinases can create reactive radicals, making them ideal for synthesizing WDHs from natural or synthesized compounds (Guebitz and Nyanhongo 2018; Nyanhongo et al. 2012; Moreira Teixeira et al. 2012).

Salehi et al. created thermoplastic hydrogels that depend on starch and zeolite nanoparticles infused with chamomile extract, a herbal medication. Zeolite-embedded wound dressing composites exhibiting appropriate antioxidant and antibacterial properties showed sustained drug-release behaviour. For example, chamomile reduces inflammation by blocking prostaglandin E2 release triggered by lipopolysaccharide and reducing EDC [COX-2] activity without compromising COX-1, the constitutive form of the enzyme. Zeolite samples have good biocompatibility, particularly for mouse fibroblast, per the MTT assay. Along with angiogenesis, histological examination of treated tissue samples revealed improved collagen production, epithelialization, and reduced inflammation. Because of the dressing's porosity, nanostructure, and controlled fluid absorption rates, burn wounds, ulcers, and bed sores healed more quickly. Notably, in the clinical pilot research, five refractory ulcers healed without hypersensitive reaction (Salehi et al. 2017).

Green synthesized silver nanoparticles from herbal tea (*Stachys lavandulifolia*) (Zangeneh et al. 2019), leaf extract of *Prosopis juliflora* (Arya et al. 2019), kiwifruit juice (Kodasi et al. 2023), *Scutellaria barbata* (Veeraraghavan et al. 2021) displayed excellent results in wound healing. Green synthesized copper oxide nanoparticles from Table Olive also showed better results in burn wound treatment (Wang et al. 2023). Green synthesized zinc oxide nanoparticles from *Calendula officinalis* flower extract showed significant antioxidant and wound-healing properties (Aydin Acar et al. 2024). Ahmed HE et al. developed a chitosan-based hydrogel incorporating Cerium oxide nanoparticles that showed antibacterial activity against *K. pneumonia* and *S. aureus*, significantly improving wound-healing progression (Ahmed et al. 2021).

9.3.6 Bioprinting

Over the last few years, there has been a growing interest in using engineered artificial tissues to mend injured tissues or as in vitro models to assess the human body's reaction to drugs. Researchers are using 3D bioprinting, a new fabrication process, to create artificial tissues that closely resemble authentic tissues. Bioengineers have expanded the applications of 3D bioprinting in biomedical fields, including preoperative surgery simulation, due to its fast and precise controllability (Li 2019). Enzymes like horse radish peroxidase and glucose peroxidase are catalyzed for crosslinking in bioprinting (Peng et al. 2021). Freeform bioprinting, in which cell-suspending ink is ejected into a support bath containing supportive elements, has become an encouraging method for creating cell-laden soft hydrogel constructions (Hinton et al. 2015; Bhattacharjee et al. 2015). The supporting materials provide physical support throughout the 3D printing process, while the gelation of the embedded ink occurs in the support bath. As a result, freeform bioprinting allows for the creation of soft constructions with higher shape accuracy than standard air printing (Shiwarski et al. 2021; Hinton et al. 2015). Sakai et al. developed a hydrogel using a freeform bioprinting process involving gelation mediated by the enzymatic cascade reaction of horseradish peroxidase (HRP) and choline oxidase (COD). COD catalyzes the conversion of choline and oxygen into betaine glycine and hydrogen peroxide. HRP further catalyzes the formation of covalent bonds between phenolic hydroxyl groups through oxidation in the presence of H_2O_2 (Sakai et al. 2021).

9.4 Methods Involved in the Synthesis of Green Catalyst

9.4.1 Green Solvent Synthesis Method of Catalyst

The conventional method used to synthesize catalysts applied at high temperatures and toxic solvents may cause hazards for both the environment and human usage. To overcome these problems and to produce a sustainable catalyst, green solvent synthesis of catalyst was introduced as an alternative method. Water, ionic liquids, and ethanol are renewable, non-toxic, biodegradable, and sustainable solvents that can be used as an alternative to conventional solvents. These synthesis processes can be carried out with low pressure and temperature, which may lead to less energy consumption (Khalaf et al. 2024; Jiang et al. 2020; Yilmaz and Soylak 2020).

Water cannot be reduced or oxidized promptly. Water's other advantages for catalyst use are high polarity and solvation ability. Moreover, water is safe, non-flammable, non-toxic, cheap and readily available. Sheldon states, "The ideal solvent is no solvent; if one is required, water is preferred". Thus, water is a promising green chemistry solvent (Sheldon 2005).

Low volatility, good dissolving capability in different solutions, high thermal stability, and non-toxicity make ionic liquids a green solvent in various production processes. Some ionic liquids immobilize non-metals or metals on polymers or

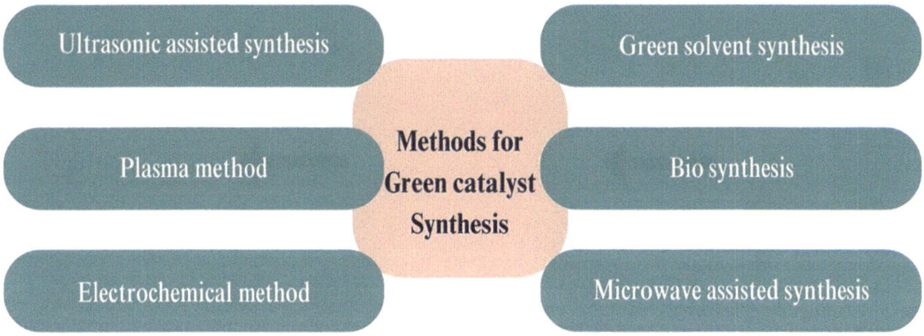

Fig. 9.3 Methods for green catalysts synthesis

minerals, allowing them to be utilized as "immobilized catalysts." or "biphasic catalysts" (Sood et al. 2023). Metal nanoparticles readily synthesized using plant extracts are the best example of green solvent synthesis. Plant extracts contain polyphenols and flavonoids, which can reduce metal ions. This process has been used to create silver nanoparticles from plant extracts, including grapefruit peel, green tea, and ginger. Another example includes the use of ionic liquids as solvents in catalytic reactions. Ionic liquids, low-volatile liquid salts at room temperature, are effective solvents for catalytic reactions such as hydrogenation and oxidation. Furthermore, these ionic liquids can be recyclable and used without losing catalytic activity. An outline of methods involved in synthesising green catalysts is shown in Fig. 9.3.

9.4.2 Biosynthesis Method of Catalyst

This method synthesizes catalysts from living organisms, including microbes, plants, and enzymes capable of synthesizing catalytic material or nanoparticles. Microbes, like bacteria, can also synthesize catalytic materials. *Shewanella oneidensis* is a gram-negative bacterial species that can synthesize magnetite Fe_3O_4 nanoparticles (Chmykhalo et al. 2021). Metal nanoparticle synthesis from microbes and plant extracts are the best examples of biosynthesis methods of catalysts.

For instance, the basic procedure associated with the biosynthesis of silver nanoparticles from plant extract is explained as follows: First, plant extracts are collected by boiling plant parts (leaf, stem, root, seed or flower) using appropriate green solvents. Then, metal ions in precursor material ($AgNO_3$) are introduced to the plant extract. Later, the reduction of metal ions (Ag^+) into atoms (Ag^0) leads to the formation of Ag nanoparticles (Fig. 9.4). The plant extract is a reducing agent and capping mediator (Zuhrotun et al. 2023).

The green synthesis of nanoparticles from microbes can be performed using intracellular and extracellular methods. For extracellular synthesis, after culturing

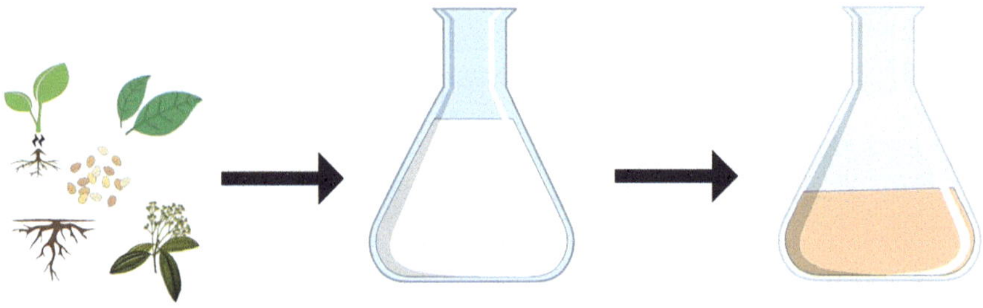

Fig. 9.4 Synthesis of nanoparticles from plant parts

the microbes in optimum growth conditions, the biomass is harvested using centrifugation, and the recovered supernatant is mixed with aqueous metallic salt solution to synthesize the nanoparticle. For the Intracellular method, the harvested biomass is rinsed well with sterile water and incubated with an aqueous metal ion. The change in the colour indicates the synthesis of nanoparticles. Later, the biomass is physically disrupted using ultrasonication to release nanoparticles from the cell wall (Soni et al. 2018). The process associated with the biogenic synthesis of nanoparticles from microbes is mentioned in Fig. 9.5.

9.4.3 Electrochemical Synthesis Method of Catalyst

Electrochemical synthesis is a potential green procedure for creating nanocatalysts due to its ease of use, scalability, and control over nanoparticle size, shape, structure, efficiency and scalability. Electrochemical synthesis uses an electrochemical cell with two electrodes submerged in an electrolyte solution. The electrode responsible for reduction or oxidation is a cathode or anode. When a voltage is applied between the electrodes, electrons are transported from one to the other via the electrolyte solution. This technique causes the formation of nanoparticles on one or more electrodes based on the reaction conditions (Budnikova et al. 2024). A rapid, green, and inventive approach for producing high-quality graphene oxide (GO) sheets using water electrolytic oxidation of graphite has been disclosed by Pei et al. At ambient temperature, the synthesis was conducted after two electrochemical actions. Mercantile pliable graphite samples submerged in H_2SO_4 solutions were first subjected to a high potential to initiate an electrochemical intercalation process. The chemical was utilized as an anode in diluted H_2SO_4 solutions to initiate an electrochemical reaction. Oxygen radicals from the anodic electrocatalytic oxygen evolution reaction create low-oxidation GO layers (Pei et al. 2018).

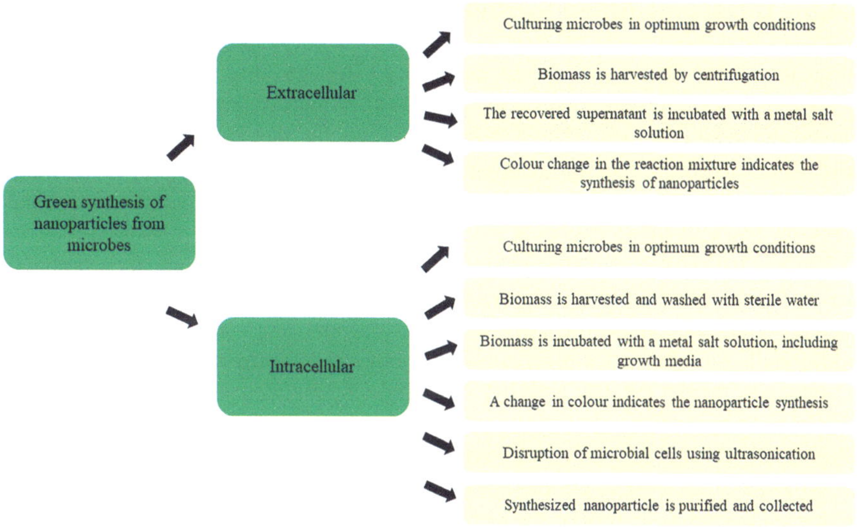

Fig. 9.5 Synthesis of nanoparticles from microbes

9.4.4 Plasma Method

Plasma techniques are a viable way to create green catalysts. Plasma is composed of charged particles. Creating plasma involves applying an electric field to a gas or liquid, which causes it to ionize and become conductive. Plasma technologies have numerous benefits for creating green catalysts. Catalysts are readily synthesized using inexpensive materials like carbon precursors. Plasma approaches can create catalysts with distinct features that are challenging to generate through regular synthesis methods. Plasmas are classified as high and low-temperature. Thermal plasma is formed by heating a gas to elevated temperatures.

In contrast, nonthermal plasma is formed by applying an electric field to a gas at low temperatures. Wang et al. found that both types of plasma can activate and change catalyst surfaces, enhancing their effectiveness in chemical processes. Plasma can affect the properties of reactants to make thermodynamically unfavourable reactions more favourable, even at normal temperatures (Wang et al. 2018; Zhao et al. 2017; Zhao et al. 2019). Guo et al. (2013) used cold plasma to create zirconia with a monoclinic structure at temperatures below 150 °C, usually accomplished at temperatures above 1000 °C. Plasma generates mostly negatively charged nanoparticles, avoiding aggregation and resulting in lower dimensions than thermal techniques. The main benefit of cold plasmas is that they operate at low temperatures. Cold plasmas may produce structural distortions due to undecomposed precursors and moisture in the surroundings, which can hydrate catalyst precursors. In many circumstances, thermal therapy must adhere to the plasma technique (Wang et al. 2018; Huang et al. 2019).

9.4.5 Ultrasonic-Aided Synthesis

Ultrasonic-aided synthesis is a promising strategy for improving catalyst characteristics. Ultrasonic-aided synthesis employs high-frequency sound waves to trigger chemical processes. The procedure follows the fact known as sonic cavitation, which happens as soon as sound waves travel through a liquid medium. It causes tiny bubbles to form and collapse, resulting in high temperatures and pressures that encourage chemical reactions (Chatel 2019).

Ultrasonic-assisted synthesis provides advantages over typical catalyst preparation methods. Initially, because cavitation bubbles have a larger surface area and mass transfer, they allow for rapid reaction rates and more remarkable outcomes. Additionally, it allows for fine control over particle form and size, which can dramatically impact catalyst activity. Ultrasonic-aided synthesis is an environmentally friendly technology that eliminates the need for complex chemicals or elevated temperatures, making it ecofriendly benign (Chatel 2019).

Alinavaz et al. used a simple sonochemical approach (ultrasonic radiation) to create nano-sized nanocomposites with high phase purity (Alinavaz et al. 2023). Yao et al. synthesized mordenite zeolite under diverse conditions, including salt addition, hydrothermal process and temperature, and ultrasonic assistance. The ultrasonic post-treatment of the zeolite sample revealed a unique structure of many mordenite nanosheets that enhanced catalytic stability and attained an elevated carbonylation rate (Yao et al. 2020).

9.4.6 Microwave-Aided Synthesis (MAS)

This technology uses microwave radiation to heat the reaction mixture, producing more rapid and potent synthesis than traditional heating methods. Rapid heating and cooling can shorten reaction times (Rao et al. 1999). Compared to conventional techniques, microwave-aided alumina and boron carbide sintering can save energy and time. It can result in increased outcomes and better control over the size and shape of catalyst particles (Roy et al. 1999). Microwave radiation can heat specific reaction mixture components based on their dielectric characteristics (Leonelli et al. 2013). It can selectively activate precursors or stimulate particular processes while limiting adverse effects. MAS can be conducted without or with minimal solvents, reducing waste, simplifying product purification and being environmentally friendly (Kokel et al. 2019; Wu et al. 2013).

Microwave ovens can operate safely depending on the technology, as the external components are not too hot. Microwave research is becoming increasingly popular due to its potential for time and energy savings and breakthroughs in industrial-scale technologies—microwave heating results in little conductive and convective heat losses. A material's catalytic activity is determined by its surface structure and capacity to speed up reactions. Microwaves improve catalytic activity by accelerating the transition of polar reactant molecules, leading to quicker reaction rates than standard heating methods. Microwave heating can create microplasmas on metal surfaces, resulting in enhanced selectivity. Changes in grain size, poisoning, and coking can all impact catalyst stability. In some situations, microwave-assisted reactions can delay catalyst degradation and extend lifespan (Muley et al. 2021). This process has been used to create metal nanoparticles on diverse materials, including zeolites and carbon nanotubes. These catalysts exhibit enhanced activity and selectivity in numerous methods, including hydrogenation and oxidation (Khalaf et al. 2024).

He et al. (2019) synthesized MCM-41-NH_2 from fly ash by applying an alkali fusion process with microwave assistance, resulting in a shorter ageing period of 30 minutes compared to the traditional hydrothermal method. Golchinvafa et al. used a solution combustion synthesis approach to produce $FeNi_3/NiFe_2O_4$ composites using conventional and microwave heating. Microwave heating produced composites with higher $FeNi_3$ phase and larger sintered particles (Golchinvafa and Masoudpanah 2019).

9.4.7 Alternative Green Methods

An appealing method for producing low-dimensional nanomaterials with an eco-friendly approach involves sonicating Ni^{2+} and Fe^{2+} with MoS_2 acting as two-dimensional templates. This technique uses weak van der Waals interactions at the 2D interfaces to create 2D/2D heterostructures made of NiFe nano-sheets deposited onto MoS_2 layers (Wang et al. 2019). Additionally, Zhang et al. proposed an outstanding method to produce high yields of 2D MOFs. Intending to exfoliate the 3D MOFs and create 2D sheets, they developed a surfactant-mediated technique in which the surfactant's interaction with MOF crystal defects produces a pseudoassembly process. Significant progress has also been made in the environmentally friendly production of MXenes nano-sheets, a relatively new and exciting 2D material (Zhang et al. 2019c). Zhang et al. have progressed in this direction by creating a straightforward template-free process involving fast liquid freezing in nitrogen. It opens the door to producing inexpensive 3D microporous structured MXenes (Zhang et al. 2019d).

Remarkably, zeolite synthesis has been accomplished without the need for solvents. ZSM-5 was synthesized using $TPABr/SiO_2$, rice husk ash, and $Na_2CO_3 \cdot 10H_2O/SiO_2$ mixture to remove the organic template, as reported by Zhang et al. The samples were processed in an autoclave at 150 °C for 72 hours and then calcined at 550 °C for 6 hours. They exhibit similar characteristics when synthesized by hydrothermal solvent-free approaches (Zhang et al. 2019a). Furthermore, it was observed that hierarchical zeolite Co particles with a large surface area and remarkable crystallinity were trapped inside ZSM-5.

Compared to wet impregnation or coprecipitation, the mechanochemical approach has become a green technology. It enables the nonsolvent manufacture of the catalyst anchored on mesoporous material, particularly metal oxide nanoparticles. To demonstrate this, Saberia et al. synthesized Fe_2O_3/SBA-15, for which propionic acid and $Fe(NO_3)_3 \cdot H_2O$ were distributed in a planetary ball mill before being calcined for 30 minutes at 300 °C. The nanomaterial had a high surface area and could be separated magnetically (Zhou et al. 2019). However, information on nonmagnetic Fe–SBA-15 modified by sulfonic groups already exists (Ostovar et al. 2018).

Additionally, Co_3O_4 was supported on SBA15 by Pineda et al. to improve the catalytic qualities, resulting in a bifunctional nanomaterial with redox and acid sites (Pineda et al. 2018). Gold and silver were similarly supported on SBA-15 by taking the same approach. Aluminium species were also held up on MOF, MCM41, and SBA-15. Compared to the impregnation approach, the green synthesis route showed reduced superficial area and less acidity (Marquez-Medina et al. 2019).

Abundant work has been done to create encapsulated NP materials using solvent-free techniques. During the sacrificial phase, the mechanochemical technique enables the nanoparticles in the MOF matrices through a solid-state reaction. Consequently, ball milling of Palladium NPs in Zeolitic imidazolate framework-8 crystal without stabilizing agents was carried out (Li et al. 2019). The nanomaterials have a high surface area and an excellent crystalline structure. Furthermore, a quick and easy solvent-free technique was used to create a mesoporous platinum-iron oxide mechanochemical structure. In a solid-state reaction lasting 0.5 hours, the triblock copolymer and iron salt enable a narrow pore size distribution, a high surface area, and well-dispersed Platinum (Zhou et al. 2019).

9.5 Conclusion

To summarize, using green synthetic techniques, considerable attempts must be made to produce green catalysts with exceptional features such as high firmness, accuracy, and activity. This chapter explains the usage of green catalysts in synthesising biomaterials for biomedical applications, as well as current advancements and practical implementations. Green catalysts are essential to sustainable chemistry, providing eco-friendly alternatives to conventional catalysts. They are prepared using a variety of ways that have a low impact on the environment. Renewable starting materials, including plants, microorganisms, and materials derived from biomass, are commonly used as precursors. Eco-friendly procedures such as hydrothermal synthesis, microwave-aided approaches, and enzymatic activation improve the sustainability of catalysts. Techniques such as impregnation, sol-gel synthesis, and ion exchange use water or toxic-free solvents as the media.

Green catalysts have numerous applications in synthesising biomaterials for biomedical applications, including tissue engineering, drug delivery, wound healing, biosensors in clinical diagnosis, polymer coatings, and inks for bioprinting. Green catalysts provide excellent selectivity, efficiency, and stability, producing hygienic manufacturing and less dependence on destructive chemicals. Using green catalysts can help the industry transition to a more organic future.

Green catalysts have a bright future in analysis that will help them tackle their challenges. As our economy becomes more sustainable, there will be a greater need for eco-friendly and efficient catalysts. Green catalysts have proven helpful in biomedical applications and are expected to improve further. Green catalysts are predicted to become more cheap, ascendable, and steady as technology advances and sustainability consciousness grows. Adopting these technologies has significant promise to create a healthier and more organic future.

Green catalysts confront obstacles in mainstream adoption, but continued analysis opens the way for services to market. Green catalysts are more environmentally friendly and efficient than conventional catalysts, making them a popular choice. Green catalysts can help create a more viable and progressive future.

References

Abdul Salam H, Sivaraj R, Venckatesh R (2014) Green synthesis and characterization of zinc oxide nanoparticles from Ocimum basilicum L. var. purpurascens Benth.-Lamiaceae leaf extract. Mater Lett 131:16–18. Available at: https://doi.org/10.1016/j.matlet.2014.05.033

Agarwal H, Venkat Kumar S, Rajeshkumar S (2017) A review on green synthesis of zinc oxide nanoparticles—an eco-friendly approach. Res Efficient Technol 3(4):406–413. Available at: https://doi.org/10.1016/j.reffit.2017.03.002

Ahmed S, Chaudhry SA, Ikram S (2017) A review on biogenic synthesis of ZnO nanoparticles using plant extracts and microbes: a prospect towards green chemistry. J Photochem Photobiol B Biol 166:272–284

Ahmed HE et al (2021) Green synthesis of CeO2 nanoparticles from the Abelmoschus esculentus extract: evaluation of antioxidant, anticancer, antibacterial, and wound-healing activities. Molecules. Available at: https://doi.org/10.3390/molecules26154659

Akolpoglu MB et al (2019) Biosensing–drug delivery systems for in vivo applications. In: Advanced biosensors for health care applications. Elsevier, pp 249–262

Ali A et al (2020) Advancements in plant and microbe-based synthesis of metallic nanoparticles and their antimicrobial activity against plant pathogens. Nano 10(6):1146. Available at: https://doi.org/10.3390/nano10061146

Alinavaz S et al (2023) Sonochemical synthesis, characterization and investigation of the electrochemical hydrogen storage properties of Ho3Fe5O12/Mg2Al(OH)7 nanocomposites. J Energy Storage 66:107369. Available at: https://doi.org/10.1016/j.est.2023.107369

Amiryaghoubi N et al (2023) Application of graphene in articular cartilage tissue engineering and chondrogenic differentiation. J Drug Deliv Sci Technol 83:104437. Available at: https://doi.org/10.1016/j.jddst.2023.104437

Anandan M et al (2019) Green synthesis of anisotropic silver nanoparticles from the aqueous leaf extract of Dodonaea viscosa with their antibacterial and anticancer activities. Process Biochem 80:80–88. Available at: https://doi.org/10.1016/j.procbio.2019.02.014

Anju S, Mohanan PV (2021) Biomedical applications of transition metal dichalcogenides (TMDCs). Synth Met 271:116610. Available at: https://doi.org/10.1016/j.synthmet.2020.116610

Arya G et al (2019) Catalytic, antibacterial and antibiofilm efficacy of biosynthesised silver nanoparticles using Prosopis juliflora leaf extract along with their wound healing potential. J Photochem Photobiol B Biol 190:50–58. Available at: https://doi.org/10.1016/j.jphotobiol.2018.11.005

Aydin Acar C et al (2024) Green and eco-friendly biosynthesis of zinc oxide nanoparticles using Calendula officinalis flower extract: wound healing potential and antioxidant activity. Int Wound J 21(1):e14413

Azizi A (2020) Green synthesis of Fe3O4 nanoparticles and its application in preparation of Fe3O4/cellulose magnetic nanocomposite: A suitable proposal for drug delivery systems. J Inorg Organomet Polym Mater 30(9):3552–3561. Available at: https://doi.org/10.1007/s10904-020-01500-1

Badali E et al (2021) Enzymatic crosslinked hydrogels for biomedical application. Polym Sci Ser A 63(1):S1–S22. Available at: https://doi.org/10.1134/S0965545X22030026

Badhe RV et al (2017) A composite chitosan-gelatin bi-layered, biomimetic macroporous scaffold for blood vessel tissue engineering. Carbohydr Polym 157:1215–1225

Bagal-Kestwal D, Kestwal RM, Chiang B-H (2015) Invertase-nanogold clusters decorated plant membranes for fluorescence-based sucrose sensor. J Nanobiotechnol 13:1–11

Berketa K et al (2022) Novel highly sensitive conductometric biosensor based on arginine deiminase from mycoplasma hominis for determination of arginine. Sensors Actuators B Chem 367:132023. Available at: https://doi.org/10.1016/j.snb.2022.132023

Berti FV et al (2017) Synthesis and characterization of electroactive gellan gum spongy-like hydrogels for skeletal muscle tissue engineering applications. Tissue Eng A 23(17–18):968–979

Bharathala S, Sharma P (2019) Biomedical applications of nanoparticles. In: Nanotechnology in modern animal biotechnology. Elsevier, pp 113–132

Bhattacharjee T et al (2015) Writing in the granular gel medium. Sci Adv 1(8):e1500655

Bhowmick S et al (2017) Biomimetic electrospun scaffolds from main extracellular matrix components for skin tissue engineering application–the role of chondroitin sulfate and sulfated hyaluronan. Mater Sci Eng C 79:15–22

Bigham A et al (2024) Inspired by nature: bioinspired and biomimetic photocatalysts for biomedical applications. Nano Mater Sci [preprint]. Available at: https://doi.org/10.1016/j.nanoms.2024.02.006

Biswal T, Badjena SK, Pradhan D (2020) Sustainable biomaterials and their applications: a short review. Mater Today Proc 30:274–282

Bollella P et al (2017) Green synthesis and characterization of gold and silver nanoparticles and their application for development of a third generation lactose biosensor. Electroanalysis 29(1):77–86

Budnikova YH et al (2024) Electrochemistry in organics: a powerful tool for "green" synthesis. J Solid State Electrochem 28(3):659–676. Available at: https://doi.org/10.1007/s10008-023-05507-9

Castro B et al (2021) Stress-induced reactive oxygen species compartmentalization, perception and signalling. Nat Plants 7(4):403–412

Chatel G (2019) Sonochemistry in nanocatalysis: the use of ultrasound from the catalyst synthesis to the catalytic reaction. Curr Opin Green Sustain Chem 15:1–6. Available at: https://doi.org/10.1016/j.cogsc.2018.07.004

Chen H et al (2024) Trilayered biomimetic hydrogel scaffolds with dual-differential microenvironment for articular osteochondral defect repair. Mater Today Bio 26:101051. Available at: https://doi.org/10.1016/j.mtbio.2024.101051

Chhangani MK (2016) Catalytic green synthesis of chemicals for sustainable future. Int Res J Manag Sci Technol 7(3):1959–2250

Chmykhalo V et al (2021) Microbial-based magnetic nanoparticles production: a mini-review. Intergr Biol 13(4):98–107. Available at: https://doi.org/10.1093/intbio/zyab005

Choi W et al (2017) Recent development of two-dimensional transition metal dichalcogenides and their applications. Mater Today 20(3):116–130

Coimbra P et al (2017) Coaxial electrospun PCL/gelatin-MA fibers as scaffolds for vascular tissue engineering. Colloids Surf B: Biointerfaces 159:7–15

Da L et al (2017) Composite elastomeric polyurethane scaffolds incorporating small intestinal submucosa for soft tissue engineering. Acta Biomater 59:45–57

Dayakar T et al (2017) Novel synthesis and characterization of pristine Cu nanoparticles for the non-enzymatic glucose biosensor. J Mater Sci Mater Med 28:1–8

de Assis S et al (2023) Review of bacterial Nanocellulose-based electrochemical biosensors: functionalization, challenges, and future perspectives. Biosensors 13(1):142

Dogutan DK, Nocera DG (2019) Artificial photosynthesis at efficiencies greatly exceeding that of natural photosynthesis. Acc Chem Res [preprint]. Available at: https://doi.org/10.1021/acs.accounts.9b00380

Driver M (2012) Coatings for cardiovascular devices: coronary stents. In: Coatings for biomedical applications. Elsevier, pp 223–250

Du C, Zhang G, Cheng Y (2019) 2D materials meet biomacromolecules: opportunities and challenges. Acta Phys -Chim Sin:1078–1089. Available at: https://doi.org/10.3866/PKU.WHXB201812057

Du C et al (2022) On the interface between biomaterials and two-dimensional materials for biomedical applications. Adv Drug Deliv Rev 186:114314. Available at: https://doi.org/10.1016/j.addr.2022.114314

Dung TT et al (2009) Structural and magnetic properties of starch-coated magnetite nanoparticles. J Exp Nanosci 4(3):259–267

Duraisamy N, Muthu S, Krishnan K (2020) Synthesis of silver nanoparticles using Lyngbya Majuscula extract and their antibacterial effects (2). Int J Pharm Sci 11:83–91

El-Kebir A, Harrane A, Belbachir M (2016) Protonated montmorillonite clay used as green non-toxic catalyst for the synthesis of biocompatible Polyglycidol. Arab J Sci Eng 41(6): 2179–2184. Available at: https://doi.org/10.1007/s13369-015-1862-z

Escobar AM et al (2021) Recent applications of heteropolyacids and related compounds in heterocycle synthesis. Contributions between 2010 and 2020. Catalysts 11(2):291

Folorunso A et al (2019) Biosynthesis, characterization and antimicrobial activity of gold nanoparticles from leaf extracts of Annona muricata. J Nanostructure Chem 9:111–117

Fooladi S, Nematollahi MH, Iravani S (2023) Nanophotocatalysts in biomedicine: cancer therapeutic, tissue engineering, biosensing, and drug delivery applications. Environ Res:116287

Gao S et al (2020) A novel degradable injectable HLC-HPA hydrogel with anti-inflammatory activity for biomedical materials: preparation, characterization, in vivo and in vitro evaluation. SCIENCE CHINA Technol Sci 63(11):2449–2463. Available at: https://doi.org/10.1007/s11431-020-1544-3

Georgakilas V et al (2016) Noncovalent functionalization of graphene and graphene oxide for energy materials, biosensing, catalytic, and biomedical applications. Chem Rev 116(9): 5464–5519

Gildner PG, Colacot TJ (2015) Reactions of the 21st century: two decades of innovative catalyst design for palladium-catalyzed cross-couplings. Organometallics 34(23):5497–5508. Available at: https://doi.org/10.1021/acs.organomet.5b00567

Gkiliopoulos D et al (2022) SBA-15 mesoporous silica as delivery vehicle for rhBMP-2 bone Morphogenic protein for dental applications. Nanomaterials 12(5). Available at: https://doi.org/10.3390/nano12050822

Gobalakrishnan S, Chidhambaram N, Chavali M (2021) Role of greener syntheses at the nanoscale. In: Handbook of greener synthesis of nanomaterials and compounds. Elsevier, pp 107–134

Golchinvafa S, Masoudpanah SM (2019) Magnetic and microwave absorption properties of FeNi3/NiFe2O4 composites synthesized by solution combustion method. J Alloys Compd 787:390–396. Available at: https://doi.org/10.1016/j.jallcom.2019.02.039

Gomes S et al (2017) Evaluation of nanofibrous scaffolds obtained from blends of chitosan, gelatin and polycaprolactone for skin tissue engineering. Int J Biol Macromol 102:1174–1185

Gómez-lópez P et al (2020) Nanomaterials and catalysis for green chemistry. Curr Opin Green Sustain Chem 24:48–55. Available at: https://doi.org/10.1016/j.cogsc.2020.03.001

Grotta C et al (2017) 3D printing of 2D atomically thin materials. arXiv preprint arXiv:1710.03956 [Preprint]

Guebitz GM, Nyanhongo GS (2018) Enzymes as green catalysts and interactive biomolecules in wound dressing hydrogels. Trends Biotechnol:1–14. Available at: https://doi.org/10.1016/j.tibtech.2018.05.006

Guo Q et al (2013) Carbon template removal by dielectric-barrier discharge plasma for the preparation of zirconia. Catal Today 211:156–161. Available at: https://doi.org/10.1016/j.cattod.2013.02.032

Hasan A et al (2018) Nanoparticles in tissue engineering: applications, challenges and prospects. Int J Nanomedicine:5637–5655
Hassaan MA, El MA, Marwa N (2023) Principles of Photocatalysts and their different applications: a review. Top Curr Chem (Z) 381. Springer International Publishing. Available at: https://doi.org/10.1007/s41061-023-00444-7
Hataminia F et al (2022) Green synthesis of oxidized starch with a novel catalyst based on Fe3O4 nanoparticles and H2O2 reagent to form thermoplastic as a stable gel coating on the cardiovascular stents. Int J Biol Macromol 219:290–303. Available at: https://doi.org/10.1016/j.ijbiomac.2022.07.119
He Y et al (2019) Microwave assistant rapid synthesis MCM-41-NH2 from fly ash and Cr (VI) removal performance. Environ Sci Pollut Res 26(30):31463–31477. Available at: https://doi.org/10.1007/s11356-019-06255-y
Hinton TJ et al (2015) Three-dimensional printing of complex biological structures by freeform reversible embedding of suspended hydrogels. Sci Adv 1(9):e1500758
Huang X et al (2019) Efficient conversion of CO2 to methane using thin-layer SiOx matrix anchored nickel catalysts. New J Chem 43(33):13217–13224. Available at: https://doi.org/10.1039/C9NJ03152A
Huang H, Feng W, Chen Y (2021) Two-dimensional biomaterials: material science, biological effect and biomedical engineering applications. Chem Soc Rev 50(20):11381–11485
Ishak NAI, Kamarudin SK, Timmiati SN (2019) Green synthesis of metal and metal oxide nanoparticles via plant extracts: an overview. Mater Res Express 6. Available at: https://doi.org/10.1088/2053-1591/ab4458
Jafari Z et al (2021) Nanotechnology-abetted astaxanthin formulations in multimodel therapeutic and biomedical applications. J Med Chem 65(1):2–36
Jastrzębska AM et al (2017) In vitro studies on cytotoxicity of delaminated Ti3C2 MXene. J Hazard Mater 339:1–8
Jastrzębska AM et al (2021) On the rapid in situ oxidation of two-dimensional V2CTz MXene in culture cell media and their cytotoxicity. Mater Sci Eng C 119:111431
Jeon EY et al (2017) Natural healing-inspired collagen-targeting surgical protein glue for accelerated scarless skin regeneration. Biomaterials 134:154–165
Ji D-K, Ménard-Moyon C, Bianco A (2019) Physically-triggered nanosystems based on two-dimensional materials for cancer theranostics. Adv Drug Deliv Rev 138:211–232. Available at: https://doi.org/10.1016/j.addr.2018.08.010
Jiang Y et al (2020) Ethylene glycol: a green solvent for visible light-promoted aerobic transition metal-free Cascade Sulfonation/cyclization reaction. Adv Synth Catal 362(13):2609–2614. Available at: https://doi.org/10.1002/adsc.202000233
Jing X et al (2017) Mussel-inspired electroactive chitosan/graphene oxide composite hydrogel with rapid self-healing and recovery behavior for tissue engineering. Carbon 125:557–570
Kawatra A, Dhankhar R, Gulati P (2022) Microbial arginine deiminase: a multifaceted green catalyst in biomedical sciences. Int J Biol Macromol 196:151–162. Available at: https://doi.org/10.1016/j.ijbiomac.2021.12.015
Keskin Z, Urkmez AS, Hames EE (2017) Novel keratin modified bacterial cellulose nanocomposite production and characterization for skin tissue engineering. Mater Sci Eng C 75:1144–1153
Khalaf HA, Shehata N, Abdelbaqi RF (2024) Green catalyst: definitions, recent development, applications, and circular economy. In: Reference module in materials science and materials engineering. Elsevier. Available at: https://doi.org/10.1016/B978-0-443-15738-7.00013-1
Khorsandi D et al (2024) Catalytic and biomedical applications of nanocelluloses: a review of recent developments. Int J Biol Macromol 268:131829. Available at: https://doi.org/10.1016/j.ijbiomac.2024.131829
Kim SG et al (2021) Mitochondria-targeted ROS-and GSH-responsive diselenide-crosslinked polymer dots for programmable paclitaxel release. J Ind Eng Chem 99:98–106

Kodasi B et al (2023) Novel jointured green synthesis of chitosan-silver nanocomposite: an approach towards reduction of nitroarenes, anti-proliferative, wound healing and antioxidant applications. Int J Biol Macromol 246:125578. Available at: https://doi.org/10.1016/j.ijbiomac.2023.125578

Kokel A, Schäfer C, Torok B (2019) Microwave-assisted reactions in green chemistry. In: Green chemistry and chemical engineering. Springer, New York, pp 573–612. Available at: https://doi.org/10.1007/978-1-4939-9060-3_1008

Kotani T et al (2023) Horseradish peroxidase-mediated bioprinting via bioink gelation by alternately extruded support material. ACS Biomater Sci Eng 9(10):5804–5812

Krishani M et al (2023a) Development of scaffolds from bio-based natural materials for tissue regeneration applications: a review. Gels 9(2):100. Available at: https://doi.org/10.3390/gels9020100

Krishani M, Suhaimi H, Sambudi NS (2023b) A review of hydroxyapatite: sustainable product development in terms of waste valorization. In What to know about hydroxyapatite. Nova Science Publishers, Inc. pp 219–243

Kumar VB et al (2019) Fluorescent metal-doped carbon dots for neuronal manipulations. Ultrason Sonochem 52:205–213

Kumar R, Kumar VB, Gedanken A (2020) Sonochemical synthesis of carbon dots, mechanism, effect of parameters, and catalytic, energy, biomedical and tissue engineering applications. Ultrason Sonochem 64:105009. Available at: https://doi.org/10.1016/j.ultsonch.2020.105009

Leonelli C, Veronesi P, Cravotto G (2013) Microwave-assisted extraction: an introduction to dielectric heating. In: Chemat F, Cravotto G (eds) Microwave-assisted extraction for bioactive compounds: theory and practice. Springer US, Boston, MA, pp 1–14. Available at: https://doi.org/10.1007/978-1-4614-4830-3_1

Levin A et al (2020) Biomimetic peptide self-assembly for functional materials. Nat Rev Chem 4(11):615–634. Available at: https://doi.org/10.1038/s41570-020-0215-y

Li YE (2019) Sustainable biomass materials for biomedical applications. ACS Biomater Sci Eng [Preprint]. Available at: https://doi.org/10.1021/acsbiomaterials.8b01634

Li X et al (2017) Recent advances in synthesis and biomedical applications of two-dimensional transition metal dichalcogenide nanosheets. Small 13(5):1602660

Li X et al (2019) Mechanochemistry-assisted encapsulation of metal nanoparticles in MOF matrices via a sacrificial strategy. J Mater Chem A 7(24):14504–14509

Lim TC et al (2012) The effect of injectable gelatin-hydroxyphenylpropionic acid hydrogel matrices on the proliferation, migration, differentiation and oxidative stress resistance of adult neural stem cells. Biomaterials 33(12):3446–3455. Available at: https://doi.org/10.1016/j.biomaterials.2012.01.037

Lim GP et al (2021) Synthesis, characterization and biophysical evaluation of the 2D Ti2CTx MXene using 3D spheroid-type cultures. Ceram Int 47(16):22567–22577

Liu X et al (2020) Injectable electrical conductive and phosphate releasing gel with two-dimensional black phosphorus and carbon nanotubes for bone tissue engineering. ACS Biomater Sci Eng 6(8):4653–4665

Liu X et al (2021) 2D phosphorene nanosheets, quantum dots, nanoribbons: synthesis and biomedical applications. Biomater Sci 9(8):2768–2803. Available at: https://doi.org/10.1039/d0bm01972k

Mahajan A, Arya A, Chundawat TS (2019) Green synthesis of silver nanoparticles using green alga (Chlorella vulgaris) and its application for synthesis of quinolines derivatives. Synth Commun 49(15):1926–1937

Marquez-Medina MD et al (2019) Post-synthetic mechanochemical incorporation of Al-species into the framework of porous materials: toward more sustainable redox chemistries. ACS Sustain Chem Eng 7(10):9537–9543

Moeini S, Mohammadi MR, Simchi A (2017) In-situ solvothermal processing of polycaprolactone/hydroxyapatite nanocomposites with enhanced mechanical and biological performance for bone tissue engineering. Bioact Mater 2(3):146–155

Mokhtar M et al (2022) New green perspective to dihydropyridines synthesis utilizing modified heteropoly acid catalysts. Catal Today 397–399:484–496. Available at: https://doi.org/10.1016/j.cattod.2021.07.006

Moreira Teixeira LS et al (2012) Enzyme-catalyzed crosslinkable hydrogels: emerging strategies for tissue engineering. Biomaterials 33(5):1281–1290. Available at: https://doi.org/10.1016/j.biomaterials.2011.10.067

Moreno-Luna FB et al (2019) Quick synthesis of gold nanoparticles at low temperature, by using Agave potatorum extracts. Mater Lett 235:254–257. Available at: https://doi.org/10.1016/j.matlet.2018.09.122

Muley PD, Wang Y, Hu J, Shekhawata D (2021) Microwave-assisted heterogeneous catalysis. In Catalysis, pp. 1–37. Available at: https://doi.org/10.1039/9781839163128-00001

Murugiah K et al (2021) Synthesis and characterisation of hydroxyapatite (HAp) from Asiatic hard clam (Meretrix meretrix) and blood cockle clam (Anadara granosa) using wet precipitation process. In National biomedical engineering conference, pp 1–6. Available at: https://doi.org/10.1109/nbec53282.2021.9618744

Muthuchamy N et al (2018) High-performance glucose biosensor based on green synthesized zinc oxide nanoparticle embedded nitrogen-doped carbon sheet. J Electroanal Chem 816:195–204. Available at: https://doi.org/10.1016/j.jelechem.2018.03.059

Nabi G, Raza W, Tahir MB (2020) Green synthesis of TiO2 nanoparticle using cinnamon powder extract and the study of optical properties. J Inorg Organomet Polym Mater 30(4):1425–1429. Available at: https://doi.org/10.1007/s10904-019-01248-3

Nadaf SJ et al (2022) Green synthesis of gold and silver nanoparticles: updates on research, patents, and future prospects. OpenNano 8:100076

Nair RS et al (2017) A gold nanoparticle coated porcine cholecyst-derived bioscaffold for cardiac tissue engineering. Colloids Surf B: Biointerfaces 157:130–137

Nwosu FO et al (2020) Colorimetric based polysorbate crosslinked cellulose-Ag–Cu nanohybrid sensor for urea sensing applications. J Polym Environ 28:1475–1483

Nyanhongo GS, Nugroho Prasetyo E, Herrero Acero E, Guebitz GM (2012) Engineering strategies for successful development of functional polymers using oxidative enzymes. Chem Eng Technol 35(8):1359–1372. Available at: https://doi.org/10.1002/ceat.201100590

Ostovar S et al (2018) Efficient mechanochemical bifunctional nanocatalysts for the conversion of isoeugenol to vanillin. Front Chem 6:77

Park H et al (2023) Bioactive inorganic compound MXene and its application in tissue engineering and regenerative medicine. J Ind Eng Chem 117:38–53. Available at: https://doi.org/10.1016/j.jiec.2022.10.014

Pei S et al (2018) Green synthesis of graphene oxide by seconds timescale water electrolytic oxidation. Nat Commun 9(1):145. Available at: https://doi.org/10.1038/s41467-017-02479-z

Peng K et al (2021) 3D bioprinting of reinforced vessels by dual-cross-linked biocompatible hydrogels. ACS Appl Bio Mater 4(5):4549–4556. Available at: https://doi.org/10.1021/acsabm.1c00283

Pérez-Amodio S, Engel E (n.d.) Bone biology and regeneration. In Bio-ceramics with clinical applications, pp 315–342. Available at: https://doi.org/10.1002/9781118406748.ch11

Pineda A et al (2018) Mechanochemical synthesis of supported cobalt oxide nanoparticles on mesoporous materials as versatile bifunctional catalysts. Microporous Mesoporous Mater 272:129–136

Pirich CL et al (2017) Piezoelectric immunochip coated with thin films of bacterial cellulose nanocrystals for dengue detection. Biosens Bioelectron 92:47–53

Rai P, Gupta D (2021) Magnetic nanoparticles as green catalysts in organic synthesis-a review. Synth Commun 51(20):3059–3083. Available at: https://doi.org/10.1080/00397911.2021.1968910

Ranjbar S, Shahrokhian S (2018) Design and fabrication of an electrochemical aptasensor using Au nanoparticles/carbon nanoparticles/cellulose nanofibers nanocomposite for rapid and sensitive detection of Staphylococcus aureus. Bioelectrochemistry 123:70–76

Ranjbarvan P et al (2018) Natural compounds for skin tissue engineering by electrospinning of nylon-Beta vulgaris. ASAIO J 64(2):261–269

Rao KJ et al (1999) Synthesis of inorganic solids using microwaves. Chem Mater 11(4):882–895

Rashad S, El-Chaghaby GA, Elchaghaby MA (2019) Antibacterial activity of silver nanoparticles biosynthesized using Spirulina platensis microalgae extract against oral pathogens. Egypt J Aquatic Biol Fish 23(5 (Special Issue)):261–266

Rasool K et al (2016) Antibacterial activity of Ti3C2T x MXene. ACS Nano 10(3):3674–3684

Rhee SG (2006) H2O2, a necessary evil for cell signaling. Science 312(5782):1882–1883

Rosales M et al (2020) Unprecedented arsenic photo-oxidation behavior of few-and multi-layer Ti3C2Tx nano-sheets. Appl Mater Today 20:100769

Roy R et al (1999) Full sintering of powdered-metal bodies in a microwave field. Nature 399(6737): 668–670. Available at: https://doi.org/10.1038/21390

Sakai S, Harada R, Kotani T (2021) Freeform 3D bioprinting involving ink gelation by Cascade reaction of oxidase and peroxidase: a feasibility study using hyaluronic acid-based ink. Biomolecules. Available at: https://doi.org/10.3390/biom11121908

Saleh TA (2020) Nanomaterials: classification, properties, and environmental toxicities. Environ Technol Innov 20:101067

Salehi H et al (2017) Effects of nanozeolite/starch thermoplastic hydrogels on wound healing. J Res Med Sci 22(1):110

Sassetti E, Clausen MH, Laraia L (2021) Small-molecule inhibitors of reactive oxygen species production. J Med Chem 64(9):5252–5275

Sattar T, Athar M (2018) Some Nano bio-Mofs evaluated for storage purpose of drugs. Biomed J Sci Tech Res 2:2348–2359

Šebesta M et al (2023) Mycosynthesis of metal-containing nanoparticles—synthesis by ascomycetes and basidiomycetes and their application. Int J Mol Sci. Available at: https://doi.org/10.3390/ijms24010304

Shariatinia Z, Pourzadi N (2021) Designing novel anticancer drug release vehicles based on mesoporous functionalized MCM-41 nanoparticles. J Mol Struct 1242:130754. Available at: https://doi.org/10.1016/j.molstruc.2021.130754

Sheldon RA (2005) Green solvents for sustainable organic synthesis: state of the art. Green Chem 7(5). Available at: https://doi.org/10.1039/b418069k

Shiwarski DJ et al (2021) Emergence of FRESH 3D printing as a platform for advanced tissue biofabrication. APL Bioeng 5(1):010904

Shoda S et al (2016) Enzymes as green catalysts for precision macromolecular synthesis. Chem Rev 116(4):2307–2413. Available at: https://doi.org/10.1021/acs.chemrev.5b00472

Soni M et al (2018) Green nanoparticles: synthesis and applications. IOSR J Biotechnol Biochem 4(3):78–83

Sood K, Saini Y, Thakur KK (2023) Ionic liquids in catalysis: a review. Mater Today Proc 81:739–744. Available at: https://doi.org/10.1016/j.matpr.2021.04.225

Stratakis E et al (2020) Laser engineering of biomimetic surfaces. Mater Sci Eng R: Rep 141: 100562. Available at: https://doi.org/10.1016/j.mser.2020.100562

Sukumar S, Rudrasenan A, Padmanabhan Nambiar D (2020) Green-synthesized rice-shaped copper oxide nanoparticles using Caesalpinia bonducella seed extract and their applications. ACS Omega 5(2):1040–1051

Szuplewska A et al (2019) Multilayered stable 2D nano-sheets of Ti 2 NT x MXene: synthesis, characterization, and anticancer activity. J Nanobiotechnol 17:1–14

Tabrizi MA, Varkani JN (2014) Green synthesis of reduced graphene oxide decorated with gold nanoparticles and its glucose sensing application. Sensors Actuators B Chem 202:475–482

Tahriri M et al (2018) Evaluation of the in vitro biodegradation and biological behavior of poly (lactic-co-glycolic acid)/nano-fluorhydroxyapatite composite microsphere-sintered scaffold for bone tissue engineering. J Bioact Compat Polym 33(2):146–159

Tao J, Xu J-H (2009) Biocatalysis in development of green pharmaceutical processes. Curr Opin Chem Biol 13(1):43–50. Available at: https://doi.org/10.1016/j.cbpa.2009.01.018

Thomas B et al (2018) Nanocellulose, a versatile green platform: from biosources to materials and their applications. Chem Rev 118(24):11575–11625

Veeraraghavan VP et al (2021) Green synthesis of silver nanoparticles from aqueous extract of Scutellaria barbata and coating on the cotton fabric for antimicrobial applications and wound healing activity in fibroblast cells (L929). Saudi J Biol Sci 28(7):3633–3640. Available at: https://doi.org/10.1016/j.sjbs.2021.05.007

Wang L-S et al (2010) Injectable biodegradable hydrogels with tunable mechanical properties for the stimulation of neurogenesic differentiation of human mesenchymal stem cells in 3D culture. Biomaterials 31(6):1148–1157. Available at: https://doi.org/10.1016/j.biomaterials.2009.10.042

Wang C, Zhao Q, Wang M (2017a) Cryogenic 3D printing for producing hierarchical porous and rhBMP-2-loaded Ca-P/PLLA nanocomposite scaffolds for bone tissue engineering. Biofabrication 9(2):25031

Wang L et al (2017b) Electrospun conductive nanofibrous scaffolds for engineering cardiac tissue and 3D bioactuators. Acta Biomater 59:68–81

Wang Z et al (2018) Catalyst preparation with plasmas: how does it work? ACS Catal 8(3):2093–2110

Wang Y et al (2019) Environmentally-friendly exfoliate and active site self-assembly: thin 2D / 2D Heterostructure amorphous nickel—iron alloy on 2D materials for efficient oxygen evolution reaction. Small 1805435:1–8. Available at: https://doi.org/10.1002/smll.201805435

Wang X et al (2021a) 2D materials for bone therapy. Adv Drug Deliv Rev 178:113970. Available at: https://doi.org/10.1016/j.addr.2021.113970

Wang X et al (2021b) Photothermally triggered biomimetic drug delivery of Teriparatide via reduced graphene oxide loaded chitosan hydrogel for osteoporotic bone regeneration. Chem Eng J 413:127413. Available at: https://doi.org/10.1016/j.cej.2020.127413

Wang F, Zheng Y, Ning J (2023) Biogenic preparation of copper oxide nanoparticles using table olive: catalytic reduction, cytotoxicity, and burn wound healing activities. Environ Res 237:116995. Available at: https://doi.org/10.1016/j.envres.2023.116995

Wu Q et al (2013) Microwave-assisted aqueous multicomponent reaction: facile synthesis of Polyfunctionalized Indoline-spiro fused Pyran derivatives. J Heterocyclic Chem 50(3):599–602. Available at: https://doi.org/10.1002/jhet.1537

Xue X et al (2022) Neutrophil-erythrocyte hybrid membrane-coated hollow copper sulfide nanoparticles for targeted and photothermal/ anti-inflammatory therapy of osteoarthritis. Compos Part B 237:109855. Available at: https://doi.org/10.1016/j.compositesb.2022.109855

Yao J et al (2020) Effects of mordenite zeolite catalyst synthesis conditions on dimethyl ether carbonylation. Microporous Mesoporous Mater 306:110431. Available at: https://doi.org/10.1016/j.micromeso.2020.110431

Yaqoob AA et al (2020) Recent advances in metal decorated nanomaterials and their various biological applications: a review. Front Chem 8:341

Yi H et al (2017) Tissue-specific extracellular matrix promotes myogenic differentiation of human muscle progenitor cells on gelatin and heparin conjugated alginate hydrogels. Acta Biomater 62:222–233

Yilmaz E, Soylak M (2020) Chapter 5 – type of green solvents used in separation and preconcentration methods. In: Soylak M, Yilmaz IS (eds) New generation green solvents for separation and preconcentration of organic and inorganic species. Elsevier, pp 207–266. Available at: https://doi.org/10.1016/B978-0-12-818569-8.00005-X

Yook H et al (2023) Design strategies for hydroxyapatite-based materials to enhance their catalytic performance and applicability. Adv Mat (Deerfield Beach, Fla) 35(43):e2204938. Available at: https://doi.org/10.1002/adma.202204938

You Q et al (2020) Persistent regulation of tumor hypoxia microenvironment via a bioinspired Pt-based oxygen nanogenerator for multimodal imaging-guided synergistic phototherapy. Adv Sci 7(17):1903341

Yu T et al (2019) Green synthesis of gold nanoclusters using papaya juice for detection of L-lysine. Chin Chem Lett 30(3):660–663
Zamarchi F, Vieira IC (2021) Determination of paracetamol using a sensor based on green synthesis of silver nanoparticles in plant extract. J Pharm Biomed Anal 196:113912
Zangeneh MM et al (2019) Green synthesis of silver nanoparticles using aqueous extract of Stachys lavandulifolia flower, and their cytotoxicity, antioxidant, antibacterial and cutaneous wound-healing properties. Appl Organomet Chem 33(9):e5016
Zarrintaj P et al (2020) Zeolite in tissue engineering: opportunities and challenges. MedComm 1(1): 5–34
Zhang C, Li S, Bao S (2019a) Sustainable synthesis of ZSM-5 zeolite from rice husk ash without addition of solvents. Waste Biomass Valor 10:2825–2835
Zhang Q et al (2019b) Recent progress on heteropolyacids for green fuels synthesis. Curr Green Chem 6(3). Available at: https://doi.org/10.2174/2213346106666191014104156
Zhang X et al (2019c) Fabrication of 2D metal–organic framework nanosheets with tailorable thickness using bio-based surfactants and their application in catalysis. Green Chem 21(1): 54–58. Available at: https://doi.org/10.1039/C8GC02835D
Zhang X et al (2019d) Template-free synthesized 3D macroporous MXene with superior performance for supercapacitors. Appl Mater Today 16:315–321. Available at: https://doi.org/10.1016/j.apmt.2019.06.013
Zhao B et al (2017) CO Methanation over Ni/SiO2 catalyst prepared by ammonia impregnation and plasma decomposition. Top Catal 60(12):879–889. Available at: https://doi.org/10.1007/s11244-017-0752-x
Zhao B et al (2019) Preparation of Ni/SiO2 catalyst via novel plasma-induced micro-combustion method. Catal Today 337:28–36. Available at: https://doi.org/10.1016/j.cattod.2019.04.068
Zheng B et al (2010) Preparation of gold nanoparticles on eggshell membrane and their biosensing application. Talanta 82(1):177–183
Zheng Y et al (2021) 2D nanomaterials for tissue engineering and regenerative nanomedicines: recent advances and future challenges. Adv Healthc Mater 10(7):2001743
Zhou P et al (2014) Enhanced bone tissue regeneration by antibacterial and osteoinductive silica-HACC-zein composite scaffolds loaded with rhBMP-2. Biomaterials 35(38):10033–10045. Available at: https://doi.org/10.1016/j.biomaterials.2014.09.009
Zhou M et al (2019) Solvent-free and rapid synthesis of mesoporous Pt–iron oxide catalysts via mechanochemical assembly. Cat Sci Technol 9(15):3907–3913
Zuhrotun A, Oktaviani DJ, Hasanah AN (2023) Biosynthesis of gold and silver nanoparticles using phytochemical compounds. Molecules 28(7):3240
Zulkifli FH et al (2017) A facile synthesis method of hydroxyethyl cellulose-silver nanoparticle scaffolds for skin tissue engineering applications. Mater Sci Eng C 79:151–160

Ms. Murugiah Krishani is a research scholar in the Department of Chemical and Process Engineering under the Faculty of Integrated Technologies, Universiti Brunei Darussalam. She obtained her B.Tech in Biotechnology (2015) and M.Tech in Biotechnology (2017) from Anna University, India. Later in 2017, Krishani joined in PABCET as an Assistant Professor and worked for three years. Her major area of research is Bone Tissue Engineering and she published several articles in Bone Tissue Engineering and Machine Learning Techniques in the field of Biomedical applications.

Dr. Nonni Soraya Sambudi is a Faculty Member at the Department of Chemical Engineering, Universitas Pertamina. She earned her Ph.D. in Chemical and Biomolecular Engineering from Korea Advanced Institute of Science and Technology (KAIST). Dr. Sambudi has over 10 years of experience in the field of material science, with a particular focus on nanotechnology. Her works on the synthesis of carbon quantum dots for various applications such as photocatalytic process, drug delivery, heavy metals detection and removal have been published in several prestigious journals. She has collaborated with many institutions globally and has been a keynote speaker at numerous international conferences. Dr. Sambudi's current projects include developing advanced nanomaterials and composite as electrodes for photoelectrochemical process and drug delivery vehicles.

Dr. Hazwani Suhaimi obtained her BEng in Chemical Engineering (2009), MSc in Advanced Chemical Engineering with IT and Management (2010), and PhD in Chemical Engineering (2015) from Loughborough University. In January 2016, Hazwani joined Universiti Brunei Darussalam as a lecturer in Chemical and Process Engineering, and was promoted to Assistant Professor in December 2021. She has been the Programme Leader of the Chemical and Process Engineering since August 2016. Her major research is on the nutrient diffusion in tissue engineering membranes and scaffolds. Over the years, she has expanded her research activities and interests including in the area of biomaterials, ankle injury, rehabilitation, waste management, and membrane science.

Utilisation of Plant Extracts for Green Synthesis of Metallic Nanoparticles

10

S. Giridhar Reddy

Abstract

This chapter focuses on making metal nanoparticles in an environmentally friendly way using plant materials. It talks about how different plants have special chemicals that can help reduce and stabilize these nanoparticles when they're being made. The mechanisms, influencing factors, and characterisation techniques are discussed in detail. Additionally, the chapter examines the environmental impact and potential applications of these green-synthesized nanoparticles across diverse fields. Overall, it provides a concise and insightful overview, emphasising the importance of sustainable practices in contemporary nanotechnology.

Keywords

Plant extracts · Green synthesis · Metallic nanoparticles · Nanoparticle green chemistry synthesis

10.1 Introduction

Plant-derived substances have garnered attention as environmentally friendly and sustainable options for creating nanoparticles (NPs). Many plants contain active compounds capable of both reducing and encasing NPs, aiding in their formation. These extracts are abundant in phytochemicals like flavonoids, phenols, terpenoids, and alkaloids, which are essential for reducing metal ions and stabilizing NPs. Utilizing plant extracts for nanoparticle synthesis brings several benefits, including being

S. G. Reddy (✉)
Department of Physical Sciences, Amrita School of Engineering, Bengaluru, Amrita Vishwa Vidyapeetham, Bengaluru, Karnataka, India
e-mail: s_giri@blr.amrita.edu

R. Malviya, S. Sundram (eds.), *Sustainable Green Biomaterials As Drug Delivery Systems*, Biomaterials, Bioengineering and Sustainability 1,
https://doi.org/10.1007/978-3-031-79062-1_10

eco-friendly, cost-effective, easily scalable, and compatible with biological systems. Furthermore, the vast diversity of plant species offers a wide range of phytochemical compositions, enabling the creation of NPs tailored to specific properties for applications in medicine, catalysis, sensing, and environmental clean-up.

In recent years, the interest has been garnered by the eco-friendly nature of plant products for the synthesis of metallic NPs, owing to their sustainable approach. This method leverages phytochemicals natural reducing and stabilising properties in different plant extracts to create metallic NPs tailored for various applications (Shankar et al. 2003a).

Plants are rich reservoirs of bioactive compounds like flavonoids, phenols, terpenoids, and alkaloids, which act as effective agents in reducing and capping metallic NPs (Iravani 2011). The green synthesis process involves the gentle reduction of metal ions by these phytochemicals in plant extracts, avoiding harsh chemicals and high temperatures common in traditional synthesis methods (Mittal and Chisti 2013).

The appeal of plant extract-based nanoparticle synthesis lies in its sustainability, affordability, scalability, and compatibility with living organisms. By embracing the principles of green chemistry, researchers can reduce the environmental impact of NP production while exploring novel applications in nanotechnology (Khan et al. 2018). This chapter will explore the mechanisms behind green synthesis utilizing plant extracts, emphasizing the contribution of phytochemicals in the formation and stability of NPs (Siddiqi and Husen 2018; Sathishkumar et al. 2009). Additionally, we will explore the wide range of metallic NPs produced through this method and their applications in medicine, catalysis, sensing, and environmental clean-up (Prasad and Elumalai 2011) (Fig. 10.1).

Through our exploration of plant extract-mediated synthesis, we acknowledge the significant potential of this approach in advancing sustainable nanotechnology and addressing global challenges in science, technology, and environmental conservation.

Plants excel at synthesizing NPs compared to other biological methods because they offer abundant resources, are non-toxic, and naturally provide capping agents. Also, using plant extracts is cheaper than isolating and culturing microorganisms, making NPs synthesis more cost-effective (Sharma et al. 2009; Singhal et al. 2011). The ease of access, safety, and diverse range of secondary metabolites make plants a favourable choice for NPs synthesis. Many plants are currently under investigation for their NPs synthesis potential (Prasad 2014).

10.1.1 Silver Oxide Nanoparticles

Different plants, like leaves, fruit, stems, and root extracts are used for green synthesis, chosen based on availability, phytochemical content, and ease of extraction. After cleaning, the plant material is processed to make an extract using methods such as grinding, blending, or maceration in solvents like water, ethanol, or methanol. In a distinct phase, a solution containing silver salt, typically silver nitrate ($AgNO_3$), is combined with the plant extract, which influences the size and shape

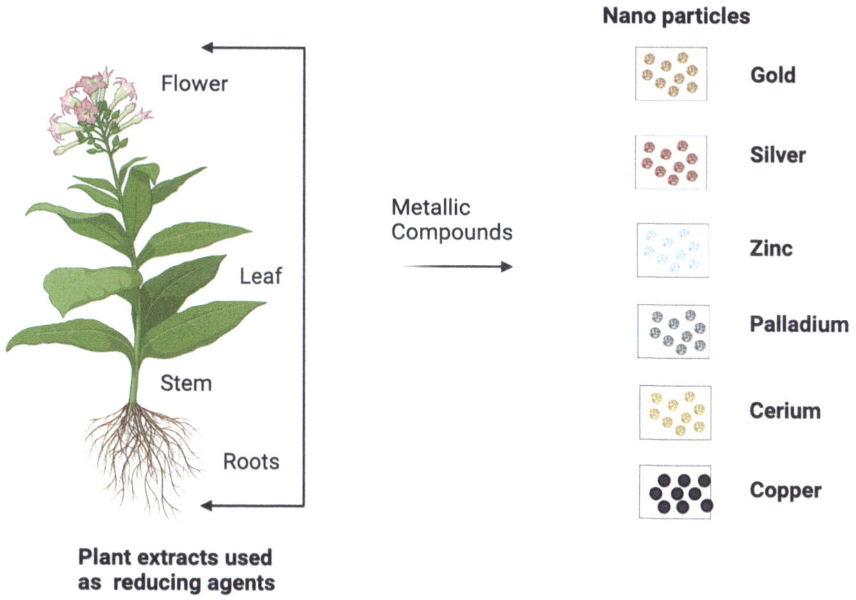

Fig. 10.1 Various parts of different plant products are used to synthesise the nanoparticles using metallic compounds

of the NPs. The phytochemicals of plants act as reducing agents, converting silver ions (Ag^+) into silver NPs (AgNP), while specific biomolecules may serve to stabilize the NPs, preventing agglomeration.

Analytical techniques like UV-vis spectroscopy confirm the AgNP formation based on absorption peaks around 400–450 nm, with further characterization using methods like TEM or SEM examining their size, shape, and distribution.

Green synthesis presents advantages over conventional chemical approaches, such as its environmental friendliness, cheaper, and potential for scalable production.

Various reports on the AgNP synthesis from plants are given in Table 10.1.

10.1.2 Synthesis of Gold Nanoparticles

Gold nanoparticles (AuNPs), as noble metal NPs, are well known for their resistance to corrosion and oxidation. Their exceptional physicochemical and biological characteristics have garnered significant attention (Sreeprasad and Pradeep 2013). These NPs offer several desirable characteristics, including high electrical, catalytic activity, chemical stability, and notable thermal conductivity. The characteristics mentioned, including their nanoscale dimensions, customizable shapes, and surface structures, are intricately tied to their properties.

Table 10.1 List of plant extracts used for AgNPs synthesis

Sl. no.	Plant extract	NP size (nm)	Characterisations	Ref
1	Salvia spinosa	19–125	Antimicrobial property, FESEM, XRD, FTIR	Pirtarighat et al. (2019)
2	Costus speciosus		Antimicrobial property	Malabadi et al. (2012)
3	Jatropha curcas	10–20	HRTEM, XRD and UV	Bar et al. (2009)
4	Euphorbia		Antibacterial characteristics	Sameena and Thoppil (2022)
5	Andrographis paniculata, Phyllanthus niruri, and Tinospora cordifolia		Antiviral agents against Chikungunya	Sharma et al. (2019)
6	Ziziphora tenuior	8 to 40	SEM, TEM, XRD and FTIR	Sadeghi and Gholamhoseinpoor (2015)
7	Pinus eldarica Bark	10–40	TEM	Iravani and Zolfaghari (2013)
8	Coleus aromaticus	38	SEM, ED, FTIR, UV-vis, bactericidal properties.	Vanaja and Annadurai (2013)
9	Capsicum annuum L.		TEM, SAED, XRD, X-ray photoemission spectroscopy, electrochemical measurements, FTIR and differential spectrum techniques.	Li et al. (2007)
10	Curcuma longa tuber-powder	6.3 ± 2.64	TEM, FT-IR, UV-vis, XRD, SEM, and EDXF	Shameli et al. (2012a, b)
11	Sorghum bran extracts	10	XRD	Njagi et al. (2011)
12	Green tea extract (Camellia sinensis)		ICP-MS, XPS, TEM, XRD, UV, FTIR, TGA, antibacterial activity (gram-positive, gram-negative)	Rolim et al. (2019)
13	Tribulus terrestris L. fruit	16–28	TEM, AFM, XRD, FTIR, UV–vis spectroscopy.	Gopinath et al. (2012)
14	Dimocarpus longan seed	40	XRD,TEM, anti-oxidant, strong catalytic and reducing properties	Khan et al. (2016)
15	Aloe barbadensis miller and Ocimum tenuiflorum	7–70 and 9–48	TEM, UV-vis, anti-bacterial activity	Sharma et al. (2021)
16	Coffea arabica seed	20–30	TEM, FTIR, XRD, UV-visible, antibacterial activity	Dhand et al. (2016)
17	Malus domestica (red apple)	145	EDS, XRD, FTIR, SEM, Zeta potential	Umoren et al. (2014)

(continued)

Table 10.1 (continued)

Sl. no.	Plant extract	NP size (nm)	Characterisations	Ref
18	Trifolium resupinatum seed exudates	17	TEM, XRD, FTIR, and UV-vis, antifungal activity against plant-pathogenic fungi	Khatami et al. (2016)
19	Ananas comosus	12	UV-vis, EDAX, SAED and HRTEM, TEM	Ahmad and Sharma (2012)
20	Aqueous garlic extract	7.3 ± 4.4	Antibacterial activity	Rastogi and Arunachalam (2011)
21	Ocimum Sanctum		Antibacterial properties	Jain and Mehata (2017)
22	Leptadenia reticulata leaf	50 to 70	XRD, UV-vis, cytotoxic effects with HCT15 cancer cell line	Kumara Swamy et al. (2015)
23	Bacillus licheniformis	40	SEM and XRD	Kalishwaralal et al. (2008)
24	Erythrina suberosa (Roxb.)		DLS, FTIR, UV-vis, antioxidant property and cytotoxicity was studied using A-431 osteosarcoma cell line	Mohanta et al. (2017)
25	Melia azedarach	12 to 46	Zeta potential analysis	Ashraf et al. (2020)
26	Moringa oleifera	23	Anti-inflammatory activity	Muhammad et al. (2023)
27	Cissus quadrangularis	37 to 44	Antibacterial activity	Vanaja et al. (2013)
28	Green tea		Luminescence property	Vilchis-Nestor et al. (2008)
29	Artemisia nilagirica (Asteraceae)	70 to 90	EDX	Vijayakumar et al. (2013)
30	Desmodium triflorum	5 to 20	XRD, antimicrobial activity against common pathogens	Ahmad et al. (2011)
31	Geranium leaf	16 to 40	TEM	Shankar et al. (2003a, b, c)
32	Mesocarp layer extract of Cocos nucifera	23 ± 2	TEM, XRD, gas chromatography-mass spectrometry.	Roopan et al. (2013)
33	Eucalyptus leucoxylon	50	XRD, SEM, TEM, and UV–vis	Rahimi-Nasrabadi et al. (2014)
34	Callicarpa maingayi Stem bark	12.4 ± 3.27	Zeta potential, and antibacterial testing.	Shameli et al. (2012a, b)
35	Saltmarsh plant Sesuvium portulacastrum L.	5 to 20	TEM and FTIR, antimicrobial activity.	Nabikhan et al. (2010)
36	Ocimum sanctum	5 to10	TEM, SAED, and XRD	Ahmad et al. (2010)

(continued)

Table 10.1 (continued)

Sl. no.	Plant extract	NP size (nm)	Characterisations	Ref
37	Sesbania grandiflora	10 to 25	Antibacterial activity	Vijay Kumar et al. (2014)
38	Boerhaavia diffusa plant	25	Antibacterial activity	Jeeva et al. (2014)
39	Caesalpinia coriaria		Antimicrobial activity	Jeeva et al. (2014)
40	Vitex negundo L. extract	18.2	TEM, XRD, and UV–visible, antimicrobial activity	Zargar et al. (2011)
41	Latex from Euphorbiaceae family		Antibacterial activity	Patil et al. (2012)
42	Moringa oleifera leaf	9 to11	Antimicrobial activity	Jerushka et al. (2018)
43	Blackberry, blueberry, pomegranate, and turmeric	5–150	SEM, TEM, HR-TEM.	Nadagouda et al. (2014)
44	Musa balbisiana, Azadirachta indica, and Ocimum tenuiflorum	200	Antibacterial efficacy and toxicity	Banerjee et al. (2014)
45	Boswellia ovalifoliolata stem bark extract	30 to 40	SEM, EDAX, UV-vis	Ankanna et al. (2010)
46	Glaucium corniculatum (L.) Curtis plant extract	53 to 45	SEM and TEM, antibacterial activity	Allafchian et al. (2018)
47	Leaf extracts from Semecarpus anacardium, Glochidion lanceolarium, and Bridelia retusa	52 and 96	UV–Vis, DLS, FTIR, anti-biofilm agents, antibacterial activity	Mohanta et al. (2020)
48	Breynia rhamnoides	64	SEM, TEM, UV-vis, FT-IR	Gangula et al. (2011)
49	Carica papaya	60 to 80	FTIR, SEM	Mude et al. (2009)
50	Aloe vera leaf extract		XRD, SEM, antimicrobial activity, MTT assay	Tippayawat et al. (2016)
51	Tectona grandis (teak) seed extract	10 to 30	Antimicrobial activity	Rautela, A et al. (2019)
52	Neem leaf (Azadirachta indica) extract	5 to 35	TEM, UV-vis, XRD	Shiv Shankar et al. (2004)

Oat (*Avena sativa*) biomass was explored by Armendariz, V et al. to synthesise AuNPs from Au(III) ions, with pH influencing binding and NPs size. Maximum adsorption occurred at pH 3, highlighting pH dependency in the process. High-resolution TEM revealed various NPs shapes, including tetrahedral and rod shapes, with nanorods particularly notable, influenced by pH. Smaller NPs were prevalent at pH 3 & 4, while larger ones predominated at pH 2, showcasing pH's impact on NPs size distribution (Armendariz et al. 2004).

Nagajyothi et al. utilized Lonicera japonica flower extract to bioreduce chloroauric acid, yielding AuNPs. Constituents like alcohols, alkanes, amines, and amides in the extract played roles in both stabilizing and reducing the AuNPs. These NPs displayed antimicrobial efficacy against various bacteria and fungi (Nagajyothi et al. 2014).

The AuNPs with a 50 nm size were synthesized by S.K. Srivastava et al. at room temperature using Escherichia coli K12 cells, with specific proteins aiding in the bioreduction process and NPs stabilization (Srivastava et al. 2013).

K. Chandran, et al. synthesised the AuNP using medicinal plants, Cucurbita pepo and Malva crispa. They are showing significant resistance for foodborne pathogens like Escherichia coli and Listeria monocytogenes (Chandran et al. 2019).

Mallikarjuna N. et al. employed antioxidants sourced from extracts of blackberry, blueberry, pomegranate, and turmeric for the preparation of AuNPs. The fruit extracts of 8.5% are prepared, while green tea extract with five green tea bags for 15 min, and turmeric extract was obtained of 1% of turmeric powder for 12 min. After filtering the green tea and turmeric extracts, they were mixed with auric chloride, respectively. The solutions were stirred vigorously overnight to complete the synthesis process (Nadagouda et al. 2014).

A novel method for synthesizing AuNPs from stem extract of Breynia rhamnoides is introduced. This eco-friendly approach offers rapid synthesis rates (approximately 7 min), with the ability to control NPs size and catalytic activity by adjusting concentrations. The phenolic glycosides and reducing sugars in the extract reduces Au^{3+} ions to AuNPs. These AuNPs shows efficient catalytic activity and reduces 4-nitrophenol to 4-aminophenol, with reaction kinetics dependent on NPs size or extract concentration (Gangula et al. 2011).

The Sesbania seedlings exposed to chloroaurate solution accumulated gold, forming stable AuNP within plant tissues. TEM results revealed monodisperse NPs, may due to reduction by secondary metabolites. XRD showed efficient biotransformation of Au^{3+} into Au by plant tissues. The NPs rich biomass exhibited catalytic activity by reducing aqueous 4-nitrophenol. (Sharma et al. 2007).

A genetically modified tobacco mosaic virus (MTMV), was engineered to show a metal-binding and reducing peptide. Unlike wild-type MTMV facilitated the formation of 10–40 nm AuNPs when mixed with potassium tetrachloroaurate. These NPs were found to be crystalline and stable, suggesting the potential of MTMV in eco-friendly metal nanomaterial production (Love et al. 2015).

G.S. Ghodake et al. present a new environmentally friendly approach for producing AuNP using pear fruit extract. When subjected to alkaline conditions, the extract prompted the creation of plate-shaped AuNPs, including triangular and hexagonal nanoplates, alongside hexagonal AuNPs. Various characterization techniques like TEM, EDAX, XRD, XPS, and AFM are used to report the size, crystal structure, purity, and shapes of these nanostructures. The nanostructures sizes ranged from 200–500 nm in edge lengths, with nanohexagons measuring 12 to 20 nm in thickness (Ghodake et al. 2010).

Leaf extracts from Magnolia kobus et al. are used in the eco-friendly synthesis of AuNP by reducing aqueous $HAuCl_4$ solution. The process was rapid, achieving

>90% conversion to AuNPs within minutes at 95 °C, showcasing comparable or even superior reaction rates to chemical methods. Characterization techniques revealed the formation of stable AuNPs with diverse morphologies, including plate and spherical structures ranging from 5 to 300 nm in size. They highlights the potential of plant-based approaches for efficient and sustainable NPs synthesis (Song et al. 2009).

Jalil, S.U et al. study aimed to synthesize AuNPs from *Nicotiana tabacum* L. cv. xanthi leaf. Exposure to bioengineered AuNPs at concentrations between 100 and 500 ppm resulted in significant enhancements in shoot growth, root, seed germination and antioxidant capacity in N. tabacum plants. Optimal growth was observed at concentrations of 200–300 ppm, while higher concentrations (400–500 ppm) had detrimental effects. Overall, AuNPs exhibited dose-dependent impacts on plant physiology and antioxidant activity (Jalil et al. 2019).

Aljabali, A.A.A et al. preparation of AuNPs using ***Ziziphus plant*** extracts is reasonable and environmentally friendly. Phytochemicals from the leaf extract serve as efficient reducing agents for AuNPs synthesis, which are characterized using various analytical techniques. The resulting AuNPs are deemed biocompatible and safe for diverse nanomedicine applications, with thermogravimetric analysis show they act as a stabilizing agent for the NPs (Aljabali et al. 2018).

The study V. Ganesh Kumar presents a synthesis of AuNPs from *Cassia auriculata* leaf extract. AuNPs were rapidly formed in 10 min, indicating a faster reaction rate compared to chemical synthesis methods. Characterizations confirm the production of stable, triangular, and spherical crystalline AuNPs with sizes of ~15–25 nm. The influence of pH on AuNPs stability are explored. The research aims to utilize the antidiabetic properties of C. auriculata, suggesting potential applications for AuNPs in managing hyperglycemia pending further investigation (Kumar et al. 2011).

Gayathiri, E., explores the utilization of *Decalepis hamiltonii methanol* root extracts for antioxidant assays, anti-cancer evaluations, and AuNPs characterization, along with ADME/T and molecular docking studies. The root extract displays potent antioxidant and enzyme properties. The AuNPs synthesized from the roots exhibit varied morphology, size, and shape, confirmed by various analytical instruments. FTIR analysis indicates biomolecule involvement in NPs reduction and stabilization. Molecular docking suggests 1TUP as a potential anti-cancer inhibitor, while 2-Hydroxy-3-Methoxybenzaldehyde shows promising anti-cancer activity, supporting cancer treatment and nanomedicine development (Kumar et al. 2023).

The *Abelmoschus esculentus* seed extract is used to synthesise the AuNPs and evaluates their antifungal activity. The AuNPs exhibit a peak at 536 nm and a size of 62 nm. FTIR analysis indicates OH functional groups' involvement from the extracts in capping the NPs. AFM and FESEM images reveal the sizes of spherical particles ranging from 45-75 nm. Antifungal assays demonstrate significant inhibition zones against Puccinia graminis and Candida albicans, suggesting the AuNPs as effective antifungal agents for drug development against fungal diseases (Jayaseelan et al. 2013).

S. S. Shankar et al. study investigates the synthesis of AuNPs with *geranium leaves* and an endophytic fungus. Both geranium leaves and the fungus showed rapid reduction of chloroaurate ions, yielding stable AuNPs of various sizes. Terpenoids act as reducing and capping agents in geranium leaves, while polypeptides/enzymes perform these functions in Colletotrichum sp. The NPs synthesized from the fungus are predominantly spherical, whereas those from the leaves exhibit diverse shapes including rods, sheets, and triangles. The potential for NPs shape control within a leaf–fungus system is highlighted, although the exact mechanisms remain unclear (Shankar et al. 2003a, b, c).

10.1.3 Synthesis Iron Oxide Nanoparticles

The plants are used in the synthesis of iron oxide NPs (IONPs) presents an eco-friendly, economical, and adaptable method, finds many applications. These NPs, produced through green synthesis methods, possess biocompatible properties suitable for biomedical uses like drug delivery and imaging. Additionally, they exhibit multifunctionality, including magnetic and antimicrobial properties, making them applicable in fields in environmental remediation and catalysis.

Ting, A.S.Y et al. investigates the using *Apple peel extract* (APE) for biosynthesizing iron nanoparticles (INPs). Characterization techniques confirmed the synthesis of APE-INPs, revealing polyphenol peaks and primary functional groups integrated into the NPs. The APE-INPs exhibited elliptical and spherical shapes with a size of 50 nm. When tested for dye decolorization, APE-INPs demonstrated 71.51% decolorization of malachite green dye within the first minute. This research underscores the potential of APE in producing INPs for efficient dye removal applications (Ting and Chin 2020).

Mehdi Fazlzadeh et al. nanoscale Fe(0) particles were prepared using plant extracts from *Rosa damascene (RD), Thymus vulgaris (TV), and Urtica dioica (UD)*. FTIR analysis confirms the presence of polyphenols, proteins, and organic acids in the extracts preventing NPs aggregation. The removal efficiency of Cr (VI) was highest at pH 2, with increased efficiency observed with longer contact times and higher doses of Fe(0). However, higher initial concentrations of Cr(VI) led to decreased removal efficiency. Application of 0.2 g/l Fe(0) for 10 min resulted in over 90% removal efficiency, with complete removal achieved within 25 min for TV-Fe and UD-Fe, and 30 min for RD-F (Fazlzadeh et al. 2017).

Eric C. Njagi successfully synthesized amorphous IONPs of ~50 nm using *sorghum extracts*. The IONPs were formed rapidly under normal environmental conditions. Their reactivity was evaluated through the degradation of contaminant bromothymol blue, catalysed by H_2O_2, indicating their promise for applications in environmental remediation (Njagi et al. 2011).

Lebogang Katata-Seru et al. explore the INPs using leaf (L) and seed (S) extracts of *Moringa oleifera* (MO). INPs synthesized with M. oleifera seed extract exhibited absorption at 240 nm, and leaf extract at 210 nm. The synthesized NPs demonstrated enhanced nitrate ion removal from water, with MOS-INPs showing an 85% removal rate compared to 26% for MOL-INPs. Additionally, MOS-INPs exhibited better antibacterial activity with Escherichia coli, suggesting the potential of M. oleifera-based INPs for water treatment applications (Katata-Seru et al. 2018).

The Fe_3O_4 (magnetite) nanoparticles were synthesised using seaweed *Kappaphycus alvarezii* as natural agents for both reduction and stabilization. XRD showed distinct patterns typical of Fe_3O_4, indicating high purity and crystal structure. FT-IR confirmed Fe_3O_4 presence with specific absorption peaks. TEM revealed mostly spherical Fe_3O_4 NPs, averaging 14.7 nm in size (Yew et al. 2016).

A method using *green tea* leaf extract to synthesize mesoporous α-IONP has been developed by Bashir Ahmmad et al. offering a simple and scalable process. The IONPs, of ~60 nm, were highly pure and well crystallized. Evaluation of their photocatalytic activity, measured by hydroxyl radical formation exposure to visible light, revealed twice the efficiency of commercial α-IONP. Additionally, they demonstrated improved performance in a photoelectrochemical cell (Ahmmad et al. 2013).

10.1.4 Cerium Oxide Nanoparticles

The synthesizing cerium oxide NPs (CeO_2 NPs) from plants involves plant extracts or biomaterials as reducing agents as well as stabilizers. This green synthesis approach offers several advantages, including environmental friendliness, cheaper, and biocompatibility. They possess unique features which makes them useful in various applications, including catalysis, energy storage, environmental remediation, and biomedical fields. CeO_2 NPs exhibit excellent redox properties, high surface area, and stability, which contribute to their versatility in diverse applications (Nadeem et al. 2020). They are commonly synthesized using green methods involving plants, microbes, or other biological products. Here's an Table 10.2 of the mentioning the various plants in synthesizing CeO_2 NPs from plants.

10.1.5 Zinc Oxide Nanoparticle

The plant extracts are used in the synthesis of zinc oxide NPs (ZnO NPs) provides a sustainable and environmental safe approach compared to conventional methods. This process utilizes plant extracts rich in bioactive compounds, resulting in biocompatible and safe NPs. The tailored properties of ZnO NPs, influenced by plant extracts, enable diverse applications such as antimicrobial, antioxidant, and photocatalytic properties. Additionally, the cost-effectiveness and scalability of plant-based synthesis make it a promising method for addressing environmental and healthcare challenges.

Raunak Saha et al. utilised phytochemicals from *Terminalia arjuna, Swertia chirayita, and Psidium guajava* phytochemicals for the synthesis of ZnO NPs of varied size and shape with distinct structural defects were synthesized. These NPs exhibit increased reactive oxygen species (ROS) production within cancer cells,

Table 10.2 List of CeO_2 NPs using plant extracts

No.	Plant extract	Size (nm)	Shape	Properties	Ref
1	*Moringa oleifera L*	100	Spherical	Antimicrobial and wound healing	Tammineni and Roopan (2016)
2	*Gloriosa Superba*			Antibacterial activity	Arumugam and Karthikeyan (2015)
3	*Hibiscus sabdariffa*	3.9	Spherical	Morphological, structural and optical	Thovhogi et al. (2015)
4	*Aloe Barbadensis*	63.6	Spherical	Crystallinity, morphological	Priya et al. (2014)
5	*Jatropha curcus* leaf	3–5		Photocatalysis	Magudieshwaran et al. (2019)
6	*Oleo Europaea*	24	Spherical	High antimicrobial activity	Maqbool et al. (2016)
7	*Rubia cordifolia* leaf fusions	26	Hexagonal	Anti-cancer potential	Sisubalan et al. (2018)
8	*Origanum majorana*	20	Pseudo spherical	FT-IR confirms the phenolic and flavonoids groups.	Nezhad et al. (2019)
9	*Pedalium murex* L.	5–55	Nano rod	Antibacterial activity	Nithya et al. (2019)
10	*China rose* petal	7	Nano sheet	Bio template for the facile fabrication	Qian et al. (2011)

enhancing their efficacy in cancer treatment. This study explores ZnO NPs induce DNA damage via ROS, leading to apoptosis in human breast cancer cells (Saha et al. 2023).

D. Suresh, P.C et al. presents the ZnO NPs synthesis of using *Cassia fistula* plant extract. The XRD, UV–vis studies, and TEM revealed hexagonal wurtzite structure with average sizes of ~5–15 nm. They displayed potent bactericidal activity against various strains. This environmentally friendly approach utilizing plant extracts showcases the potential of multifunctional ZnO NPs in diverse applications (Suresh et al. 2015).

An aqueous leaf extract of *Tulbaghia violacea*, Mbenga, Y et al. utilized to stabilize the ZnO NPs synthesis and assessed their *in vitro* cytotoxic potential. Characterization techniques confirmed the spherical morphology of the ZnO NPs with an ~45.26 nm and single-phase crystallinity. *In vitro* assays revealed high cytotoxicity of the biogenic ZnO NPs, with an IC50 value of 4.04×10^{-1} μg/mL against human liver cells, suggesting their potential in modulating proliferation and inducing apoptosis (Mbenga et al. 2022).

Li Fu et al. describe the fabrication of ZnO NPs employing leaf extract from *Plectranthus amboinicus*. SEM examination displayed rod-shaped NPs with an ~88 nm. UV–vis spectroscopy revealed a band gap of 3.07 eV for the produced

NPs. Photocatalytic effectiveness was assessed by decomposing methyl red under UV light exposure, demonstrating enhanced performance compared to ZnO NPs synthesized via hydrothermal methods and P25 (Li and Zhuxian 2015).

Bioaugmented ZnO-NPs were prepared using *Myristica fragrans* fruit and extensively characterized through various techniques. The NPs exhibited an size of ~41.23 nm. The NPs shown significant antibacterial activity against strains, potent antidiabetic effects, and remarkable larvicidal activity against Aedes aegypti mosquitoes. Additionally, they showed promising inhibitory potential against enzymes and exhibited excellent antioxidant properties. Moreover, the ZnO-NPs served as efficient photocatalytic agents, degrading methylene blue dye by 88% in 140 min. These environmentally friendly and biocompatible NPs hold immense potential for biomedical and environmental applications (Faisal et al. 2021).

The ZnO NPs synthesized using leaf extracts from *Cassia fistula* and *Melia azadarach* plants exhibited strong antimicrobial activity, indicating the potential of plant-based synthesis for eco-friendly biomedical products (Naseer et al. 2020). Additionally, a novel approach presented by Raliya and Tarafdar (2013) utilized extracellular secretions of Aspergillus fumigatus TFR-8 to synthesize ZnO NPs, which were applied to clusterbean plants, enhancing various plant parameters and gum production. Moreover, Niranjan Bala et al. synthesized ZnO NPs using Hibiscus subdariffa leaf, which exhibited antibacterial properties against E. coli and S. aureus, along with superior anti-diabetic effects on streptozotocin-induced diabetic mice and the potential to regulate gene expression associated with diabetes (Bala et al. 2015).

10.1.6 Copper Oxide Nanoparticle

The copper oxide nanoparticles (CuO NPs) with plant extracts appears to be eco friendly. This method harnesses the reducing and stabilizing properties of plant compounds, eliminating the need for harsh chemicals. The CuO NPs have diverse applications, including catalysis, sensing, electronics, and biomedicine, due to their strange properties such as optical, electrical, and catalytic activity. Additionally, their eco-friendly synthesis aligns with sustainable practices, making them attractive for green technology initiatives and reducing environmental impact.

Soheyla Honary et al. synthesized CuO NPs extracellularly using *Penicillium aurantiogriseum*, from soil. SEM analysis shows uniform spherical NPs, with size and polydispersity confirmed by DLS using a nano zetasizer. The presence of secreted proteins facilitated metal precursor hydrolysis for metal oxide formation. Parameters affecting particle size and polydispersity index were explored, suggesting rapid and scalable biological synthesis potential (Honary et al. 2012).

Jagpreet Singh et al. in his study, prepared CuO NPs with *Psidium guajava leaf.* The resulting CuO NPs were monodispersed and spherical, with an size ~2 to 6 nm and a BET surface area of 52.6 m^2/g. These NPs demonstrated very high degradation for industrial dyes Nile blue (NB) and reactive yellow 160 (RY160), with 93% and 81% removal rates, respectively, within 120 min. The CuO catalyst showed high reusability, retaining its photocatalytic activity over five cycles. With low detection limits and environmentally friendly synthesis, biogenic CuO NPs offer a better solution for treating water polluted with industrial dyes (Singh et al. 2019).

The investigation by H. Raja Naika highlights the eco-friendly synthesis of CuO NPs with *Gloriosa superba* leaf, demonstrating its efficacy as a green fuel. XRD and UV-vis absorption analyses confirm the NPs characteristics, while SEM and TEM imaging provide insights into their morphology and size distribution. Remarkably, the prepared CuO NPs exhibit antibacterial activity against various strains (Raja Naika et al. 2015).

10.1.7 Palladium Nanoparticles

The Palladium NPs synthesized using plant extracts have diverse applications across fields. They contribute to environmental remediation by effectively removing pollutants from air and water. In electronics, these NPs find use in sensors, conductive coatings, and electronic devices because of their superior electrical conductivity. Furthermore, they hold promise in biomedical uses towards drug delivery and imaging, owing to their biocompatibility. Lastly, in energy storage, palladium NPs enhance the efficiency of fuel cells and batteries, making them valuable in sustainable energy technologies.

The Palladium NPs were produced using various plant leaf /bark extracts are listed in Table 10.3.

10.2 Conclusion

The synthesis of NPs using plants presents a fascinating intersection of botanical science and nanotechnology, offering a wealth of opportunities for scientific exploration and practical application. Through leveraging the unique biochemical properties of plants, researchers have developed innovative methods to produce NPs with tailored characteristics, ranging from size and shape to surface properties and functionality.

This synthesis approach holds great promise for numerous fields, including biomedicine, agriculture, environmental remediation, and materials science. By utilizing plants as natural sources of reducing and stabilizing agents, NPs synthesis becomes not only environmentally friendly but also cheaper and scalable. Additionally, the biocompatibility and biodegradability of plant-derived NPs make them particularly appealing for biomedical applications in a drug delivery, imaging, and therapeutics.

As our comprehension of plant-NPs interactions grows and synthesis methodologies evolve, we foresee additional advancements in the creation and production of innovative nanomaterials boasting enhanced properties and functionalities. Nevertheless, obstacles persist, such as the need for standardized synthesis protocols, scalability for industrial manufacturing, and comprehensive evaluation of potential environmental repercussions.

Table 10.3 List of Palladium NPs using plant extracts

S. no.	Plants extract	Size	Shape	Properties	Ref
1	*Hippophae rhamnoides Linn leaf*	5 ± 2.5 nm	Spherical	Catalytic property of Pd NPs to the Suzuki–Miyaura	Mahmoud Nasrollahzadeh et al. (2015)
2	*Allium fistulosum, Basella alba, and Tabernaemontana divaricate*	500 nm, 2 μm and 2 μm	Spherical	Antioxidant, antifungal and antibacterial activity.	Vinodhini et al. (2022)
3	*Filicium decipiens leaf*	2–22 nm		Antibacterial activity	Sharmila et al. (2017)
4	*Cocculus hirsutus leaf*	10 ± 3	Spherical, FCC	Antimicrobial and anti-inflammatory activities	Sandhya et al. (2024)
5	*Zaleya decandra leaf*	14 nm	Spherical	Eco-safe assessment, toxicity bioassay	Naveenkumar et al. (2024)
6	*Turmeric extract*	100	Spherical	MTT assay and flow cytometry tests	Golestannezhad et al. (2024)
7	*Rosa damascena (r. damascena) leaf*	20–50 nm	Spherical	Anti-inflammatory and analgesic activity	Shagufta and Srivastava (2024)
8	*Rosmarinus officinalis L.*	4.91 nm	Spherical	Antimicrobia, photocatalytic activities	Tiri et al. (2024)
9	*Indian cork extract*		–	Hydrogen evolution study	Limaye et al. (2024)
10	*Aegle marmelos* fruit	5 nm	–	Antibacterial activity, molecular docking studies	Vijeata et al. (2024)

References

Ahmad N, Sharma S (2012) Green synthesis of silver nanoparticles using extracts of Ananas comosus. Green Sustain Chem 2(4):Article ID: 24781

Ahmad N, Sharma S, Alam MK, Singh VN, Shamsi SF, Mehta BR, Fatma A (2010) Rapid synthesis of silver nanoparticles using dried medicinal plant of basil. Colloids Surf B: Biointerfaces 81(1):81–86

Ahmad N, Sharma S, Singh VN, Shamsi SF, Fatma A, Mehta BR (2011) Biosynthesis of silver nanoparticles from Desmodium triflorum: a novel approach towards weed utilization. Biotechnol Res Int 2011:Article ID 454090

Ahmmad B, Leonard K, Shariful Islam M, Kurawaki J, Muruganandham M, Ohkubo T, Kuroda Y (2013) Green synthesis of mesoporous hematite (α-Fe_2O_3) nanoparticles and their photocatalytic activity. Adv Powder Technol 24(1):160–167

Aljabali AAA, Akkam Y, Al Zoubi MS, Al-Batayneh KM, Al-Trad B, Abo Alrob O, Alkilany AM, Benamara M, Evans DJ (2018) Synthesis of gold nanoparticles using leaf extract of *Ziziphus zizyphus* and their antimicrobial activity. Nanomaterials *8*:174

Allafchian AR, Jalali SAH, Aghaei F, Farhang HR (2018) Green synthesis of silver nanoparticles using Glaucium corniculatum (L.) Curtis extract and evaluation of its antibacterial activity. IET Nanobiotechnol 12:574–578

Ankanna S, Prasad TNVKV, Elumalaib EK, Savithramma N (2010) Production of biogenic silver nanoparticles using boswellia ovalifoliolata stem bark. Digest J Nanomater Biostruct 5(2): 369–372

Armendariz V, Herrera I, Peralta-videa JR (2004) Size controlled gold nanoparticle formation by Avena sativa biomass: use of plants in nanobiotechnology. J Nanopart Res 6:377–382

Arumugam, A., Karthikeyan, C., Hameed, A.S; Synthesis of cerium oxide nanoparticles using Gloriosa superba L. leaf extract and their structural, optical and antibacterial properties. Mater Sci Eng C. 49, 408-415 (2015)

Ashraf H, Anjum T, Riaz S, Naseem S (2020) Microwave-assisted green synthesis and characterization of silver nanoparticles using Melia azedarach for the management of Fusarium Wilt in tomato. Front Microbiol 10(11):228

Bala N, Saha S, Chakraborty M, Maiti M, Das S, Basu R, Nandy P (2015) Green synthesis of zinc oxide nanoparticles using Hibiscus subdariffa leaf extract: effect of temperature on synthesis, anti-bacterial activity and anti-diabetic activity. RSC Adv 5:4993–5003

Banerjee P, Satapathy M, Mukhopahayay A (2014) Leaf extract mediated green synthesis of silver nanoparticles from widely available Indian plants: synthesis, characterization, antimicrobial property and toxicity analysis. Bioresour Bioprocess 1:3

Bar H, Bhui DKR, Sahoo GP, Sarkar P, De SP, Misra A (2009) Green synthesis of silver nanoparticles using latex of Jatropha curcas. Colloids Surf A Physicochem Eng Asp 339(1–3):134–139

Chandran K, Song S, Yun S (2019) Effect of size and shape controlled biogenic synthesis of gold nanoparticles and their mode of interactions against food borne bacterial pathogens. Arab J Chem 12(8):1994–2006

Dhand V, Soumya L, Bharadwaj S, Chakra S, Bhatt D, Sreedhar B (2016) Green synthesis of silver nanoparticles using Coffea arabica seed extract and its antibacterial activity. Mater Sci Eng C 58:36–43

Faisal S, Jan H, Shah SA, Shah S, Khan A, Akbar MT, Rizwan M, Jan F, Akhtar N, Khattak A, Syed S (2021) Green synthesis of zinc oxide (ZnO) nanoparticles using aqueous fruit extracts of Myristica fragrans: their characterizations and biological and environmental applications. ACS Omega 6(14):9709–9722

Mehdi Fazlzadeh., Kourosh Rahmani., Ahmad Zarei., Hossein Abdoallahzadeh., Fakhraddin Nasiri., Rasoul Khosravi; A novel green synthesis of zero valent iron nanoparticles (NZVI) using three plant extracts and their efficient application for removal of Cr(VI) from aqueous solutions, Adv Powder Technol, 28(1)122–130 (2017)

Gangula A, Podila R, Ramakrishna M, Karanam L, Janardhana C, Apparao MR (2011) Catalytic reduction of 4-Nitrophenol using biogenic gold and silver nanoparticles derived from Breynia rhamnoides. Langmuir 27(24):15268–15274

Ghodake GS, Deshpande NG, Lee YP, Jin ES (2010) Pear fruit extract-assisted room-temperature biosynthesis of gold nanoplates. Colloids Surf B: Biointerfaces 75(2):584–589

Golestannezhad N, Divsalar A, Badalkhani-Khamseh F (2024) Oxali-palladium nanoparticle synthesis, characterization, protein binding, and apoptosis induction in colorectal cancer cells. J Mater Sci Mater Med 35:4

Gopinath V, MubarakAli D, Priyadarshini S, Meera Priyadharsshini N, Thajuddin N, Velusamy P (2012) Biosynthesis of silver nanoparticles from Tribulus terrestris and its antimicrobial activity: a novel biological approach. Colloids Surf B: Biointerfaces 96:69–74

Honary S, Barabadi H, Gharaei-Fathabad E, Naghibi F (2012) Green synthesis of copper oxide nanoparticles using penicillium aurantiogriseum, penicillium citrinum and penicillium waksmanii. Dig J Nanomater Biostruct 7(3):999–1005

Iravani S (2011) Green synthesis of metal nanoparticles using plants. Green Chem 13(10): 2638–2650

Iravani S, Zolfaghari B (2013) Green synthesis of silver nanoparticles using Pinus eldarica bark extract. Biomed Res Int 5:Article ID 639725, 5

Jain S, Mehata MS (2017) Medicinal plant leaf extract and pure flavonoid mediated green synthesis of silver nanoparticles and their enhanced antibacterial property. Sci Rep 7:15867

Jalil SU, Zahera M, Khan MS, Ansari MI (2019) Biochemical synthesis of gold nanoparticles from leaf protein of Nicotiana tabacum L. cv. xanthi and their physiological, developmental, and ROS scavenging responses on tobacco plant under stress conditions. IET Nanobiotechnol 13:23–29

Jayaseelan C, Ramkumar R, Rahuman AA (2013) Pachiappan Perumal; green synthesis of gold nanoparticles using seed aqueous extract of Abelmoschus esculentus and its antifungal activity. Ind Crop Prod 45:423–429

Jeeva K, Thiyagarajan M, Elangovan V, Geetha N, Venkatachalam P (2014) Caesalpinia coriaria leaf extracts mediated biosynthesis of metallic silver nanoparticles and their antibacterial activity against clinically isolated pathogens. Ind Crop Prod 52:714–720

Kalishwaralal K, Deepak V, Ramkumarpandian S, Nellaiah H, Sangiliyandi G (2008) Extracellular biosynthesis of silver nanoparticles by the culture supernatant of Bacillus licheniformis. Mater Lett 62(29):4411–4413

Katata-Seru L, Moremedi T, Aremu OS, Bahadur I (2018) Green synthesis of iron nanoparticles using Moringa oleifera extracts and their applications: removal of nitrate from water and antibacterial activity against Escherichia coli. J Mol Liq 256:296–304

Khan FU, Chen Y, Khan NU, Khan ZUH, Ahmad AUKA, Tahir K, Wang L, Khan MR, Wan P (2016) Antioxidant and catalytic applications of silver nanoparticles using Dimocarpus longan seed extract as a reducing and stabilizing agent. J Photochem Photobiol B Biol 164:344–351

Khan M, Khan ST, Khan M, Adil SF, Tahir MN, Tremel W (2018) Plant extracts as green reductants for the synthesis of silver nanoparticles: lessons from chemical synthesis. Dalton Trans 47(35):11988–12004

Khatami M, Nejad MS, Salari S, Almani PGN (2016) Plant-mediated green synthesis of silver nanoparticles using Trifolium resupinatum seed exudate and their antifungal efficacy on Neofusicoccum parvum and Rhizoctonia solani. IET Nanobiotechnol 10:237–243

Kumar VG, Gokavarapu SD, Rajeswari A, Dhas TS, Karthick V, Kapadia Z, Shrestha T, Barathy IA, Roy A, Sinha S (2011) Facile green synthesis of gold nanoparticles using leaf extract of antidiabetic potent Cassia auriculata. Colloids Surf B Biointerfaces 87(1):159–163

Kumar VG, Prakash P, Selvam K et al (2023) Exploring the therapeutic potential of Decalepis hamiltonii root extract: synthesis of gold nanoparticles and assessment of antimicrobial, antioxidant, and anti-proliferative activities. Appl Nanosci 13:5967–5981

Kumara Swamy M, Sudipta KM, Jayanta K (2015) The green synthesis, characterization, and evaluation of the biological activities of silver nanoparticles synthesized from Leptadenia reticulata leaf extract. Appl Nanosci 5:73–81

Li F, Zhuxian F (2015) Plectranthus amboinicus leaf extract–assisted biosynthesis of ZnO nanoparticles and their photocatalytic activity. Ceram Int Part A 41(2):2492–2496

Li S, Shen Y, Xie A, Xuerong Y, Qiu L, Zhang L, Zhang Q (2007) Green synthesis of silver nanoparticles using *Capsicum annuum* L. extract. Green Chem. 9:852–858

Limaye AS, Yhobu Z, Thrilokraj R, Budagumpi S, Ramesh BD (2024) Biogenically synthesized palladium nanoparticles for hydrogen evolution study: an efficient catalyst for 4-nitrophenol reduction and C-C coupling reactions. Colloids Surf A Physicochem Eng Asp 688:133555

Love AJ, Makarov VV, Sinitsyna OV, Shaw J, Yaminsky IV, Kalinina NO, Taliansky M (2015) A genetically modified tobacco mosaic virus that can produce gold nanoparticles from a metal salt precursor. Front Plant Sci 6:984

Magudieshwaran R, Ishii J, Raja KCN (2019) Green and chemical synthesized CeO2 nanoparticles for photocatalytic indoor air pollutant degradation. Mater Lett 239:40–44

Malabadi RB, Meti NT, Mulgund GS, Nataraja K, Vijaya Kumar S (2012) Synthesis of silver nanoparticles from in vitro derived plants and callus cultures of Costus speciosus (Koen.); assessment of antibacterial activity. Res Plant Biol 2(4):32–42

Maqbool Q, Nazar M, Naz S (2016) Antimicrobial potential of green synthesized CeO_2 nanoparticles from Olea europaea leaf extract. Int J Nanomedicine 11:5015–5025

Mbenga Y, Mthiyane MN, Botha TL (2022) Nanoarchitectonics of ZnO nanoparticles mediated by extract of Tulbaghia violacea and their cytotoxicity evaluation. J Inorg Organomet Polym 32: 3249–3259

Mittal, A.K., Chisti, Y., Banerjee, U.C; Synthesis of metallic nanoparticles using plant extracts. Biotechnol Adv 31(2), 346–356(2013)

Mohanta YK, Panda SK, Jayabalan R, Sharma N, Bastia AK, Mohanta TK (2017) Antimicrobial, antioxidant and cytotoxic activity of silver nanoparticles synthesized by leaf extract of Erythrina suberosa (Roxb.). Front Mol Biosci 17(4):14

Mohanta YK, Biswas K, Jena SK, Hashem A, Allah EFA, Mohanta TK (2020) Anti-biofilm and antibacterial activities of silver nanoparticles synthesized by the reducing activity of Phytoconstituents present in the Indian medicinal plants. Front Microbiol 11:1143

Moodley JS, Krishna SBN, Pillay K, Sershen, Govender P (2018) Green synthesis of silver nanoparticles from Moringa oleifera leaf extracts and its antimicrobial potential. Adv Nat Sci Nanosci Nanotechnol 9:015011

Mude N, Ingle A, Gade A (2009) Synthesis of silver nanoparticles using callus extract of Carica papaya— a first report. J Plant Biochem Biotechnol 18:83–86

Muhammad S, Ali A, Shah J, Hamza M, Kashif M, Ajat Khel BK, Iqbal A (2023) Using Moringa oleifera stem extract for green synthesis, characterization, and anti-inflammatory activity of silver oxide nanoparticles. Nat Appl Sci Int J 4(1):80–97

Nabikhan A, Kandasamy K, Raj A, Alikunhi NM (2010) Synthesis of antimicrobial silver nanoparticles by callus and leaf extracts from saltmarsh plant, Sesuvium portulacastrum L. Colloids Surf B: Biointerfaces 79(2):488–493

Nadagouda MN, Iyanna N, Lalley J, Han C, Dionysiou DD, Varma RS (2014) Synthesis of silver and gold nanoparticles using antioxidants from blackberry, blueberry, pomegranate, and turmeric extracts. ACS Sustain Chem Eng 2(7):1717–1723

Nadeem M, Khan R, Afridi K, Nadhman A, Ullah S, Faisal S, Abbasi BH (2020) Green synthesis of cerium oxide nanoparticles (CeO_2 NPs) and their antimicrobial applications: a review. Int J Nanomedicine 15:5951–5961

Nagajyothi PC, Lee KD, Sreekanth TVM (2014) Biogenic synthesis of gold nanoparticles (quasi-spherical, triangle, and hexagonal) using *Lonicera japonica* flower extract and its antimicrobial activity. Synth React Inorganic Met Nano-Metal Chem 44:1011–1018

Naseer M, Aslam U, Khalid B (2020) Green route to synthesize zinc oxide nanoparticles using leaf extracts of Cassia fistula and Melia azadarach and their antibacterial potential. Sci Rep 10:9055

Nasrollahzadeh M, Sajadi SM, Maham M (2015) Green synthesis of palladium nanoparticles using Hippophae rhamnoides Linn leaf extract and their catalytic activity for the Suzuki–Miyaura coupling in water. J Mol Catal A Chem 396:297–303

Naveenkumar S, Kamaraj C, Prem P, Karthik Raja R, Priyadharsan A, Alrefaei AF, Govindarajan RK, Thamarai R, Subramaniyan V (2024) Eco-friendly synthesis of palladium nanoparticles using Zaleya decandra: assessing mosquito larvicidal activity, zebrafish embryo developmental toxicity, and impacts on freshwater sludge worm Tubifex tubifex. J Environ Chem Eng 12(2): 111912

Nezhad S, Haghi A, Homayouni M (2019) Green synthesis of cerium oxide nanoparticle using Origanum majorana L. leaf extract, its characterization and biological activities: green synthesis of nanoparticle. Appl Organomet Chem 34:e5314

Nithya P, Murugesan B, Sonamuthu J, Samayanan S, Mahalingam S (2019) [BMIM] PF6 ionic liquid mediated green synthesis of ceramic SrO/CeO2 nanostructure using Pedalium murex leaf extract and their antioxidant and antibacterial activities. Ceram Int 45(9):12138–12148

Njagi EC, Huang H, Stafford L, Genuino H, Galindo HM, Collins JB, Hoag GE, Suib SL (2011) Biosynthesis of iron and silver nanoparticles at room temperature using aqueous sorghum bran extracts. Langmuir 27(1):264–271

Patil SV, Borase HP, Patil CD (2012) Biosynthesis of silver nanoparticles using latex from few Euphorbian plants and their antimicrobial potential. Appl Biochem Biotechnol 167:776–790

Pirtarighat S, Ghannadnia M, Baghshahi S (2019) Green synthesis of silver nanoparticles using the plant extract of Salvia spinosa grown in vitro and their antibacterial activity assessment. J Nanostructure Chem 9:1–9

Prasad R (2014) Synthesis of silver nanoparticles in photosynthetic plants. J Nanoparticles 8:Article ID 963961

Prasad TNVKV, Elumalai EK (2011) Biofabrication of iron nanoparticles using Moringa oleifera seed extract and their antibacterial activity. Mater Lett 65(15–16):2524–2526

Priya G, Kanneganti A, Kumar KA, Rao KV, Bykkam S (2014) Bio synthesis of cerium oxide nanoparticles using aloe barbadensis miller gel. Int J Sci Res Publ 4:1–4

Qian J, Chen F, Zhao X (2011) China rose petal as biotemplate to produce two-dimensional ceria nanosheets. J Nanopart Res 13(12):7149–7158

Rahimi-Nasrabadi M, Pourmortazavi SM, Shandiz SAS, Ahmadi F, Batooli H (2014) Green synthesis of silver nanoparticles using Eucalyptus leucoxylon leaves extract and evaluating the antioxidant activities of extract. Nat Prod Res 28(22):1964–1969

Raja Naika H, Lingaraju K, Manjunath K, Kumar D, Nagaraju G, Suresh D, Nagabhushana H (2015) Green synthesis of CuO nanoparticles using Gloriosa superba L. extract and their antibacterial activity. J Taibah Univ Sci 9(1):7–12

Raliya R, Tarafdar JC (2013) ZnO nanoparticle biosynthesis and its effect on phosphorous-mobilizing enzyme secretion and gum contents in Clusterbean (Cyamopsis tetragonoloba L.). Agric Res 2:48–57

Rastogi L, Arunachalam J (2011) Sunlight based irradiation strategy for rapid green synthesis of highly stable silver nanoparticles using aqueous garlic (Allium sativum) extract and their antibacterial potential. Mater Chem Phys 129(1–2):558–563

Rautela A, Rani J, Debnath (Das) M (2019) Green synthesis of silver nanoparticles from Tectona grandis seeds extract: characterization and mechanism of antimicrobial action on different microorganisms. J Anal Sci Technol 10:5

Rolim WR, Pelegrino MT, de Araújo Lima B, Ferraz LS, Costa FN, Bernardes JS, Rodigues T, Brocchi M, Seabra AB (2019) Green tea extract mediated biogenic synthesis of silver nanoparticles: characterization, cytotoxicity evaluation and antibacterial activity. Appl Surf Sci 463:66–74

Roopan SM, Rohit, Madhumitha G, Rahuman AA, Kamaraj C, Bharathi A, Surendra TV (2013) Low-cost and eco-friendly phyto-synthesis of silver nanoparticles using Cocos nucifera coir extract and its larvicidal activity. Ind Crop Prod 43:631–635

Sadeghi B, Gholamhoseinpoor F (2015) A study on the stability and green synthesis of silver nanoparticles using Ziziphora tenuior (Zt) extract at room temperature. Spectrochim Acta A Mol Biomol Spectrosc 134:310–315

Saha R, Subramani K, Dey S, Sikdar S, Incharoensakdi A (2023) Physicochemical properties of green synthesised ZnO nanoparticles and utilisation for treatment of breast cancer. Process Biochem 129:170–184

Sameena VP, Thoppil JE (2022) Green synthesis of silver nanoparticles from Euphorbia and its biological activities. Nanotechnol Environ Eng 7:377–392

Sandhya K, Bhagavanth Reddy G, Ayodhya D, Venkatesh B, Kondaiah S, Noorjahan M, Yadagiri Swamy P, Girija Mangatayaru K (2024) Ultrasound assisted biogenic synthesis of palladium nanoparticles using Cocculus hirsutus leaf extract: antimicrobial, anti-inflammatory and catalytic activities. J Mol Struct 1306:137848

Sathishkumar M, Sneha K, Won SW, Cho CW, Kim S, Yun YS (2009) Cinnamon zeylanicum bark extract and powder mediated green synthesis of nano-crystalline silver particles and its bactericidal activity. Colloids Surf B Biointerfaces 73(2):332–338

Shagufta B, Srivastava R (2024) Rosa damascena leaf extract mediated palladium nanoparticles and their anti-inflammatory and analgesic applications. Inorg Chem Commun 162:112122

Shameli K, Ahmad MB, Zamanian A, Sangpour P, Shabanzadeh P, Abdollahi Y, Zargar M (2012a) Green biosynthesis of silver nanoparticles using Curcuma longa tuber powder. Int J Nanomedicine 7:5603–5610

Shameli K, Bin Ahmad M, Jaffar Al-Mulla EA, Ibrahim NA, Shabanzadeh P, Rustaiyan A, Abdollahi Y, Bagheri S, Abdolmohammadi S, Usman MS (2012b) Green biosynthesis of silver nanoparticles using *Callicarpa maingayi* stem bark extraction. Molecules 17:8506–8517

Shankar SS, Ahmad A, Pasricha R, Sastry M (2003a) Bioreduction of chloroaurate ions by geranium leaves and its endophytic fungus yields gold nanoparticles of different shapes. J Mater Chem 13:1822

Shankar SS, Ahmad A, Sastry M (2003b) Geranium leaf assisted biosynthesis of silver nanoparticles. Biotechnol Progress 19:1627–1631

Shankar SS, Ahmad A, Sastry M (2003c) Geranium leaf assisted biosynthesis of silver nanoparticles. Biotechnol Prog 19(6):1627–1631

Sharma NC, Sahi SV, Nath S, Parsons JG, Gardea-Torresde JL, Pal T (2007) Synthesis of plant-mediated gold nanoparticles and catalytic role of biomatrix-embedded nanomaterials. Environ Sci Technol *41*(14):5137–5142

Sharma VK, Yngard RA, Lin Y (2009) Silver nanoparticles: green synthesis and their antimicrobial activities. Adv Colloid Interf Sci 145(1–2):83–96

Sharma V, Kaushik S, Pandit P (2019) Green synthesis of silver nanoparticles from medicinal plants and evaluation of their antiviral potential against chikungunya virus. Appl Microbiol Biotechnol 103:881–891

Sharma S, Kumar K, Thakur N (2021) Green synthesis of silver nanoparticles and evaluation of their anti-bacterial activities: use of *Aloe barbadensis miller* and *Ocimum tenuiflorum* leaf extracts. Nano 6(1):52–67

Sharmila G, Farzana Fathima M, Haries S, Geetha S, Manoj Kumar N, Muthukumaran C (2017) Green synthesis, characterization and antibacterial efficacy of palladium nanoparticles synthesized using Filicium decipiens leaf extract. J Mol Struct 1138:35–40

Shiv Shankar S, Rai A, Ahmad A, Sastry M (2004) Rapid synthesis of Au, Ag, and bimetallic Au core–Ag shell nanoparticles using neem (Azadirachta indica) leaf broth. J Colloid Interface Sci 275(2):496–502

Siddiqi KS, Husen A (2018) Plant-mediated synthesis of nanoparticles: synthesis, characterization, and applications. In: Siddiqi KS, Husen A (eds) Nanotechnology: an introduction for the food and bioprocessing industries, 1st edn. Academic, London, pp 135–169

Singh J, Kumar V, Kim K-H, Rawat M (2019) Biogenic synthesis of copper oxide nanoparticles using plant extract and its prodigious potential for photocatalytic degradation of dyes. Environ Res 177:108569

Singhal G, Bhavesh R, Kasariya K, Sharma AR, Singh RP (2011) Biosynthesis of silver nanoparticles using Ocimum sanctum (Tulsi) leaf extract and screening its antimicrobial activity. J Nanopart Res 13(7):2981–2988

Sisubalan N, Ramkumar VS, Pugazhendhi A (2018) ROS-mediated cytotoxic activity of ZnO and CeO2 nanoparticles synthesized using the Rubia cordifolia L. leaf extract on MG-63 human osteosarcoma cell lines. Environ Sci Pollut Res. 25(11):10482–10492

Song JY, Jang H-K, Kim BS (2009) Biological synthesis of gold nanoparticles using Magnolia kobus and Diopyros kaki leaf extracts. Process Biochem 44(10):1133–1138

Sreeprasad TS, Pradeep T (2013) Noble metal nanoparticles. In: Vajtai R (ed) Springer handbook of nanomaterials, Springer handbooks. Springer, Berlin/Heidelberg, Germany, pp 303–388

Srivastava SK, Yamada R, Ogino C, Kondo A (2013) Biogenic synthesis and characterization of gold nanoparticles by *Escherichia coli*K12 and its heterogeneous catalysis in degradation of 4-nitrophenol. Nanoscale Res Lett 8:70

Suresh D, Nethravathi PC, Udayabhanu, Rajanaika H, Nagabhushana H, Sharma SC (2015) Green synthesis of multifunctional zinc oxide (ZnO) nanoparticles using Cassia fistula plant extract and their photodegradative, antioxidant and antibacterial activities. Mater Sci Semicond Process 31:446–454

Tammineni S, Roopan S (2016) Photocatalytic and antibacterial properties of phytosynthesized of CeO_2 NPs using Moringa oleifera peel extract. J Photochem Photobiol B 161:122–128

Thovhogi N, Diallo A, Gurib-Fakim A (2015) Nanoparticles green synthesis by Hibiscus sabdariffa flower extract: main physical properties. J Alloys Compd 647:392–396

Ting ASY, Chin JE (2020) Biogenic synthesis of iron nanoparticles from apple peel extracts for decolorization of malachite green dye. Water Air Soil Pollut 231:278

Tippayawat P, Phromviyo N, Boueroy P, Chompoosor A (2016) Green synthesis of silver nanoparticles in aloe vera plant extract prepared by a hydrothermal method and their synergistic antibacterial activity. Peer J 4:e2589

Tiri RNE, Aygun A, Bekmezci M (2024) Environmental energy production and waste water treatment using synthesized Pd nanoparticles with biological and photocatalytic activity. Top Catal 67:714–724

Umoren SA, Obot IB, Gasem ZM (2014) Green synthesis and characterization of silver nanoparticles using red apple (Malus domestica) fruit extract at room temperature. J Mater Environ Sci 5(3):907–914

Vanaja M, Annadurai G (2013) Coleus aromaticus leaf extract mediated synthesis of silver nanoparticles and its bactericidal activity. Appl Nanosci 3:217–223

Vanaja M, Gnanajobitha G, Paulkumar K et al (2013) Phytosynthesis of silver nanoparticles by Cissus quadrangularis: influence of physicochemical factors. J Nanostructure Chem 3:17

Vijay Kumar PPN, Pammi SVN, Kollu P, Satyanarayana KVV, Shameem U (2014) Green synthesis and characterization of silver nanoparticles using Boerhaavia diffusa plant extract and their anti bacterial activity. Ind Crop Prod 52:562–566

Vijayakumar M, Priya K, Nancy FT, Noorlidah A, Ahmed ABA (2013) Biosynthesis, characterisation and anti-bacterial effect of plant-mediated silver nanoparticles using Artemisia nilagirica. Ind Crop Prod 41:235–240

Vijeata A, Garg A, Chaudhary GR, Chaudhary S, Bhalla A (2024) Nanocatalytic and antimicrobial potency of CQD@Pd nanoparticles: experimental and theoretical approach. Appl Surf Sci 653: 159394

Vilchis-Nestor AR, Sanchez-Mendieta V, Camacho-Lopez MA, Gomez-Espinosa RM, Camacho-Lopez MA, Arenas-Alatorre JA (2008) Solventless synthesis and optical properties of Au and Ag nanoparticles using *Camellia sinensis* extract. Mater Lett 62(17–18):3103–3105

Vinodhini S, Scholastica Mary Vithiya B, Augustine Arul Prasad T (2022) Green synthesis of palladium nanoparticles using aqueous plant extracts and its biomedical applications. J King Saud Univ Sci 34(4):102017

Yew YP, Shameli K, Miyake M (2016) Green synthesis of magnetite (Fe_3O_4) nanoparticles using seaweed (Kappaphycus alvarezii) extract. Nanoscale Res Lett 11:276

Zargar M, Hamid AA, Bakar FA, Shamsudin MN, Shameli K, Jahanshiri F, Farahani F (2011) Green synthesis and antibacterial effect of silver nanoparticles using *Vitex Negundo* L. Molecules 16:6667–6676

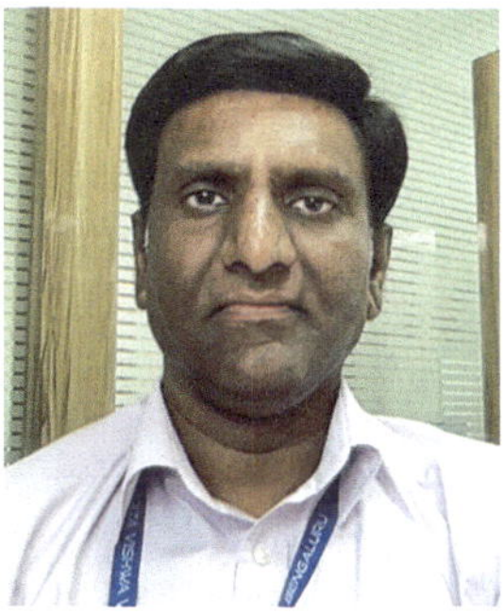

Dr. S. Giridhar Reddy an Assistant Professor (Selection grade) working at Amrita Vishwa Vidyapeetham as a Chairman of Physical sciences, is a skilled researcher who has significantly contributed to biopolymers for controlled drug delivery, nanomaterials and graft polymers. With a background in materials science and engineering, he worked extensively in drug delivery, designing biopolymers and nanomaterials that can be used to control the release of drugs in a targeted manner. His research also focuses on developing new graft polymers with unique properties, which can be used in various applications for transdermal and controlled drug delivery. An impressive track record of academic publications, with more than 25 peer-reviewed journal articles and book chapters from Elsevier, Springer Nature and Taylor & Francis publishers to his name. His work has been recognised and cited by numerous researchers in the field, and they are considered leading experts in biopolymers and nanomaterials. Beyond his research work, he is also a dedicated educator, committed to mentoring and guiding the next generation of researchers. He supervises several graduate and doctoral researchers and is actively involved in teaching and curriculum development at Amrita Vishwa Vidyapeetham.

Green and Sustainable Synthesis of Silver Nanoparticles Using Wastes of Crude Drugs for Traditional Medicinal Use in Nara, Japan

11

Kimihiro Tani, Suguru Sakamoto, Yukie Tatsumoto, Masanao Imai, and Kazumitsu Naoe

Abstract

Recent topics on eco-friendly metal nanoparticle preparation using plant extracts and sustainable metal nanoparticle preparation using waste materials was introduced. Furthermore, original experimental data on the preparation of antimicrobial silver nanoparticles using the waste leaves of *Yamato* peony *(Paeonia lactiflora*), a traditional medicinal plant cultivated in Nara, Japan, which is a World Heritage site, were presented. In the silver nanoparticle preparation using waste leaves of *Yamato* peony, after a 1-hour reaction at 90 °C of a mixture of silver nitrate solution with the extract obtained by hot water extraction of the leaves, UV-vis spectra of the solution showed a surface plasmon resonance peak at ca. 440 nm, indicating the formation of silver nanoparticles. The dynamic light scattering (DLS) measurement of the solution also showed peaks at around 10 and 100 nm, respectively. SEM observations and EDS analysis of the dried sample confirmed the presence of agglomerates of nanoparticles with a silver signal. The higher the reaction temperature, the narrower the particle size

K. Tani
Department of Chemical Engineering, National Institute of Technology, Nara College, Yamato-Koriyama, Nara, Japan

Applied Chemistry, Kyoto Institute of Technology, Kyoto, Japan

S. Sakamoto · K. Naoe (✉)
Department of Chemical Engineering, National Institute of Technology, Nara College, Yamato-Koriyama, Nara, Japan
e-mail: naoe@chem.nara-k.ac.jp

Y. Tatsumoto
Nara Prefecture Institute of Industrial Development, Nara, Japan

M. Imai
Institute of Science and Technology, Kanto Gakuin University, Yokohama, Kanagawa, Japan

R. Malviya, S. Sundram (eds.), *Sustainable Green Biomaterials As Drug Delivery Systems*, Biomaterials, Bioengineering and Sustainability 1,
https://doi.org/10.1007/978-3-031-79062-1_11

distribution was obtained. In addition, antimicrobial tests of the prepared silver nanoparticles against the model microorganisms *Escherichia coli* and *Bacillus subtilis* showed that the silver nanoparticles have antibacterial activity against both bacteria. In conclusion, antimicrobial silver nanoparticles can be prepared using the waste parts of crude drug plants for traditional medicinal use in Nara, Japan, without the emission of effluents containing organic solvents or strong chemical reducing agents.

Keywords

Nanoparticles · Silver · Antibacterial · Medicinal plant · Waste · Plant extract · *Paeonia lactiflora*

11.1 Introduction

Nanomaterials, mainly metal nanoparticles, are expected to be applied in various fields, such as biomedicine (Hang et al. 2024), energy (Tony 2021), and catalysis for material synthesis (Ishida et al. 2020), due to their excellent properties. However, their preparation often involves chemical reactions in the liquid phase (Park et al. 2007), which are often not eco-friendly because they use organic solvents and highly reactive chemical reagents at high temperatures, resulting in the discharge of large amounts of liquid waste after preparation and high energy consumption. In order for our society to continue to take advantage of the wonderful properties of nanoparticles, it is essential to establish sustainable and eco-friendly nanoparticle preparation methods.

This paper first outlines the conventional preparation of metal nanoparticles by chemical reactions in liquid-phase systems and then introduces recent topics on the preparation of eco-friendly metal nanoparticles using plant extracts. Furthermore, examples of sustainable preparation of metal nanoparticles using waste materials are presented. In the last section, we present our experimental data on the preparation of antibacterial Ag nanoparticles using waste products from traditional medicinal plants in Nara, a world heritage site.

11.2 Eco-Friendly Synthesis of Metal Nanoparticles Using Plant Extracts

In general, metal nanoparticles are prepared by reducing metal ions with a strong chemical reducing agent in the presence of a surfactant as a phase transfer agent in an oil/water two-phase system. The most representative method is that reported by Brust et al. at Liverpool University in 2004 and is called the Brust–Schiffrin method (Brust et al. 1994). Using two-phase (water-toluene) reduction of $AuC1^{4-}$ by sodium borohydride in the presence of an alkanethiol after phase transfer by ammonium surfactant, solutions of 1–3 nm gold nanoparticles bearing a surface coating of thiol were prepared. The method enabled the room-temperature synthesis of highly stable functionalized metal nanoparticles of small size (2–2.5 nm). The proposed method significantly impacted the pace of subsequent metal nanoparticle synthesis. A wide variety of functionalized metal nanoparticles have been synthesized using this method over the years, which has also led to the development of new scientific fields (Templeton et al. 2000; Pileni 2023). The method has the advantage of accurate preparation control, but it is not eco-friendly because it produces a large amount of liquid waste including organic solvents, chemical reducing agents, and surfactants.

On the other hand, for the synthesis of metal nanoparticles using a single phase, the Turkevich method (Turkevich et al. 1951) is a relatively simple and reproducible technique for the synthesis of spherical Au nanoparticles between 10 and 30 nm. However, above 30 nm size the nanoparticles become less spherical, the size distribution becomes broader, and the results were less reproducible for the synthesis of Au nanoparticles. In contrast to Au nanoparticles obtained by the Turkevich method, Ag nanoparticles obtained by this method are much larger (60–200 nm) with high polydispersity and different morphologies (Wuithschick et al. 2015).

For centuries, traditional extracts from plants, spices, and culinary herbs have been added to enhance the flavor and extend the shelf life of foods. These extracts are rich in compounds with high antioxidant activity, such as polyphenols, carotenoids, and vitamins E and C (Embuscado 2015; López-Pedrouso et al. 2022). Numerous studies have been conducted on the preparation of metallic nanomaterials by reducing metal ions using the reducing properties of plant components. The preparation methods of nanomaterials using plant components are eco-friendly because they produce no liquid waste including organic solvents and chemical reducing agents.

In the preparation of nanomaterials using plant components, Easy availability of plants Not competing with food use are important. A similar example is the issue of competition with food production in bioethanol production. Bioethanol is made from sugars and starches, such as corn and sugarcane which are used as food for people and livestock. As the demand for bioethanol increases, crops that would normally be used as food are used for bioethanol production, resulting in food shortages and higher crop prices.

Therefore, the preparation of metal nanoparticles using easily available plant extracts has been reported (Song et al. 2009; Song and Kim 2009; Huang et al. 2011; Lu et al. 2014). Kim's group in Cheongju, Korea prepared Au nanoparticles using the extracts of *Magnolia kobus* and *Diopyros kaki* leaves (Song et al. 2009). *Magnolia kobus* is a species of Magnolia native to Japan and Korea, and is widely planted for ornamental purposes. *Diopyros kaki*, the oriental persimmon, is native to East Asia and is grown in Japan, Korea, and China. Because the ripe fruits are edible, the tree has been widely grown as a fruit tree in Asia. In modern times, it is also grown and consumed outside East Asia, such as Europe. The authors proposed a method for preparing Au nanoparticles using the aqueous extracts of common tree leaves and reported more than 90% conversion within a few minutes and a higher reaction rate than conventional chemical methods. The size and shape of the nanoparticles were controlled to some extent by changing the reaction conditions.

The aqueous extract of *the Acacia ehrenbergiana* plant cortex from Saudi Arabia was used to prepare Ag nanoparticles (Alamier et al. 2023). It is commonly found in the Sahara, northern Sahel, parts of East Africa, the Arabian Peninsula, and Iran. The prepared Ag nanoparticles with a particle size distribution ranging from 1 to 40 nm were used as heterogeneous catalysts to reduce Rhodamine B dye from aqueous solutions in the presence of sodium borohydride, and 96% catalytic reduction was accomplished within 32 min. The Ag nanoparticles also exhibited high antimicrobial activity against various Gram-positive, Gram- negative, and fungal microorganisms.

11.3 Sustainable Synthesis of Metal Nanoparticles Using Waste

Using waste to prepare value-added materials is important not only because it reduces the cost of disposal but also because it converts waste into useful materials. Preparation of metal nanoparticles using waste discharged in various fields has been reported. Wastes are classified according to the sector in which it is generated as follows; Agri-wastes, E-wastes, and Industrial wastes.

11.3.1 Agri-Wastes

Agri-wastes include not only agricultural wastes but also food wastes. While food shortages are severe in developing countries because of population growth and disasters caused by climate change, large amounts of food waste are generated every year in economically wealthy countries because of overconsumption by their societies. Many of these Agri wastes contain a large amount of moisture, and incineration of such wastes consumes a large amount of energy, which is undesirable from the perspectives of energy conservation and CO_2 emission control.

Italian researchers in Rome prepared Ag nanoparticles using extracts from bilberry waste or spent coffee grounds as raw materials (Baiocco et al. 2016). The wastes are commonly found in Italy. The authors focused on them as raw materials containing relatively high levels of phenolic compounds. They extracted the components from them using an aqueous ethanol solution and used them to prepare Ag nanoparticles. Ag nanoparticles with a spherical shape and an average size of 10–20 nm were successfully prepared. This is a truly local production of raw materials for nanomaterials.

Using the aqueous extract of cotton boll peel waste, SnO_2 nanoparticles were successfully prepared (Narasaiah et al. 2022). The powder of dried cotton boll peel was boiled to obtain an aqueous extract. The mixture of the extract and $SnCl_2$ aqueous solution was thoroughly stirred at 80 °C for 3 h, and the resulting powder after centrifugation was calcinated at 200 °C for 3 h to obtain SnO_2 nanoparticles. The SnO_2 nanoparticles exhibited superior photocatalytic efficiency for the degradation of methylene blue and methyl orange dyes under UV light exposure.

In India, the preparation of Fe nanoparticles using the aqueous extract of waste peels of *Aegle marmelos* (bael) fruit was reported (Srivastava et al. 2023). Bael is native to India and Bangladesh, and it is cultivated throughout India, Sri Lanka, Thailand, and Malaysia. Fruits are eaten either fresh from trees or dried and produced into candy, toffee, pulp powder, or nectar. Thus, a large amount of peel is discarded after eating and cooking. Peels are rich in polyphenols, which can act both as a reducing agent and capping agent. Fe nanoparticles were easily prepared by simply mixing 0.1 M $FeCl_2$ solution with waste peel extract at room temperature and adjusting the pH. The nanoparticles as a Fenton catalyst showed catalytic degradation of the anionic dye Eosin Yellow and the cationic dye Fuchsin Basic, with removal efficiencies as high as 97% within 20 min and 95% within 25 min at 318 K.

In the Czech Republic, bimetallic nanoparticles were produced using extracts derived from *Cannabis sativa* and *Vitis vinifera* waste (Michailidu et al. 2024). *Cannabis sativa* is cultivated and used as a source of industrial fiber, seed oil, food, and medicine. As for *Vitis vinifera* grape vines, which are common, grapes are eaten fresh or dried to produce raisins, etc. The fresh grapes are also processed into juice that is fermented to make wine and vinegar. The extracts were obtained from dried *Vitis vinifera* canes collected during the dormancy period and waste material extracted from C. *sativa* cannabinoid extracts. The prepared bimetallic nanoparticles exhibited strong antimicrobial activity and biofilm inhibition against several strains of *Pseudomonas aeruginosa*, a common pathogen responsible for hospital-acquired infections and major contributor to antibiotic resistance. It is expected that the toxicity of the bimetallic nanoparticles will be lowered while their antibacterial activity is preserved, as Ag nanoparticles are characterized by higher toxicity for mammalian cells compared to Au nanoparticles. The authors also prepared Ag nanoparticles using the extract of *Vitis vinifera* canes (Miskovska et al. 2024).

Solmaz et al. used the waste of tropical fruit *Citrus fortunella* for fruit juice, liqueur, or jam production and used it to synthesize selenium nanoparticles (Solmaz et al. 2024). Recently, Se nanoparticles are expected to be applied in the biomedical field due to their antibacterial (Zhang et al. 2021), biofilm-inhibiting (Ullah et al. 2023), anticancer (Geoffrion et al. 2020), and anti-inflammatory (Sun et al. 2023) effects. The authors investigated the adsorption behavior of the Se nanoparticles onto a crystal violet dye. A higher maximum adsorption capacity of 23.55 mg/g was achieved compared with the other nanoparticles.

11.3.2 E-Wastes

E-waste is becoming one of the fastest-growing waste categories globally. In 2019, 53.6 million tons of E-waste were generated worldwide. Only 17% of the E-waste was collected and recycled (Forti et al. 2020). This means that gold, silver, copper, platinum, and other high-value, recoverable materials were mostly dumped or burned rather than collected for treatment and reuse. However, it is important to consider the safety of recovering and recycling processes. E-waste recycling processes themselves are also not necessarily sustainable. Typically, they use toxic substances such as acids and organic solvents, emit greenhouse gases, and often require significant energy.

In the case of the preparation of copper oxide nanoparticles from copper recovered from discarded electrical circuits (Gautam et al. 2023) and Fe_2O_3 nanoparticles from waste toner and their application as catalyst support (Kouser et al. 2022), the use of organic solvents, treatment with strong acids and alkalis, and sintering at high temperatures are required.

11.3.3 Industrial-Wastes

In addition, nanomaterials have also been created in recent years from large quantities of industrial waste. Ag nanoparticles were prepared using wastes from paper and textile industries, and their antibacterial, antioxidant, and cytotoxic potential were evaluated (Paosen et al. 2021). Using *Eucalyptus camaldulensis* leaves collected as waste from a paper manufacturer and sericin extracted from silkworm cocoons. The prepared Ag nanoparticles exhibited antibacterial activity against foodborne pathogens (*Bacillus cereus*, *Listeria monocytogenes*, *Staphylococcus aureus*, *Escherichia coli* O157:H7, *Klebsiella pneumoniae*, *Salmonella Typhimurium*, *Shigella sonnei*, *Vibrio cholerae*, and *Vibrio parahaemolyticus*). In addition, the Ag nanoparticles provided antioxidant and antibiofilm activities without cytotoxic effects on Caco-2 and human erythrocytes.

Waste pickling acid, an iron-rich effluent produced worldwide as a result of pretreatment of steel surfaces in galvanizing plants, is a potential source of iron ions for many industrial processes despite its high pollutant loads. Approximately 300,000 tons of waste pickling acid are generated annually in the EU alone, resulting in a large amount of sludge containing metals for conventional treatment, which usually involves precipitation with limestone. It is important to make effective use of large amounts of waste metals to produce new nanomaterials. Polenz et al. (2022) prepared magnetic Fe_3O_4 nanoparticles using waste pickling acid as an iron source. The nanoparticles were then silica-coated and tested for biocompatibility. The tests showed no detection of hazardous contaminants (heavy metals) from waste-based precursors and a high ability to immobilize spermatozoon cells, indicating the biocompatibility of the magnetic nanoparticles for applications in biomedical fields. However, the conventional chemical coprecipitation and sol-gel methods were used to prepare the magnetic nanoparticles, which are not environmentally friendly.

In the above sections, we review recent studies on eco-friendly and sustainable preparations of metal nanoparticles using plant extracts and waste. The optical properties and catalytic activity of metal nanoparticles vary greatly depending on their size and shape (Sardar et al. 2009; Ishida et al. 2020; Hang et al. 2024). Therefore, precise control of particle size and shape during preparation is important for nanoparticle applications. On the other hand, in the eco-friendly and sustainable preparation of metal nanoparticles, it is difficult to precisely control the size and shape of the nanoparticles during preparation. This is due to the large variation in the components of the raw material, which makes the composition of the extract unstable. This is especially true for the preparation of nanoparticles using bio-waste materials.

Therefore, any measures on the eco-friendly methods are needed to overcome this drawback. The measures include the following:

Pretreatment (e.g., purification) of the raw materials or their extracts.

Post-treatment by size selection of the generated nanoparticles (Taleb et al. 1997) or by direct chemical modification of the size and shape of the generated nanoparticles (Naoe et al. 2007; Shimpi et al. 2017).

However, using only such measures, it is difficult to achieve the quality of nanoparticles produced by eco-friendly methods to reach the level of nanoparticles prepared by chemical methods. In other words, using current technology, it is difficult to replace chemical preparation methods with eco-friendly preparation methods completely. Therefore, it is better to select a preparation method that yields nanoparticles that are suitable for the given application. It is effective to use chemically prepared nanoparticles and eco-friendly prepared nanoparticles separately or mixed together (hybrid use) depending on the application. For example, in the aviation industry, Sustainable Aviation Fuel (SAF) is mixed into jet fuel to reduce CO_2 emissions (SAF blending limit: up to 50%) (ICAO 2020).

In the preparation of nanoparticles using plant extracts, it is also necessary to evaluate the shelf life of the extract and to improve its shelf life. In addition, it is important to establish methods for selecting plants and bio-waste suitable for nanoparticle synthesis. For plant parts and bio-waste, it is necessary to develop simple evaluation methods to determine the following

1. Does it have reducing properties? (containing reducing substances such as polyphenols?)
2. Does it have a coating capability for nanoparticles? (containing biomolecules such as proteins acting as coating agents?)

11.4 Preparation of Antibacterial Silver Nanoparticles Using Wastes of Traditional Medicinal Plants from the World Heritage Site Nara

Nara, the ancient capital of Japan approximately 1300 years ago, has many valuable historical heritages in the prefecture and has three UNESCO World Heritage sites: "Buddhist buildings in the Horyu-ji area", "Historic Monuments of Ancient Nara", and "Sacres Sites and Pilgrimage Routes in the Kii Mountain Range" (Nara Prefecture's World Heritage Properties HP 2024). At the time, Nara enjoyed great prosperity, emerging as the fountainhead of Japanese culture.

Kampo is a traditional Japanese herbal medicine that originated in ancient China. With its roots in ancient Chinese medicine, this antecedent form of empirical medicine was introduced to Japan in approximately the fifth to sixth century. Subsequently, it developed into a unique form of medicine by adapting to the climate and culture of Japan and was further refined to suit the constitutions of the Japanese people before evolving into a distinct form of traditional medicine (Watanabe et al. 2011).

In Nara, formerly known as "*Yamato*", where Japan's oldest imperial court was established, there had been a close relationship with crude drugs for traditional medicinal use since ancient times. To prepare for epidemics, medicinal plants were cultivated in the Kinki region centering on *Yamato*, and crude drugs from China and other countries were also gathered in *Yamato*. In addition to these historical factors, Nara is surrounded by mountains with ample rainfall, hot summers, cold winters, and little snowfall, making it geologically suitable for the cultivation of various plants for crude drugs. In the Edo period (1603–1867), demand for Kampo medicines increased, and as a measure for self-sufficiency in Japan, many attempts were made to import seeds and seedlings of medicinal plants from China, survey and collect medicinal herbs and trees growing wild in the mountains and fields in Japan, and cultivate them. In particular, the eighth Shogun, Yoshimune TOKUGAWA, encouraged the cultivation of medicinal plants throughout Japan. Due to these circumstances, Nara has been one of the most important areas where medicinal plants have been cultivated since ancient times. Even nowadays, there are many small pharmaceutical companies in Nara that manufacture Kampo medicines. For example, *Daranisuke* (陀羅尼助) is a gastrointestinal drug produced from the inner bark extract of yellow cedar (*Phellodendron amurense* Rupr) and other herbal extracts. It improves unpleasant conditions, such as weak stomach, overeating, loss of appetite, excessive drinking, heavy stomach, stomach or abdominal bloating, heartburn, pressure on the chest, and nausea and vomiting (ex. Sankogan Co., Ltd. (HP) 2024).

The peony is a perennial plant of the genus *Paeonia*. It is native to Siberia, China, and Mongolia. Its scientific name is *Paeonia lactiflora*. The peony came to Japan from China during the Nara period (710–794). The beautiful looking peony has a large number of varieties as horticultural species. Its roots have sedative, pain-relieving, antitussive, and antihypertensive effects and are used as an ingredient in Kampo medicine. The varieties suitable for medicinal use are currently being studied. In Japan, *Yamato* peony from Nara, which has been used for medicinal purposes for many years, is considered the highest quality (Fig. 11.1).

For medicinal plants, dried roots and bark are used as raw materials for crude drugs, whereas other parts are discarded. In the case of the peony, the roots of the peony are used as crude drugs, and the flowers are distributed as ornamental, whereas the leaves and buds are discarded because they are no longer used. In Japan, more than 1600 tons of peony are used annually as raw material for crude drugs (Yamamoto et al. 2021). In other words, a large number of non-root parts of the peony are discarded. However, it has been reported that discarded parts of peony, such as leaves, contain reducing components (Tatsumoto 2023), and if these discarded parts can be used as reducing agents, an eco-friendly large-scale production method for metallic nanoparticles can be developed. In this study, we focused on peony (*Paeonia lactiflora*) grown in Nara as a crude drug plant for traditional medicinal use and prepared antimicrobial silver nanoparticles using discarded parts (leaves) of the plant.

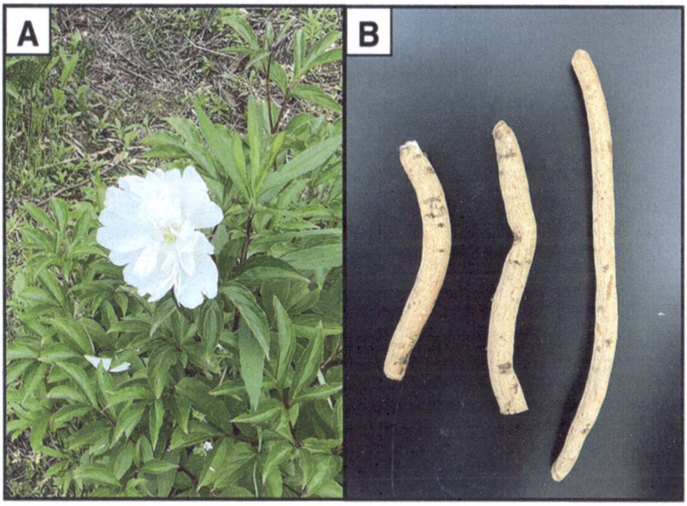

Fig. 11.1 Photos of (**a**) flower and (**b**) roots (after washing and drying) of *Yamato* peony (*Paeonia lactiflora*) in Nara

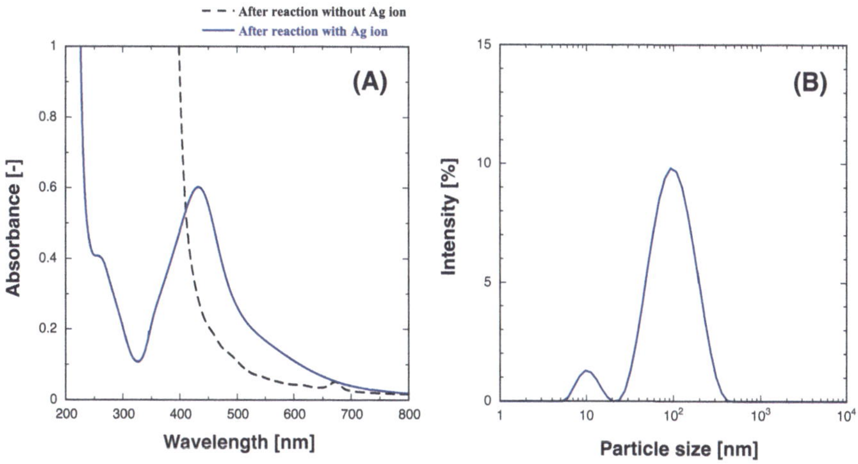

Fig. 11.2 (**a**) UV-Vis spectra of the mixture of *Paeonia lactiflora* leaf extract and Ag nitrate solution after heat treatment and (**b**) particle size distribution of the reaction mixture after heating. The spectrum of the mixture after the reaction without Ag ions is also shown for comparison. Reaction temp. = 90 °C, Reaction time = 1 h

Dried *Paeonia lactiflora* leaves were ground in a grinder, and hot water extraction was performed for 10 min. After centrifugation, the supernatant was filtered through a filter with a pore diameter of 4 μm, and the resulting filtrate was used as the extract. A silver nitrate solution was added to the extract, and the reduction reaction was carried out under stirring at a given temperature. After the reaction, the UV-Vis spectrum of the reaction mixture was recorded. Figure 11.2a shows the UV-Vis spectrum of a mixture of silver nitrate solution and *Paeonia lactiflora* leaf extract after the reaction at 90 °C for 1 h. The spectrum of the reaction solution without Ag ions is also shown for comparison. The UV-Vis spectrum in the presence of Ag ions shows an absorption peak at around 440 nm. As surface plasmon resonance (SPR) absorption due to Ag nanoparticles is observed at approximately 400 nm (Solomon et al. 2007), indicating the formation of Ag nanoparticles using the extract. On the other hand, no absorption peak was observed near the corresponding wavelength in the reaction without Ag ions. In the following sections, the absorption peak near 440 nm is used as an indicator of Ag nanoparticles. In addition, the DLS measurement of the solution after the reaction revealed peaks at approximately 10 and 100 nm, respectively (Fig. 11.2b). The results show that the size of primary nanoparticles was 10 nm, and they aggregated into aggregates of approximately 100 nm.

Next, the effect of reaction time on the formation of Ag nanoparticles was investigated. The absorbance at the maximum absorption wavelength can be used as an indicator of the number of nanoparticles produced. The absorbance at the maximum absorption wavelength λ_{max} increased with the reaction time and became almost constant after a reaction time of 1 h (Fig. 11.3). The DLS profile also changed little with reaction times greater than 1 h (data not shown). Under the reaction conditions, the reduction reaction was considered to have reached equilibrium after a reaction time of 1 h. In the following sections, the reaction time is set at 1 h).

The effect of reaction temperature on the formation of Ag nanoparticles was also investigated. Reduction of aqueous Ag nitrate solution by the extract from *Paeonia lactiflora* leaves at 70–90 °C. The reaction time was 1 h. The absorbance at the maximum absorption wavelength λ_{max} increased with the reaction

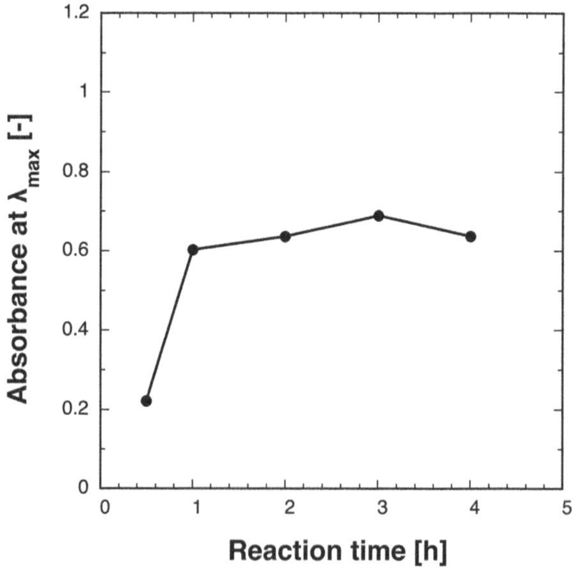

Fig. 11.3 Effect of reaction time on absorbance at λ_{max} of the mixture of *Paeonia lactiflora* leaf extract and Ag nitrate solution. Reaction temperature = 90 °C

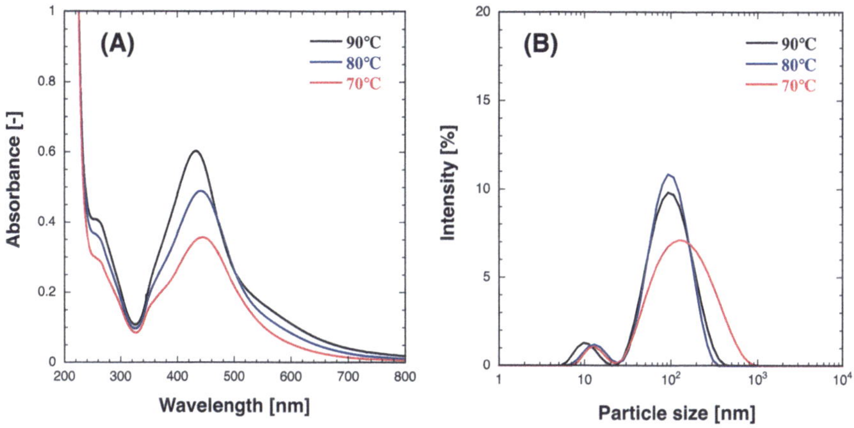

Fig. 11.4 Effect of reaction temperature on (**a**) UV-Vis spectra and (**b**) dynamic light scattering profiles of a mixture of *Paeonia lactiflora* leaf extract and Ag nitrate solution after heat treatment. Reaction time = 1 h

temperature (Fig. 11.4a), indicating that the number of Ag nanoparticles produced increased with the reaction temperature. The peak wavelength shifted to lower wavelengths as the reaction temperature increased. This result implies that the higher the reaction temperature, the smaller the size of the Ag nanoparticles. In the preparation of nanoparticles using other plant extracts, nanoparticles with smaller sizes have been reported to form at higher reaction temperatures (Song and Kim 2009; Song et al. 2010). In fact, in this study, the DLS profile under higher temperature conditions exhibited peaks at smaller particle sizes and a narrower particle size distribution (Fig. 11.4b). In the chemical synthesis of metal nanoparticles, reactions are often conducted under reflux at fairly high temperatures to obtain monodisperse nanoparticles (Shimizu et al. 2003; Park et al. 2004).

SEM observation and EDS analysis of the prepared Ag nanoparticle powders obtained after lyophilization were performed. Figure 11.5 shows SEM image and EDS analysis results for the Ag nanoparticle powder. The red dots in the figure represent silver. Aggregates of nanoparticles were observed. The elemental mapping of Ag shows that Ag is homogenously distributed only on the aggregates of the nanoparticles. The EDS spectrum (Fig. 11.5b) also confirmed the presence of Ag peaks, confirming the formation of Ag nanoparticles.

Thermogravimetry (TG) measurements of the prepared Ag nanoparticles were performed. Figure 11.6 shows the thermogravimetric profile of the Ag nanoparticles in a nitrogen gas atmosphere. The weight of the sample remained constant above 450 °C. There was a degradation of 53.2% of the sample attributed to the bio-organic part of the material, indicating 46.8% of the nanoparticle is Ag component. TG measurements of Ag nanoparticles prepared by other plant extracts have reported that the silver content is approximately 80% in Ag nanoparticles by propolis extract (Barbosa et al. 2019), 61% in Ag nanoparticles by tea leaf extract (Sun et al. 2014),

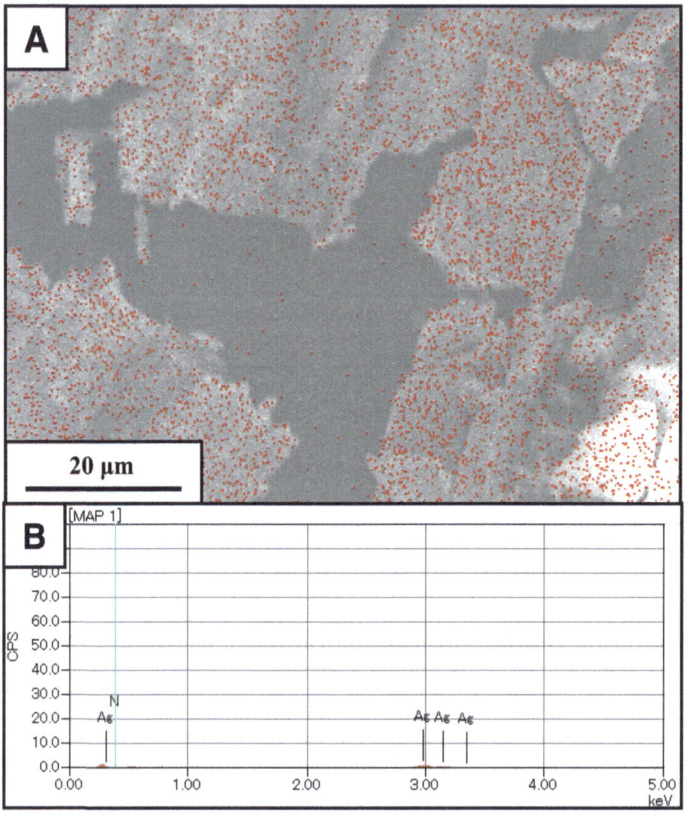

Fig. 11.5 (**a**) SEM image and EDS map (Ag) of the lyophilized Ag nanoparticles prepared using *Paeonia lactiflora* leaf extract and (**b**) its EDS analysis

and 66.7% in Ag nanoparticles by *Acacia ehrenbergiana* exocarp extract (Alamier et al. 2023). The Ag nanoparticles prepared using *Paeonia lactiflora* leaf extract in this study appear to contain more biomolecular coating components.

The antibacterial activity of the Ag nanoparticles against typical Gram-negative bacteria *Escherichia coli* and Gram-positive bacteria *Bacillus subtilis* was evaluated. The number of viable cells after incubating with Ag nanoparticles was determined by measuring colony formation using the standard plate count technique. Figure 11.7 shows the ratio of viable cells to the initial cell count for each Ag nanoparticle concentration. The ratio of viable cells decreased with increasing Ag nanoparticle concentration, indicating that the Ag nanoparticles have antibacterial activity against both bacteria. In the low-concentration region of the nanoparticles, the antibacterial activity against *E. coli* was slightly greater than that against *B. subtilis*. In general, Gram-positive bacteria have a thick peptidoglycan layer, whereas Gram-negative bacteria have a thin peptidoglycan layer that is easily permeable to Ag ions (Marcus et al. 2012). The antibacterial effect of the Ag nanoparticles was also greater against

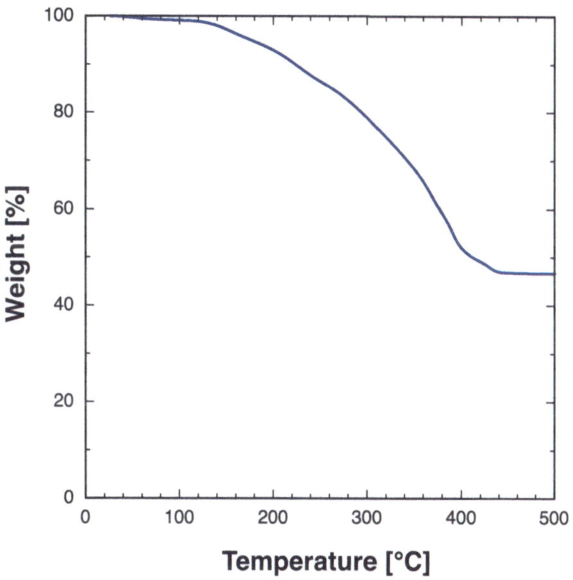

Fig. 11.6 TG analysis of Ag nanoparticles prepared using *Paeonia lactiflora* leaf extract

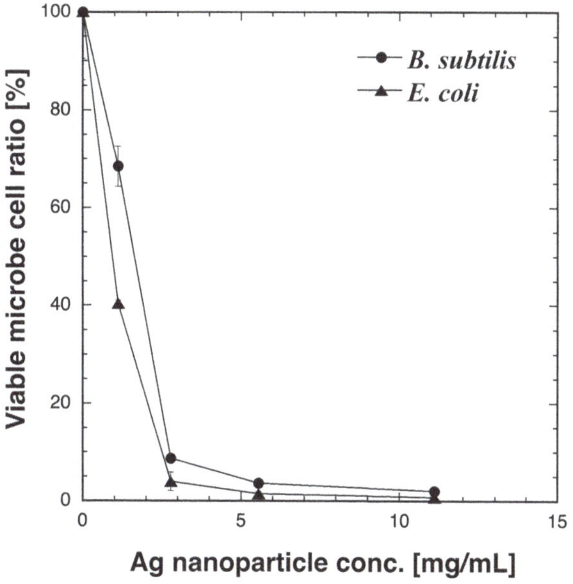

Fig. 11.7 Viable microbe cell ratios of *B. subtilis* and *E. coli* in the presence of Ag nanoparticles prepared using *Paeonia lactiflora* leaf extract

E. coli with a thin peptidoglycan layer in our experiment. Proteins, terpenoids, and flavonoids in plants function as coating agents in metal nanoparticle preparation using plant extracts (Shankar et al. 2004; Elumalai et al. 2014). Therefore, this result is related to the surface chemistry of Ag nanoparticles due to the components in *Paeonia lactiflora* leaves.

In addition, the Ag nanoparticles prepared using *Paeonia lactiflora* leaf extract have sufficient antimicrobial properties even though their Ag content is lower than that of Ag nanoparticles prepared using other plant extracts, as mentioned in the TG measurement results.

The Trolox equivalent antioxidant capacity (TEAC) values of the extracts of medicinal plant parts were measured as an indicator of reducing ability. The TEAC assay measures the antioxidant capacity of a sample compared with the standard Trolox. The TEAC assay is often used to measure the antioxidant capacity of foods (Huang et al. 2005). In this study, the 2,2-Diphenyl-1-picrylhydrazyl (DPPH) assay was used to measure the TEAC value of the samples. The samples measured were extracts of *Paeonia lactiflora* leaves and flowers from different collection sites, Gose and Shimoichi in Nara, and *Phellodendron amurense* fruit, respectively. The TEAC values of the samples are shown in Fig. 11.8a. For *Paeonia lactiflora*, differences in TEAC levels were observed depending on the plant part. Flowers had higher TEAC values than leaves. The TEAC values also differed depending on the location of the *Paeonia lactiflora* collection. Differences by sampling location may be due to the soil conditions in which the plants grew. The fruit of *Phellodendron amurense*, one of the plants used in Kampo medicine, which has reducing components similar to *Paeonia lactiflora*, had a lower TEAC value than that of *Paeonia lactiflora* leaves.

Figure 11.8b shows the absorbances at the maximum absorption wavelength of aqueous solution of Ag nanoparticles prepared using these plant extracts. The absorbance at the maximum absorption wavelength can be used as an indicator of the number of nanoparticles produced. The relative difference in absorbance for each Ag nanoparticle sample (Fig. 11.8b) corresponds to the relative difference in TEAC values for each extract sample (Fig. 11.8a). This indicates that if the TEAC values of plant part extracts are known, the nanoparticle-forming ability of the plant parts can be determined. However, as mentioned above, in the preparation of metal nanoparticles using plant extracts, it is important not only whether the extract has

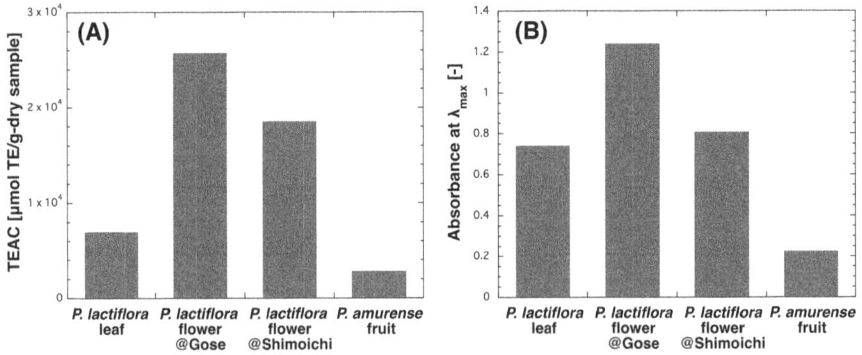

Fig. 11.8 (**a**) TEAC values of extracts from various parts of medicinal plants in Nara and (**b**) absorbances at the maximum absorption wavelength of aqueous solutions of Ag nanoparticles prepared using these extracts. Used medicinal plants: *Paeonia lactiflora* leaves, *Paeonia lactiflora* flowers from Gose and Shimoichi in Nara, and *Phellodendron amurense* fruit

reducing components, but also whether it has components with coating ability on the nanoparticles. It is necessary to evaluate the ability of plant extract biomolecules to act as coating agents in the future.

11.5 Conclusion

In this chapter, first, recent topics on eco-friendly metal nanoparticle preparation using plant extracts and sustainable metal nanoparticle preparation using waste materials were introduced. The overview of recent topics demonstrates that the preparation of metal nanoparticles using plant extracts is an eco-friendly preparation method that does not discharge hazardous waste solutions and that the preparation of metallic nanoparticles using Agri-wastes is a sustainable method that reduces waste and creates high-value-added functional materials. On the other hand, the drawback of the preparation methods using plant extracts and Agri-wastes were also highlighted, and measures to overcome the drawback were proposed. It was also noted the importance of establishing a method for selecting plant parts and bio-wastes suitable for nanoparticle synthesis. Finally, the preparation of antibacterial silver nanoparticles using the extract of waste leaves of *Yamato* peony (*Paeonia lactiflora*), which is cultivated in Nara, a World Heritage site, as a traditional medicinal plant, was presented. The conditions for the preparation of silver nanoparticles using the extract of waste leaves were experimentally investigated, and the antibacterial properties of the prepared silver nanoparticles were evaluated. In addition, the relationship between the reducing ability of plant extracts and the production of silver nanoparticles was also investigated. Thus, it is demonstrated that antimicrobial silver nanoparticles can be prepared using the waste parts of crude drug plants for traditional medicinal use in Nara, Japan, without the emission of effluents containing organic solvents or strong chemical reducing agents.

References

Alamier WM, Oteef MDY, Bakry AM, Hasan N, Ismail KS, Awad FS (2023) Green synthesis of silver nanoparticles using *acacia ehrenbergiana* plant cortex extract for efficient removal of Rhodamine B cationic dye from wastewater and the evaluation of antimicrobial activity. ACS Omega 8:18901–18914. https://doi.org/10.1021/acsomega.3c01292

Baiocco D, Lavecchia R, Natali S, Zuorro A (2016) Production of metal nanoparticles by agro-industrial wastes: a green opportunity for nanotechnology. Chem Eng Trans 47:67–72. https://doi.org/10.3303/CET1647012

Barbosa VT, Souza JKC, Alvino V, Meneghetti MR, Florez-Rodriguez PP, Moreira RE, Paulino GVB, Landell MF, Basílio-Júnior ID, do Nascimento TG, LAM G, Dornelas CB (2019) Biogenic synthesis of silver nanoparticles using Brazilian propolis. Biotechnol Prog 35: e2888. https://doi.org/10.1002/btpr.2888

Brust M, Walker M, Bethell D, Schiffrin DJ, Whyman R (1994) Synthesis of thiol-derivatised gold nanoparticles in a two-phase liquid–liquid system. J Chem Soc Chem Commun 4:801–802. https://doi.org/10.1039/C39940000801

Elumalai EK, Kayalvizhi K, Silvan S (2014) Coconut water assisted green synthesis of silver nanoparticles. J Pharm Bioallied Sci 6:241–245. https://doi.org/10.4103/0975-7406.142953

Embuscado ME (2015) Herbs and spices as antioxidants for food preservation. In: Shahidi F (ed) Handbook of antioxidants for food preservation. Woodhead Publishing, Cambridge, pp 251–283. https://doi.org/10.1016/B978-1-78242-089-7.00011-7

Forti V, Baldé CP, Kuehr R, Bel G (2020) The global E-waste monitor 2020: quantities, flows and the circular economy potential, United Nations University (UNU)/United Nations Institute for Training and Research (UNITAR) –co-hosted SCYCLE Programme. International Telecommunication Union (ITU) & International Solid Waste Association (ISWA), Bonn/Geneva/ Rotterdam

Gautam P, De AK, Sinha I, Behera CK, Singh KK (2023) Genesis of copper oxide nanoparticles from waste printed circuit boards and evaluation of their photocatalytic activity. Environ Res 229:115951. https://doi.org/10.1016/j.envres.2023.115951

Geoffrion LD, Hesabizadeh T, Medina-Cruz D, Kusper M, Taylor P, Vernet-Crua A, Chen J, Ajo A, Webster TJ, Guisbiers G (2020) Naked selenium nanoparticles for antibacterial and anticancer treatments. ACS Omega 5:2660–2669. https://doi.org/10.1021/acsomega.9b03172

Hang Y, Wang A, Wu N (2024) Plasmonic silver and gold nanoparticles: shape- and structure-modulated plasmonic functionality for point-of-caring sensing, bio-imaging and medical therapy. Chem Soc Rev 53:2932–2971. https://doi.org/10.1039/d3cs00793f

Huang D, Ou B, Prior RL (2005) The chemistry behind antioxidant capacity assays. J Agric Food Chem 53:1841–1856. https://doi.org/10.1021/jf030723c

Huang J, Zhan G, Zheng B, Sun D, Lu F, Lin Y, Chen H, Zheng Z, Zheng Y, Li Q (2011) Biogenic silver nanoparticles by *cacumen platycladi* extract: synthesis, formation mechanism, and antibacterial activity. Ind Eng Chem Res 50:9095–9106. https://doi.org/10.1021/ie200858y

Ishida T, Murayama T, Taketoshi A, Haruta M (2020) Importance of size and contact structure of gold nanoparticles for the genesis of unique catalytic processes. Chem Rev 120:464–525. https://doi.org/10.1021/acs.chemrev.9b00551

Kouser M, Chowhan B, Sharma N, Gupta M (2022) Transformation of waste toner powder into valuable Fe2O3 nanoparticles for the preparation of recyclable Co(II)-NH2-SiO2@Fe2O3 and its applications in the synthesis of polyhydroquinoline and quinazoline derivatives. ACS Omega 7:47619–47633. https://doi.org/10.1021/acsomega.2c04512

López-Pedrouso M, Lorenzo JM, Franco D (2022) Advances in natural antioxidants for food improvement. Antioxidants 11:1825. https://doi.org/10.3390/antiox11091825

Lu F, Sun D, Huang J, Du M, Yang F, Chen H, Hong Y, Li Q (2014) Plant-mediated synthesis of Ag–Pd alloy nanoparticles and their application as catalyst toward selective hydrogenation. ACS Sustain Chem Eng 2:1212–1218. https://doi.org/10.1021/sc500034r

Marcus IM, Herzberg M, Walker SL, Freger V (2012) Pseudomonas aeruginosa attachment on QCM-D sensors: the role of cell and surface hydrophobicities. Langmuir 28:6396–6402. https://doi.org/10.1021/la300333c

Michailidu J, Miskovska A, Jarosova I, Cejkova A, Matatkova O (2024) Bimetallic nanoparticle production using *Cannabis sativa* and *Vitis vinifera* waste extracts. RSC Adv 14:5309–5318. https://doi.org/10.1039/d3ra07134k

Miskovska A, Michailidu J, Kolouchova IJ, Barone L, Gornati R, Montali A, Tettamanti G, Berini F, Marinelli F, Masak J, Cejkova A, Matatkova O (2024) Biological activity of silver nanoparticles synthesized using viticultural waste. Microb Pathog 190:106613. https://doi.org/10.1016/j.micpath.2024.106613

Naoe K, Petit C, Pileni MP (2007) From worm like to spherical palladium nanocrystals: digestive ripening. J Phys Chem C 111:16249–16254. https://doi.org/10.1021/la702908y

Nara Prefecture's World Heritage Properties. https://www.pref.nara.jp/miryoku/ikasu-nara/en/sekai/. Accessed 29 May 2024

Narasaiah BP, Banoth P, Sohan A, Mandal BK, Dominguez AGB, Valladares LDLS, Kollu P (2022) Green biosynthesis of tin oxide nanomaterials mediated by agro-waste cotton boll peel extracts for the remediation of environmental pollutant dyes. ACS Omega 7:15423–15438. https://doi.org/10.1021/acsomega.1c07099

Paosen S, Wunnoo S, Bilhman S, Lethongkam S, Septama AW, Voravuthikunchai SP (2021) Inhibition of biofilm formation, adhesion and invasion in Caco-2 cells by foodborne pathogens using phyto-mediated synthesised silver nanoparticles from industrial wastes. Int J Food Sci Technol 56:5881–5892. https://doi.org/10.1111/ijfs.15329

Park J, An K, Hwang Y, Park J-G, Noh H-J, Kim J-Y, Park J-H, Hwang N-M, Hyeon T (2004) Ultra-large-scale syntheses of monodisperse nanocrystals. Nature Mater 3:891–895. https://doi.org/10.1038/nmat1251

Park J, Joo J, Kwon SG, Jang Y, Hyeon T (2007) Synthesis of monodisperse spherical nanocrystals. Angew Chem Int Ed 46:4630–4660. https://doi.org/10.1002/anie.200603148

Pileni MP (2023) Metal nano 3D superlattices: synthesis, properties, and applications. Wiley, Weinheim

Polenz MF, Sante LGG, Malschitzky E, Bail A (2022) The challenge to produce magnetic nanoparticles from waste containing heavy metals aiming at biomedical application: new horizons of chemical recycling. Sustain Chem Pharm 27:100678. https://doi.org/10.1016/j.scp.2022.100678

Sankogan Co., Ltd. https://sankogan.co.jp/en/. Accessed 29 May 2024

Sardar R, Funston AM, Mulvaney P, Murray RW (2009) Gold nanoparticles: past, present, and future. Langmuir 25:13840–13851. https://doi.org/10.1021/la9019475

Shankar SS, Rai A, Ahmad A, Sastry M (2004) Rapid synthesis of Au, Ag, and bimetallic Au core-Ag shell nanoparticles using neem (*Azadirachta indica*) leaf broth. J Colloid Interface Sci 275:496–502. https://doi.org/10.1016/j.jcis.2004.03.003

Shimizu T, Teranishi T, Hasegawa S, Miyake M (2003) Size evolution of alkanethiol-protected gold nanoparticles by heat treatment in the solid state. J Phys Chem B 107:2719–2724. https://doi.org/10.1021/jp026920g

Shimpi JR, Sidhaye DS, Prasad BLV (2017) Digestive ripening: a fine chemical machining process on the nanoscale. Langmuir 33:9491–9507. https://doi.org/10.1021/acs.langmuir.7b00193

Solmaz A, Turna T, Baran A (2024) Ecofriendly synthesis of selenium nanoparticles using agricultural *citrus fortunella* waste and decolourization of crystal violet from aqueous solution. Can J Chem Eng 102:2051–2067. https://doi.org/10.1002/cjce.25179

Solomon SD, Bahadory M, Jeyarajasingam AV, Rutkowsky SA, Boritz C, Mulfinger L (2007) Synthesis and study of silver nanoparticles. J Chem Educ 84:322–325. https://doi.org/10.1021/ed084p322

Song JY, Kim BS (2009) Rapid biological synthesis of silver nanoparticles using plant leaf extracts. Bioprocess Biosyst Eng 32:79–84. https://doi.org/10.1007/s00449-008-0224-6

Song JY, Jang HK, Kim BS (2009) Biological synthesis of gold nanoparticles using *Magnolia kobus* and *Diopyros kaki* leaf extracts. Process Biochem 44:1133–1138. https://doi.org/10.1016/j.procbio.2009.06.005

Song JY, Kwon E-Y, Kim BS (2010) Biological synthesis of platinum nanoparticles using *Diopyros kaki* leaf extract. Bioprocess Biosyst Eng 33:159–164. https://doi.org/10.1007/s00449-009-0373-2

Srivastava A, Dutta S, Jain R (2023) Waste mediated synthesis of iron nanoparticles using *Aegle marmelos* peel extract for catalytic degradation of toxic dye contaminants. RSC Sustainab 1:2261–2269. https://doi.org/10.1039/d3su00246b

Sun Q, Cai X, Li J, Zheng M, Chen Z, Yu C-P (2014) Green synthesis of silver nanoparticles using tea leaf extract and evaluation of their stability and antibacterial activity. Colloids Surf A Physicochem Eng Aspects 444:226–231. https://doi.org/10.1016/j.colsurfa.2013.12.065

Sun Y, Liang L, Yi Y, Meng Y, Peng K, Jiang X, Wang H (2023) Synthesis, characterization and anti-inflammatory activity of selenium nanoparticles stabilized by aminated yeast glucan. Int J Biol Macromol 245:125187. https://doi.org/10.1016/j.ijbiomac.2023.125187

Taleb A, Petit C, Pileni MP (1997) Synthesis of highly monodisperse silver nanoparticles from AOT reverse micelles: a way to 2D and 3D self-organization. Chem Mater 9:950–959. https://doi.org/10.1021/jp972807s

Tatsumoto Y (2023) Evaluation of antioxidant active components and free amino acid content of peonies from Nara prefecture. Res Rep Nara Prefect Industr Promot Center 49:9–18

Templeton A, Wuelfing W, Murray R (2000) Monolayer-protected cluster molecules. Acc Chem Res 33:27–36. https://doi.org/10.1021/ar9602664

The Business Aviation Coalition for Sustainable Aviation Fuel of The International Civil Aviation Organization (ICAO) (2020) Sustainable Aviation Fuel Guide Edition 2. https://www.futureofsustainablefuel.com. Accessed 29 May 2024

Tony MA (2021) Recent frontiers in solar energy storage via nanoparticles enhanced phase change materials: succinct review on basics, applications, and their environmental aspects. Energy Storage 3:e238. https://doi.org/10.1002/est2.238

Turkevich J, Stevenson PC, Hillier J (1951) A study of the nucleation and growth processes in the synthesis of colloidal gold. Discuss Faraday Soc 11:55–75. https://doi.org/10.1039/DF9511100055

Ullah A, Mirani ZA, Binbin S, Wang F, Chand MWH, Aslame S, Yonghong L, Hasan N, Naveed M, Hussaing S, Khatoon Z (2023) An elucidative study of the anti-biofilm effect of selenium nanoparticles (SeNPs) on selected biofilm producing pathogenic bacteria: a disintegrating effect of SeNPs on bacteria. Process Biochem 126:98–107. https://doi.org/10.1016/j.procbio.2022.12.031

Watanabe K, Matsuura K, Gao P, Hottenbacher L, Tokunaga H, Nishimura K, Imazu Y, Reissenweber H, Witt CM (2011) Traditional Japanese Kampo medicine: clinical research between modernity and traditional medicine—the state of research and methodological suggestions for the future. Evid Based Complement Alternat Med 2011:513842. https://doi.org/10.1093/ecam/neq067

Wuithschick M, Birnbaum A, Witte S, Sztucki M, Vainio U, Pinna N, Rademann K, Emmerling F, Kraehnert R, Polte J (2015) Turkevich in new robes: key questions answered for the most common gold nanoparticle synthesis. ACS Nano 9:7052–7071. https://doi.org/10.1021/acsnano.5b01579

Yamamoto Y, Kasahara R, Taira M, Takeda O, Higuchi Y, Yamaguchi Y, Shiratori M, Sasaki H (2021) Survey on crude drug usage in Japan (2). Shoyakugaku Zasshi 75:89–105. https://doi.org/10.24684/jspharm.75.2_89

Zhang H, Li Z, Dai C, Wang P, Fan S, Yu B, Qu Y (2021) Antibacterial properties and mechanism of selenium nanoparticles synthesized by *Providencia sp*. DCX Environ Res 194:110630. https://doi.org/10.1016/j.envres.2020.110630

Kimihiro Tani was a student at the Department of Chemical Engineering, National Institute of Technology, Nara College. He worked on the eco-friendly synthesis of metal nanoparticles using medicinal plant extracts in Nara at Prof. Naoe's laboratory. At present, he is a bachelor course student of Applied Chemistry program at Kyoto Institute of Technology.

Suguru Sakamoto is a student at the Department of Chemical Engineering, National Institute of Technology, Nara College. He is working on the eco-friendly synthesis of metal nanoparticles using extracts from the wastes of traditional medicinal plants in Nara at Prof. Naoe's laboratory.

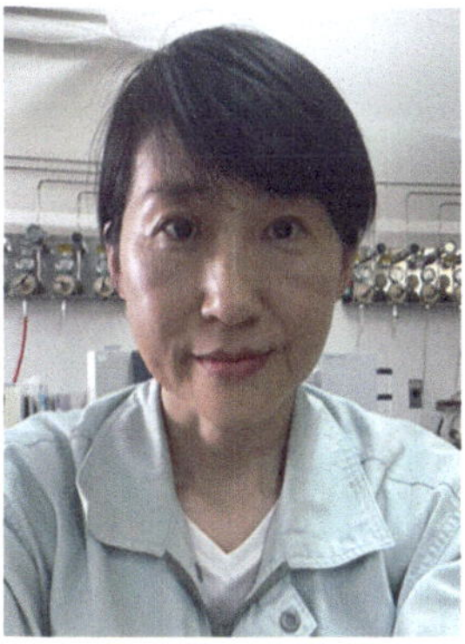

Yukie Tatsumoto is the chief researcher at the Nara Prefecture Institute of Industrial Development. She graduated from Kyoto Pharmaceutical University in 1993. She is working on the evaluation of the functional ingredients of traditional medicinal plants in Nara and their application to foods, and the development of alcoholic beverages and foods using unique yeasts in Nara.

Masanao Imai is a Research Fellow at the Institute of Science and Technology, Kanto Gakuin University. In 1976, he Graduated from Dept, of Industrial Chemistry, National Institute of Technology, Toyama College. In1980, he graduated from Dept. of Chemical Engineering, The University of Tokyo. In 1987, he has received PhD from The University of Tokyo under supervisory Professor Shintaro Furusaki. In April 1985, he entered the Fuji Oil Co. Ltd. In 1987, he moved to Tokyo University of Agriculture and Technology. In 1998, he moved to Nihon University and work as Professor of Chemical Engineering (2012–2022)) and Director of Food Manufacturing and Educational Center (2017–2021)). In 2021, he moved to Institute of Science and Technology, Kanto Gakuin University and worked as Research Fellow of Chemical Engineering. He has received honorable the award "Academic Cultural Award of Takaoka City" (Mayor Prize) in Feb, 2001. His research activities are presented as followed; 22 Books, 109 refereed papers, 249 international presentations and 73 other academic publications. He specializes in enzyme reaction in microemulsion phase and supercritical carbon dioxide concerned about separation of functional component from natural resources. Biopolymer sciences involved in characteristics of gel and membrane are currently investigated.

Kazumitsu Naoe is Professor, Department of Chemical Engineering, National Institute of Technology, Nara College. He graduated from Tokyo University of Agriculture and Technology and received his PhD from Kyushu University in 2001 under the supervision of Prof. Shintaro Furusaki. He studied applications of amphiphile aggregates in protein extraction and enzymatic reactions. He was awarded the Research Encouragement Award of The Membrane Society of Japan in 2004. After spending time as a visiting scientist at Université Pierre et Marie Curie (Paris VI), France, he began the preparation of metal nanoparticles using amphiphile aggregates and their applications. Recently, his laboratory focuses on emulsions and droplets stabilized by food-grade microparticles and eco-friendly synthesis of metal nanoparticles using local plant extract.

Metal Framework in Biosensor

12

Stephen Rathinaraj Benjamin, Eli José Miranda Ribeiro Júnior, Rosa Fireman Dutra, and Geanne Matos de Andrade

Abstract

Metal-organic frameworks (MOFs) are permeable substances composed of inorganic metal-based elements and naturally occurring ligands. Their unique properties, including tunable pores and diverse functional sites, have positioned them as promising candidates for immobilizing biomolecules in various environments. Recent advancements have enabled biomolecules like antibodies, enzymes, peptides, nucleic acids, and phages to be incorporated into or within the cavities of MOFs. This integration has opened up new possibilities for sensing applications, offering enhanced sensitivity, specificity, and a broader range of target detection. Researchers have shown significant interest in developing MOF

S. R. Benjamin (✉)
Laboratory of Behavioral Neuroscience, Drug Research and Development Center (NPDM), Department of Physiology and Pharmacology, Federal University of Ceará, Fortaleza, Ceará, Brazil

Faculty of Medicine, Department of Physiology and Pharmacology, Federal University of Cearà, Fortaleza, Ceará, Brazil

Department of Science and Engineering Materials, Federal University of Ceará, Pici Campus, Fortaleza, CE, Brazil

E. J. M. R. Júnior
Department of Pharmacy, Faculty of CGESP (Centro Goiano de Ensino Superior), Goiânia, GO, Brazil

R. F. Dutra
Biomedical Engineering Laboratory, Federal University of Pernambuco, Recife, PE, Brazil

G. M. de Andrade
Laboratory of Behavioral Neuroscience, Drug Research and Development Center (NPDM), Department of Physiology and Pharmacology, Federal University of Ceará, Fortaleza, Ceará, Brazil

Faculty of Medicine, Department of Physiology and Pharmacology, Federal University of Cearà, Fortaleza, Ceará, Brazil

R. Malviya, S. Sundram (eds.), *Sustainable Green Biomaterials As Drug Delivery Systems*, Biomaterials, Bioengineering and Sustainability 1,
https://doi.org/10.1007/978-3-031-79062-1_12

biocomposites with synergistic properties by combining biomolecules with MOFs. In sensing applications, voltammetric immunosensors based on electrochemical analytical techniques have emerged as a sensitive approach for quantitative virus detection using current measurements. This review examines the detection platforms for nucleic acids, antigens, and antibodies, evaluating the underlying sensing processes. Furthermore, the review addresses future directions for developing MOFs in virus detection. This review provides a comprehensive review of existing research on disease diagnosis and decision-making platforms, making it useful for those interested in advancing this field.

Keywords

Metal-organic framework · Immunosensors · Nucleic acid · Virus detection

12.1 Introduction

In recent years, depending on the advances in physical science, important studies have been carried out in the research of new materials for the development of sensors with potential information in nature. Carbon-based nanomaterials such as graphene and carbon nanotubes are frequently used in modern technology. In addition, many other interesting and new elements have gained importance, such as metal-based and metal oxides, quantum dots (QDs), and metal-organic frameworks (MOFs) (Bhadra et al. 2019). MOFs are interesting because they contain metal ions or clusters and organic molecules that give them a unique shape (Doonan et al. 2017). They have crystalline glass and cavities, making them useful in some industries, such as sensor technology. MOFs can control their chemical composition and size, which is a significant advantage. Organic linkers cause changes in these structures. This makes MOFs independent and unique features that have attracted the attention of many people over the years. MOFs are made using various combinations, each modified to achieve different goals and designs.

As the number of sensors increases, solid metallurgy (SMFs), especially MOFs and other 2D materials such as graphene, are gaining more attention due to their special products. Surface modified membranes (SMFs) allow you to carefully control their chemical composition and porous structure. This allows the creation of sensitive sensors that can detect specific chemicals or ions. Their excellent porosity, as well as their thickness and value-bearing ability, make them more sensitive and stable, allowing them to operate reliably for a long time. The addition of biological material to surface modified fiber (SMF) makes biomonitoring technology more effective, especially in identifying health problems and focusing on the environment. Using advanced manufacturing techniques, scientists have developed composite biosensors that can detect a variety of things, including biomolecules, antigens, viruses and bacteria. These biosensors are promising for use in laboratory and other important applications because of their ability to mimic living things. This makes them powerful and effective in many cases. Metal-organic frameworks

(MOFs) are known for their stability and transparency, high permeability, and the ability to modify the chemicals used to make them (Jahangiri-Dehaghani et al. 2020). Due to these properties, they are ideal for applications that require water absorption. Metal-organic frameworks (MOFs) can be easily modified by adding different groups such as -NH_2 and -COOH during or after the formation process (Ling et al. 2015). Due to the way they are made and their natural porous structure, they easily absorb objects and recognize living things. MOFs have been widely investigated as they can be used in many fields, including biology, fuel storage, catalysis, drug delivery, and photobiology (Daglar et al. 2021). MOFs are useful in these areas because they are tailored to specific needs. Researchers have discovered that MOFs have unique properties and can perform in many different ways. In the fabrication of sensors, MOFs are chosen because they are flexible in design and have many advantages. An important advantage of MOF-based sensors is that their framework is flexible and can create specific regions with their properties. Scientists can create special sensors that can be customized to detect specific chemicals or ions. Due to their thin structure, MOFs are ideal as sensors. An important feature of MOFs is the π conjugation system, which allows charges to move efficiently. These properties make MOF-based devices more sensitive to changes in their environment. Due to its large surface area and high porosity, the sensor is sensitive to small analytes. This is because it provides more space for the target molecule to interact with itself (Wang et al. 2021a, b; Cao et al. 2019; Banerjee et al. 2020). path. Using synthetic techniques to create advanced biosensors. Bioactive substances can be added during the synthesis of the MOF or introduced after synthesis. The method improves analyte detection ability by ensuring the even distribution of biomolecules in the MOF role. MOF composite biosensors can detect a variety of substances, including biomolecules, antibody-antigen complexes, enzymes, bacteria, and viruses (Safaei et al. 2019). Due to their sensitivity and versatility, biosensors are important in many fields such as environment, food safety, and diagnostics (Robison et al. 2019; Miller et al. 2023; Cheng et al. 2021). The purpose of this chapter is to provide a detailed and in-depth introduction to biosensors using MOFs. This study investigates the formation process, properties, and utilization of MOFs in biosensor applications. This section discusses how MOFs can help in the development of biosensor technology due to their optical and electrical properties. This section presents an in-depth study of MOF-based biosensors, focusing on recent developments and potential areas for further research.

12.2 Synthesis of MOFs

Metal-organic frameworks Frameworks develop by changing several factors, including the fluid used, the length and temperature of the response, and the chemical features of the metallic ions and organic ligands that are used. Aside from that, the size and shape of the morphological nodes, the presence of rivals, and the complexity of crystallization dynamics all have big effects. What makes MOF crystals grow and form depends on the way each of these components are placed along with how well they work together. To make MOFs, solutions with natural ligands and metallic salts are mixed to make a liquid. It is very important for picking the right solvent because it requires to be able to react and dissolve properly to allow the chemical processes to happen. Additionally, the solvent's composition has a significant impact on the thermodynamic characteristics and energetic stimulation of each reaction, which are crucial for attaining the desired structural as well as operational attributes of the MOFs. Despite facing difficulties in the development of single crystals, solid-state synthetic techniques have been used in specific cases. The extended evaporation procedure is commonly employed in the development of MOFs, which are crystal structures that serve as a mode of operation for the reaction solution. Typically, MOFs have been produced by solvothermal methods or hydrothermal techniques (Arul et al. 2021), which involve carrying out the reaction at high temperatures and pressures. The above technique embodies the fundamental approach to the synthesis of MOFs. Recently, several other synthetic approaches have arisen, including mechanochemical, electrochemical, microwave for heating, and sonochemical processes (Beamish-Cook et al. 2021). These techniques demonstrate impressive cost-effectiveness, accelerated completion times, and a tendency to provide high-quality results (Safaei et al. 2019).

12.3 Sensors

Electrochemical sensors have gained popularity for their ability to quickly and inexpensively analyze many substances with sensitivity and accuracy (Sun and Zhou 2015). Several electrocatalysts possess constraints that impede their ability to perform. These constraints encompass inadequate design of the structure, a sparse distribution of active spots, and a limited surface area (Sun and Zhou 2015). Therefore, it is imperative to explore novel materials that can enhance electrochemical efficiency. This part introduces a novel and advanced point on material science, providing a strategy that is different from traditional methods that mainly concentrate on the process of transduction (Ni and Masel 2006). It investigates the use of MOFs as essential elements in sensors and biological sensors, considering the specifics of the manufacturing processes, and emphasizes many convincing examples that might result in significant advancements in electrochemical sensing.

12.3.1 Various Types of Biosensors

12.3.1.1 Electrochemical Biosensors

Electrochemical immunosensors are devices that incorporate biological recognition components like a pair of electrodes or field-effect transistors (FET) with transduction components (Carson et al. 2011). Electrochemical immunosensors are ideal for detecting diseases early on, thanks to their sensitivity, affordability, small size, repeatability, and biocompatibility (Benjamin et al. 2023). Electrochemical immunosensors must be categorized based on their sensing principle: label-free (without labels) and sandwich-type (with labels). Immunosensors without labels can directly identify physical alterations resulting from the development of immunocomplexes, such as the Antigen-Antibodies complex. The utilization of multiple labels, including enzymes, liposomes, antibodies, chromophores, or antigens immobilized on conducting membranes, has substantially enhanced the responsiveness as well as the specificity of immune sensor technologies that have been marked. It is essential to include the electromechanical propagation element to increase the efficacy and selectivity of an electromechanical immunosensor. This is accomplished by converting the biological signal produced by interactions between antibodies and antigens into an electrochemical signal.

The electrochemical transduction section usually involves an electrochemical cell that functions either in a three-electrode setup within a potentiostat/galvanostat or a two-electrode mode in conductometry. Electrochemical immunosensors use electrodes from different conductive and semi-conductive materials, forming various configurations. Due to their excellent conductivity, label-free electrochemical immunosensors often use metallic electrodes like Au and Pt. These electrodes are synthesized by applying a thin metal film onto insulating substrates. In addition, scientists have studied different materials as potential substrates for detecting tumors, such as ceramic electrodes like indium tin oxide (ITO) and titanium dioxide (TiO_2), carbon electrodes, and polymer electrodes. Although ceramic electrodes have high conductivity, they are commonly used for the Confirmation of analytical substances in their original location due to their openness. Polymeric electrodes, on the other hand, are renowned for their exceptional mechanical properties, which enable them to store and transfer energy in various forms, and their compatibility with living organisms. This makes them incredibly valuable in a wide range of scenarios.

Researchers have developed a groundbreaking biological sensor that utilizes gold nanoparticles and copper-based metalorganic frameworks to detect microRNA in blood samples. By employing this novel approach, gold nanoparticles were created and positioned on top of the cumofs. Then, the dna tags that were modified by adding a thiol group were chemically attached to the gold nanoparticles. Mirna155 was identified in the sample through a deliberate design. In the realm of electrical immunological sensors, there are three primary methods to measure them: amperometric, potentiometric, and impedimetric. Frequently, these assessment methods are combined to obtain a thorough understanding of how the immunosensor functions. This comprehensive assessment enables you to track the progress of the stability process and determine the sensitivity of the sensor towards target antigens, as well as its limit of detection.

12.3.1.2 Amperometric and Voltammetric Immunosensor

Amperometric and voltammetric immunosensors are common components of analytical instruments that use various electrical methods to measure current and voltage in an electrolytic system. This list includes cyclic voltammetry (CV), differential pulse voltammetry (DPV), square wave voltammetry (SWV), anodic stripping voltammetry (ASV), and linear sweep voltammetry (LSV). By utilizing specific potentials to a reference electrode and monitoring the electrical signals passing through it, it is possible to study the redox characteristics of different substances. Two widely used methods to evaluate the redox capabilities of immunosensors are amperometry and voltammetry. When employing these techniques, it is crucial to establish distinct potentials between the reference electrode and the conducting membrane, and then meticulously analyze the currents flowing through them. Due to their precision and comprehensive nature, amperometric methods are highly recommended for accurate and detailed measurements. They can precisely identify redoxactive chemicals within a specific range, guaranteeing accurate and dependable outcomes.

Most of the time, micro-fabrication methods using small materials like glass, silicon, and printed circuit boards are used to add all three the electrodes to these systems. Based materials are often made from gold, titanium, or copper, which are then used in other production steps. There are numerous substances, like plastics with conducting properties, that are put on the backing material to improve the operating electrode's electromagnetic properties and sensitivity. Conductive polymers like Polypyrrole (Ppy), Polyaniline (PANI), Poly(3-thiophene methyl acetate) (PTMA), and Poly(3,4-ethylenedioxythiophene) (PEDOT) are often used (Wang et al. 2023). These materials have been selected because they are very good at conducting electricity, can last in a variety of environments, and are safe for living things. These characteristics render them ideal for improving the signals that are changed in immunosensor uses. The materials that are made are joined with semi-conductive or conductive resins to make the electrode functions more conductive.

MOFs have been mixed with protein and genetic material in novel studies to make electrochemical sensors that can identify target proteins and DNA signals (Yang et al. 2019). Yang et al. (2019), for example, made a positive ions MOF-5 that binds strongly to negative-charged DNA through both π-π stacking and electrostatic bonds. In addition to high accuracy, this MOF-5 found particular sequences of DNA. Researchers (https://www.sciencedirect.com/science/article/pii/S0026265X23006264) have made a unique electrochemical detector that uses Au@Cu-MOFs. They made Au NPs right on top of the Cu-MOFs and used covalent contact to strongly attach thiol-modified probe DNA to the AuNPs. MiRNA-155 can be found by this probe (Sun et al. 2020) made BPNSs/Mn-MOF nanocomposites to find SIP1, which is a stress-induced protein. A DNA aptamer was attached to the nanocomposites' surface. MOFs are perfect for making Point-of-Care (POC) devices for disease evaluation because they have large specific surface areas, can be used in numerous of different ways, and are easy to use. It has historically been possible to make carbon cloth paper (CCP) film electrodes that are cheaply made from paper. When it came to identifying HIV DNA, these sensors were very accurate and precise. They were able to recognize HIV DNA in a wide range of concentrations, from 10 nM to 1 μM, and at a very low level (0.13 nM) (Du et al. 2020). Incorporating functional molecules or nanomaterials to Metal-Organic Frameworks (MOFs) can make electrochemical instruments a lot more effective. These improvements can make the equipment better at identifying things, which makes it better for a variety of research assignments.

12.3.1.3 Electrochemiluminescence (ECL) Biosensor

Over the past few years, electrochemiluminescence (ECL) has become more useful using very strong luminophores and co-reactant boosters. As a result of its high responsiveness and low cost, ECL is widely used in pharmaceutical and therapeutic fields. To ensure that biosensors are accurate and reliable, they require a high-quality ECL signal reader. Improving the electrolytic contact between the luminophore and the co-reactant is an effective method to enhance the signal output in electrochemiluminescence (ECL). ECL biosensors for markers of neurodegenerative diseases such as cardiac troponin (Du et al. 2020; Yan et al. 2019; https://doi.org/10.1016/j.bios.2021.113532), amyloid-beta protein (Wang et al. 2019a, b, c; Zhao et al. 2019) and tumor markers such as CyFRA 21–1. Recently, an ECL assay that was very good at identifying CYFRA 21-1, which is a biomarker for certain types of cancer, the luminophore Sr(HCOO)2 +/Eu3+, and the co-reactant boosters silver chloride (AgCl) reduced-graphene nanocomposites (RRNCs) show significant limitations in detection.LOD) of 2.73 femtograms per milliliter (fg/mL) and a broad linear detection range spanning 5 to 100 nanograms per milliliter (ng/mL), rendering it very suitable for practical diagnostic applications (Zhao et al. 2023).

12.3.1.4 Aptamers

The Aptamer, sometimes referred to as Apt, is an extraordinary biological receptor due to its remarkable affinity and enduring stability, even under elevated temperatures. The importance of aptamers in the foreseeable future of receptors cannot be exaggerated. These brief, single-stranded DNA molecules are remarkable and can potentially transform the field (Khan et al. 2023). Adding functional compounds like -SH and -NH2 can significantly enhance the binding capacity of materials and electrodes (Miao et al. 2022a, b). The aptamer sensors exhibit promising potential in biomedical diagnosis due to their superior thermal stability compared to conventional antibody-based biosensors, simple manufacture and modification, cost-effectiveness, and no toxicity and immunogenicity (Zhang et al. 2023). Recently, Shu et al., (Su et al. 2023) have introduced a revolutionary wearable and flexible electrochemical aptasensor that allows for Efficient and accurate detection of cortisol in perspiration. The sensor is composed of a polyurethane (PU) film coated with CNTs (carbon nanotubes) and nickel-copper-MOF nanosheets. The cortisol detection tool has a broad range of values that are arranged in a straight line of 0.1–100 ng/mL and can accurately detect cortisol levels as low as 0.032 ng/mL. The development of bioactive Zr-MOF composites has resulted in highly efficient and ultrasensitive biosensors for detecting targeted biomolecules. These composite materials have shown a strong attraction to the specific proteins they are designed to target, making them outstanding applicants for maximizing the effectiveness of biosensors. The biosensors developed may change the way biomolecules are detected and provide the basis for more specific and efficient diagnostic tools (Zhang et al. 2017). Sun et al. (2020) recently developed a novel electrochemical aptasensor to measure STIP1 (stress-related phosphoprotein 1) levels. The researchers used a 3-D manganese-doped metal-organic framework (MOF) that resembled flower-shaped black phosphorus nanosheets. This unique design facilitates STIP1 detection and reliable measurement. Regarding STIP1 detection, the aptasensor is highly sensitive with a linear measurement range from 2×10^{-3} to 1×10^4 ng/mL. It also reached the effective limit of detection (LOD) of 1 picogram per milliliter (pg/mL), indicating that it can accurately and sensitively measure the amount of STIP1.

12.3.1.5 Field-Effect-Transistor-Based Sensors (FET)

Field-effect transistors also have a very quick output signal response, are easy to calibrate and they can measure concentrations in the orders of single digits nM. These sensors monitor for one or more molecules of interest and relay this information in the easy-to-read form of electrical signals (Zhang et al. 2022a, b, c). Recently, the Keum team developed c-MOF-gated FET biosensors that can distinguish between various neurotransmitters found in body fluids (Keum et al. 2023). The biosensor array utilized a new conductivity MOF-gated FET to distinguish neural transmitters. The inclusion of catalytic c-MOF Facilitated the identification of chemical messengers with exceptional sensitivity, even at concentrations relevant to clinical applications. By incorporating catalytic c-MOF, it becomes possible to detect neurotransmitters with remarkable sensitivity, even at concentrations that are significant in clinical applications (Sayyad et al. 2021) successfully developed a FET using microporous ZIF-67 MOFs, demonstrating high mobility compared to other ZIFs. ZIF-67 displayed The material exhibits conventional p-type characteristics, with an internal hole accessibility of around 0.85 cm2/Vs and an Ion/off ratio of 10^2; among MOFs, this material has the potential to be a promising channel for FET applications. A newly proposed gas sensor for SO_2 operates at room temperature and uses a buried-gate FET and interdigitated electrode structure. The sensor uses a composite film made of TiO2 nanoparticles and rGO, which makes it highly sensitive and compatible with traditional MEMS processes. The results showed a sensitivity of up to 3.46% at 20 ppm SO2, using a 2 V source-drain voltage and zero gate voltage (Zeng et al. 2022).

12.3.1.6 MOF-Nanomaterials-Based Biosensors

Various nanomaterials, including Gold (AU), silver (AG), silicon (SI), and copper (Cu) micron-sized particles together with carbon-based compounds such as graphite (Petit et al. 2011), graphene (Zhang et al. 2022a, b, c), and carbon nanotubes, are utilized in the fabrication of biological sensor immobilization (Benjamin et al. 2022). The use of nanoparticle-based materials is highly effective in developing electrochemical and other biosensors due to their sensitivity and specificity. Gold nanoparticles are more stable and less toxic than other metallic nanoparticles, like silver, that can oxidize and have toxic effects (Luo et al. 2019)There are benefits and downsides to using nanoparticles as signal mechanisms, but these tools are essential for improving single-molecule recognition sensitivity and limit. Electrochemical amplification using a single-label response has been developed using platinum-based nanoparticles to detect low glucose concentrations (Uzak et al. 2020). Tumor antigens may be targeted with excellent specificity and affinity by linking semiconducting quantum dots and iron oxide nanocrystals to tumor-targeting compounds (Gaviria-Arroyave et al. 2020). The potential use of biosensors ranging from millimeter- to nanometer-scale cantilevers is thoroughly evaluated.

12.3.1.7 Food Quality Monitoring

Electrochemical biosensors utilizing MOFs as integrated analytical devices equipped with electrochemical transducers have emerged as advanced tools for analytical responses. The reduction catalyst and analyte-MOF interaction in the redox process determine electrochemical signals. These biosensors possess advanced capabilities, facilitating the development of portable and rapid signal-responsive materials for cost-effective and highly sensitive food quality monitoring, such as food freshness (Dey et al. 2023). Consequently, they present a promising alternative to traditional techniques in the field.

Recently, sensors based on MOF technology have been effectively ensuring food safety (Cheng et al. 2021; Yao et al. 2020; Wang et al. 2019a, b, c; Zhang et al. 2021). An effective method for detecting The detection of aflatoxin M1 that is present in powdery and pasteurized milk specimens may be achieved using a label-free electromechanical aptasensor. This sensor utilizes platinum nanoparticles embedded within Fe-based MOFs (3) (Liu et al. 2021) have developed a highly sensitive electrochemical biosensor for detecting Hg^{2+}. In developing this biosensor, they used clustered copper-based MOFs that resemble peonies. They utilized a method of dual amplification signal, which relied on a DNAzyme-powered DNA Walker. This innovative approach allows for efficient and accurate detection of Hg^{2+} ions. The aptasensor's results are unequivocally remarkable as it has exhibited high accuracy in detecting the presence of Hg^{2+}. The current shift correlates directly with the negative logarithm of the Hg^{2+} concentration within the range of 0.001–100 nM and LOD of 0.52 pM.One example involves using Pb-MOF phosphorescence, which incorporates terephthalic acid, to determine PV in luminescence studies. The luminescence spectra in Pb-MOF can be quenched by iodide ions (I^-), which act as reductants for peroxides. This allows for the evaluation of PV in edible oils.The developed model produces quick results in just 10 minutes, with a LOD value of 30 μmol/kg and a linearity of 0.35–25.62 mmol/kg for PV detection, which can be applied to various oil samples (Wu et al. 2022a, b; Wang et al. 2022) developed a composite material called MOF@MOF, specifically ZIF-8@PCN–128Y (Zn-MOF@Zr-MOF), using internal extension growth. The tetracycline detection process was facilitated by this substance, which served as a turn-off sensor. Throughout the boundaries of linear ranges, it demonstrated a low limit of detection (LOD) of 60 nanometers from 0.4–80 μM. Moreover, when the sensor was utilized to test milk samples, it exhibited a remarkable recovery rate ranging from 94.2% to 106.3%.

12.3.1.8 Environmental Analysis

Peng et al. (2022) introduced an original and efficient method for the rapid detection of streptomycin (STR) based on its interaction with kanamycin aptamer immobilized onto 96 interferon wells prepared using TiO2 layers modified by gold nanoparticles modified zinc oxide quantum dots. This approach was based on the fabrication of a

Ti3C2 aptasensor and zirconium-based metal-organic frameworks (Zr-MOFs). The formation of nanocomposites within MOF structure through the impregnation method to get an improvement on electrochemical activity for STR detection. The LOD was as low as 0.0033 nM within the broad concentration range of 0.01–200 nM, achieved by such a specialized sensor Milk sample was a very good matrix for STR and the sensor worked well with milk samples, demonstrating high recovery ranging between 94.0% to 105.0%. It was very specific, consistent, and reliable (Miller et al. 2023) made an electrochemical sensor electrode with graphene oxide (GO) and epoxy-functionalized BEA nano-zeolite to find bisphenol E. With a correlation value of 0.995 and a low limit of detection (LOD) of 0.056 μM, this nanocomposite sensor was capable of picking up things from 0.07 to 4.81 micromolar (μM).

Also, (Yang et al. 2019) led a group of researchers who believe made an electrochemical immunosensor with NH2-UiO-66 that can identify both triazophos and thiacloprid at the same time. This method shows the versatility and useful it is to combine MOFs with electrochemical detecting technologies to find more than one analyte.For triazophos, the LOD (limit of discovery) is very low—0.07 ng·mL − 1—while for thiacloprid, it is 0.1 ng·mL − 1 (Tu et al. 2019) made an electrochemical detector that uses sensors made of Ce-MOF and reduced graphene oxide. The sensor has exceptionally high sensitivity for finding dichlorophen, with a lowest limit of detection as low as 0.007 μM. It had been demonstrated that MWCNTs had very high conductivity, which made them a good alternative to MOFs. For example, (Wang et al. 2021a, b) have successfully created an electrochemical sensor that combines MWCNTs and MOF to rapidly detect cadmium ions in meat samples. The sensor has demonstrated a strong linear correlation in the linearity of 0.5 to 170 μg/L and achieved an incredibly low detection value of 0.2 μg/L.

12.3.1.9 Pesticide

Xie et al. (2019) conducted an investigation where they developed an electrochemical gauge that does not require enzymes and instead uses a combined material of copper-cerium (CuO-CeO2). This hybrid material was formed by subjecting a Cu (II)/Ce(III) MOF to calcination together with a CuO-CeO2 nanostructure. The electrode exhibited a detection spectrum that was linear and spanned from 0.1 × 10–-7 to 0.1 μmol L − 1. Additionally, it achieved a low limit of detection (LOD) of 0.33 × 10–8 ± 0.03 μmol L − 1. The detection device identified trace quantities of malathion (organophosphorus pesticides) in the reservoir's water (He et al. 2019) effectively created a Zr luminous pore-filled MOF that is resistant to water and used it to directly identify MP (methyl parathion) in different water supplies and cowpeas. This sensor is capable of detecting a linear value ranging from 70 to 5.0 milligrammes per litre (mg L − 1) and has a relatively small limit of detection of 0.115 times 10 to the power of negative 3 (0.115 × 10–3) mg L − 1 for MP (Peng et al. 2022) came up with a new way to determine streptomycin (STR) that worked really well, with returns ranging from 77.6% to 106.5% and relative standard deviations (RSD) ranging from 3.8% to 24.0%. The Zr-MOF monitor shows a lot of promise for finding STR accurately and reliably.

Wang et al. (2020) created a special light sensor that only recognizes parathion in water used for farming. The sensor could pick up numerous amounts of various concentrations, from 0.005 mg/L to 1 mg/L, and it reached a limit of detection (LOD) of 1.95 μg/L. Over a 15-day period, the system achieved 97.23% accuracy and 93.76% repeatability. This shows that it can be used to track real agriculture (Liu et al. 2020) developed a porphyrin MOF probe that does two things: It effectively blocks fluorescence to capture nitenpyram. The probe has a lower limit of detection (LOD) of 0.03 mg/L and a linear range of 0.05 to 10.0 mg/L in groundwater and soil samples. The probe's ability to manipulate a variety of natural patterns shows how flexible and useful it is for agriculture. By combining MOF nanocubes with platinum nanoparticles (PtNPs), Guo and colleagues created an immune system that can capture Salmonella by amplifying the signal. In their approach, they used Fe-MOF/PtNPs as peroxidase analogs in anti-inflammatory drugs. This allows collecting oxygen (O2) from temperature changes measured by thermometers, including smartphones. and co-deposited Fe-MOF/PtNPs into a microfluidic device. The most important step to make sure it works is to isolate and repair the components. This highly integrated system can detect contamination easily and quickly, making it a powerful tool for food and environmental protection. Fe-MOF/PtNP then decomposes H2O2 and reacts with water and calcium oxide powder, causing heat and high pressure. This reaction can be easily detected by the thermal monitor of the smartphone, making it easier to determine whether there is an infection (Guo et al. 2021).

12.3.1.10 Gas Sensors

MOF-based composites are ideal for gas chemo-resistive sensors due to their excellent selectivity, sensitivity, and low detection value. A team of researchers led by (Zhang et al. 2019) A sensor that detects SO2 gas using MOFs-derived TiO2/rGO nanocomposites has been successfully created. This sensor outperforms pure TiO_2 sensors and can detect SO_2 concentrations ranging from 250 ppb to 20 ppm, with a 43% responsivity to 1 ppm SO_2 gas. Researchers have recently focused on designing ECL biosensors and MOFs functionalized with ECL molecules (Yang et al. 2019) fabricated 2D Ru-MOF nanosheets using ECL properties to detect Cardiac Troponin I. Furthermore, as contrasted with large MOF crystals, 2D MOF nanosheets provide a greater number of active regions that are easily accessible on their extensive surfaces. This facilitates the interaction with substrate molecules, enhancing MOFs' sensing capabilities. The detection range for cTnI by the immunosensor was between 1 fg/mL and 10 ng/mL, with an impressive LOD of 0.48 fg/mL with good sensitivity and excellent selectivity.

12.3.1.11 Temperature Sensor

The easy approach Naderi et al., (https://doi.org/10.1016/j.matchemphys.2023.127775) proposed to produce 2D MOF-5 sheets over ZnO nanorods for enhancing MOF-5 stability and properties for light sensing devices is very promising. This approach enhances mechanical strength, improves flow distribution, and increases surface accessibility and porosity. The photoresponse characteristics of a Metal-

Semiconductor-Metal (MSM) photodetector made from a hybrid film of MOF-5/ ZnO nanorods were studied to change in temperature. The distinctive charge transfer processes of organic and inorganic semiconductors reflect the significant improvement in photodetector performance at lower temperatures. The photodetector's sensitivity increased 5.7-fold when the temperature was reduced by 15°. The produced SnO_2/ Co-MOF derived Co3O4 films exhibited nanometer-scale thickness control, 3D porous and undulating designs, and significantly enhanced conductivity. The optimized SnO_2/Co_3O_4 sensor demonstrated rapid response and recovery times of 11 Sec and 16 Sec, respectively (Zhang et al. 2019). It also exhibited excellent selectivity and achieved a satisfactory sensitivity of 150% towards TEA at room temperature (https://www.sciencedirect.com/science/article/pii/S0272884223008775). The successful method for chemo-resistive toluene sensing has been showcased by applying a composite electrode composed of UiO-66-NH_2 and PEDOT: PSS, illustrating its effectiveness. This method can be performed in toluene with high accuracy, efficiency and reliability (Garg et al. 2023). Analysis was performed to measure the response to different concentrations of toluene ranging from 1 to 20 parts per million (ppm). The answer to this scale is 1.42 ± 0.14. The mean reaction time was 362 ± 4 s, and the mean recovery time was 106 ± 1 s.

12.4 Diagnosis of Diseases

Metal-organic frameworks (MOFs) have received significant attention as a diagnostic approach in recent years. The next section discusses how metal-organic frameworks (MOFs) can help identify diseases that pose a serious threat to humans worldwide.

12.4.1 Cancer

Cancers such as stomach, breast, ovarian, and medullary thyroid tumors can be detected with a blood-converted biomarker called carcinoembryonic antigen (CEA). (https://pubmed.ncbi.nlm.nih.gov/31995543/) developed a state-of-the-art electrochemical immunosensor using a novel nanoprobe called MOF-Ce@HA/Ag-HRP-Ab2 to improve CEA detection. This nanoprobe is specifically designed for the direct detection and understanding of CEA. This is the immune system sandwich stack and allows it to see CEA better. Details of this electrolytic sensor include that it is biocompatible and very good at catalysis. The entire discovery process takes less than 30 minutes, so results can be evaluated quickly. Regardless of the CEA value, the examiner responded to a range of 1 pg/mL to 80 ng/mL. Due to the relationship between the generated signal and the amount of CEA, the sensor can detect the amount of CEA quickly and accurately. As little as 0.2 pg/ml of CEA might be enough to be detected. Also, there is a way to find the biomarker HER2 in breast cancer using a voltammetric immunosensor with a hybrid structure. This sensor is made of a composite material with gold nanoparticles on top of a Cu-MOF and C3N4 (Yola 2021).

The LOD was found to be 3.00 fg/mL, and the calibration curve that formed was between 0.01 and 1.00 pg/mL. TNF-α is a factor that plays a role in both healthy and sick body processes. People who have higher amounts of TNF-α are more likely to have insulin resistance, gout, and cancer. To find TNF-α, a new sandwich-style voltammetric immunosensor has been made. This immunosensor utilized gold nanoparticles combined with thiol-functionalized MWCNT and bimetallic Ni/Cu-MOFs. The immunosensor produced showed a detection limit of 2.00 fg/mL and exhibited excellent selectivity, even in other chemicals (Yola 2021). Additionally, MIL-96 is a biosensor that can detect alpha-fetoprotein with high sensitivity. It utilizes copper for better conductivity and has a low LOD of 0.12 pg/mL. The dynamic range for detection is between 0.001 to 0.5 ng/mL (Gu et al. 2020).

12.4.2 Glucose Sensor

Recently, Copper (Zhang et al. 2020; Zhang et al. 2022a, b, c; Ma et al. 2021; Wu et al. 2020; Su et al. 2023), nickel (Ma et al. 2022; Xiao et al. 2022; Wei et al. 2023), bimetallic (Ma et al. 2023; Wang et al. 2021a, b), conducting polymer (Chen et al. 2022) and cobalt MOFs enhance enzymatic-free detection of glucose using electrochemical sensors. In addition, a newly developed 3D-printed gadget incorporates a water-resistant and harmless Fe(II)-MOF enzyme for detecting GLU using voltammetry in synthetic sweat (Koukouviti et al. 2023). The device created using MOF and 3D-printing technology showed a linear response to DPV and a calibration range within 100–600 μmol/L. Additionally, it had a low LOD value of 17.6 μmol/L. It has been established that the metal-enzyme compound GOx-Eu^{3+}@UMOF exhibits high selectivity and sensitivity toward glucose in serum and urine samples while utilized at room temperature (Zhang et al. 2019). A sandwich-style electrochemical immunosensor was created with a signaling molecule consisting of a nanocomposite of a zinc-based MOF/Fe_3O_4-COOH/thionine labeled anti-CTnI -moAb. It has been demonstrated that a plastic film made of cetyltrimethylammonium bromide in an intense eutectic solution including the compounds choline chloride and urea may be developed to detect CTnI through a sensor. The use of an anti-CTnI polyclonal antibody facilitated immobilization. The nanocomposite's precise arrangement amplifies electrochemical reduction, thereby reducing the accurate evaluation of CTnI via DPV. The performance of the immunosensor was exceptional, showing a concentration range from 0.04–50 ng/mL and a LOD of 0.0009 ng/mL (Ahmadi et al. 2022). In addition, a combination of copper-MOF ($Cu3(BTC)_2$) and PANI was utilized to create an impedimetric sensor that can detect cTnI. The Cu3(BTC)2/PANI composite was applied as a thin layer on SPE and linked to anti-cTnI antibodies. Impedimetric analysis of cTnI can be swiftly accomplished using the immunosensor, with a linear range of 1–400 ng/mL. The process can be completed within 5 minutes, ensuring speedy and efficient detection (Gupta et al. 2021).

12.4.3 Parkinson's Disease (PD)

The global incidence of Parkinson's disease (PD) is progressively rising as the population undergoes ageing. PD affects a global population exceeding 6 million individuals, according to a data published by the WHO. Recently, a new biosensor that combines electrochemical and colorimetric techniques has been developed using Cu-MOF@Ag to detect nitro-α-syn with high sensitivity. Based on the assay, the ranges of detection are 10–175 and 20–350 ng/mL. The detection limits are 0.23 and 7.63 ng/mL, respectively. These results demonstrate the sensitivity of the biosensor (Xu et al. 2023).

12.4.4 HIV Sensor

Lu et al. (2021) conducted a study that created a highly sensitive and selective electrode made of CCP film for detecting HIV DNA. This electrode is also cost-effective and versatile. By adding a conjugated π-electron framework, a hybrid material composed of nickel and gold can efficiently expand the outermost area of the electrode, hence enhancing its ability to sense (Li et al. 2021). With a low limit of detection (LOD) value of 0.13 nM, the electrode exhibited a wide range of linearity ranging from 10 nM to 1 μM (84). and excellent stability even after storage, making it a promising POC sensing platform for pathogen diagnosis. Additionally, the electrode showed good selectivity against mismatches and could successfully detect HIV DNA in complex serum samples.

12.4.5 MOF Used for Optical Sensors

An optical biosensor is an analytical tool consisting of a receptor, such as peptides, antibodies, or nucleic acids, an optical transducer, and a detector. The receptor is combined with optical probes like AuNPs (Li et al. 2021), AgNPs (Wu et al. 2022a, b), QDs, etc., in the transducer component. The fundamental principle behind optical biosensors is modifying the optical probe's characteristic signal when an analyte is present. The dynamic interplay between the receptor and the analyte instigates biorecognition phenomena, encompassing intricate processes, including hybridization and binding of proteins or peptides (Ly et al. 2018), and metabolic reactions. These events inevitably result in physical or chemical changes within the system. The transducer converts these modifications into optical or electronic signals, which are then detected by the detector. Different categories of optical biosensors use various optical principles, some of which include SPR (Han et al. 2020), absorbance, fluorescence, chemiluminescence (Li et al. 2019), photoluminescence (Basaleh and Sheta 2021), refractive index, surface-enhanced Raman spectroscopy (SERS) (Huang et al. 2020), scattering, reflectance (Pander et al. 2021), diffraction, and others. These principles encompass various electromagnetic spectra, such as UV-visible, infrared (IR), and near-infrared (NIR) regions (Law et al. 2020).

Recently, (Miao et al. 2022a, b) proposed a biomimetic mineralization technique as a reasonable design approach for increasing the thermostability and activity of biosensors. The process of Fe-MOF mineralization creates a protective layer on HRP, which shields it from high temperatures and successfully addresses the biosensor's low thermostability. The biosensor can regenerate, thanks to the biomimetic mineralization of HRP@Fe-MOF and the aptamer's remarkable stability even at high temperatures. The new optical biosensor has been created to measure AβO levels using a spectrum transformation accurately. This biosensor has a low LOD of 0.03 pM and can measure within a linearity value of 0.0001–10 nM. However, a novel optical biosensor was developed using aptamer-functionalized Cu-MOF to detect CRP. The study utilized immobilized RNA that binds to CRP on substrates that trigger florescent and peroxidase Cu-MOF, which enables dual-mode visual recognition using colorimetry and fluorometry. This biosensor exhibited excellent CRP detection in COVID-19 patient serum, achieving a low LOD of 40 pg/mL in the fluorescence-based mode and 240 pg/mL in the colorimetric technique (Ali and Omer 2022). The U-shaped surface plasmon resonance (SPR) sensor's sensitivity has improved by MOF surface modification. Subsequently, To facilitate reliable identification of the low-concentration MUC1 protein, the outermost layer of the U-shaped fiber probe was immobilized with the nucleic acid aptamer. The results of the experiments show that the sensor's response is linear throughout a wide range of MUC1 protein concentrations, including 1 pg/ml to 100 g/ml. Notably, the sensor demonstrates a sensitivity of 5.33 nm/log(μg/ml) and achieves LOD value of 0.16 pg/ml (Wang et al. 2023).

12.5 Conclusion and Future Perspective

The sensing applications of MOFs have been widely addressed, especially in electrochemical sensing. MOF sensors facilitate the detection of organic ions and molecules. This overview focuses on MOF-based composites utilized for detecting disease biomarkers through electrochemical sensing. It explores nanocomposite components, including metal/metal oxide nanoparticles and graphene derivatives. Future studies should focus on using various functionalized MOFs to enhance the identification ability of composite-based sensors. It has been discovered that integrating separate NPs is crucial for improving the effectiveness of the sensing region. There is a need for further study to improve the analytical sensitivity, and stability, MOF-based compounds have features that include specificity and interference suppression behavior. MOFs' inability to detect viruses at low enough concentrations in clinical samples reduces their sensitivity. Building MOF structures with lower particle sizes, including the application of viral RNA in the development of innovative biological sensors that make use of metallic nanoparticles, and creating automated, highly sensitive, highly selective MOF-based biosensors are all promising areas for further study. High sensitivity would significantly improve if conductive functional components were added to MOFs. It is recommended that future researchers focus on automating ultrasensitive MOF-based immunosensors to enable rapid detection on a larger scale. This can be achieved by incorporating electrochemistry technology and developing structures using MOF by the use of micron-sized particles, which will contribute to the enhancement of luminescent signals and the enhancement of quenched operations and selection. Cutting-edge research approaches in this field demand using aptamers, affibodies, peptide arrays, and molecularly imprinted polymers. These methods offer unparalleled potential for groundbreaking studies in the future.

Acknowledgments The authors express their gratitude to the Coordination for the Improvement of Higher Education Personnel (CAPES)-Brazil, the Post Graduate Program in Medical Sciences, Drug Research and Development Centre (NPDM) of the Federal University of Ceará (UFC) in Fortaleza, Brazil for their invaluable support.

Declaration of Competing Interest There are no conflicts of interest to declare.

References

Ahmadi A, Khoshfetrat SM, Mirzaeizadeh Z, Kabiri S, Rezaie J, Omidfar K (2022) Electrochemical immunosensor for determination of cardiac troponin I using two-dimensional metal-organic framework/Fe3O4–COOH nanosheet composites loaded with thionine and pCTAB/DES modified electrode. Talanta 237:122911

Ali GK, Omer KM (2022) Ultrasensitive aptamer-functionalized Cu-MOF fluorescent nanozyme as an optical biosensor for detection of C-reactive protein. Anal Biochem [Internet] 658:114928. Available from: https://linkinghub.elsevier.com/retrieve/pii/S0003269722003888

Arul P, Huang ST, Gowthaman NSK, Mani G, Jeromiyas N, Shankar S et al (2021) Electrocatalyst based on Ni-MOF intercalated with amino acid-functionalized graphene nanoplatelets for the determination of endocrine disruptor bisphenol A. Anal Chim Acta 1150:338228

Banerjee S, Lollar CT, Xiao Z, Fang Y, Zhou H-C (2020) Biomedical integration of metal–organic frameworks. Trends Chem [Internet] 2(5):467–479. Available from: https://linkinghub.elsevier.com/retrieve/pii/S2589597420300320

Basaleh AS, Sheta SM (2021) Manganese metal–organic framework: chemical stability, photoluminescence studies, and biosensing application. J Inorg Organomet Polym Mater 31(4):1726–1737

Beamish-Cook J, Shankland K, Murray CA, Vaqueiro P (2021) Insights into the mechanochemical synthesis of MOF-74. Cryst Growth Des [Internet] 21(5):3047–3055. Available from: https://pubs.acs.org/doi/10.1021/acs.cgd.1c00213

Benjamin SR, Miranda Ribeiro Júnior EJ (2022) Graphene-based electrochemical sensors for detection of environmental pollutants. Curr Opin Environ Sci Heal [Internet] 29:100381. Available from: https://linkinghub.elsevier.com/retrieve/pii/S2468584422000563

Benjamin SR, Júnior EJMR, Geanne Matos de Andrade A, Oriá RB (2023) Metal-organic framework–based electrochemical immunosensors for virus detection. In: Advanced functional metal-organic frameworks, pp 269–288

Bhadra BN, Vinu A, Serre C, Jhung SH (2019) MOF-derived carbonaceous materials enriched with nitrogen: preparation and applications in adsorption and catalysis. Mater Today 25:88–111

Cao J, Li X, Tian H (2019) Metal-Organic Framework (MOF)-based drug delivery. Curr Med Chem 27(35):5949–5969

Carson CG, Brown AJ, Sholl DS, Nair S (2011) Sonochemical synthesis and characterization of submicrometer crystals of the metal–organic framework Cu[(hfipbb)(H 2 hfipbb) 0.5]. Cryst Growth Des [Internet] 11(10):4505–4510. Available from: https://pubs.acs.org/doi/10.1021/cg200728b

Chen S, Liu D, Song N, Wang C, Lu X (2022) Promoting non-enzymatic electrochemical sensing performance toward glucose by the integration of conducting polypyrrole with metal-organic framework. Compos Commun [Internet] 30:101074. Available from: https://linkinghub.elsevier.com/retrieve/pii/S2452213922000195

Chen X, Qin C, Zheng Z, Chi H, Ye Z, Zhu L (2023) Constructing bilayer sensors with co-MOF-derived Co3O4 porous sensing films and SnO2 catalytic overlayers to enhance room-temperature triethylamine sensing performance. Ceram Int [Internet] 49(13):21455–21464. Available from: https://linkinghub.elsevier.com/retrieve/pii/S0272884223008775

Cheng W, Tang X, Zhang Y, Wu D, Yang W (2021) Applications of metal-organic framework (MOF)-based sensors for food safety: enhancing mechanisms and recent advances. Trends Food Sci Technol 112:268–282

Daglar H, Gulbalkan HC, Avci G, Aksu GO, Altundal OF, Altintas C et al (2021) Effect of Metal–Organic Framework (MOF) database selection on the assessment of gas storage and separation potentials of MOFs. Angew Chemie Int Ed 60(14):7828–7837

Dey B, Ahmad MW, Kim BH, Kamal T, Yang D-J, Patra CN et al (2023) Manganese cobalt-MOF@carbon nanofiber-based non-enzymatic histamine sensor for the determination of food freshness. Anal Bioanal Chem [Internet]. 415(17):3487–3501. Available from: https://link.springer.com/10.1007/s00216-023-04737-0

Doonan C, Riccò R, Liang K, Bradshaw D, Falcaro P (2017) Metal–organic frameworks at the biointerface: synthetic strategies and applications. Acc Chem Res [Internet] 50(6):1423–1432. Available from: https://pubs.acs.org/doi/10.1021/acs.accounts.7b00090

Du D, Shu J, Guo M, Haghighatbin MA, Yang D, Bian Z et al (2020) Potential-resolved differential Electrochemiluminescence Immunosensor for cardiac troponin I based on MOF-5-wrapped CdS quantum dot Nanoluminophores. Anal Chem [Internet]. 92(20):14113–14121. Available from: https://pubs.acs.org/doi/10.1021/acs.analchem.0c03131

Garg G, Garg N, Deep A, Soni D (2023) Zr-MOF and PEDOT:PSS composite sensor for chemoresistive sensing of toluene at room temperature. J Alloys Compd [Internet] 956: 170309. Available from: https://linkinghub.elsevier.com/retrieve/pii/S0925838823016122

Gaviria-Arroyave MI, Cano JB, Peñuela GA (2020) Nanomaterial-based fluorescent biosensors for monitoring environmental pollutants: a critical review. Talanta Open [Internet] 2:100006. Available from: https://www.sciencedirect.com/science/article/pii/S2666831920300060

Gu C, Peng Y, Li J, Liu CS, Pang H (2020) Controllable synthesis of copper ion guided MIL-96 octadecahedron: highly sensitive aptasensor toward alpha-fetoprotein. Appl Mater Today 20: 100745

Guo R, Xue L, Cai G, Qi W, Liu Y, Lin J (2021) Fe-MIL-88NH2Metal-organic framework nanocubes decorated with Pt nanoparticles for the detection of salmonella. ACS Appl Nano Mater 4(5):5115–5122

Gupta A, Sharma SK, Pachauri V, Ingebrandt S, Singh S, Sharma AL et al (2021) Sensitive impedimetric detection of troponin I with metal-organic framework composite electrode. RSC Adv 11(4):2167–2174

Han H, Hou D, Luan N, Bai Z, Song L, Liu J et al (2020) Surface plasmon resonance sensor based on dual-side polished microstructured optical fiber with dual-core. Sensors (Switzerland) 20: 3911

He K, Li Z, Wang L, Fu Y, Quan H, Li Y et al (2019) A water-stable luminescent metal-organic framework for rapid and visible sensing of organophosphorus pesticides. ACS Appl Mater Interfaces 11(29):26250–26260

Huang C, Li A, Chen X, Wang T (2020) Understanding the role of metal–organic frameworks in surface-enhanced raman scattering application. Small 16:e2004802

Jahangiri-Dehaghani F, Zare HR, Shekari Z (2020) Measurement of aflatoxin M1 in powder and pasteurized milk samples by using a label–free electrochemical aptasensor based on platinum nanoparticles loaded on Fe–based metal–organic frameworks. Food Chem [Internet] 310: 125820. Available from: https://linkinghub.elsevier.com/retrieve/pii/S0308814619319545

Keum C, Park S, Kim H, Kim H, Lee KH, Jeong Y (2023) Modular conductive MOF-gated field-effect biosensor for sensitive discrimination on the small molecular scale. Chem Eng J [Internet] 456:141079. Available from: https://linkinghub.elsevier.com/retrieve/pii/S1385894722065603

Khan S, Cho WC, Sepahvand A, Haji Hosseinali S, Hussain A, Nejadi Babadaei MM et al (2023) Electrochemical aptasensor based on the engineered core-shell MOF nanostructures for the detection of tumor antigens. J Nanobiotechnol [Internet] 21(1):136. Available from: https://jnanobiotechnology.biomedcentral.com/articles/10.1186/s12951-023-01884-5

Koukouviti E, Plessas AK, Pagkali V, Economou A, Papaefstathiou GS, Kokkinos C (2023) 3D-printed electrochemical glucose device with integrated Fe(II)-MOF nanozyme. Microchim Acta [Internet] 190(7):274. Available from: https://link.springer.com/10.1007/s00604-023-05860-6

Law CS, Marsal LF, Santos A (2020) Electrochemically engineered nanoporous photonic crystal structures for optical sensing and biosensing. In: Handbook of nanomaterials in analytical chemistry [Internet]. Elsevier, pp 201–226. Available from: https://linkinghub.elsevier.com/retrieve/pii/B9780128166994000098

Li D, Zhang S, Feng X, Yang H, Nie F, Zhang W (2019) A novel peroxidase mimetic Co-MOF enhanced luminol chemiluminescence and its application in glucose sensing, vol 296. B Chem, Sensors Actuators, p 126631

Li Y, Yuan J, Zhan S, Hu J, Guo Y, Ding L et al (2021) Dynamic light scattering immunosensor based on metal-organic framework mediated gold growth strategy for the ultra-sensitive detection of alpha-fetoprotein. Sensors Actuators B Chem 341:130030

Ling P, Lei J, Zhang L, Ju H (2015) Porphyrin-encapsulated metal–organic frameworks as mimetic catalysts for electrochemical DNA sensing via allosteric switch of hairpin DNA. Anal Chem [Internet]. 87(7):3957–3963. Available from: https://pubs.acs.org/doi/10.1021/acs.analchem.5b00001

Liu J, Liu J, Xiong WH, Ye LY, Zhang WS, Yang H et al (2020) Developing a novel nanoscale porphyrinic metal-organic framework: a bifunctional platform with sensitive fluorescent detection and elimination of Nitenpyram in agricultural environment. J Agric Food Chem 68(20):5572–5578

Liu H, Wang J, Jin H, Wei M, Ren W, Zhang Y et al (2021) Electrochemical biosensor for sensitive detection of Hg2+ baesd on clustered peonylike copper-based metal-organic frameworks and DNAzyme-driven DNA Walker dual amplification signal strategy. Sensors Actuators B Chem 329:129215

Lu Q, Su T, Shang Z, Jin D, Shu Y, Xu Q et al (2021) Flexible paper-based Ni-MOF composite/AuNPs/CNTs film electrode for HIV DNA detection. Biosens Bioelectron [Internet] 184:113229. Available from: https://linkinghub.elsevier.com/retrieve/pii/S0956566321002669

Luo G, Xie H, Niu Y, Liu J, Huang Y, Li B et al (2019) Electrochemical myoglobin biosensor based on magnesium metal-organic frameworks and gold nanoparticles composite modified electrode. Int J Electrochem Sci 14(3):2405–2413

Ly HGT, Fu G, Kondinski A, Bueken B, De Vos D, Parac-Vogt TN (2018) Superactivity of MOF-808 toward peptide bond hydrolysis. J Am Chem Soc 140(20):6325–6335

Ma Z-Z, Ma Y, Liu B, Xu L, Jiao H (2021) A high-performance Co-MOF non-enzymatic electrochemical sensor for glucose detection. New J Chem [Internet] 45(45):21350–21358. Available from: http://xlink.rsc.org/?DOI=D1NJ04480J

Ma Z-Z, Wang Y-S, Liu B, Jiao H, Xu L (2022) A non–enzymatic electrochemical sensor of Cu@Co–MOF composite for glucose detection with high sensitivity and selectivity. Chem Int 10(10):416. Available from: https://www.mdpi.com/2227-9040/10/10/416

Ma Y, Leng Y, Huo D, Zhao D, Zheng J, Yang H et al (2023) A sensitive enzyme-free electrochemical sensor based on a rod-shaped bimetallic MOF anchored on graphene oxide nanosheets for determination of glucose in huangshui. Anal Methods [Internet] 15(20):2417–2426. Available from: http://xlink.rsc.org/?DOI=D2AY01977A

Miao YB, Ren HX, Zhong Q, Song FX (2022a) Tailoring a luminescent metal – organic framework precise inclusion of Pt-Aptamer nanoparticle for noninvasive monitoring Parkinson's disease. Chem Eng J 441:136009

Miao YB, Zhong Q, Ren HX (2022b) Engineering a thermostable biosensor based on biomimetic mineralization HRP@Fe-MOF for Alzheimer's disease. Anal Bioanal Chem 414(29–30): 8331–8339

Miller AH, Nguyen HT-T, Nery JG, Fielding AJ (2023) Electrochemical sensor based on epoxy-functionalized BEA nanozeolite and graphene oxide modified glassy carbon electrode for bisphenol E determination. J Appl Electrochem [Internet] 53(8):1715–1725. Available from: https://link.springer.com/10.1007/s10800-023-01875-2

Ni Z, Masel RI (2006) Rapid production of metal-organic frameworks via microwave-assisted solvothermal synthesis. J Am Chem Soc 128(38):12394–12395

Pander M, Janeta M, Bury W (2021) Quest for an efficient 2-in-1 MOF-based catalytic system for cycloaddition of CO2 to epoxides under mild conditions. ACS Appl Mater Interfaces 13(7): 8344–8352

Peng H, Hui Y, Zhang L, Zhang F, Liu Y, Zheng J et al (2022) A novel electrochemical aptasensor based on Ti3C2-MOFs nanocomposites for rapid streptomycin detection in milk samples. Sensors Actuators B Chem 368:132119

Petit C, Mendoza B, O'Donnell D, Bandosz TJ (2011) Effect of graphite features on the properties of metal–organic framework/graphite hybrid materials prepared using an in situ process. Langmuir [Internet] 27(16):10234–10242. Available from: https://pubs.acs.org/doi/10.1021/la2017424

Robison L, Zhang L, Drout RJ, Li P, Haney CR, Brikha A et al (2019) A bismuth metal-organic framework as a contrast agent for X-ray computed tomography. ACS Appl Bio Mater 2(3): 1197–1203

Safaei M, Foroughi MM, Ebrahimpoor N, Jahani S, Omidi A, Khatami M (2019) A review on metal-organic frameworks: synthesis and applications. TrAC Trends Anal Chem [Internet] 118: 401–425. Available from: https://linkinghub.elsevier.com/retrieve/pii/S0165993619301840

Sayyad PW, Farooqui AA, Ingle NN, Al-Gahouari T, Bodkhe GA, Mahadik MM et al (2021) High carrier mobility and environmentally stable microporous zeolite imidazolate framework (ZIF-67): a field-effect transistor (FET) approach. Chem Phys Lett [Internet] 776:138690. Available from: https://linkinghub.elsevier.com/retrieve/pii/S0009261421003730

Su T, Mi Z, Xia Y, Jin D, Xu Q, Hu X et al (2023) A wearable sweat electrochemical aptasensor based on the Ni–Co MOF nanosheet-decorated CNTs/PU film for monitoring of stress biomarker. Talanta [Internet] 260:124620. Available from: https://linkinghub.elsevier.com/retrieve/pii/S0039914023003715

Sun Y, Zhou HC (2015) Recent progress in the synthesis of metal-organic frameworks. Sci Technol Adv Mater 16. https://doi.org/10.1088/1468-6996/16/5/054202

Sun Z, Jin H, Sun Y, Jiang X, Gui R (2020) Mn-Doping-induced hierarchical petal growth of a flower-like 3D MOF assembled with black phosphorous nanosheets as an electrochemical aptasensor of human stress-induced phosphoprotein 1. Nanoscale [Internet] 12(27): 14538–14548. Available from: http://xlink.rsc.org/?DOI=D0NR02342F

Tu X, Xie Y, Ma X, Gao F, Gong L, Wang D et al (2019) Highly stable reduced graphene oxide-encapsulated Ce-MOF composite as sensing material for electrochemically detecting dichlorophen. J Electroanal Chem [Internet]. 848:113268. Available from: https://linkinghub.elsevier.com/retrieve/pii/S1572665719305363

Uzak D, Atiroğlu A, Atiroğlu V, Çakıroğlu B, Özacar M (2020) Reduced graphene oxide/Pt nanoparticles/Zn-MOF-74 nanomaterial for a glucose biosensor construction. Electroanalysis 32(3):510–519

Wang C, An B, Lin W (2019a) Metal-organic frameworks in solid-gas phase catalysis. ACS Catal 9(1):130–146

Wang Y, Zhang Y, Sha H, Xiong X, Jia N (2019b) Design and biosensing of a Ratiometric Electrochemiluminescence resonance energy transfer Aptasensor between a g-C 3 N 4 Nanosheet and Ru@MOF for amyloid-β protein. ACS Appl Mater Interfaces [Internet] 11(40):36299–36306. Available from: https://pubs.acs.org/doi/10.1021/acsami.9b09492

Wang P-L, Xie L-H, Joseph EA, Li J-R, Su X-O, Zhou H-C (2019c) Metal–organic frameworks for food safety. Chem Rev [Internet] 119(18):10638–10690. Available from: https://pubs.acs.org/doi/10.1021/acs.chemrev.9b00257

Wang L, He K, Quan H, Wang X, Wang Q, Xu X (2020) A luminescent method for detection of parathion based on zinc incorporated metal-organic framework. Microchem J 153:104441

Wang X, Xu Y, Li Y, Li Y, Li Z, Zhang W et al (2021a) Rapid detection of cadmium ions in meat by a multi-walled carbon nanotubes enhanced metal-organic framework modified electrochemical sensor. Food Chem [Internet] 357:129762. Available from: https://linkinghub.elsevier.com/retrieve/pii/S0308814621007688

Wang B, Luo Y, Gao L, Liu B, Duan G (2021b) High-performance field-effect transistor glucose biosensors based on bimetallic Ni/Cu metal-organic frameworks. Biosens Bioelectron 171:112736

Wang X, Zhang L, Ye N, Xiang Y (2022) Synthesis of a dual metal–organic framework heterostructure as a fluorescence sensing platform for rapid and sensitive detection of tetracycline in milk and beef samples. Food Anal Methods 15(10):2801–2809

Wang R, Liu C, Wei Y, Ran Z, Jiang T, Liu C et al (2023) Fiber SPR biosensor sensitized by MOFs for MUC1 protein detection. Talanta [Internet]. 258:124467. Available from: https://linkinghub.elsevier.com/retrieve/pii/S0039914023002187

Wei Y, Hui Y, Lu X, Liu C, Zhang Y, Fan Y et al (2023) One-pot preparation of NiMn layered double hydroxide-MOF material for highly sensitive electrochemical sensing of glucose. J Electroanal Chem [Internet] 933:117276. Available from: https://linkinghub.elsevier.com/retrieve/pii/S1572665723001364

Wu L, Lu Z-W, Ma Y, Zhang J-J, Mo G-Q, Du H-J et al (2020) Cu(II) metal-organic framework encapsulated in carbon paste electrode for high-performance non-enzymatic glucose sensing. Chinese J Anal Chem [Internet] 48(3):e20038–e20046. Available from: https://linkinghub.elsevier.com/retrieve/pii/S1872204020600068

Wu Z, Wei J, Jiao T, Chen Q, Oyama M, Chen Q et al (2022a) A lead-based room-temperature phosphorescent metal–organic framework sensor for assessing the peroxide value of edible oils. Food Chem 385:132710

Wu Y, Chen J-Y, He W-M (2022b) Surface-enhanced Raman spectroscopy biosensor based on silver nanoparticles@metal-organic frameworks with peroxidase-mimicking activities for ultrasensitive monitoring of blood cholesterol. Sensors Actuators B Chem [Internet] 365:131939. Available from: https://linkinghub.elsevier.com/retrieve/pii/S0925400522005810

Xiao L, Yang K, Duan J, Zheng S, Jiang J (2022) The nickel phosphate rods derived from Ni-MOF with enhanced electrochemical activity for non-enzymatic glucose sensing. Talanta [Internet] 247:123587. Available from: https://linkinghub.elsevier.com/retrieve/pii/S0039914022003836

Xie Y, Yang C, Chen P, Yuan D, Guo K (2019) MnO2-decorated hierarchical porous carbon composites for high-performance asymmetric supercapacitors. J Power Sources [Internet] 425:1–9. Available from: https://linkinghub.elsevier.com/retrieve/pii/S0378775319303830

Xu X, Chen J, Hu R, Zhang Y, Chen H, Hu X et al (2023) A dual-modality immunosensor for simple and reliable detection of nitrated alpha-synuclein in serum based on silver-coated MOF. Microchim Acta [Internet]. 190(5):196. Available from: https://link.springer.com/10.1007/s00604-023-05783-2

Yan M, Ye J, Zhu Q, Zhu L, Huang J, Yang X (2019) Ultrasensitive Immunosensor for cardiac troponin I detection based on the Electrochemiluminescence of 2D Ru-MOF Nanosheets. Anal Chem [Internet] 91(15):10156–10163. Available from: https://pubs.acs.org/doi/10.1021/acs.analchem.9b02169

Yang Y, Cheng J, Wang B, Guo Y, Dong X, Zhao J (2019) An amino-modified metal-organic framework (type UiO-66-NH2) loaded with cadmium(II) and lead(II) ions for simultaneous electrochemical immunosensing of triazophos and thiacloprid. Microchim Acta [Internet] 186(2):101. Available from: http://link.springer.com/10.1007/s00604-018-3201-z

Yao C-X, Zhao N, Liu J-C, Chen L-J, Liu J-M, Fang G-Z et al (2020) Recent progress on luminescent metal-organic framework-involved hybrid materials for rapid determination of contaminants in environment and food. Polymers (Basel) [Internet] 12(3):691. Available from: https://www.mdpi.com/2073-4360/12/3/691

Yola ML (2021) Sensitive sandwich-type voltammetric immunosensor for breast cancer biomarker HER2 detection based on gold nanoparticles decorated Cu-MOF and Cu2ZnSnS4 NPs/Pt/g-C3N4 composite. Microchim Acta [Internet]. 188(3):78. Available from: http://link.springer.com/10.1007/s00604-021-04735-y

Zeng S, Zhang Y, Zhang Y, Li Y, Tang C, Li K et al (2022) A novel room temperature SO2 gas sensor based on TiO2/rGO buried-gate FET. Microelectron Eng [Internet] 263:111841. Available from: https://linkinghub.elsevier.com/retrieve/pii/S0167931722001356

Zhang Z-H, Duan F-H, Tian J-Y, He J-Y, Yang L-Y, Zhao H et al (2017) Aptamer-embedded zirconium-based metal–organic framework composites prepared by de novo bio-inspired approach with enhanced biosensing for detecting trace analytes. ACS Sensors [Internet] 2(7): 982–989. Available from: https://pubs.acs.org/doi/10.1021/acssensors.7b00236

Zhang D, Wu D, Zong X, Yang Z (2019) Enhanced SO2 gas sensing properties of metal organic frameworks-derived titanium dioxide/reduced graphene oxide nanostructure. J Mater Sci Mater Electron 30(12):11070–11078

Zhang L, Wang N, Cao P, Lin M, Xu L, Ma H (2020) Electrochemical non-enzymatic glucose sensor using ionic liquid incorporated cobalt-based metal-organic framework. Microchem J [Internet]. 159:105343. Available from: https://linkinghub.elsevier.com/retrieve/pii/S0026265X20313606

Zhang Z, Lou Y, Guo C, Jia Q, Song Y, Tian JY et al (2021) Metal–organic frameworks (MOFs) based chemosensors/biosensors for analysis of food contaminants. Trends Food Sci Technol 118:569–588

Zhang X, Liu T, Boyle A, Bahreman A, Bao L, Jing Q et al (2022a) Dielectric-modulated biosensing with ultrahigh-frequency-operated graphene field-effect transistors. Adv Mater [Internet] 34(7):2106666. Available from: https://onlinelibrary.wiley.com/doi/10.1002/adma.202106666

Zhang M, Shan Y, Kong Q, Pang H (2022b) Applications of metal–organic framework–graphene composite materials in electrochemical energy storage. FlatChem 32:100332

Zhang Y, Huang Y, Gao P, Yin W, Yin M, Pu H et al (2022c) Bimetal-organic frameworks MnCo-MOF-74 derived Co/MnO@HC for the construction of a novel enzyme-free glucose sensor. Microchem J [Internet] 175:107097. Available from: https://linkinghub.elsevier.com/retrieve/pii/S0026265X21011838

Zhang Y, Xu Y, Li N, Ma W, Yang M, Hou C et al (2023) An ultra-sensitive dual-signal ratio electrochemical aptasensor based on functionalized bimetallic MOF nanocomplexes by the in-situ electrochemical synthesis for detect HER2. Int J Hydrogen Energy [Internet] 48:24548–24558. Available from: https://linkinghub.elsevier.com/retrieve/pii/S0360319923012399

Zhao G, Wang Y, Li X, Yue Q, Dong X, Du B et al (2019) Dual-quenching electrochemiluminescence strategy based on three-dimensional metal–organic frameworks for ultrasensitive detection of amyloid-β. Anal Chem [Internet] 91(3):1989–1996. Available from: https://pubs.acs.org/doi/10.1021/acs.analchem.8b04332

Zhao L, Wang B, Wang C, Fan D, Liu X, Wei Q et al (2023) Dual-strategy ECL biosensor based on rare Eu(II,III)-MOF as probe with antenna effect and sensitization for CYFRA 21-1 trace analysis. Sensors Actuators B Chem [Internet] 377:133101. Available from: https://linkinghub.elsevier.com/retrieve/pii/S0925400522017440

Dr. Stephen Rathinaraj Benjamin is a distinguished researcher and engineer renowned for his profound knowledge and expertise in biosensor technology. He is primarily dedicated to advancing the field of biosensors to create state-of-the-art technology for medical diagnostics and the monitoring of environmental conditions. His work encompasses the design of sensors capable of detecting antibodies, the integration of nanomaterials, and the advancement of surface modification techniques to optimize biomolecule interactions. His research is primarily directed towards medical applications, with a strong emphasis on early disease detection and the exploration of innovative treatment approaches. Notably, he is currently dedicated to the development of biosensors tailored for immunosensors/Point of Care (POC) systems, specifically targeting the diagnosis and treatment of neurodegenerative diseases.

Eli José Miranda Ribeiro Júnior is a highly accomplished professional with a strong academic background in Environmental Management and Business Administration. He has excelled as a university professor, sharing his expertise at prestigious institutions and teaching a wide range of subjects including Administration, Accounting, and Civil Engineering. With his diverse skill set, including expertise in team leadership, process improvement, and adaptability to complex environments, he continues to make significant contributions to education as a Master Professor at Faculdade Cgesp.

Professor, Dr. Rosa Fireman Dutra is a distinguished leader in the field of Biodevices and Bioelectronics at the Laboratory of Biomedical Engineering and holds the position of Associate Professor at the Federal University of Pernambuco in the Department of Biomedical Engineering, Brazil. She has a solid educational background, having graduated in Electronic Engineering, obtained an MSc in Biomedical Engineering, and earned a PhD in Biosensors and Immunosensors, followed by a Postdoctoral in Analytical Chemistry. Her main area of expertise involves improving the analytical capabilities of a wide range of sensor types, such as amperometric, capacitive, impedimetric, piezoelectric, and optical transducer-based sensors.

Professor, Dr. Geanne Matos de Andrade holds a degree in Pharmacy, a master's, and a doctorate in Pharmacology from the Federal University of Ceará (UFC), with a post-doctorate in Neuroscience from the University of Coimbra, Portugal. She is a Full Professor in the Department of Physiology and Pharmacology at UFC, Fortaleza, Ceará, a 1C research productivity fellow, and serves on several key committees, including the Central Analytical Management Committee, UFC PRINT Program, and the Chamber of Biological and Environmental Sciences of FUNCAP. Currently, she is the Assistant Coordinator of Professional Postgraduate Programs at CBII of CAPES. She has held prominent roles, such as Coordinator of the Pharmacology Postgraduate Program and the Graduate Program in Medical Sciences at UFC, as well as UFC's Research Coordinator. She teaches Neurophysiology in Medicine, Pharmacy, Nursing, Biological Sciences, Psychology, and Biology, while also mentoring and leading courses in the Postgraduate Programs in Pharmacology and Medical Sciences. Her research focuses on experimental Neuropsychopharmacology, specializing in neurodegenerative diseases like Parkinson's and Alzheimer's, cerebral ischemia, neuroprotection with natural compounds, and purinergic signaling.

Cellulose, Chitin, and Chitosan Composite-Based Sustainable Biomaterials

13

Sultan Gul, Yesim Karahan, Ozan Baris Kurtur, and Yasemin Budama-Kilinc

Abstract

The escalating environmental concerns highlight fears about the sustainability of our planet's livability. Synthetic polymers, widely used across industries, are major contributors to these issues due to their poor biocompatibility, potential toxicity from degradation, and sustainability challenges in production and disposal. To address these issues, researchers are turning to green biomaterials derived from renewable natural sources, crucial for developing bioactive and biodegradable materials.

Among natural polymers, cellulose, chitin, chitosan, and their derivatives stand out in drug delivery systems due to their biocompatibility, biodegradability, accessibility, modifiability, non-toxicity, and stability against environmental factors. Cellulose, abundant in plant cell walls and sourced from various natural origins, offers structural integrity through intermolecular hydrogen bonding. Chitin, found in crustacean shells, and its derivative chitosan, known for its cationic nature and unique properties like antimicrobial activity, they are also prominent.

Chemical modifications enhance these biopolymers' solubility and interactions, enabling their integration into innovative composites. Combining cellulose and chitosan has yielded materials with enhanced functionality, suitable for diverse biomedical applications such as films, aerogels, and membranes.

O. B. Kurtur
Graduate School of Natural and Applied Science, Bioengineering Department, Yildiz Technical University, Istanbul, Turkey

Y. Budama-Kilinc (✉)
Faculty of Chemical and Metallurgical Engineering, Department of Bioengineering, Yildiz Technical University, Istanbul, Turkey

Health Biotechnology Joint Research and Application Center of Excellence, Istanbul, Turkey

R. Malviya, S. Sundram (eds.), *Sustainable Green Biomaterials As Drug Delivery Systems*, Biomaterials, Bioengineering and Sustainability 1,
https://doi.org/10.1007/978-3-031-79062-1_13

These biopolymer composites exhibit 1D to 3D structures, leveraging cellulose's hydrophilicity and mechanical strength alongside chitosan's antimicrobial properties.

The growing interest in these sustainable biomaterials stems from their biodegradability and eco-friendliness, offering viable alternatives to conventional petroleum-based products. This chapter provides an overview of cellulose, chitin, and chitosan, discusses their production, and explores recent advancements in their application as green polymers in drug delivery systems, showcasing their potential to address environmental challenges while advancing biomedical technology.

Keywords

Sustainable Green Biomaterials · Cellulose · Chitin · Chitosan · Drug delivery

13.1 Introduction

In today's world of material science and engineering, there is a growing need for sustainable solutions (Apelian 2012). In particular, while cost-effective and consistent in composition, synthetic polymers often lack biocompatibility and pose toxicity risks from degradation products. Moreover, the production sustainability of synthetic polymers and increasing concerns about the environmental impacts of synthetic polymers, including microplastic problems and disposal, have led researchers to investigate alternative polymer sources (Kostag and El Seoud 2021; Muhamad et al. 2014). For these reasons, the exploration and use of biopolymers derived from renewable resources have become a significant focus in the search for green sustainable materials. Among these biopolymers, cellulose, chitin, and chitosan stand out due to their natural abundance and desirable properties. Cellulose, as the predominant biopolymer found on Earth, plays a crucial role in the construction of plant cell walls, enhancing their remarkable strength and stability (Heinze 2016).

On the other hand, chitin and its derivative, chitosan, are present in the exoskeletons of crustaceans and insects. They have unique properties like biocompatibility, biodegradability, antimicrobial properties, and wound-healing capabilities, making them invaluable in various biomedical, pharmaceutical, and environmental applications (Tao et al. 2020). Moreover, they are renewable and have minimal environmental impact, making them highly desirable for sustainable material development (Nadeem et al. 2020).

Blending different polymers is a technique employed to produce innovative composites with tailored properties (Kulshreshtha and Vasile 2002). This technique offers a technological avenue to create customized composites, enhancing processability and yielding superior properties compared to individual polymers (Díez-Pascual 2019). For instance, cellulose, chitin, and chitosan can be combined with composite materials, and this new composite offers many advantages. These composites inherit each biopolymer's individual properties while exhibiting enhanced mechanical strength, flexibility, and biodegradability (Kestur 2010; Kochkina and Lukin 2020; Yadav et al. 2023). Furthermore, these composites can be improved by incorporating additives, nanomaterials, and functionalization strategies. In this way, they can be used in various fields, from biodegradable packaging materials to tissue engineering scaffolds, biomedical implants, drug delivery systems, water purification membranes, and environmental monitoring sensors (Pal et al. 2021; Rizvi et al. 2022). This approach fosters the production of multifunctional, biodegradable composites for various uses like films, foams, fibers, and filters, including biomedical research and potential *in vivo* applications (Kostag and El Seoud 2021). Each application leverages the unique properties of cellulose, chitin, and chitosan composites to address societal and environmental concerns while advancing the paradigm of green, sustainable materials (Anwer et al. 2023). In this context, integrating cellulose, chitin, and chitosan into composite materials represents a pioneering approach in sustainable biomaterials research. In the past few years, there has been rapid advancement in the utilization of composites made from sustainable biomaterials extracted from cellulose, chitin, and chitosan, serving as eco-friendly polymers in drug delivery systems. By harnessing the abundance and versatility of these biopolymers, researchers and engineers can unlock new frontiers in material science, ushering in a future characterized by environmentally friendly and socially responsible innovations.

In this chapter, we will briefly overview the molecular structures of cellulose, chitin, and chitosan. We will then discuss the manufacturing process of composite-based biomaterials utilizing these three substances. Lastly, we will discuss the latest advancements in applying cellulose, chitin, and chitosan composite-based eco-friendly biomaterials as sustainable polymers in drug delivery systems.

13.2 General Structures of Cellulose, Chitin and Chitosan

13.2.1 Cellulose

Cellulose is a natural material found in various sources such as plants, animals, algae, fungi, and minerals (McNamara et al. 2015). Nevertheless, plant fiber is the primary origin of cellulose. Cellulose is an important component of the cell walls of plants, making up about 40% of the carbon fraction. It provides structural support and is often found with other substances, such as hemicelluloses, lignin, and minor quantities of extractives. While cellulose can exist in a pure form in plants, it is more commonly found in conjunction with other substances (Heinze 2016).

Cellulose is a polymer formed by "polycondensation" (Abrahamsen-Mills and Small 2021). It consists of D-glucopyranose ring units in the 4C1-chair configuration, the conformation with the lowest energy. These units are connected by β-1,4-glycosidic bonds, which lead to a 180° alternation in the direction of the cellulose chain (Gopi et al. 2019). Each anhydroglucose unit (AGU) in the cellulose chain contains three reactive hydroxyl groups: a primary group at C6 and two secondary groups at C2 and C3, all located in the ring's plane (Heinze 2016). Its

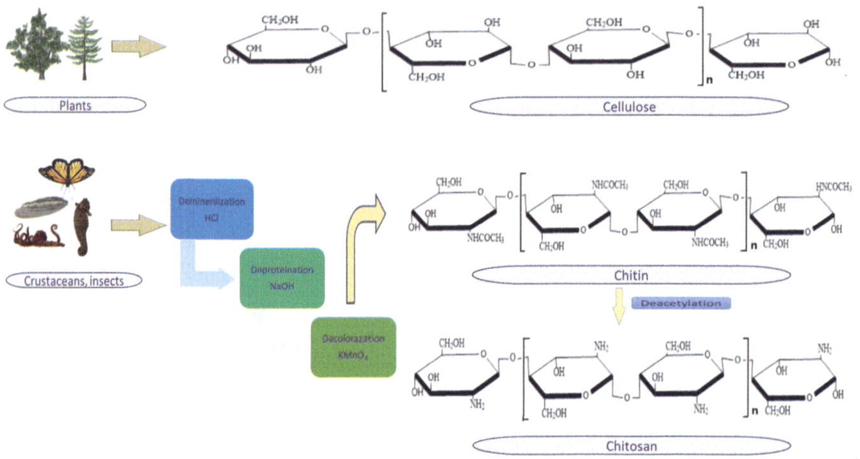

Fig. 13.1 Chemical structures of chitin-chitosan-cellulose and their natural source. Deacetylation of chitin to chitosan

molecule has two chemically different chain ends. One end contains an anomeric C atom linked by the glycosidic bonds, also known as the non-reducing end. The reducing and non-reducing ends of cellulose chains make them directionally asymmetric molecules. At the other end, a D-glucopyranose unit is in equilibrium with the aldehyde function, serving as the reducing end group (Fig. 13.1). Cellulose's molecular structure can undergo modification through reactions such as hydrolysis or oxidation of the cellulose chain. These reactions primarily take place either on the surface of the fibrils or within the amorphous regions (Heinze 2016).

The chemical reactivity of cellulose is dependent on its polymorphic structure. The less-ordered amorphous regions exhibit higher reactivity compared to the highly-ordered crystallite regions (Heinze 2016). Inaugural chemical reactions occur primarily on the surfaces of the less-ordered fibrils, while the impenetrable crystalline structure has minimal impact. Despite its moisture-absorbing ability, cellulose remains insoluble, yet it can undergo distensioning in water, diluted acid, and some solvents (Etale et al. 2023; Wohlert et al. 2022). Solutions with low pH values can render cellulose soluble, albeit at the cost of significant degradation via acetal (glycosidic) hydrolysis. Alkaline solutions cause hemicelluloses to swell and dissolve within cellulose (Przypis et al. 2023). Strong alkaline solutions can penetrate cellulose through swelling and capillary attraction, a mercerization process activating it for cellulose ether production. In basic media at elevated temperatures (>150 °C), cellulose undergoes hydrolysis, and oxidation may occur (Stephen and Phillips 2016). Microbiological degradation happens through enzymatic hydrolysis, leading to the cleavage of the Β-1,4-glucosidic linkages (Lakhundi et al. 2015). Chemically modified cellulose ethers, due to steric hindrance, are less susceptible to enzymatic degradation, exhibiting significantly improved stability. Furthermore, amorphous cellulose is more prone to enzymatic hydrolysis than highly crystalline cellulose. Breaking down natural cellulose with a combination of enzymes leads to products that offer specific practical benefits, such as enhanced solubility (Stephen and Phillips 2016).

13.2.2 Chitin

Chitin is the second most abundant natural polymer after cellulose (Barbosa et al. 2019; Elieh-Ali-Komi and Hamblin 2016). The term "chitin" originates from the Greek word "chiton," which refers to a coat of mail or armor. Chitin is mainly derived from insects like cuticles, beetle cocoons, and ovipositors, as well as from crustaceans such as shrimp shells and crab shells (Kostag and El Seoud 2021). Squids like Loligo stomach, centric diatoms such as algae and *Thalassiosira fluviatilis*, and fungi like *Aspergillus nidulans* and *Mucor rouxii* are also significant sources of chitin (Kaur et al. 2023). Chitin is characterized by its chemical structure (1→4)-2-acetamido-2-deoxy-β-d-glucan, and composed of N-acetylglucosamine units that form a linear array with a molecular weight of 2–3 million Daltons (Ahmad et al. 2020). These N-acetylglucose units in chitin are rotated 180° with respect to each other, forming the disaccharide N,N′-diacetylchitobiose, similar to cellulose. Despite the enzymatic hydrolysis of chitin yielding N-acetylglucosamine and glucosamine, solid-state NMR analysis suggests little to no presence of glucosamine (Elieh-Ali-Komi and Hamblin 2016). Glycosidic bonds in chitin that form between modified monomeric glucose molecules are responsible for its structure. This long chain of chitin can form a solid structure in an insect's wings or a clam's shell (Nicolay et al. 2020).

Chitin tends to form microfibrils stabilized by hydrogen bonds, with lengths sometimes exceeding 0.5μm and bundled in groups of 10 or more. X-ray diffraction analysis reveals three crystalline modifications of chitin—*α*-, *β*-, and γ-chitin—differing in hydration, unit cell size, and chitin chain arrangement (Blackwell 1988). *α*-chitin exhibits antiparallel chain orientation, *β*-chitin features parallel chain arrangement, and γ-chitin is characterized by two parallel strands alternating with a single antiparallel strand. Furthermore, non-crystalline transient states have been noted in fungal chitin (Hou et al. 2021). Alpha chitin's dense arrangement of antiparallel chains contributes to its higher resistance to hydrogen bond disruption. Beta-isoforms, as observed in squid, display enhanced solubility, affinity, and reactivity towards solvents due to the weaker hydrogen bonding between chains. The third chitin isoform is γ-chitin, which is a variant of *α*-chitin (Ahmad et al. 2020). In the crystalline structures of *α*- and *β*-chitin, *α*-chitin's unit cell is orthorhombic with chitobiose sections forming helical conformations. The unit cell of *β*-chitin is monoclinic, exhibiting a helical conformation similar to *α*-chitin but packed in a parallel arrangement. Unlike *α*-chitin, *β*-chitin chains form H-bonded sheets along the a-axis, leading to increased swelling by solvents and enhanced reactivity towards derivatization. This difference explains why *β*-chitin exhibits more swelling and reactivity compared to *α*-chitin (Pillai et al. 2009; Salavati 2023).

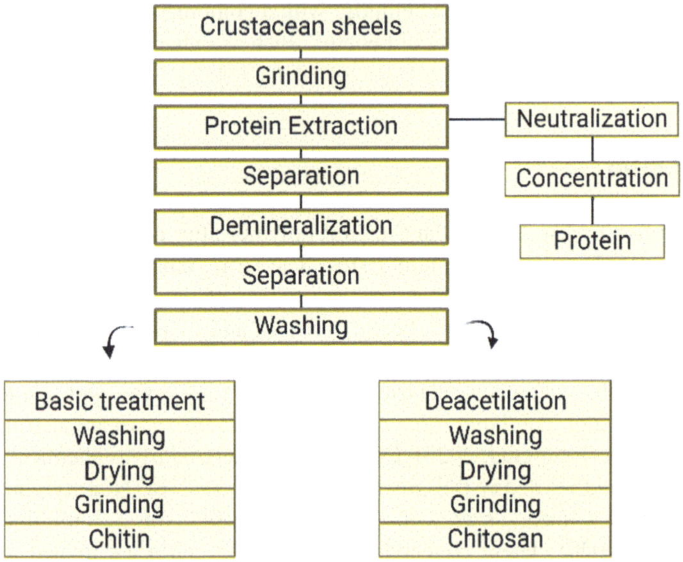

Fig. 13.2 Schematic representation for the biological and chemical routes for producing chitin

In industrial chitin extraction, there are typically three primary stages: (i) deproteinization, which involves the removal of proteins using aqueous solutions of chemicals such as sodium hydroxide or potassium hydroxide; (ii) demineralization, where acids like hydrochloric acid are used to eliminate calcium carbonate; and (iii) decolorization, achieved through solvent extraction with substances like ethanol or bleaching with hydrogen peroxide to remove pigments. Alternatively, a biological route employing bacteria that produce protease, followed by treatment with lactic acid, can be utilized for the first two steps, offering reduced environmental impact (Fig. 13.2) (Younes and Rinaudo 2015).

13.2.3 Chitosan

Chitosan is a substance that can be obtained from through biological processes and chemical methods (Rehman et al. 2023). For example, the microbial synthesis of chitosan is a biological pathway that involves various organisms such as *Phycomyces blakesleeanus*. Chitosan is produced through the utilization of cell cultures from these organisms, with the addition of *Aspergillus niger* to the culture medium to facilitate the production process. This production mechanism also causes deacetylation of chitin, and chitosan is obtained after an incubation period of 96 h (Kuzgun and İnanlı 2013).

Chitosan, a linear polysaccharide, is created through the deacetylation process of chitin, where the acetamide groups are hydrolyzed via potent alkaline treatment and is a naturally occurring substance found in only trace amounts within the ecosystem, notably prevalent in select fungal species. It consists of 2-amino-2-deoxy-β-D-glucopyranose linked through β (1–4) with 2-acetamino-2-deoxy-β-D-glucopyranose (Brasselet et al. 2019; Shariatinia 2019). It is shown that their turn to each other by chitin's deacetylation and chitosan's acetylation in (Fig. 13.1). The difference between chitin and chitosan lies in the degree of deacetylation (DA), with chitosan having a DA higher than 50 mol% and being soluble in dilute acid solutions (Younes and Rinaudo 2015). Initially, chitin undergoes deacetylation using strong alkalis, which removes acetamide groups. This process involves several chemical steps, including demineralization, deproteinization, and deacetylation, which are essential for extracting chitosan from chitin (Fig. 13.1) (Kumar et al. 2023). In heterogeneous deacetylation, chitin is suspended in hot, concentrated alkali, whereas homogeneous deacetylation involves suspending chitin in a concentrated base followed by rapid cooling (Methacanon et al. 2003). Ionic liquids have also found application in the dissolution of chitin and in homogeneous deacetylation processes (Ma et al. 2020). On the other hand, environmentally friendly enzymatic methods employ chitin deacetylases to catalyze the hydrolysis of N-acetamido bonds in chitin, leading to well-defined chitosan structures without degradation (Tsigos et al. 2000).

Researchers have been interested in chitosan because of the amines present at the C-2 position of glucosamine residues, which contribute key functional characteristics to chitosan (Stephen and Phillips 2016). Hydroxyl at C3 and C6 and primary amine at C2 in every sequence make it highly susceptible to chemical transformation (Nicolle et al. 2021). Due to their beneficial biological and film-forming properties, chitosan and its derivatives have emerged as an excellent eco-friendly substitute for synthetic plastic polymers, which have found various applications in the environment, packaging, and medical sectors (Jiang et al. 2023). Furthermore, chitosan can be converted into a polycation with protonated amino groups, enabling it to interact with various natural or synthetic anionic species such as lipids, proteins, DNA, and negatively charged synthetic polymers like poly (acrylic acid). It is worth noting that chitosan is the sole naturally occurring polysaccharide that carries a positive charge (Pavinatto et al. 2010). Chitosan molecules possess both amino and hydroxyl groups, enabling the formation of stable covalent bonds through reactions such as etherification, esterification, and reductive amination (Ibrahim and El-Zairy 2015).

The physico-mechanical attributes of chitosan, which are toxicity, solubility, viscosity, reactivity of proteinaceous material coagulation, and heavy metal ion chelation, are intricately linked to its molecular weight and the extent of deacetylation. Notably, chitosan with heightened deacetylation levels displays a nuanced toxicity profile contingent upon its molecular weight: it demonstrates reduced toxicity at lower molecular weight (Román-Doval et al. 2023). It still exhibits heightened toxicity as molecular weight escalates. Conversely, chitosan with diminished deacetylation levels is an absorption facilitator across a spectrum of molecular weights. Furthermore, chitosan's solubility and degradation kinetics are intricately tied to its molecular weight profile, with higher molecular weight variants displaying decreased solubility and a more protracted degradation process compared to their lower molecular weight counterparts (Román-Doval et al. 2023; Zhang and Neau 2001).

Moreover, the solubility of a substance in water is affected by the acidity or alkalinity of the solution, which is measured by its pH level. Therefore, the solubility of a substance in water can vary depending on the pH of the solution (Ibrahim and El-Zairy 2015). Additionally, the solubility and degradation kinetics of chitosan are closely linked to its molecular weight profile. Higher molecular weight variants exhibit reduced solubility and undergo a slower degradation process compared to their lower molecular weight counterparts (Román-Doval et al. 2023; Zhang and Neau 2001). Furthermore, a substance's solubility in water is influenced by the acidity or alkalinity of the solution, as indicated by its pH level. Thus, the solubility of a substance in water may vary depending on the pH of the solution (Ibrahim and El-Zairy 2015). Chitosan readily dissolves in acidic aqueous solutions but remains insoluble in both water and alkaline solutions (Blagodatskikh et al. 2023).

13.3 Preparation of Cellulose, Chitin, and Chitosan Composite-Based Biomaterials

Polymer composites can be classified as either fully or partially renewable, depending on the origin and nature of the polymer matrix and reinforcement (Thakur et al. 2014). Both reinforcement and matrix materials are incorporated and sourced from bio-renewable resources in completely renewable polymer composites. Alternatively, only one element in partially renewable composites is sourced from bio-renewable resources. For instance, the polymer matrix may be derived from bio-renewable resources. At the same time, the reinforcement comes from non-renewable resources, or the polymer matrix may be synthetic with the reinforcement originating from bio-renewable resources (Thakur et al. 2014). Chemical modifications of cellulose/chitin/chitosan composites biopolymers often enhance solubility and biological interactions (Kostag and El Seoud 2021). Three techniques can be used to obtain a composite based on cellulose and chitin-chitosan (Silva-Castro et al. 2017).

1. Physical ways (Ultrasonication, thermal treatment, microwave, electrospinning, etc.)
2. Chemical ways (Graft polymerization, catalysis, nucleophilic substitutions, etc.)
3. Biological ways (Biocatalysis, enzymatic hydrolysis, etc.)

13.3.1 Cellulose Composite-Based Biomaterials

Cellulose composites are made using diverse manufacturing techniques like compression moulding, injection moulding, resin transfer moulding, and vacuum bagging. The quality of composites is influenced by processing factors such as

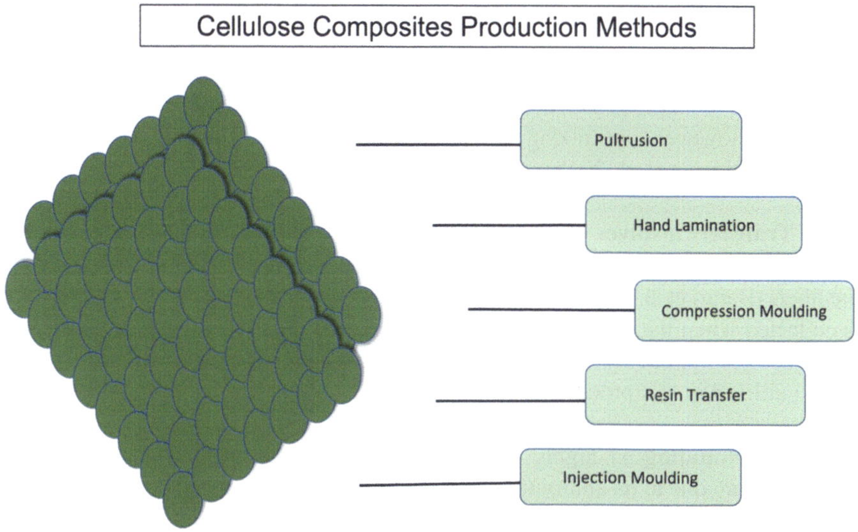

Fig. 13.3 Methods of fabrication of cellulose composite materials

temperature, pressure, and moulding time (Jaafar et al. 2019). Preheating natural fibers to reduce moisture content is expected, though cellulose degradation at high temperatures can harm mechanical properties (Vaca-Medina et al. 2013). Improper fiber dispersion may lead to agglomeration, lowering tensile strength. The choice of process technology is determined by factors like the product's design, anticipated performance, cost-effectiveness, and manufacturing simplicity. Different methods for fabricating cellulose composites are shown in Fig. 13.3 (Pattnaik and Swain 2022).

Pultrusion is an ongoing method for manufacturing composites. It involves pulling resin-impregnated fibers through a die shaped to match the aspired cross-sectional form of the creation. This method facilitates the production of slender formations and various segmental configurations (X. Peng et al. 2012). Additionally, pultrusion offers the potential for high levels of automation, making it advantageous for manufacturing (Volk et al. 2022).

Hand Lamination involves applying resin to fibers in a mould using rollers, often assisted by vacuum bags to eliminate excess air and compact the part with atmospheric pressure. While this method offers simplicity, low tooling costs, and design flexibility, it is time-consuming and lacks automation potential, which are its primary limitations (Vallittu 2018).

Compression Moulding stands as a prevalent technique for crafting composites (Sahu et al. 2024). This method utilizes a partially processed composite sheet or sheet moulding compound, which is then shaped into final parts through compression. The process involves rolling a resin film with added fibers, compressing it into a composite sheet, and then processing it in a press to obtain the intended shape (Mao et al. 2018). The compression moulding method offers advantages like ample production capability, consistent part quality and rapid cycle durations. However, optimizing pressure is crucial to prevent fiber or structural damage (Mokhena et al. 2018).

Resin Transfer involves placing fibers inside a mould with two solid components while a tube delivers liquid resin into the mould under low pressure, impregnating the fibers (Barari et al. 2016). After curing at or above room temperature, the product is collected when the mould is opened. This method efficiently produces intricate, high-quality components on a large scale. Various types of low-viscosity resins like can be utilized in this process (Quero et al. 2010).

Injection Moulding is highly regarded as the optimal method for composite manufacturing due to its ability to produce intricate geometries and precise features. It results in items boasting exceptional surface quality and exact dimensional precision, all achieved with swift production rates and minimal labor expenses (Wang et al. 2017). Studies in the literature related to the preparation methods, advantages, and use areas of cellulose composite-based biomaterials are shown in Table 13.1.

13.3.2 Chitin-Chitosan Composite-Based Biomaterials

Despite being scientifically proven to be biodegradable, biocompatible, non-toxic, non-allergenic, and bioadhesive (Tao et al. 2020), its low mechanical strength and solubility, depending on pH, are essential disadvantages. To overcome these issues, thanks to its abundant active hydroxyl and amino groups, researchers have used various methods to change the solubility and activity of chitosan by modifying it (Qin and Li 2020). Biomaterials prepared with chitosan have been used in many studies in the literature. Li et al. (2019) focused on preparing and characterizing antimicrobial films made of chitosan, polyvinyl alcohol, and methacryloyloxyethyl trimethyl ammonium chloride (DMC) through ultrasonic treatment, demonstrating that ultrasonication improved the mechanical properties, light transmittance, and barrier properties of the films, with a 2% methacryloyloxyethyl trimethyl ammonium chloride concentration showing strong bactericidal effects on *S. aureus* and *E. coli* (Li et al. 2019). Using microwave technology, Kumar et al. (2023) synthesized chitosan and polyethylene glycol hydrogel membranes containing curcumin. The membrane had high swelling, low degradation, and sustained drug release, leading to enhanced mechanical properties and faster wound healing *in vivo*, suggesting its suitability for therapeutic approaches for wound healing (Kumar et al. 2023). Beer et al. (2020) investigated the enzymatic breakdown of chitosan into a water-soluble form that interacts with lignin, using cellobiose hydrolase to hydrolyze chitosan and

Table 13.1 Some studies on cellulose composite based biomaterials

Composite type	Preparation	Advantages	Usage areas of the method	References
Chitosan-cellulose composite	Dissolving the components in an acidic solution & completely dispersed.	Better biological compatibility and antibacterial properties	Food, medicine, chemical, and other related industries	Kargarzadeh et al. (2018)
Folate-decorated reductive-responsive carboxymethylcellulose-based nanocapsules (FA-RCNCs)	The dimensions and shape of drug-loaded FA-RCNCs were assessed using DLS, SEM, and CLSM, revealing the features of the fabricated nano-capsules.	Targeted and triggered drug delivery	Precise delivery and controlled release of hydrophobic anticancer medications for cancer treatment in biomedical contexts.	He et al. (2021)
Electrospinning cellulose acetate nanofibers	Tetrabutylammonium bromide salts and sophorolipid-based bio-surfactants were used as a solvent which increased the spinnability of cellulose acetate.	Evaporation process and improved the uniformity of the nanofibers	Drug delivery systems, support structure for tissue regeneration, air purification, and selective membrane configurations.	Oldal et al. (2023)
Folic acid conjugated nano cellulose	Fabricated a drug administration system utilizing nano cellulose to enhance the delivery efficiency of curcumin, employing acid-alkali treatment on cellulose.	Regulated release of curcumin for cancer therapy.	Drug delivery systems	Anirudhan et al. (2021)

laccases to oxidize lignin, forming stable covalent lignin cross-linked chitosan hydrogels (Beer et al. 2020).

13.4 Drug Delivery Applications of Cellulose, Chitin, and Chitosan Composite-Based Biomaterials

Cellulose, a long-standing presence in pharmaceuticals, is often blended with excipients to create tablets for oral drug administration (Wilson et al. 2017). Recent researches have brought to light the versatile and promising role of cellulose in drug delivery applications (Ramezani et al. 2024). Notably, cellulose composite biomaterials have been successfully employed to enhance the oral bioavailability of drugs, develop controlled delivery systems, increase drug loading capacity, program pH-sensitive release behavior, and target drug and controlled release systems (Ul-Islam et al. 2023). Chitosan composite biomaterials, on the other hand, have been thoroughly researched as drug delivery systems by virtue of their pH-responsive release mechanism, which enables specific drug administration and controlled release systems (Omrani et al. 2023). A summary of the studies in the literature on the implementation of cellulose, chitin, and chitosan composite-based biomaterials in drug delivery utilization is provided in Table 13.2.

13.5 Advantages and Disadvantages

Despite its abundance, cellulose has limitations in processing; it is not amenable to methods like injection moulding technique or melt extrusion due to its high melting temperature, likely above its thermal decomposition temperature (Amiri et al. 2020). Moreover, many solvents cause swelling but do not dissolve cellulose, necessitating strong dipolar characteristics for solvent dissolution (Kostag and El Seoud 2021). Chitosan, on the other hand, with its distinctive traits, is a versatile material with wide applications in diverse areas, for instance, biotechnology, water purification, cosmetics, food, pharmacy, and agricolation (Mıyazakı et al. 1981). However, its limited solubility and mechanical properties constrain its biomedical application (Desai et al. 2023). Chitosan is esteemed in the biomedical domain for its low immunogenicity, robust biological compatibility, and widespread availability (Dimassi et al. 2018). Modification techniques can enhance its solubility and generate derivatives with improved properties, expanding its potential applications. Chitosan-based nanocomposites demonstrate enhanced mechanical durability and heat transfer capabilities, and antimicrobial attributes in contrast to chitosan (Kumar et al. 2023). However, its limited water solubility hampers its biological functionality, exclusively under corporal surroundings, where it is weakly dissolved and poorly absorbed (Husain et al. 2017; Snyman et al. 2002). This can be handled, by the chitosan is combined with another compound enhances its water solubility and broadens its therapeutic applications (E. M. Ahmed 2015).

Table 13.2 A precis of utilization of cellulose, chitin and chitosan based-composites in therapeutic administration

Composites	Drug Used	Role	Results	References
Fe_3O_4@C/carboxymethyl cellulose/chitosan beads	Diclofenac sodium	Controlled-release drug carrier	The beads exhibited increased swelling and release rates in phosphate buffer solutions(pH 6.8 and 7.4) compared to pH 1.2, effectively preventing abrupt releases.	Sun et al. (2019b)
Alfuzosin hydrochloride	Ethyl cellulose-polyvinyl pyrrolidone	Dermal delivery	When it is higher than body temperature more release	Pattnaik et al. (2015)
Hydroxypropyl methylcellulose, hydroxypropyl cellulose, hydroxyethyl cellulose and methyl cellulose	Theophylline (99%)	Controlled-release drug carrier	Especially hydroxypropyl cellulose (HPC) and methyl cellulose (MC) well drug release properties	Wilson et al. (2017)
Microcrystalline cellulose	Enalapril-maleate and hydrochlorothiazide	Creating pellet for Tablet coating	Better densification and deformability	Hegyesi et al. (2017)
Amphifilic bacterial cellulose-Alginate beads	Alfacalcidol	Controlled-release drug carrier	Well loading and sustained release	Yan et al. (2019)
Carboxymethylated nanocellulose	Ciprofloxacin	Controlled-release drug carrier	Highly effective in prolonging the antibiotic release.	Hong et al. (2019)
Cellulose nanocrystal and chitosan composite hydrogel	Theophylline	Controlled-release drug carrier	Well release behavior depends on pH	Xu et al. (2019)
Nano cellulose glycidyl methacrylate/ hydroxyethyl methacrylate (HEMA) /ethylene glycol dimethacrylate (EGDMA)	Curcumin	Controlled-release drug carrier for cancer treatment	Resulting in high drug-loading capacity, sustained release, and pH-responsive behavior	Anirudhan et al. (2021)
Bacterial cellulose–polyvinyl alcohol based complex composite membranes	Curcumin	Carrier on drug delivery, transdermal drug delivery	High swelling degrees, well stability and antibacterial activity	Busuioc et al. (2023)

(continued)

Table 13.2 (continued)

Composites	Drug Used	Role	Results	References
Cellulose/alginate beads	Metformin hydrochloride	Carrier in drug delivery	Evaluated stability and drug-release properties, increased mechanical properties and stability, and more sustained drug release	Guo et al. (2017)
Cellulose-PVA blended hydrogels	Streptomycin/ curcumin	Dressings tailored to address diverse wound types, including those stemming from cancer treatment.	Excellent capacity to retain moisture, biodegradable nature, and compatibility with living tissues.	Sankarganesh et al. (2024)
Carboxymethyl cellulose/starch/ reduced graphene oxide composite hydrogel	Curcumin	A pH-responsive nano-carrier	More biocompatible, cheaper, and environmentally friendly; sustained release of anticancer drugs	Parvaneh et al. (2023)
Alginate-chitosan core-shell hydrogel beads	Insulin	A carrier to administer through the mouth	Resulting sustained release, and pH-responsive behavior	Phan et al. (2022)
Chitosan/polyvinyl alcohol porous composite	Metformin nanoparticles	A carrier for oral drug delivery	Increased porosity significantly enhances drug encapsulation and effectively regulates controlled release in some body fluids via diffusion kinetics; pH-responsive behavior	Yari et al. (2023)
Montmorillonite-Famotidine/ Chitosan Bio-nanocomposite Hydrogels	Famotidine	A carrier for oral administration	Having the ability to release drugs consistently and sustainably, excellent adhesion to the stomach lining and prolonged retention time in the stomach.	Farhadnejad et al. (2022)
Carboxymethyl chitosan and oxidized pullulan	Methotrexate	Drug delivery for cancer treatment	Good biocompatibility and pronounced cytotoxicity	Rui et al. (2023)
Chitosan-folate hybrid magnetic nanoparticles	Doxorubicin	Drug delivery systems	pH-sensitiveness; induced toxicity	Amiryaghoubi et al. (2022)

ZnO/carboxymethyl cellulose/ chitosan bio-nanocomposite beads	5-fluorouracil	A carrier for colon-specific release	Excellent pH-sensitivity and biodegradability	Sun et al. (2019a)
Polyethylene oxide nanofibers immobilized with silver nanoparticles stabilized by chitosan	Silver	Wound dressing	Good antibacterial effect; suitable mechanic properties	Aljohani et al. (2023)
Chitosan-PEO nanofibers	Teicoplanin	A carrier drug delivery and wound healing	Higher antibacterial activity; decreased cytotoxicity	Amiri et al. (2020)

Interestingly, blending also introduces new characteristics to chitosan, facilitated by its unique structure, which allows for various reactions like phosphorylation, crosslinking, and complexation (Kumar et al. 2023; Qin et al. 2020). These modifications lead to derivatives with adaptable biological and chemical properties, including enhanced bioactivity and DNA complexing capabilities (Sarmento et al. 2011). Chitosan also exhibits antibacterial properties, damaging bacterial cell membranes and preventing plaque formation (Dilamian et al. 2013). While generally safe for mammals, it demonstrates a diverse array of antibacterial functions. Chitosan and its derivatives hold promise for biomedical innovations, benefiting from their diverse characteristics and potential utility in healthcare (Husain et al. 2017).

Despite being crucial to tissue engineering, scaffolds often have weak mechanical properties, which can be significantly improved. Chitosan is extensively utilized as a wound-healing material to promote wound mending and prevent contamination, although some cytotoxic effects have been observed, necessitating further research to minimize toxicity (Chen et al. 2008; de Sousa Victor et al. 2020). Despite thousands of research studies in recent decades, the complete capability of chitosan on biomedical utilization still needs to be explored, with many unresolved challenges. Chitosan-based nanoparticles show promise as multifunctional drug carriers with good biocompatibility and easy modifiability (de Sousa Victor et al. 2020). However, a systematic approach is needed to address issues related to drug delivery selectivity, adverse effects on living organisms, safety, and synthesis methods for cellulose, chitin-chitosan-based biomaterials. Using the advantages of these natural biopolymers to prepare their composites suitable for drug delivery may be the best way to overcome all these challenges. Undoubtedly, these will have some benefits and drawbacks that should not be ignored (Table 13.3).

In summary, cellulose, chitin, and chitosan composite-based sustainable biomaterials offer numerous advantages for drug delivery applications. However, they also present certain limitations and challenges that must be addressed through further research and development efforts. With ongoing advancements in material science and engineering, these biomaterials are promising to revolutionize drug delivery and sustainably improve patient outcomes.

13.6 Future Perspectives for Cellulose, Chitin, and Chitosan Composite-Based Sustainable Biomaterials

Drug delivery has witnessed remarkable advancements over the years, with a growing emphasis on sustainability and biocompatibility. In this context, cellulose, chitin, and chitosan composite-based biomaterials have shown great potential as candidates for revolutionizing therapeutic delivery systems. This chapter delves into the future perspectives of utilizing these biomaterials in drug delivery, exploring their potential, challenges, and opportunities for sustainable healthcare solutions.

Cellulose, chitin, and chitosan possess unique properties conducive to drug delivery applications. As a renewable and biocompatible material, cellulose provides

Table 13.3 Advantages and disadvantages of cellulose, chitin-chitosan based composites

Advantages		References
Biocompatibility	Cellulose, chitin, and chitosan are biologically suitable, indicating that they are well-received by organisms and exhibit minimal adverse reactions. They give biocompatible composites.	Petrova et al. (2023) and Sanmugam et al. (2023)
Renewable and sustainable	Their sustainable sourcing coordinates to the flourishing requirement for environmental substances in healthcare.	Parvaneh et al. (2023) and Song et al. (2022)
Biodegradability	This property reduces the peril of longstanding accumulation and toxicity, enhancing the safety profile of drug delivery systems.	Kim et al. (2023) and Song et al. (2022)
Tailored drug release	This allows for precise modulation of drug release kinetics, ensuring optimal therapeutic outcomes and minimizing side effects.	Leonard et al. (2023)
Targeted delivery	Chitin and chitosan exhibit mucoadhesive characteristics, facilitating the accurate administration of medications to particular areas within the human body. This enhances drug localization and uptake at the desired site of action, improving treatment efficacy and reducing systemic exposure.	Leonard et al. (2023) and Parvaneh et al. (2023)
Disadvangates		References
Limited mechanical strength	Pure cellulose, chitin, and chitosan materials may have limited mechanical strength. Reinforcement strategies or composite formulations may be necessary to overcome this limitation.	Said et al. (2023)
Complex fabrication process	The fabrication of cellulose, chitin, and chitosan-based drug delivery systems often involves complex processing techniques such as solvent casting, electrospinning, or 3D printing. These processes may require specialized equipment and expertise, leading to higher production costs and scalability challenges.	Murugan et al. (2021)
Variable drug loading capacity	Achieving consistent and high drug loading efficiencies may require optimization of formulation parameters.	Khalili et al. (2024) and Trzeciak et al. (2021)
Potential immunogenicity	While cellulose, chitin, and chitosan are generally considered biocompatible, there is a possibility of immune responses or allergic reactions in some individuals. Pre-trial and clinical investigations are imperative to evaluate the antigenicity and protection profiles of these biomaterials for drug delivery purposes.	Morishima and Bokuda (2021) and Peng et al. (2022)
Regulatory hurdles	Compliance with regulatory standards and guidelines can pose challenges for product development and commercialization.	Beheshtizadeh et al. (2022) and Peng et al. (2022)

a stable matrix for drug encapsulation and sustained release (Zhang et al. 2021). Chitin and chitosan offer inherent mucoadhesive properties, facilitating targeted delivery to specific sites within the body (Bhattarai et al. 2010). Moreover, their biodegradability ensures minimal adverse effects and enhanced biocompatibility, aligning with the principles of sustainable healthcare (Mushtaq et al. 2017).

Future Directions and Opportunities

1. **Enhanced Biocompatibility:** Future research endeavours aim to augment the biological compatibility and bioactivity of cellulose, chitin, and chitosan composites through surface modification and functionalization. These composites can exhibit improved cellular interactions and tissue-specific targeting by incorporating bioactive molecules and targeting ligands, thereby optimizing drug delivery efficacy (Gaur et al. 2023).
2. **Controlled Release Systems:** Developing advanced drug delivery systems based on cellulose, chitin, and chitosan composites holds promise for achieving precise control over drug release kinetics (Leonard et al. 2023). Future efforts may focus on designing stimuli-responsive materials capable of triggered discharge in replay to significant physiologic clues, for instance, acidity levels, thermal conditions, or enzymatic performance. Controlled release systems can optimize therapeutic results, mitigate side effects, and enhance patient adherence.
3. **Personalized Medicine:** Cellulose, chitin, and chitosan composites' adaptability allows for the customization of drug delivery systems to meet individual patient needs (Yang et al. 2022). Future developments may involve 3D printing's assimilation to create personalized implants, scaffolds, or drug-eluting devices (Ahmed et al. 2021). By tailoring drug delivery systems to patient-specific requirements, personalized medicine can be realized, offering improved treatment outcomes and healthcare efficiency.

Challenges and Considerations

Despite the promising prospects, the full realization of cellulose, chitin, and chitosan composite-based drug delivery systems hinges on effectively addressing several challenges. These include optimizing composite formulations for specific drug properties, ensuring long-term stability and efficacy, and addressing regulatory issues for their development into products for clinical use. Moreover, scalability and cost-effectiveness remain critical factors in the widespread adoption of these sustainable biomaterials in drug delivery applications.

13.7 Conclusion

Cellulose, chitin, and chitosan are inexhaustible and worthwhile biopolymers gaining significance as sustainable alternatives to synthetic polymers. Their structural similarities and functional groups allow for the formation of nanocomposites with diverse shapes like fibers, films, and hydrogels, primarily through biopolymer solution mixing for drug delivery. The composition and properties of nanocomposites can be tailored by adjusting precursor concentrations, with cellulose often reinforcing mechanical strength. Chitin and chitosan, prized for their biological activity, find applications in biomedicine and food packaging. Cross-linking and derivatization further modify nanocomposites properties, though challenges like polymer degradation and reduced dissolution must be managed for quality maintenance. Collaborative research across disciplines can leverage the unique properties of these biopolymers to develop innovative drug-delivery solutions that address unmet medical needs and contribute to a more sustainable healthcare system. Green chemistry principles like process optimization, eco-friendly solvents and chemicals, solvent recyclability, and academia-industry collaboration should be embraced in a composite to promote sustainability.

References

Abrahamsen-Mills L, Small JS (2021) Organic-containing nuclear wastes and national inventories across Europe. In: The microbiology of nuclear waste disposal. Elsevier, pp 1–20

Ahmad A, Mubarak N, Naseem K, Tabassum H, Rizwan M, Najda A et al (2020) Recent advancement and development of chitin and chitosan-based nanocomposite for drug delivery: critical approach to clinical research. Arab J Chem 13(12):8935–8964

Ahmed EM (2015) Hydrogel: preparation, characterization, and applications: a review. J Adv Res 6(2):105–121

Ahmed TA, Felimban RI, Tayeb HH, Rizg WY, Alnadwi FH, Alotaibi HA et al (2021) Development of multi-compartment 3d-printed tablets loaded with self-nanoemulsified formulations of various drugs: A new strategy for personalized medicine. Pharmaceutics 13(10):1733

Aljohani MM, Abu-Rayyan A, Elsayed NH, Alatawi FA, Al-Anazi M, Mustafa SK et al (2023) One-pot microwave synthesis of chitosan-stabilized silver nanoparticles entrapped polyethylene oxide nanofibers, with their intrinsic antibacterial and antioxidant potency for wound healing. Int J Biol Macromol 235:123704

Amiri N, Ajami S, Shahroodi A, Jannatabadi N, Darban SA, Bazzaz BSF et al (2020) Teicoplanin-loaded chitosan-PEO nanofibers for local antibiotic delivery and wound healing. Int J Biol Macromol 162:645–656

Amiryaghoubi N, Abdolahinia ED, Nakhlband A, Aslzad S, Fathi M, Barar J, Omidi Y (2022) Smart chitosan–folate hybrid magnetic nanoparticles for targeted delivery of doxorubicin to osteosarcoma cells. Colloids Surf B: Biointerfaces 220:112911

Anirudhan T, Manjusha V, Sekhar VC (2021) A new biodegradable nano cellulose-based drug delivery system for pH-controlled delivery of curcumin. Int J Biol Macromol 183:2044–2054

Anwer AH, Ahtesham A, Shoeb M, Mashkoor F, Ansari MZ, Zhu S, Jeong C (2023) State-of-the-art advances in nanocomposite and bio-nanocomposite polymeric materials: a comprehensive review. Adv Colloid Interf Sci 318:102955

Apelian D (2012) Materials science and engineering's pivotal role in sustainable development for the 21st century. MRS Bull 37(4):318–323

Barari B, Omrani E, Moghadam AD, Menezes PL, Pillai KM, Rohatgi PK (2016) Mechanical, physical and tribological characterization of nano-cellulose fibers reinforced bio-epoxy composites: an attempt to fabricate and scale the 'Green'composite. Carbohydr Polym 147: 282–293

Barbosa AI, Lima SAC, Reis S (2019) Development of methotrexate loaded fucoidan/chitosan nanoparticles with anti-inflammatory potential and enhanced skin permeation. Int J Biol Macromol 124:1115–1122

Beer B, Bartolome MJ, Berndorfer L, Bochmann G, Guebitz GM, Nyanhongo GS (2020) Controlled enzymatic hydrolysis and synthesis of lignin cross-linked chitosan functional hydrogels. Int J Biol Macromol 161:1440–1446

Beheshtizadeh N, Gharibshahian M, Pazhouhnia Z, Rostami M, Zangi AR, Maleki R et al (2022) Commercialization and regulation of regenerative medicine products: promises, advances and challenges. Biomed Pharmacother 153:113431

Bhattarai N, Gunn J, Zhang M (2010) Chitosan-based hydrogels for controlled, localized drug delivery. Adv Drug Deliv Rev 62(1):83–99

Blackwell J (1988) Physical methods for the determination of chitin structure and conformation. Methods Enzymol 161:435–442

Blagodatskikh IV, Vyshivannaya OV, Tishchenko NA, Orlov VN, Tikhonov VE, Bezrodnykh EA et al (2023) Complexation between chitosan and carboxymethyl cellulose in weakly acidic, neutral, and weakly alcaline media. Int J Biol Macromol 243:125277

Brasselet C, Pierre G, Dubessay P, Dols-Lafargue M, Coulon J, Maupeu J et al (2019) Modification of chitosan for the generation of functional derivatives. Appl Sci 9(7):1321

Busuioc C, Isopencu GO, Deleanu I-M (2023) Bacterial cellulose–polyvinyl alcohol based complex composites for controlled drug release. Appl Sci 13(2):1015

Chen J-P, Chang G-Y, Chen J-K (2008) Electrospun collagen/chitosan nanofibrous membrane as wound dressing. Colloids Surf A Physicochem Eng Asp 313:183–188

de Sousa Victor R, da Cunha M, Santos A, Viana de Sousa B, de Araújo Neves G, de Lima N, Santana L, Rodrigues Menezes R (2020) A review on Chitosan's uses as biomaterial: tissue engineering, drug delivery systems and cancer treatment. Materials 13(21):4995

Desai N, Rana D, Salave S, Gupta R, Patel P, Karunakaran B et al (2023) Chitosan: A potential biopolymer in drug delivery and biomedical applications. Pharmaceutics 15(4):1313

Díez-Pascual AM (2019) Synthesis and applications of biopolymer composites, vol 20. MDPI, p 2321

Dilamian M, Montazer M, Masoumi J (2013) Antimicrobial electrospun membranes of chitosan/poly (ethylene oxide) incorporating poly (hexamethylene biguanide) hydrochloride. Carbohydr Polym 94(1):364–371

Dimassi S, Tabary N, Chai F, Blanchemain N, Martel B (2018) Sulfonated and sulfated chitosan derivatives for biomedical applications: a review. Carbohydr Polym 202:382–396

Elieh-Ali-Komi D, Hamblin MR (2016) Chitin and chitosan: production and application of versatile biomedical nanomaterials. Int J Adv Res 4(3):411–427

Etale A, Onyianta AJ, Turner SR, Eichhorn SJ (2023) Cellulose: a review of water interactions, applications in composites, and water treatment. Chem Rev 123(5):2016–2048

Farhadnejad H, Mortazavi SA, Jamshidfar S, Rakhshani A, Motasadizadeh H, Fatahi Y et al (2022) Montmorillonite-famotidine/chitosan bio-nanocomposite hydrogels as a mucoadhesive/gastroretentive drug delivery system. Iran J Pharm Res: IJPR 21(1):e127035

Gaur M, Maurya S, Akhtar MS, Yadav AB (2023) Synthesis and evaluation of BSA-loaded PLGA–Chitosan composite nanoparticles for the protein-based drug delivery system. ACS omega 8(21):18751–18759

Gopi S, Balakrishnan P, Chandradhara D, Poovathankandy D, Thomas S (2019) General scenarios of cellulose and its use in the biomedical field. Mater Today Chem 13:59–78

Guo T, Pei Y, Tang K, He X, Huang J, Wang F (2017) Mechanical and drug release properties of alginate beads reinforced with cellulose. J Appl Polym Sci 134(8):44495

He S, Zhong S, Meng Q, Fang Y, Dou Y, Gao Y, Cui X (2021) Sonochemical preparation of folate-decorated reductive-responsive carboxymethylcellulose-based nanocapsules for targeted drug delivery. Carbohydr Polym 266:118174

Hegyesi D, Thommes M, Kleinebudde P, Sovány T, Kása P Jr, Kelemen A et al (2017) Preparation and physicochemical characterization of matrix pellets containing APIs with different solubility via extrusion process. Drug Dev Ind Pharm 43(3):458–464

Heinze T (2016) Cellulose: structure and properties. In: Rojas OJ (ed) Cellulose chemistry and properties: fibers, nanocelluloses and advanced materials. Springer, pp 1–52

Hong H-J, Kim J, Kim D-Y, Kang I, Kang HK, Ryu BG (2019) Synthesis of carboxymethylated nanocellulose fabricated ciprofloxacine–montmorillonite composite for sustained delivery of antibiotics. Int J Pharm 567:118502

Hou J, Aydemir BE, Dumanli AG (2021) Understanding the structural diversity of chitins as a versatile biomaterial. Phil Trans R Soc A 379(2206):20200331

Husain S, Al-Samadani KH, Najeeb S, Zafar MS, Khurshid Z, Zohaib S, Qasim SB (2017) Chitosan biomaterials for current and potential dental applications. Materials 10(6):602

Ibrahim H, El-Zairy E (2015) Chitosan as a biomaterial—structure, properties, and electrospun nanofibers. In: Bobbarala V (ed) Concepts, compounds and the alternatives of antibacterials. InTech, pp 81–101

Jaafar J, Siregar JP, Tezara C, Hamdan MHM, Rihayat T (2019) A review of important considerations in the compression molding process of short natural fiber composites. Int J Adv Manuf Technol 105:3437–3450

Jiang A, Patel R, Padhan B, Palimkar S, Galgali P, Adhikari A et al (2023) Chitosan based biodegradable composite for antibacterial food packaging application. Polymers 15(10):2235

Kargarzadeh H, Huang J, Lin N, Ahmad I, Mariano M, Dufresne A et al (2018) Recent developments in nanocellulose-based biodegradable polymers, thermoplastic polymers, and porous nanocomposites. Prog Polym Sci 87:197–227

Kaur M, Sharma A, Puri V, Aggarwal G, Maman P, Huanbutta K et al (2023) Chitosan-based polymer blends for drug delivery systems. Polymers 15(9):2028

Kestur GS (2010) Biodegradable polymer composites based on Brazilian lignocellulosic. Matéria (Rio de Janeiro) 15:88–95

Khalili Z, Kazemi NM, Azar ZJ, Mosavi Z, Hasanzadeh M (2024) Fabrication and characterization of a Bi2O3-modified chitosan@ ZIF-8 nanocomposite for enhanced drug loading-releasing efficacy. Int J Biol Macromol 263:130295

Kim J, Kim Y, Jeong J-P, Kim J-M, Kim MS, Jung S (2023) A pH-sensitive drug delivery using biodegradable succinoglycan/chitosan hydrogels with synergistic antibacterial activity. Int J Biol Macromol 242:124888

Kochkina NE, Lukin ND (2020) Structure and properties of biodegradable maize starch/chitosan composite films as affected by PVA additions. Int J Biol Macromol 157:377–384

Kostag M, El Seoud OA (2021) Sustainable biomaterials based on cellulose, chitin and chitosan composites-A review. Carbohydr. Polym. Technol. Appl. 2:100079

Kulshreshtha AK, Vasile C (2002) Handbook of polymer blends and composites, vol vol. 1. iSmithers Rapra Publishing

Kumar A, Yadav S, Pramanik J, Sivamaruthi BS, Jayeoye TJ, Prajapati BG, Chaiyasut C (2023) Chitosan-based composites: development and perspective in food preservation and biomedical applications. Polymers 15(15):3150

Kuzgun NK, İnanlı AG (2013) Kitosan üretimi ve özellikleri ile kitosanın kullanım alanları. Türk Bilimsel Derlemeler Dergisi 6(2):16–21

Lakhundi S, Siddiqui R, Khan NA (2015) Cellulose degradation: a therapeutic strategy in the improved treatment of Acanthamoeba infections. Parasit Vectors 8:1–16

Leonard TE, Liko AF, Gustiananda M, Putra ABN, Juanssilfero AB, Hartrianti P (2023) Thiolated pectin-chitosan composites: Potential mucoadhesive drug delivery system with selective cytotoxicity towards colorectal cancer. Int J Biol Macromol 225:1–12

Li R, Wang Y, Xu J, Ahmed S, Liu Y (2019) Preparation and characterization of ultrasound treated polyvinyl alcohol/chitosan/DMC antimicrobial films. Coatings 9(9):582

Ma Q, Gao X, Bi X, Han Q, Tu L, Yang Y et al (2020) Dissolution and deacetylation of chitin in ionic liquid tetrabutylammonium hydroxide and its cascade reaction in enzyme treatment for chitin recycling. Carbohydr Polym 230:115605

Mao D, Li Q, Bai N, Dong H, Li D (2018) Porous stable poly (lactic acid)/ethyl cellulose/hydroxyapatite composite scaffolds prepared by a combined method for bone regeneration. Carbohydr Polym 180:104–111

McNamara JT, Morgan JL, Zimmer J (2015) A molecular description of cellulose biosynthesis. Annu Rev Biochem 84:895–921
Methacanon P, Prasitsilp M, Pothsree T, Pattaraarchachai J (2003) Heterogeneous N-deacetylation of squid chitin in alkaline solution. Carbohydr Polym 52(2):119–123
Mıyazakı S, Ishıı K, Nadaı T (1981) The use of chitin and chitosan as drug carriers. Chem Pharm Bull 29(10):3067–3069
Mokhena TC, Sefadi JS, Sadiku ER, John MJ, Mochane MJ, Mtibe A (2018) Thermoplastic processing of PLA/cellulose nanomaterials composites. Polymers 10(12):1363
Morishima R, Bokuda K (2021) Hydroxypropyl cellulose-induced fixed drug eruption as an adverse effect of generic drugs. Contact Derm 85(1):90–91
Muhamad II, Selvakumaran S, Lazim NAM (2014) Designing polymeric nanoparticles for targeted drug delivery system. Nanomed 287:287
Murugan SS, Anil S, Sivakumar P, Shim MS, Venkatesan J (2021) 3D-printed chitosan composites for biomedical applications. In: Jayakumar R, Prabaharan M (eds) Chitosan for biomaterials IV: biomedical applications, pp 87–116
Mushtaq S, Khan JA, Rabbani F, Latif U, Arfan M, Yameen MA (2017) Biocompatible biodegradable polymeric antibacterial nanoparticles for enhancing the effects of a third-generation cephalosporin against resistant bacteria. J Med Microbiol 66(3):318–327
Nadeem H, Naseri M, Shanmugam K, Dehghani M, Browne C, Miri S et al (2020) An energy efficient production of high moisture barrier nanocellulose/carboxymethyl cellulose films via spray-deposition technique. Carbohydr Polym 250:116911
Nicolay VK, Nina S, Yuliya K, Galina B (2020) Formation of polyelectrolyte complexes from chitosan and alkaline gelatin. KnE Life Sci 2020:109–119
Nicolle L, Journot CM, Gerber-Lemaire S (2021) Chitosan functionalization: covalent and non-covalent interactions and their characterization. Polymers 13(23):4118
Oldal DG, Topuz F, Holtzl T, Szekely G (2023) Green electrospinning of biodegradable cellulose acetate nanofibrous membranes with tunable porosity. ACS Sustain Chem Eng 11(3):994–1005
Omrani Z, Pourmadadi M, Yazdian F, Rashedi H (2023) Preparation and characterization of pH-sensitive chitosan/starch/MoS2 nanocomposite for control release of curcumin macromolecules drug delivery; application in the breast cancer treatment. Int J Biol Macromol 250:125897
Pal K, Bharti D, Sarkar P, Anis A, Kim D, Chałas R et al (2021) Selected applications of chitosan composites. Int J Mol Sci 22(20):10968
Parvaneh S, Pourmadadi M, Abdouss M, Pourmousavi SA, Yazdian F, Rahdar A, Díez-Pascual AM (2023) Carboxymethyl cellulose/starch/reduced graphene oxide composite as a pH-sensitive nanocarrier for curcumin drug delivery. Int J Biol Macromol 241:124566
Pattnaik S, Swain K (2022) Cellulose-Based composites and their biomedical applications. Cellul Chem Technol 56(1–2):115–122
Pattnaik S, Swain K, Rao JV, Varun T, Mallick S (2015) Temperature influencing permeation pattern of alfuzosin: an investigation using DoE. Medicina 51(4):253–261
Pavinatto FJ, Caseli L, Oliveira ON Jr (2010) Chitosan in nanostructured thin films. Biomacromolecules 11(8):1897–1908
Peng X, Fan M, Hartley J, Al-Zubaidy M (2012) Properties of natural fiber composites made by pultrusion process. J Compos Mater 46(2):237–246
Peng S, Liang Y, Xiao W, Liu Y, Yu M, Liu L (2022) Anaphylaxis induced by intra-articular injection of chitosan: a case report and literature review. Clin Case Reports 10(12):e6596
Petrova VA, Gofman IV, Dubashynskaya NV, Golovkin AS, Mishanin AI, Ivan'kova EM et al (2023) Chitosan composites with bacterial cellulose nanofibers doped with nanosized cerium oxide: characterization and cytocompatibility evaluation. Int J Mol Sci 24(6):5415
Phan VG, Mathiyalagan R, Nguyen M-T, Tran T-T, Murugesan M, Ho T-N et al (2022) Ionically cross-linked alginate-chitosan core-shell hydrogel beads for oral delivery of insulin. Int J Biol Macromol 222:262–271

Pillai CK, Paul W, Sharma CP (2009) Chitin and chitosan polymers: chemistry, solubility and fiber formation. Prog Polym Sci 34(7):641–678
Przypis M, Wawoczny A, Gillner D (2023) Biomass and Cellulose Dissolution—The Important Issue in Renewable Materials Treatment. Appl Sci 13(2):1055
Qin Y, Li P (2020) Antimicrobial chitosan conjugates: current synthetic strategies and potential applications. Int J Mol Sci 21(2):499
Qin Y, Li P, Guo Z (2020) Cationic chitosan derivatives as potential antifungals: a review of structural optimization and applications. Carbohydr Polym 236:116002
Quero F, Nogi M, Yano H, Abdulsalami K, Holmes SM, Sakakini BH, Eichhorn SJ (2010) Optimization of the mechanical performance of bacterial cellulose/poly (L-lactic) acid composites. ACS Appl Mater Interfaces 2(1):321–330
Ramezani G, Stiharu I, van de Ven TG, Nerguizian V (2024) Advancements in hybrid cellulose-based films: Innovations and applications in 2D nano-delivery systems. J Funct Biomater 15(4):93
Rehman KU, Hollah C, Wiesotzki K, Heinz V, Aganovic K, Rehman RU et al (2023) Insect-derived chitin and chitosan: A still unexploited resource for the edible insect sector. Sustain For 15(6):4864
Rizvi M, Gerengi H, Gupta P (2022) Functionalization of nanomaterials: synthesis and characterization. In: Functionalized nanomaterials for corrosion mitigation: synthesis, characterization, and applications. ACS Publications, pp 1–26
Román-Doval R, Torres-Arellanes SP, Tenorio-Barajas AY, Gómez-Sánchez A, Valencia-Lazcano AA (2023) Chitosan: properties and its application in agriculture in context of molecular weight. Polymers 15(13):2867
Rui Q, Gao J, Yin Z-Z, Li J, Cai W, Wu D, Kong Y (2023) A biodegradable pH and glutathione dual-triggered drug delivery system based on mesoporous silica, carboxymethyl chitosan and oxidized pullulan. Int J Biol Macromol 224:1294–1302
Sahu R, Singh MK, Singh G, Agarwal N, Yadav AS, Chincholkar P et al (2024) Optimization techniques in compression moulding: a comprehensive review. Mater Sci Forum 1112:41–48
Said HA, Noukrati H, Mahdi I, Oudadesse H, Barroug A (2023) In situ precipitated hydroxyapatite-chitosan composite loaded with ciprofloxacin: formulation, mechanical, in vitro antibiotic uptake, release, and antibacterial properties. Mater Chem Phys 294:127008
Salavati M (2023) Mechanical properties of α-Chitin and chitosan biocomposite: a molecular dynamic study. Journal of Composites Science 7(11):464
Sankarganesh P, Parthasarathy V, Ganesh Kumar A, Ragu S, Saraniya M, Udayakumari N, Anbarasan R (2024) Preparation of cellulose-PVA blended hydrogels for wound healing applications with controlled release of the antibacterial drug: an in vitro anticancer activity. Biomass Convers Biorefinery 14(3):3385–3395
Sanmugam A, Abbishek S, Kumar SL, Sairam AB, Palem VV, Kumar RS et al (2023) Synthesis of chitosan based reduced graphene oxide-CeO2 nanocomposites for drug delivery and antibacterial applications. J Mech Behav Biomed Mater 145:106033
Sarmento B, Goycoolea FM, Sosnik A, das Neves J (2011) Chitosan and chitosan derivatives for biological applications: chemistry and functionalization. Int J Carbohydr Chem. https://doi.org/10.1155/2011/802693
Shariatinia Z (2019) Pharmaceutical applications of chitosan. Adv Colloid Interf Sci 263:131–194
Silva-Castro I, Martín-Ramos P, Matei PM, Fernandes-Correa M, Hernánez-Navarro S, Martín-Gil J (2017) Eco-friendly nanocomposites of chitosan with natural extracts, antimicrobial agents, and nanometals. In: Thakur VK, Thakur MK, Kessler MR (eds) Handbook of composites from renewable materials. Wiley, pp 35–60
Kumar P, Khan N, Kumar D, Ali M, Tandon R (2023) Microwave-assisted synthesis of curcumin-encapsulated silica-chitosan composite for drug release study. Bulg Chem Commun 55, Special Issue A:102–108

Snyman D, Hamman J, Kotze J, Rollings J, Kotze A (2002) The relationship between the absolute molecular weight and the degree of quaternisation of N-trimethyl chitosan chloride. Carbohydr Polym 50(2):145–150

Song Y, Wu T, Bao J, Xu M, Yang Q, Zhu L et al (2022) Porous cellulose composite aerogel films with super piezoelectric properties for energy harvesting. Carbohydr Polym 288:119407

Stephen AM, Phillips GO (2016) Food polysaccharides and their applications. CRC Press

Sun X, Liu C, Omer A, Lu W, Zhang S, Jiang X et al (2019a) pH-sensitive ZnO/carboxymethyl cellulose/chitosan bio-nanocomposite beads for colon-specific release of 5-fluorouracil. Int J Biol Macromol 128:468–479

Sun X, Shen J, Yu D, Ouyang X-K (2019b) Preparation of pH-sensitive Fe3O4@ C/carboxymethyl cellulose/chitosan composite beads for diclofenac sodium delivery. Int J Biol Macromol 127: 594–605

Tao F, Cheng Y, Shi X, Zheng H, Du Y, Xiang W, Deng H (2020) Applications of chitin and chitosan nanofibers in bone regenerative engineering. Carbohydr Polym 230:115658

Thakur VK, Thakur MK, Gupta RK (2014) Raw natural fiber–based polymer composites. Int J Polym Anal Charact 19(3):256–271

Trzeciak K, Chotera-Ouda A, Bak-Sypien II, Potrzebowski MJ (2021) Mesoporous silica particles as drug delivery systems—the state of the art in loading methods and the recent progress in analytical techniques for monitoring these processes. Pharmaceutics 13(7):950

Tsigos I, Martinou A, Kafetzopoulos D, Bouriotis V (2000) Chitin deacetylases: new, versatile tools in biotechnology. Trends Biotechnol 18(7):305–312

Ul-Islam M, Alabbosh KF, Manan S, Khan S, Ahmad F, Ullah MW (2023) Chitosan-based nanostructured biomaterials: synthesis, properties, and biomedical applications. Adv Ind Eng Polym Res 7:79–99

Vaca-Medina G, Jallabert B, Viet D, Peydecastaing J, Rouilly A (2013) Effect of temperature on high pressure cellulose compression. Cellulose 20:2311–2319

Vallittu PK (2018) An overview of development and status of fiber-reinforced composites as dental and medical biomaterials. Acta biomaterialia odontologica Scandinavica 4(1):44–55

Volk M, Yuksel O, Baran I, Hattel JH, Spangenberg J, Sandberg M (2022) Cost-efficient, automated, and sustainable composite profile manufacture: A review of the state of the art, innovations, and future of pultrusion technologies. Compos Part B 246:110135

Wang L, Ishihara S, Hikima Y, Ohshima M, Sekiguchi T, Sato A, Yano H (2017) Unprecedented development of ultrahigh expansion injection-molded polypropylene foams by introducing hydrophobic-modified cellulose nanofibers. ACS Appl Mater Interfaces 9(11):9250–9254

Wilson MR, Jones DS, Andrews GP (2017) The development of sustained release drug delivery platforms using melt-extruded cellulose-based polymer blends. J Pharm Pharmacol 69(1):32–42

Wohlert M, Benselfelt T, Wågberg L, Furó I, Berglund LA, Wohlert J (2022) Cellulose and the role of hydrogen bonds: not in charge of everything. Cellulose 29:1–23

Xu Q, Ji Y, Sun Q, Fu Y, Xu Y, Jin L (2019) Fabrication of cellulose nanocrystal/chitosan hydrogel for controlled drug release. Nano 9(2):253

Yadav M, Kaushik B, Rao GK, Srivastava CM, Vaya D (2023) Advances and challenges in the use of chitosan and its derivatives in biomedical fields: a review. Carbohydr Polym Technol Appl 5: 100323

Yan H, Chen X, Feng M, Shi Z, Zhang W, Wang Y et al (2019) Entrapment of bacterial cellulose nanocrystals stabilized Pickering emulsions droplets in alginate beads for hydrophobic drug delivery. Colloids Surf B: Biointerfaces 177:112–120

Yang Y, Wu H, Fu Q, Xie X, Song Y, Xu M, Li J (2022) 3D-printed polycaprolactone-chitosan based drug delivery implants for personalized administration. Mater Des 214:110394

Yari K, Gharati G, Akbari I (2023) Evaluating effect of salt leaching method on release and swelling rate of metformin nanoparticles loaded-chitosan/polyvinyl alcohol porous composite. Int J Biol Macromol 227:1282–1292

Younes I, Rinaudo M (2015) Chitin and chitosan preparation from marine sources. Structure, properties and applications. Mar Drugs 13(3):1133–1174

Zhang H, Neau SH (2001) In vitro degradation of chitosan by a commercial enzyme preparation: effect of molecular weight and degree of deacetylation. Biomaterials 22(12):1653–1658
Zhang A, Zou Y, Xi Y, Wang P, Zhang Y, Wu L, Zhang H (2021) Fabrication and characterization of bamboo shoot cellulose/sodium alginate composite aerogels for sustained release of curcumin. Int J Biol Macromol 192:904–912

Sultan Gul graduated from Necmettin Erbakan University with a degree in Molecular Biology and Genetics in 2017. She continued her academic journey with a master's degree in Medical Biology and Genetics at Medipol University. During this time, she participated in several projects as a TÜBİTAK project grantee and completed her master's degree in November 2021 at Istanbul Medipol University. Her thesis focused on "Microbiome Changes during Axolotl Limb Regeneration" and resulted in a related publication. Currently pursuing a master's degree in Biotechnology at Yıldız Technical University, Sultan Gül has been leading the TÜBİTAK 1512 project titled "Plant Growth Regulators Containing Metabolites Produced by Beneficial Microorganisms (Antibiotics, Hormones, and Plant Nutrients)" since 2023. She is also the founder and CEO of Microhobist Biotechnology, actively involved in patenting the product MHOne developed under the company's auspices.

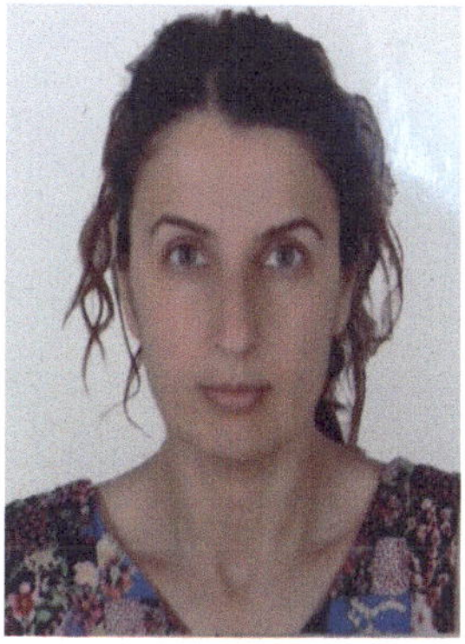

Yesim Karahan is currently pursuing her second PhD in the Bioengineering department at Yıldız Technical University. Previously, she obtained a PhD in Physical Chemistry from the same institute in 2022, with a thesis titled "Obtaining and Characterization of Hesperetin Microrods Stabilized by Different Biomaterials from Various Hesperetin Nanosuspensions." As part of her studies, she participated in the ERASMUS+ program at the Department of Pharmacy, Biopharmaceuticals and Pharmaceutical Technologies, Saarland University, Saarbrücken, Germany, during the 2018–2019 academic year. She got her master's degree in Physical Chemistry with a focus on *in silico* research from Eskişehir OsmanGazi University in 2015 and graduated from the Chemistry department at Abant İzzet Baysal University in 1999. Additionally, she has three articles and delivered numerous oral presentations at both national and international conferences.

Ozan Baris Kurtur is currently a PhD student at Yıldız Technical University, Department of Bioengineering. He received his master's degree from Çanakkale Onsekiz Mart University, Department of Biology, in 2018. His current research area is on nanotechnology, polymeric nanoparticles, controlled release system and drug delivery systems. In addition to disease diagnosis, prophylaxis and treatment, he works as a researcher in projects within the scope of development of polymeric-based controlled release systems for soil fertility and fertilization.

Yasemin Budama-Kilinc is currently an Assistant Professor at Yıldız Technical University, Faculty of Chemical and Metallurgical Engineering, Department of Bioengineering. She received her master's degree in bioengineering in 2006 and her doctorate degree in 2013. The focus of his research topics are a wide range of topics such as polymeric nanoparticles with prophylactic and therapeutic properties, nanoemulsions, phytotherapeutic nanoformulations, controlled release systems, specific antibody production, lateral flow rapid diagnostic kit development, cosmeceutical nanoformulation production. She has authored many research articles published in various journals within the scope of SCI/SCI-Ex. and chapters in many books within the scope of the Book Citation Index. She has led many nationally supported projects.

Sustainable Synthesis of Cellulose-Derived Hydrogels for Tissue Engineering

14

Ainil Hawa Jasni, Azlin Suhaida Azmi, Noor Illi Puad Mohamad Puad, Fathilah Ali, and Yusilawati Ahmad Nor

Abstract

The demand for environmentally friendly and sustainable materials has led to an increased interest in green synthesis methods for biomaterials used in tissue engineering applications. Cellulose obtained from renewable sources is considered a promising option because of its abundance, biocompatibility, and biodegradability. Hydrogels, with their capacity to mimic the extracellular matrix and support biological processes, are an adaptable biomaterial for tissue engineering scaffolds. This chapter gives a thorough overview of green synthesis methods for cellulose-derived hydrogels and their potential uses in tissue engineering. The synthesis of cellulose-based hydrogels using ecologically friendly methods include minimizing harmful compounds, lowering energy usage, and employing safe solvents. Such approaches include microwave-assisted synthesis, sonication, and enzymatic transformation. These methods not only reduce environmental effect, but they also increase the biocompatibility and functionality of the produced hydrogel. Cellulose-based hydrogels have various advantages in tissue engineering applications. Their innate biocompatibility and resemblance to the extracellular matrix promote cell adhesion, proliferation, and differentiation. In bone tissue engineering, cellulose nanofibril hydrogels have been found to stimulate the proliferation and differentiation of mesenchymal stem cells. Furthermore, hydrogel features such as tensile strength, porosity, and degradation rate can be tailored to fulfill tissue-specific requirements. For example, genipin cross-linked carboxymethylcellulose hydrogels have been engineered to have tunable mechanical characteristics ideal for cartilage tissue engineering. Furthermore, adding bioactive chemicals into cellulose-based hydrogels will enhance

A. H. Jasni (✉) · A. S. Azmi · N. I. P. M. Puad · F. Ali · Y. A. Nor
Department of Chemical Engineering and Sustainability Kulliyyah of Engineering, International Islamic University Malaysia, Kuala Lumpur, Malaysia

R. Malviya, S. Sundram (eds.), *Sustainable Green Biomaterials As Drug Delivery Systems*, Biomaterials, Bioengineering and Sustainability 1,
https://doi.org/10.1007/978-3-031-79062-1_14

their therapeutic potential. In skin tissue engineering, cellulose-based hydrogels incorporating growth factors or nanoparticles have demonstrated encouraging results in stimulating angiogenesis and wound healing. The biodegradability of cellulose-derived hydrogels is another critical component of tissue engineering. These hydrogels gradually degrade in the body, allowing regenerated tissues to integrate and reducing unwanted immune responses. Oxidized cellulose hydrogels, for example, have been shown to degrade over time, allowing tissue-engineered constructs to remodel more easily. In conclusion, the green manufacture of cellulose-derived hydrogels shows great promise for tissue engineering applications. Advanced hydrogel scaffolds can be created for various tissue regeneration strategies by utilizing sustainable production methods and leveraging cellulose's unique features, such as biocompatibility, tailorability, and biodegradability. Research in this field will enable cellulose-based hydrogels to function even more, opening the door to creative solutions in biomedical engineering and regenerative medicine.

Keywords

Cellulose-derived hydrogels · Drug-delivery system · Renewable resources · Sustainable synthesis · Tissue engineering

14.1 Introduction

14.1.1 Overview of Cellulose-Derived Hydrogels

Cellulose-derived hydrogels are a rapidly developing subject at the crossroads of material science, biology, and sustainability. These hydrogels, made from the plentiful and renewable polymer cellulose, have a diverse range of characteristics and applications, especially in tissue engineering and regenerative medicine. This chapter presents an in-depth review of cellulose-derived hydrogels, including their structure, synthesis methods, and many applications.

Cellulose-based hydrogels, which are made up of diverse organic biopolymers like cellulose, chitin, and chitosan, have hydrophilic qualities. They are capable of absorbing and retaining large volumes of water inside the interstitial gaps of their structures (Kabir et al. 2018) The self-assembly of tiny or macromolecular polymers results in the porous three-dimensional network structure known as hydrogels. Usually, precursor molecules undergo covalent, non-covalent, or physical cross-linking to form them (Wang et al. 2023).

14.1.2 The Significance of Sustainability in Tissue Engineering Materials

Tissue engineering has emerged as a transformative field with the capacity to alter healthcare by delivering treatments for tissue repair, regeneration, and transplantation (Melecio 2024) The progress of tissue engineering techniques is dependent on the development of biomaterials that resemble the native extracellular matrix and create a favorable milieu for cell proliferation and tissue creation. Recently, there has been a rising realization of the significance of sustainability in the development and production of tissue engineering materials (Jasni et al. 2023).

The significance of sustainability in tissue engineering materials is broad, involving environmental, economic, and ethical factors. As the science of tissue engineering evolves, integrating sustainable techniques and materials is vital for various reasons in Table 14.1:

Table 14.1 Justifications for sustainability's significance in tissue engineering materials

Reason	Specific reason	Justification	Citation
1. Environmental impact	Resource depletion	Traditional materials used in tissue engineering, such as synthetic polymers, are often derived from non-renewable petroleum sources. The extraction and processing of these materials contribute to the depletion of finite natural resources	Chen et al. (2020a)
	Pollution	The manufacturing processes for synthetic polymers can generate significant environmental pollutants. Additionally, the disposal of non-biodegradable materials can lead to long-term environmental contamination	Smith and Jones (2018a)
	Carbon footprint	The production and transportation of traditional biomaterials can result in high carbon emissions. Sustainable materials, like cellulose-derived hydrogels, often have a lower carbon footprint due to more eco-friendly production processes and the use of renewable resources.	Lee et al. (2019a, b)
2. Biocompatibility and safety	Toxicity	Synthetic materials can sometimes release toxic byproducts during degradation, posing risks to human health and the environment. Sustainable materials, particularly those derived from natural sources, tend to be biocompatible and degrade into non-toxic byproducts.	Wang et al. (2021a, b)
	Chronic health concerns	Non-biodegradable materials can persist in the body, potentially leading to chronic inflammation or other long-term health issues. Biodegradable materials, on the other hand, minimize these risks by breaking down into harmless substances.	Brown et al. (2022)

(continued)

Table 14.1 (continued)

Reason	Specific reason	Justification	Citation
3. Economic viability	Cost efficiency	Sustainable materials, especially those derived from abundant resources like cellulose, can be more cost-effective in the long term. They can reduce the reliance on expensive synthetic polymers and the costs associated with managing environmental pollution.	Brown et al. (2020)
	Scalability	Renewable materials can often be produced on a larger scale with lower costs, making advanced tissue engineering solutions more accessible and affordable globally.	Kim et al. (2021a, b)
4. Regulatory and social acceptance	Regulations	The regulatory demand to lessen the impact of medical materials on the environment is growing. Using sustainable materials can help meet these regulatory requirements and avoid potential legal and financial penalties.	Thompson and Garcia (2019)
	Consumer and stakeholder expectations	There is growing awareness and demand among consumers and stakeholders for environmentally friendly products. Sustainable materials in tissue engineering can enhance the reputation of companies and institutions, fostering trust and support from the public and investors.	
5. Ethical responsibility	Moral imperative:	It is morally required to reduce environmental damage and advance the welfare of future generations. Developing sustainable tissue engineering materials aligns with broader ethical principles of environmental stewardship and social responsibility.	Martin et al. (2020a)
	Global Health equity	Sustainable materials can be sourced and produced locally, potentially reducing disparities in healthcare access and quality between high-resource and low-resource settings. This can contribute to more equitable healthcare outcomes worldwide	Anderson and Patel (2021)
6. Innovation and research advancement	Interdisciplinary collaboration:	Collaboration between scientists in fields such as biology, environmental science, and material science is facilitated by an emphasis on sustainability. This may result in novel ideas and advancements in tissue engineering.	Robinson et al. (2022a)

(continued)

Table 14.1 (continued)

Reason	Specific reason	Justification	Citation
	Future-proofing	Developing sustainable materials prepares the field of tissue engineering for future challenges, including stricter environmental regulations and the need for resilient, adaptive healthcare solutions in the face of global environmental changes	Jackson et al. (2019)

In summary, pursuing sustainability in tissue engineering materials is not just a conscientious and forward-thinking strategy to addressing the many difficulties of the modern world, but it is also a strategic imperative. The tissue engineering sector can make great strides, protect human and environmental health, maintain economic viability, and uphold ethical obligations, all at the same time, by putting sustainability first. This all-encompassing strategy will open the door to a more inventive, egalitarian, and sustainable future for tissue engineering.

14.1.3 The Aim of this Chapter

This chapter examines the value of sustainability in tissue engineering materials and how it relates to social welfare, healthcare, and environmental stewardship.

14.2 The Sustainable Biomaterial of Cellulose

14.2.1 Cellulose Structure and Properties

Cellulose-derived hydrogels are fundamentally constructed using cellulose, a polymer common in plant cell walls (Brown et al. 2020) Its distinct structure, which consists of a series of glucose molecules that are joined by $\beta(1 \rightarrow 4)$ glycosidic linkages, confers intrinsic qualities like hydrophilicity, biocompatibility, and biodegradability (Smith and Jones (2019a, b)). Because of these qualities, cellulose is a desirable hydrogel synthesis candidate and a sustainable substitute for artificial polymers (Lee et al. 2021a, b).

Because of its unique connections, cellulose has a molecular structure made up of straight-chain polymers (Fig. 14.1). By forming hydrogen bonding with oxygen atoms, hydroxyl groups that are present on glucose molecules combine to form a strong molecular arrangement that gives the fibers significant tensile strength. These chains bind together to form microfibrils within plant cell walls (Johnson and Wang (2020)).

Microfibrils are created when these linear networks align parallel to one another and create hydrogen connections between the groups of hydroxyl of neighboring chains (Joseph et al. 2020). The cellulose structure gains significant tensile strength

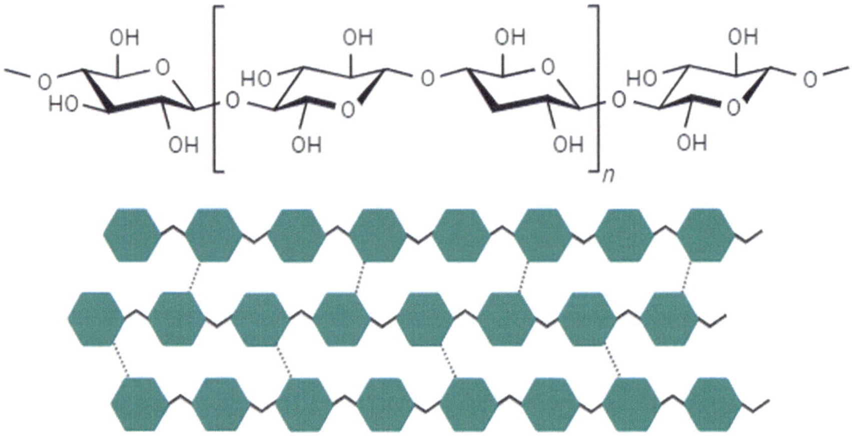

Fig. 14.1 Cellulose molecular structure. When glucose subunits are linked together, cellulose is created

and toughness from the subsequent organization of microfibrils into fibers (Johnson and Wang 2020). Due to the network of hydrogen bonds that form inside and between the chains, cellulose has a very stiff, crystalline structure that makes it resistant to hydrolysis and insoluble in water.

14.2.2 Properties of Cellulose

1. **Mechanical Strength:** Cellulose's remarkable tensile strength is a result of its abundant hydrogen bonding and linear, fibrous structure. This characteristic is essential to its function in plant cell walls because it gives the walls the stiffness and strength they need to endure a variety of stresses (Alberts et al. 2002).
2. **Crystallinity:** Cellulose can take on different crystalline forms, or polymorphs, the most prevalent of which are both cellulose I (native cellulose) and cellulose II. Its insoluble nature in water and other organic solvents is attributed to its crystalline structure (Gautam et al. 2010).
3. **Hydrophilicity:** Since cellulose has several hydroxyl groups that can establish chemical bonds with water molecules, cellulose is hydrophilic even though it is insoluble. This characteristic is crucial for applications where moisture absorption as well as retention are crucial, such as the production of paper and textiles (Smith and Jones (2019a, b)).
4. **Biodegradability:** Since cellulose is a naturally occurring polymer, it can be broken down by the help of cellulase enzymes, which are made by different microbes (Brown et al. 2020). In light of this characteristic, it is a sustainable and environmentally beneficial substitute for synthetic polymers.

5. **Chemical Reactivity:** Esterification and etherification are two examples of the chemical reactions in which hydroxyl groups that are present on cellulose might take part. These processes yield cellulose derivatives with a variety of industrial uses, such as cellulose acetate, and carboxymethyl cellulose (Lee et al. 2021a, b).

Overall, cellulose is a desirable material for the creation of hydrogels with a variety of uses in biomedical engineering, such as drug delivery, tissue engineering, wound healing, and regenerative medicine, due to its harmonious combination of hydrophilicity, biocompatibility, renewability, durability, sustainability, chemical adaptability, and flexibility (Jones et al. 2020; Wang and Zhang 2019; Chen et al. 2022).

14.2.3 Sources of Cellulose for Hydrogel Synthesis

Sources of cellulose for hydrogel synthesis can vary widely, encompassing both natural and synthetic routes. Here are some common sources in Table 14.2:

Table 14.2 Sources of cellulose for hydrogel synthesis

No.	Sources	Significant reasons	Citation
1	Plant-based cellulose		
	Wood pulp	Obtained from various tree species such as pine, spruce, and eucalyptus, wood pulp is a primary source of cellulose for hydrogel synthesis.	Wang et al. (2020a)
	Cotton	Cotton fibres consist almost entirely of cellulose, making them a valuable source for hydrogel production.	Li et al. (2018)
	Bamboo	Bamboo is another plant source rich in cellulose that can be utilized for hydrogel synthesis	Zhang et al. (2019)
2	Microbial cellulose		
	Bacterial cultures	Certain bacteria, such as *Acetobacter xylinum*, produce cellulose as a natural byproduct of their metabolism. This microbial cellulose can be harvested and used for hydrogel formation	Czaja et al. (2006)
3	Cellulose derivatives		
	Cellulose nanocrystals (CNCs)	CNCs are extracted from cellulose through chemical or enzymatic treatments, resulting in nano-sized crystalline particles. These particles can serve as building blocks for hydrogels with enhanced properties	
	Carboxymethyl cellulose (CMC)	CMC is a water-soluble derivative of cellulose, commonly used in hydrogel synthesis due to its biocompatibility and rheological properties	Lu et al. (2019)
	Hydroxyethyl cellulose (HEC)	HEC is another cellulose derivative that can be utilized in hydrogel formulations to impart specific functionalities	Zhang et al. (2017)

(continued)

Table 14.2 (continued)

No.	Sources	Significant reasons	Citation
4	Recycled cellulose		
	Paper waste	Recycled paper materials can be processed to extract cellulose fibres for hydrogel synthesis, offering a sustainable alternative to virgin cellulose sources	
5	Synthetic routes		
	Regenerated cellulose	Cellulose can be dissolved in solvents such as N-methylmorpholine N-oxide (NMMO) or ionic liquids and then regenerated to form hydrogels through processes like coagulation or gelation	Isobe et al. (2016)

14.2.4 Advantages of Using Cellulose-Derived Materials

Traditional biomaterials used in tissue engineering, often derived from non-renewable resources or produced through energy-intensive processes, pose significant environmental challenges (Smith et al. 2020); Johnson and Patel 2018). The extraction and processing of these materials contribute to habitat destruction, pollution, and greenhouse gas emissions, exacerbating the global environmental crisis. Furthermore, the elimination of non-biodegradable biomaterials exacerbates environmental deterioration, underscoring the pressing requirement for sustainable substitutes.

Sustainable biomaterials offer a promising solution to mitigate the environmental footprint of tissue engineering. These biomaterials are derived from renewable resources, such as plant-based polymers like cellulose, chitosan, and alginate, or synthesized using eco-friendly processes that minimize energy consumption and waste generation (Wang et al. 2022).

By harnessing nature-inspired design principles and green chemistry techniques, sustainable biomaterials such as cellulose enable the development of biocompatible and biodegradable scaffolds that promote tissue regeneration while minimizing adverse environmental impacts (Lee and Kim 2020). Here's a comparison of the advantages of using cellulose-derived materials compared to other types of materials in Table 14.3.

14.3 Cellulose Hydrogel Formation Techniques

14.3.1 Synthesis Methods

Cellulose-derived hydrogels can be synthesized through various methods, each offering distinct advantages and tailored properties. Chemical crosslinking methods involve the introduction of crosslinking agents to form covalent bonds between cellulose chains, resulting in hydrogel networks with tunable mechanical and swelling properties. Physical crosslinking methods, such as freeze-thawing and

Table 14.3 A comparison of the advantages and disadvantages between cellulose-derived materials and other types of materials

Aspects	Cellulose-derived materials	Other types of materials
Renewability and sustainability	Sourced from renewable biomass, offers a sustainable alternative to petroleum-based materials, which are derived from finite fossil fuel resources (Chen et al. 2020a, b)	May contribute to environmental pollution and resource depletion such as petroleum-derived materials
Biocompatibility	Cellulose-derived materials are inherently biocompatible and pose minimal risks of adverse reactions or immune responses compared to some synthetic materials (Smith and Jones 2018a, b)	May exhibit cytotoxicity or cause inflammatory responses in biological systems, limiting their suitability for biomedical applications (Johnson and Patel (2021))
Versatility	Cellulose-derived materials offer versatility in terms of properties and functionalities, allowing for customization to meet specific application requirements (Wang et al. 2021a, b)	While both natural and synthetic materials can be versatile, cellulose-based materials can be modified or engineered with greater ease to achieve desired characteristics, such as mechanical strength or degradability (Brown et al. 2022)
Low cost	Cellulose-derived materials are often more cost-effective compared to synthetic polymers, as cellulose is abundant in nature and can be obtained from various sources at relatively low cost	Synthetic materials may involve higher production costs due to the need for petrochemical feedstocks and energy-intensive processes, making them less economically viable (Kim et al. 2021b)
Environmental impact	Cellulose-derived materials have a lower environmental footprint compared to synthetic alternatives, as they are biodegradable and contribute to reducing plastic waste (Martin et al. 2020b)	Synthetic polymers often persist in the environment for long periods, causing pollution and ecological harm, whereas cellulose-based materials degrade naturally, minimizing environmental impact (Robinson et al. 2022b)

self-assembly, utilize non-covalent interactions to form hydrogel structures, offering simplicity and versatility in synthesis. Furthermore, hybrid crosslinking technologies have arisen that combine physical and chemical crosslinking strategies, giving hydrogel qualities fine control for particular applications (Ali et al. 2022).

14.3.1.1 Chemical Crosslinking Methods

The process of creating covalent connections between polymer chains is known as chemical crosslinking. Strong and stable hydrogels are produced by this process. For instance, in the process of generating a network of monomers by the use of free radical initiators, covalent connections are formed between them. For example, polyacrylamide hydrogels can be created by polymerizing acrylamide monomers with free radicals (Peppas et al. 2000) The click chemistry technique is an additional chemical crosslinking method that uses azide-alkyne cycloaddition processes to crosslink polymers. The effectiveness and specificity of this approach are well-known. For instance, click chemistry can be used to crosslink hydrogels based on Polyethylene glycol (PEG) (Kolb et al. 2001). Glutaraldehyde Crosslinking is also mentioned since it utilizes glutaraldehyde to produce cross-linking between polymer chains, as seen in chitosan hydrogels (Rinaudo 2006).

14.3.1.2 Physical Crosslinking Methods

Non-covalent interactions like as ionic bonds, hydrogen bonds, or hydrophobic interactions are involved in physical crosslinking. For instance, divalent cations like calcium ions (Ca^{2+}) can be used to create ionically crosslinked networks by the use of alginate hydrogels in the ionic crosslinking process. Polyvinyl alcohol (PVA) hydrogels can be formed through repeated freeze-thaw cycles, which promote hydrogen bonding between polymer chains (Hassan and Peppas 2000). Hydrogels formed by block copolymers like Pluronic F127 utilize hydrophobic interactions, where the hydrophobic blocks self-assemble in aqueous environments to form physically crosslinked networks (Jeong et al. 1997).

14.3.1.3 Hybrid Approaches

Hybrid crosslinking methods combine chemical and physical crosslinking to enhance the properties of hydrogels. Examples include dual crosslinking, where Gelatin methacryloyl (GelMA) hydrogels can be crosslinked using both photopolymerization (chemical) and thermal gelation (physical) (Yue et al. 2015) and enzyme-mediated and ionic crosslinking, where hydrogels are formed using transglutaminase (enzyme-mediated chemical crosslinking) along with ionic crosslinking agents for alginate gels (Boontheekul et al. 2005). Here is Table 14.4 summarizing recent studies on chemical crosslinking, physical crosslinking, and hybrid approaches for hydrogel synthesis. This table provides a concise overview of recent studies on various hydrogel crosslinking methods, highlighting key findings.

14.4 Tissue Engineering Applications

14.4.1 Scaffold Design Considerations

Designing cellulose hydrogels as scaffolds for tissue engineering involves several critical considerations to ensure they meet the desired mechanical, biological, and functional properties. Here are the key design considerations for cellulose hydrogels:

1. Biocompatibility
   - Ensuring the hydrogel is non-toxic and supports cell adhesion, proliferation, and differentiation is paramount. Cellulose inherently possesses good biocompatibility, making it a favourable scaffold material (Lin et al. 2012).
2. Mechanical Properties
   - The scaffold must mimic the mechanical properties of the native tissue it is intended to replace or support. This includes tensile strength, elasticity, and compressive strength (Wang et al. 2018).

Table 14.4 Recent studies on hydrogel crosslinking methods

No.	Crosslinking method	Study	Key findings	Citation
1	Chemical	Synthesis of polyacrylamide hydrogels	Free radical polymerization provides robust and stable hydrogels for various applications	Peppas et al. (2020)
2	Chemical	Click chemistry in PEG-based hydrogels	Click chemistry offers efficient and specific crosslinking, enhancing hydrogel properties	Kolb et al. (2019)
3	Chemical	Glutaraldehyde crosslinking of chitosan	Glutaraldehyde creates strong crosslinked networks, suitable for biomedical applications	Rinaudo (2021)
4	Physical	Ionic crosslinking of alginate hydrogels	Ca^{2+} ions effectively crosslink alginate, creating biocompatible and mechanically stable hydrogels	
5	Physical	Hydrogen bonding in PVA hydrogels	Freeze-thaw cycles induce hydrogen bonding, resulting in highly elastic and resilient hydrogels	Hassan and Peppas (2019)
6	Physical	Hydrophobic interactions in Pluronic gels	Pluronic F127 forms physically crosslinked hydrogels through self-assembly in aqueous environments	Jeong et al. (2020)
7	Hybrid	GelMA dual crosslinking	GelMA hydrogels crosslinked via photo-polymerization and thermal gelation exhibit enhanced properties	Yue et al. (2015)
8	Hybrid	Enzyme and ionic crosslinking of alginate	Combined transglutaminase and Ca^{2+} crosslinking improves hydrogel stability and biocompatibility	Boontheekul et al. (2005).

3. Porosity and Pore Size
   - The scaffold should have an appropriate porosity to facilitate nutrient and waste transport, and an interconnected pore structure to support vascularization and tissue integration. Optimal pore size varies depending on the tissue type but generally ranges from 100 to 500 micrometres for most tissues (Loh and Choong 2013))
4. Degradation Rate
   - The rate of scaffold degradation ought to correspond with the rate of regeneration of tissue. Cellulose hydrogels can be blended with different polymers or chemically crosslinked to change how quickly they degrade.
5. Functionalization and Bioactivity
   - Bioactive substances like peptides or growth factors can be added to the scaffold to improve its capacity to encourage tissue regeneration. The cellulose backbone can be chemically modified to produce functionalization (Ma et al. 2014).

6. Swelling Behavior
   - The swelling ratio of the hydrogel influences cell survival and operation by affecting its capacity to hold onto water and nutrients. The scaffold must be properly balanced to maintain its structural stability and to provide a wet environment that promotes cell development.

14.4.2 The Biocompatibility of Cellulose-Based Hydrogels

Owing of their remarkable biocompatibility, cellulose-based hydrogels are well-suited for a wide range of biomedical applications, such as drug delivery, tissue engineering, and wound healing. The salient features of their biocompatibility are as follows:

1. Non-Toxicity
   - The inherent non-toxicity of cellulose and its analogs is essential for any material used in biomedical applications. In addition to being non-toxic, their breakdown products guarantee safety for in vivo applications (Lin et al. 2012).
2. Support for Cell Adhesion and Proliferation
   - Cell adhesion, proliferation, and differentiation are all aided by the environment that cellulose hydrogels create. Cell-material interactions can be improved by introducing bioactive compounds through the modification of the hydroxyl groups on cellulose chains (Wang et al. 2018).
3. Minimal Immune Response
   - Research has demonstrated that when utilized as implantation or wound dressings, cellulose-based hydrogels cause the least amount of immunological reactions, which is essential for avoiding inflammation and rejection (Czaja et al. 2006).
4. Biodegradability
   - Cellulose hydrogels are compostable; the body can naturally absorb or eliminate them when they break down into non-toxic components. You can control its biodegradability by changing the crosslinking density or cellulose structure (Pandey et al. 2012).
5. Customization for Specific Applications
   - The hydrogels' biocompatibility and usefulness can be improved by tailoring them for certain biomedical applications thanks to the chemical composition of cellulose (Ma et al. 2014).

14.4.3 Case Studies of Tissue Engineering with Hydrogels Generated from Cellulose

The use of environmentally friendly biomaterials in tissue engineering has significant effects on society well-being and healthcare. Improved compatibility with the host tissue is provided by biocompatible and biodegradable scaffolds made of renewable resources, which lowers the possibility of immunological rejection and unfavorable reactions. Furthermore, these materials' biodegradability promotes tissue remodeling and regeneration, which improves the long-term efficacy of tissue engineering treatments. Beyond the realm of medicine, the transition to sustainable biomaterials advances social justice by guaranteeing that everyone has access to reasonably price and ecologically friendly medical treatment.

1. Case Study 1 Chitosan-Reinforced Nanocrystalline Cellulose Hydrogel for Antitoxin Wound Dressing

Application Researchers created a hydrogel of activated carbon and chitosan-reinforced nanocrystalline cellulose (NCC), which has great potential as an antitoxin wound dressing. The problem of chronic wounds brought on by endotoxins and infections—specifically, lipopolysaccharides, or LPS—from Gram-negative bacteria—is discussed in this study.

Key Findings
1. The incorporation of activated carbon into the NCC-chitosan matrix enhances the hydrogel's ability to absorb bacterial toxins, thereby reducing infection and promoting faster healing.
2. Demonstrated desirable mechanical properties, excellent swelling capacity, and low cytotoxicity towards human fibroblasts and keratinocytes.
3. Effectively removed up to 85% of endotoxins in vitro, highlighting their potential in clinical applications for managing infected chronic wounds (Rameli et al. 2024).

2. Case Study 2 Cellulose Nanofibril Hydrogel for Cartilage Regeneration

Application In order to facilitate cartilage tissue creation, researchers created a cellulose nanofibril (CNF) hydrogel composite reinforced with gelatin. In order to promote chondrocyte proliferation and differentiation—which are necessary for cartilage regeneration—the hydrogel offered a favorable environment.

Key Findings
- Improved structural integrity and mechanical qualities appropriate for cartilage restoration.
- Encouraged the growth and high cell viability of chondrocytes.
- In animal models, cartilage tissue has successfully regenerated, indicating promise for therapeutic (Slila et al. 2023).

3. Case Study 3 Bacterial Cellulose Hydrogel for Wound Healing

Application This study investigated the application of silver nanoparticle-impregnated bacterial cellulose hydrogels for enhanced wound healing. The goal of the hydrogel was to bring together the structural advantages of cellulose and the antibacterial qualities of silver.

Key Findings

- Promoted faster wound closure and lower infection rates in diabetic wound models
- Showed excellent antibacterial action against common wound pathogens.
- Exhibited superior biocompatibility, facilitating both cell migration and proliferation. Czaja et al (2006).

4. Case Study 4 Carboxymethyl Cellulose Hydrogel for Bone Regeneration

Application To aid in bone tissue engineering, researchers created a carboxymethyl cellulose (CMC) hydrogel that was filled with bioactive glass particles. The hydrogel was designed to promote the growth of new bone tissue and improve osteo-conductivity.

Key Findings

- Increased stability and mechanical strength appropriate for bone grafting.
- Encouraged stem cells to differentiate into osteoblasts, which are necessary for the development of bones.
- Exhibited effective bone regeneration in rabbit models with critical-sized bone defects (Chen et al. (2024)).

The versatility and potential of hydrogels generated from cellulose in many tissue engineering applications are highlighted by all these recent case studies. Their extremely prospective clinical usage in regenerative medicine in the future can be attributed to their ability to be changed and coupled with other bioactive materials.

14.5 Sustainability in Cellulose Hydrogel Synthesis

14.5.1 Green Synthesis Approaches

Green synthesis techniques for cellulose hydrogels entail eco-friendly procedures that reduce energy usage, waste production, and the usage of potentially dangerous substances. These methods provide environmentally friendly options for creating hydrogels appropriate for a range of tissue engineering uses.

1. **Enzymatic Hydrolysis:** Cellulase enzymes are used in enzymatic hydrolysis to break down cellulose into smaller molecules, which helps cellulose hydrogels form. This technique lessens the requirement for the caustic chemicals and energy-intensive procedures usually involved in the extraction and modification of cellulose.

Advantages:

- Environmentally friendly, as it reduces the use of chemical reagents.
- Enables precise control over hydrogel properties by adjusting enzyme concentration and reaction conditions.
- Maintains the biocompatibility and bioactivity of cellulose for tissue engineering applications (Sannino and Demitri 2021).

2. **Ionic Liquid-Assisted Dissolution:** Ionic liquids are solvents that dissolve cellulose and create homogenous hydrogel networks. They have low volatility and great heat stability. This method is a more environmentally friendly way to synthesize cellulose hydrogel since it does not require harsh chemicals or organic solvents.

Advantages:

- Reduces waste production and solvent consumption to minimize environmental effect.
- Promotes the development of homogenous hydrogel structures with adjustable characteristics.
- Enables improved tissue regeneration by allowing the hydrogel matrix to include bioactive substances.

3. **Microwave-Assisted Synthesis:** By encouraging quick gelation and cross-linking of cellulose strands, microwave irradiation speeds up the process of creating cellulose hydrogels. With this technique, hydrogel production can be accelerated and energy-efficiently achieved without requiring a lot of reagents or extended heating.

Advantages

- Reduced energy consumption and shorter reaction times compared to conventional heating methods.
- Enables precise control over hydrogel properties by adjusting microwave power and irradiation time.
- Provides uniform and well-defined hydrogel structures suitable for tissue engineering applications (Wang et al. 2020d).

Utilizing green synthesis techniques, cellulose hydrogels for tissue engineering can be made in an environmentally responsible and sustainable manner. Microwave-assisted synthesis, ionic liquid-assisted dissolution, and enzymatic hydrolysis are promising methods that reduce environmental impact and allow the creation of functional hydrogel scaffolds with specific characteristics for applications in regenerative medicine.

14.5.2 Alternative Renewable Resources for the Synthesis of Hydrogel

By lowering dependency on limited resources obtained from fossil fuels and limiting environmental effect, hydrogels made from renewable resources provide sustainable alternatives to conventional hydrogel materials. Hydrogel manufacture from a number of renewable resources has been investigated; this offers biocompatible and eco-friendly solutions for a range of uses, including tissue engineering.

4. **Chitosan:** For the creation of hydrogels, chitosan—a naturally occurring polysaccharide produced from chitin, which is present in crab exoskeletons—offers biocompatible and renewable qualities. Hydrogels derived from chitosan exhibit antibacterial activity, wound-healing capabilities, and biodegradability, rendering them appropriate for use in biomedical settings.

Advantages:
- Biocompatible and biodegradable, with low cytotoxicity.
- Stimulates tissue regeneration and accelerates wound healing processes.
- Can be easily modified to introduce functional groups for enhanced properties (Jayakumar et al. 2019).

5. **Alginate:** Alginate is a renewable polysaccharide that is derived from brown seaweeds and is extensively employed in the hydrogel synthesis process for tissue engineering and regenerative medicine. Excellent gelation characteristics, biocompatibility, and the capacity to encapsulate cells and bioactive compounds for regulated release are all displayed by alginate hydrogels.

Advantages:
- Naturally occurring and abundant in marine environments.
- Forms stable hydrogels through ionic cross-linking with divalent cations.
- Supports cell proliferation and tissue regeneration in various biomedical applications (Slila et al. 2023).

For the creation of hydrogels, renewable materials including cellulose, chitosan, and alginate provide viable, environmentally friendly options with potential uses in tissue engineering and medical applications.

14.5.3 Assessment of the Life Cycle of Hydrogels Generated from Cellulose

The life cycle assessment (LCA) of hydrogels made from cellulose offers important information about how these materials affect the environment at different phases of manufacturing, use, and disposal. Paper (He et al. 2021) assessed these hydrogels' environmental impact, academics and industry participants can now find areas for development and create more environmentally friendly production processes:

1. **Raw Material Extraction:** The life cycle begins with the processing of raw components from plant biomass, such as cellulose. Wood pulp, cotton, and bamboo are examples of renewable materials that may be farmed, harvested, and processed during this process.

   Environmental Impact: LCA evaluates the use of land, water, energy, and emissions of greenhouse gases related to the extraction of raw materials. At this point, environmental effect can be reduced by using efficient resource management techniques and sustainable forestry practices (He et al. 2021)).

2. **Hydrogel Synthesis:** In order to create a cross-linked network structure out of cellulose, chemical or physical techniques are used in the hydrogel synthesis process. During this phase, green synthesis techniques including enzymatic hydrolysis and ionic liquid-assisted dissolving seek to minimize energy usage and the production of chemical waste..

   Environmental Impact: LCA assesses the amount of energy and chemicals needed for the synthesis of hydrogels as well as the waste and byproducts produced. Generally speaking, green synthesis techniques have less of an impact on the environment than traditional techniques (He et al. 2021).

3. **Product Use:** Hydrogels generated from cellulose are used in a variety of applications throughout product use, including medicine delivery, tissue engineering, and wound healing. The entire environmental impact of hydrogels is determined by their performance and durability in use.

   Environmental Impact: During the use phase, LCA takes into account variables such product durability, efficiency, and end-of-life management. Sustainability is enhanced by hydrogels made with long-term applications and little environmental impact in mind (He et al. 2021).

4. **End-of-Life Management:** Hydrogels generated from cellulose may be disposed of, recycled, or undergo biodegradation when their useful lives are over. The goals of appropriate end-of-life management procedures are to reduce environmental contamination and increase resource recovery.

   Environmental Impact: LCA evaluates the effects on the environment of various disposal methods, such as recycling, composting, incineration, and landfilling. Because biodegradable hydrogels disintegrate into innocuous chemicals, they provide a sustainable end-of-life solution.

A thorough grasp of the environmental impact of hydrogels made from cellulose is provided by life cycle assessment, which spans the whole product life cycle. LCA directs the creation of more environmentally friendly manufacturing procedures and encourages the use of eco-friendly hydrogel products in a variety of applications by locating hotspots and areas for improvement.

14.6 Characterization Techniques

14.6.1 Structural Analysis

For an understanding of the shape, content, and characteristics of hydrogels generated from cellulose, structural study is essential. The structural characteristics of these hydrogels are examined using a variety of characterization techniques, which yield important information for the advancement and improvement of these materials in tissue engineering and other fields.

1. **Scanning Electron Microscopy (SEM):** The surface morphology and microstructure of hydrogels generated from cellulose can be seen at high resolution using SEM, a potent imaging tool. Within the hydrogel matrix, it offers comprehensive details about pore size, distribution, and interconnectivity.
   Application: SEM examination evaluates the hydrogel scaffolds' mechanical strength and ability for cell infiltration in tissue engineering by examining their structural integrity, porosity, and interfacial interactions (Li et al. 2021).
2. **Transmission Electron Microscopy (TEM):** Compared to SEM, TEM has even greater resolution imaging capabilities, making it possible to see the interior structures and nanoscale characteristics of hydrogels made of cellulose. It sheds light on how cellulose nanofibrils and other nanostructures are arranged inside the hydrogel matrix.
   Application: Understanding the mechanical and functional characteristics of hydrogels requires the ability to characterize nanoscale interactions, such as polymer chain alignment, crystallinity, and cross-linking density, which are made possible by TEM investigation (Li et al. 2021).
3. **X-ray Diffraction (XRD):** The crystalline structure and orientation of cellulose molecules inside hydrogel materials can be examined using XRD. X-ray diffraction (XRD) measures the diffraction patterns of X-rays that interact with the sample to determine the phase composition and degree of crystallinity.
   Application: The mechanical characteristics and swelling behavior of hydrogels are influenced by the structural organization of cellulose chains, which is evaluated by XRD analysis. This includes the packing density, orientation, and crystallite size (Liu et al. 2019).
4. **Fourier Transform Infrared Spectroscopy (FTIR):** FTIR spectroscopy measures the absorption of infrared light at particular wavelengths in order to evaluate the chemical content and molecular structure of hydrogels generated

from cellulose. It offers details on the hydrogel matrix's functional groups, chemical bonds, and intermolecular interactions.
Application: Identification of cellulose derivatives, cross-linking agents, and other additives added to hydrogel formulations is made easier by FTIR analysis, which also helps characterize the hydrogel formulations' chemical makeup and suitability for biological tissues (Liu et al. 2019).

For the purpose of tissue engineering, structural analysis methods including SEM, TEM, XRD, and FTIR are essential for describing the structure, makeup, and characteristics of cellulose-derived hydrogels. These methods aid in the creation of hydrogel formulations that are optimized and have attributes that are specifically suited for particular biological applications by offering thorough insights into their structural characteristics.

14.6.2 Mechanical Properties

It is crucial to comprehend the mechanical characteristics of hydrogels generated from cellulose in order to evaluate their appropriateness for different tissue engineering uses. The mechanical behavior and performance of these hydrogels are assessed using a variety of characterization techniques, which offer information about their structural stability, reactivity to external forces, and integrity.

1. **Compression Testing:** Compression testing is the process of exerting compressive forces on hydrogel samples made of cellulose and observing how the samples deform. Through the evaluation of variables like compressive modulus, strength, and strain at failure, this method sheds light on the hydrogel's ability to support loads and withstand compression.
   Application: Compression testing is a useful tool for assessing the stability and structural integrity of hydrogel scaffolds, which assists with design decisions for load-bearing tissue engineering applications like bone or cartilage regeneration (Wang et al. 2020c).
2. **Tensile Testing:** In tensile testing, hydrogel specimens made of cellulose are subjected to axial stress till failure. This method provides information about the mechanical strength, flexibility, and resistance to stretching of the hydrogel by measuring parameters like tensile strength, modulus, and elongation at break.
   Application: Tensile testing determines whether hydrogel scaffolds are appropriate for uses like soft tissue engineering or wound dressings that call for tensile strength and flexibility (Wang et al. 2020c).
3. **Swelling Studies:** Studies on swelling measure how cellulose-derived hydrogels change in size and mass when they are hydrated or submerged in aqueous solutions. This method assesses variables like the swelling ratio, equilibrium swelling, and kinetics of water uptake, offering insights into the network structure and swelling behavior of the hydrogel.

Application: Studies on swelling aid in describing the hydrogel's capacity to take in and hold onto water, which affects the material's mechanical characteristics, porosity, and suitability for tissue engineering applications that call for the exchange of nutrients and hydration (Yang et al. 2019).

4. **Rheological Analysis:** The viscoelastic behavior of hydrogels generated from cellulose under shear or oscillatory deformation is evaluated using rheological analysis. This method provides information about the mechanical response of the hydrogel to flow and deformation by measuring parameters like viscosity, loss modulus, and storage modulus.
   Application: Hydrogel-based materials for tissue engineering and drug delivery applications are formulated and processed with the assistance of rheological analysis, which helps define the hydrogel's viscoelastic characteristics, gelation kinetics, and shear-thinning behavior (Yang et al. 2019).

Understanding the mechanical characteristics of hydrogels formed from cellulose through methods including tensile testing, compression testing, swelling research, and rheological analysis offers important information on the hydrogels' flexibility, structural integrity, and reaction to mechanical stimuli. Through comprehension of these characteristics, scientists and engineers can enhance hydrogel compositions for particular tissue engineering uses, guaranteeing their mechanical appropriateness and efficacy in living organisms.

14.6.3 Biodegradability Studies

Studies on biodegradability are crucial for assessing how hydrogels made from cellulose degrade and their effects on the environment. These investigations shed light on the hydrogel's biocompatibility and sustainability by revealing how it decomposes into innocuous chemicals in response to biological or environmental stimuli. In order to facilitate the creation of cellulose-based hydrogels for a range of biomedical and environmental applications, many approaches are utilized to evaluate the hydrogels' biodegradability:

1. **In Vitro Degradation Assays:** Assays for in vitro degradation mimic how hydrogels made of cellulose break down in carefully regulated lab settings. In these tests, hydrogel samples are submerged in enzymatic solutions or physiologically modeled fluids, and changes in mass, shape, and chemical composition are tracked over time.
   Application: In vitro degradation experiments assess the hydrogel's sensitivity to enzymatic, hydrolytic, or oxidative breakdown, offering insights into degradation kinetics, processes, and degradation products. The development of biodegradable hydrogel scaffolds for tissue engineering and medication delivery is influenced by these investigations (Sannino and Demitri 2021).
2. **Enzymatic Degradation Studies:** The goal of enzymatic degradation research is to assess how a hydrogel degrades when particular enzymes, including cellulases, proteases, or lipases, are present. These investigations evaluate the hydrogel's

vulnerability to breakdown into smaller fragments and polymer chain cleavage by enzymes.

Application: Studies on enzymatic degradation imitate biological breakdown processes and shed light on the hydrogel's biocompatibility, sensitivity to enzymes, and possible uses in tissue regeneration, wound healing, and controlled drug release (Sannino and Demitri 2021).

3. **Environmental Degradation Studies:** Studies on environmental degradation evaluate how a hydrogel degrades in real or artificial environments, including soil or water. These investigations assess the effects of variables like pH, temperature, and microbial activity on the rate and mechanism of hydrogel deterioration.

   Application: Environmental degradation studies provide insights into the eco-friendliness and sustainability of cellulose-derived hydrogels, guiding their application in environmental remediation, agriculture, or wastewater treatment (Wang et al. 2020b, d).

14.7 Challenges and Future Directions

14.7.1 Current Limitations in Cellulose-Based Hydrogel Technology

Despite the significant progress in sustainable biomaterials for tissue engineering, several challenges remain to be addressed. These include optimizing material properties, scalability of production processes, and regulatory considerations. Additionally, interdisciplinary collaboration between scientists, engineers, policymakers, and stakeholders is essential to drive innovation and accelerate the translation of sustainable biomaterials from the laboratory to clinical applications. Future research directions should focus on advancing sustainable biomaterials with enhanced functionality, bioactivity, and biocompatibility, paving the way for the next generation of green tissue engineering solutions (Smith et al. 2023a, b).

14.7.2 Opportunities for Further Research and Development

Opportunities for further research and development in sustainable biomaterials for tissue engineering abound (Brown et al. 2024). These include:

1. **Advanced Functionalization:** Exploring novel methods to functionalize biomaterials with bioactive molecules, growth factors, or nanomaterials to enhance their therapeutic efficacy and tissue regeneration potential.
2. **Biofabrication Techniques:** Advancing fabrication technologies such as 3D bioprinting, electrospinning, and self-assembly to create complex tissue constructs with precise control over the structure, mechanical properties, and functionality.

3. **Biomimetic Design:** Drawing inspiration from nature to design biomaterials that mimic the extracellular matrix's composition, architecture, and mechanical cues to better support cell adhesion, proliferation, and differentiation.
4. **Smart Materials:** Developing smart biomaterials capable of responding to external stimuli such as pH, temperature, or biochemical signals to enable controlled drug release, cell recruitment, or tissue regeneration in situ.
5. **In Vivo Evaluation:** Conducting comprehensive preclinical and clinical studies to evaluate the safety, efficacy, and long-term performance of sustainable biomaterials in relevant animal models and human patients.
6. **Regulatory Compliance:** Addressing regulatory challenges and standardization requirements to ensure the translation of sustainable biomaterials into approved clinical products and their widespread adoption in healthcare settings.
7. **Global Collaboration:** Encouraging global cooperation and information exchange between scientists, physicians, business associates, and authorities to quicken the creation and exploitation of sustainable biomaterials for the world's medical requirements (Brown et al. 2024).

Sustained funding for research and development in these domains possesses the capability to stimulate ingenuity, tackle unfulfilled clinical requirements, and transform the tissue engineering sector.

14.7.3 Incorporating Sustainable Design Ideas Into Tissue Engineering Procedures

For tissue engineering procedures to address environmental issues and guarantee the long-term viability of biomedical innovations, sustainability concepts must be incorporated (Brown et al. 2024). This integration can occur through several key strategies:

1. **Material Selection:** Selecting eco-friendly, renewable, and biodegradable materials for drug delivery systems and scaffolds used in tissue engineering, such as chitosan, alginate, or hydrogels made from cellulose. When compared to conventional polymers made from non-renewable resources, these materials provide sustainable substitutes.
2. **Green Synthesis Methods:** creating solvent-free, enzymatic, or green chemical technologies as environmentally benign synthesis approaches for biomaterials. During the material production process, these methods reduce energy usage, chemical waste, and environmental contamination.
3. **Recycling and Waste Reduction:** By putting techniques in place to recover or repurpose laboratory waste and biomaterials, tissue engineering research and development can have a smaller environmental impact. Sustainable waste management techniques help reduce pollution and conserve resources.
4. **Energy-Efficient Manufacturing:** Streamlining production procedures to cut down on energy use, greenhouse gas emissions, and dependency on fossil fuels. Utilizing energy-efficient machinery, renewable energy sources, and process

optimization strategies can reduce the environmental effect of tissue engineering production.

5. **Life Cycle Assessment (LCA):** Carrying out life cycle analyses to determine how tissue engineering products affect the environment from birth to death. LCA finds areas for improvement over the whole life cycle of a product, which helps with resource allocation and sustainable design choices.
6. **Regulatory Compliance:** Guaranteeing adherence to sustainability benchmarks, ethical standards, and environmental laws in tissue engineering research, development, and commercialization. Regulations encourage ethical behavior and reduce the possible negative effects that biomedical technologies may have on the environment and society (Jones et al. 2023). Figure 14.2 is the LCA illustration of the hydrogels.

Researchers, physicians, and industry stakeholders can encourage environmental stewardship, improve resource efficiency, and aid in the creation of sustainable healthcare solutions by incorporating sustainability principles into tissue engineering procedures.

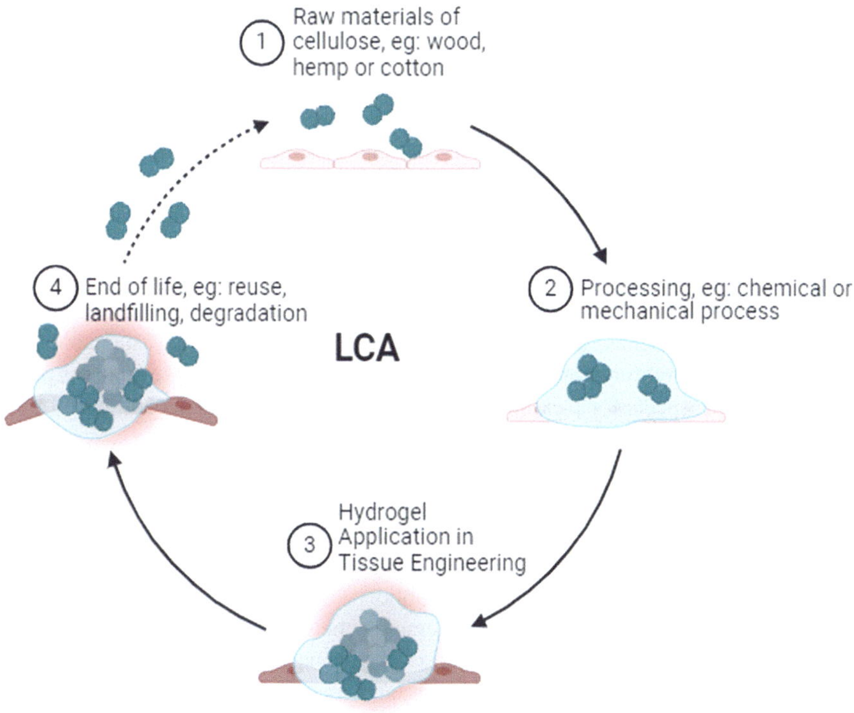

Fig. 14.2 Life cycle assessment of hydrogels in tissue engineering

14.8 Conclusion

14.8.1 Summary of Key Points

To sum up, cellulose-derived hydrogels, which take advantage of cellulose's special qualities to create sustainable and biocompatible materials, provide a flexible platform for tissue engineering and regenerative medicine. These hydrogels have great promise to address important issues in environmental sustainability and healthcare through a mix of customized synthesis techniques and functionalization tactics. The field of cellulose-derived hydrogels will develop toward revolutionary breakthroughs in biomedical engineering and other fields with continued research efforts focused on improving their characteristics and broadening their uses.

14.8.2 Implications for the Field of Tissue Engineering

Tissue engineering will be shaped by the incorporation of sustainability concepts, which will have a significant impact on environmental sustainability and global healthcare (Garcia et al. 2024). These implications include:

1. **Environmental Responsibility:** By minimizing resource depletion, pollution, and ecosystem damage associated with traditional materials and manufacturing techniques, the environmental footprint of biomedical research and development is reduced when sustainable approaches are included into tissue engineering.
2. **Healthcare Accessibility:** By offering accessible, environmentally acceptable methods for tissue regeneration, repair, and customized therapy, sustainable tissue engineering techniques hold the potential to democratize healthcare. Global access to and affordability of healthcare can be enhanced by these strategies by utilizing green technologies and renewable resources.
3. **Innovative Solutions:** Adopting sustainability encourages innovation in tissue engineering, which propels the creation of cutting-edge fabrication methods, treatment approaches, and biomaterials. Biomaterials that are sustainable and possess improved functionality, biocompatibility, and regenerative qualities present novel prospects for tackling unfulfilled clinical requirements and promoting biomedical research.
4. **Regulatory Compliance:** Sustainability enables tissue engineering procedures to be in line with changing legal requirements as well as public expectations. Adherence to environmental, social, and ethical norms bolsters tissue engineering technologies' legitimacy, dependability, and public acceptance among regulators, healthcare providers, and the general public.
5. **Interdisciplinary Collaboration:** Interaction between sustainability and tissue engineering promotes multidisciplinary cooperation between scientists, engineers, environmentalists, legislators, and medical practitioners. In order to address complex healthcare and environmental concerns, this collaborative

strategy promotes knowledge exchange, cross-disciplinary innovation, and holistic problem-solving.

6. **Global Impact:** The application of sustainable tissue engineering has the potential to significantly improve both environmental and human health on a global scale. Tissue engineers support environmental preservation, global quality of life enhancement, and sustainable development goals by emphasizing sustainability in research, development, and clinical translation.

To summarize, the incorporation of sustainability principles into tissue engineering holds significant consequences for the fields of healthcare, environmental management, innovation, and international cooperation. Tissue engineering can move closer to a future that is more robust, equitable, and sustainable for both people and the environment by embracing sustainability.

14.8.3 Recommendations for Future Work

Suggestions for further research and development in the domain of tissue engineering and sustainability principles cover a variety of topics (Smith et al. 2023a, b). These recommendations include:

1. **Material Innovation:** Extended investigation into the creation of biocompatible, mechanically strong, and slowly degrading sustainable biomaterials. We can increase the range of environmentally acceptable materials available for tissue engineering applications by investigating new natural polymers, bio-based composites, and bioactive additives.
2. **Bio-fabrication Technologies:** The development of manufacturing methods like electrospinning, self-assembly, and 3-dimension (3D) bioprinting to allow for the accurate creation of intricate tissue structures from sustainable biomaterials. Production of tissue engineering can be made more sustainable and scalable by using green manufacturing concepts, such as additive manufacturing techniques and bio-ink compositions.
3. **Regenerative Therapies:** Investigation of customized medicine techniques and regenerative therapy that take advantage of sustainable biomaterials' capacity for regeneration and tissue repair. Examining how biomaterials, cell treatments, and growth factors work in concert can hasten tissue repair and functional recovery in a variety of therapeutic contexts.
4. **Education and Outreach:** To foster a new generation of ecologically conscious researchers and practitioners, tissue engineering courses and professional development initiatives should incorporate sustainability principles. Raising people's knowledge of tissue engineering best practices, ethical issues, and environmental issues can enable them to make wise decisions and advance the industry.

By concentrating on these suggestions, practitioners and researchers can promote the incorporation of sustainability concepts into tissue engineering and help create novel, eco-friendly medical and regenerative medicine solutions.

References

Alberts B, Johnson A, Lewis J et al (2002) The plant cell wall. In: Molecular biology of the cell. Garland Science, New York

Ali F, Khan I, Chen J, Akhtar K, Bakhsh EM, Khan SB (2022) Emerging fabrication strategies of hydrogels and its applications. Gels 8:205. https://doi.org/10.3390/gels8040205

Anderson P, Patel N (2021) Promoting global health equity through sustainable biomaterials. Glob Health J 5:112–120

Boontheekul T, Kong HJ, Mooney DJ (2005) Controlling alginate gel degradation utilizing partial oxidation and bimodal molecular weight distribution. Biomaterials 26:2455–2465. https://doi.org/10.1016/j.biomaterials.2004.06.044

Brown T, Williams J, Davis H (2020) Biodegradability of cellulose and its role in sustainability. J Environ Sci Technol 44:567–578

Brown T, Williams J, Davis H (2022) Chronic health implications of non-biodegradable biomaterials. J Clin Med 11:1456

Brown D, White E, Garcia M (2024) Opportunities for further research and development in sustainable biomaterials for tissue engineering. J Tissue Eng Regen Med. https://doi.org/10.1002/term.3416

Chen Y, Li R, Zhang H (2020a) Sustainability in biomaterial development: renewable sources and green chemistry approaches. Aust J Biol Sci 31:1234–1247

Chen Y, Li R, Zhang H (2020b) Renewable biomass as a sustainable source for cellulose-derived materials. J Sustain Mater 28:567–578

Chen Y, Li R, Zhang H (2022) Recent advances in cellulose hydrogel research: a comprehensive review. Adv Mater 34:1234–1256

Chen M et al (2024) Carboxymethyl cellulose-bioactive glass composite hydrogels for bone tissue engineering. Biomaterials. https://doi.org/10.1016/j.biomaterials.2024.120418

Czaja W et al (2006) Production of microbial cellulose hydrogel by Acetobacter xylinum. Biotechnol Adv 34:567–578

Garcia M, Brown D, White E (2024) Implications for the field of tissue engineering. J Sustain Healthcare Eng. https://doi.org/10.1080/20475423.2023.2020568

Gautam SP, Bundela PS, Pandey AK, Jamaluddin, Awasthi MK, Sarsaiya S (2010) A review on systematic study of cellulose. J Appl Nat Sci 2:330–343

Hassan CM, Peppas NA (2000) Structure and applications of poly(vinyl alcohol) hydrogels produced by conventional crosslinking or by freezing/thawing methods. Adv Polym Sci 153: 37–65. https://doi.org/10.1007/3-540-46414-1_2

Hassan CM, Peppas NA (2019) Structure and applications of poly(vinyl alcohol) hydrogels produced by hydrogen bonding. Adv Polym Sci 153:37–65. https://doi.org/10.1007/3-540-46414-1_2

He Y et al (2021) Life cycle assessment of biodegradable cellulose hydrogels for agricultural applications. J Clean Prod. https://doi.org/10.1016/j.jclepro.2020.124547

Isobe N et al (2016) Regenerated cellulose-based hydrogels for biomedical applications. J Appl Polym Sci 42:672–684

Jackson R, White A, Green D (2019) Preparing for future challenges in biomaterial science. Future Mater 63:233–245

Jasni AH, Azmi AS, Puad NIM, Ali F, Nor YA (2023) Naturally derived biomaterials: advances and opportunities. In: Malviya R, Sundram S (eds) Engineered biomaterials. Engineering materials. Springer, Singapore

Jayakumar R et al (2019) Chitosan-based hydrogels for drug delivery and tissue engineering applications. Int J Biol Macromol. https://doi.org/10.1016/j.ijbiomac.2019.04.022

Jeong B, Bae YH, Kim SW (1997) Thermoreversible gelation of PEG–PLGA–PEG triblock copolymer aqueous solutions. Macromolecules 30:3500–3502. https://doi.org/10.1021/ma970144d

Jeong B et al (2020) Thermoreversible gelation of PEG–PLGA–PEG triblock copolymer aqueous solutions. Macromolecules 30:3500–3502. https://doi.org/10.1021/ma970144d

Johnson L, Patel R (2018) Sustainable alternatives in biomaterial development. Front Environ Sci 6:123

Johnson L, Patel R (2021) Cytotoxicity assessment of synthetic polymers versus cellulose-derived materials. J Biomed Mater Res A 42:672–684

Johnson L, Wang X (2020) Molecular architecture of cellulose and its role in hydrogel formation. Biomacromolecules 17:901–915

Jones A, Smith B, Davis C (2020) Cellulose-based hydrogels for biomedical applications: a review. J Biomater Sci Polym Ed 31:567–578

Jones S, Smith A, Johnson E (2023) Integration of sustainability principles into tissue engineering practices. J Sustain Biomed Eng. https://doi.org/10.1080/17568947.2022.2019643

Joseph B, Sagarika VK, Sabu C, Kalarikkal N, Thomas S (2020) Cellulose nanocomposites: fabrication and biomedical applications. J Bioresour Bioprod 5(4):223–237

Kabir SMF, Sikdar PP, Haque B, Bhuiyan MAR, Ali A, Islam MN (2018) Cellulose-based hydrogel materials: chemistry, properties and their prospective applications. Prog Biomater 7:153–174

Kim H, Lee K, Choi S (2021a) Scaling up production of renewable biomaterials. Biotechnol Adv 49:107751

Kim J et al (2021b) Economic viability of cellulose-based materials for large-scale production. J Ind Econ 28:789–802

Kolb HC, Finn MG, Sharpless KB (2001) Click chemistry: diverse chemical function from a few good reactions. Angew Chem Int Ed 40:2004–2021. https://doi.org/10.1002/1521-3773(20010601)40:11<2004::AID-ANIE2004>3.0.CO;2-X

Kolb HC et al (2019) Click chemistry in PEG-based hydrogels for biomedical applications. Angew Chem Int Ed 40:2004–2021. https://doi.org/10.1002/anie.2009039

Lee S, Kim J (2020) Nature-inspired design principles for sustainable biomaterials. Adv Mater Res 44:789–802

Lee S, Kim J, Park Y (2019a) Eco-friendly biomaterials for medical applications. Mater Sci Eng C 98:872–879

Lee S, Kim J, Park Y (2019b) Environmental impact of petroleum-based materials compared to cellulose-derived alternatives. J Environ Sci Technol 45:789–802

Lee S, Kim J, Park Y (2021a) Sustainable hydrogel synthesis using cellulose derivatives. J Sustain Mater 28:789–802

Lee S, Kim J, Park Y (2021b) Chemical modification of cellulose and its applications. J Chem Eng 73:1234–1247

Li R et al (2018) Cotton-derived cellulose for hydrogel formation. Mater Lett 36:345–356

Li Y et al (2021) Structural characterization of cellulose-based hydrogels for tissue engineering applications: a review. Carbohydr Polym. https://doi.org/10.1016/j.carbpol.2020.116504

Lin N, Huang J, Dufresne A (2012) Preparation, properties, and applications of polysaccharide nanocrystals in advanced functional materials. J Nanosci Nanotechnol 12:3003–3012. https://doi.org/10.1166/jnn.2012.5242

Liu F et al (2019) Characterization and applications of cellulose-based hydrogels: a review. Cellulose. https://doi.org/10.1007/s10570-019-02369-8

Loh QL, Choong C (2013) Three-dimensional scaffolds for tissue engineering applications: role of porosity and pore size. Tissue Eng Part B Rev 19:485–502. https://doi.org/10.1089/ten.teb.2012.0437

Lu Y et al (2019) Carboxymethyl cellulose-based hydrogels for biomedical applications. J Biomater Sci Polym Ed 31:567–578

Ma Z, He Y, Li L, Zhang Y (2014) Preparation and characterization of cellulose-based hydrogels with tunable mechanical properties. J Mater Chem B 2:5160–5168. https://doi.org/10.1039/C4TB00425A

Martin G, Richards E, Thomas K (2020a) Ethical considerations in biomaterial development. J Bioeth 14:89–97

Martin R et al (2020b) Environmental impact assessment of cellulose-derived materials. J Environ Manag 44:567–578

Melecio N (2024) Medtech. Center for MedTech Excellence, Atlanta

Pandey A et al (2012) Cellulose-based hydrogels for biomedical applications. Chem Rev 112:163–193. https://doi.org/10.1021/cr2004005

Peppas NA, Hilt JZ, Khademhosseini A, Langer R (2000) Hydrogels in biology and medicine: from molecular principles to bionanotechnology. Adv Mater 18:1345–1360. https://doi.org/10.1002/1521-4095(200011)18:18<1345::AID-ADMA1345>3.0.CO;2-8

Peppas NA et al (2020) Synthesis, properties, and biomedical applications of polyacrylamide hydrogels via free radical polymerization. Adv Mater 18:1345–1360. https://doi.org/10.1002/adma.200800702

Rameli N, Lim BY, Leong PY et al (2024) Chitosan-reinforced nanocrystalline cellulose hydrogels containing activated carbon as antitoxin wound dressing. Macromol Res. https://doi.org/10.1007/s13233-024-2083-2

Rinaudo M (2006) Chitin and chitosan: properties and applications. Prog Polym Sci 31:603–632. https://doi.org/10.1016/j.progpolymsci.2006.06.001

Rinaudo M (2021) Glutaraldehyde crosslinking of chitosan hydrogels: properties and applications. Prog Polym Sci 31:603–632. https://doi.org/10.1016/j.progpolymsci.2021.100611

Robinson D, Lee Y, Kumar S (2022a) Interdisciplinary approaches to sustainable tissue engineering. Front Bioeng Biotechnol 10:859123

Robinson A et al (2022b) Ecological consequences of synthetic polymers versus cellulose-derived alternatives. J Ecol Sci 17:901–915

Sannino A, Demitri C (2021) Biodegradable cellulose-based hydrogels: design and applications. Polymers. https://doi.org/10.3390/polym13060991

Slila H, Soufivand A, Faber J, Budday S, Lohwongwatana B, Boccaccini A (2023) Reinforcing Tissue-Engineered Cartilage: Nanofibrillated Cellulose Enhances Mechanical Properties of Alginate Dialdehyde–Gelatin Hydrogel. Adv Eng Mater 26. https://doi.org/10.1002/adem.202370013

Smith A, Jones B (2018a) Environmental impact of synthetic polymers in tissue engineering. Environ Sci Technol 52:3056–3063

Smith A, Jones B (2018b) Biocompatibility of cellulose-based materials for biomedical application. J Biomater Sci Polym Ed 31:567–578

Smith A, Jones B (2019a) Biocompatibility and biodegradability of cellulose-based hydrogels. Adv Healthc Mater 12:456–468

Smith A, Jones B (2019b) Hydrophilicity of cellulose and its applications in industry. J Appl Polym Sci 135:47890

Smith A et al (2020) Environmental impact of traditional biomaterials in tissue engineering. J Environ Sci Technol 45:567–578

Smith A, Johnson B, Garcia M (2023a) Recommendations for future work in sustainable tissue engineering. J Sustainable Biomed Eng. https://doi.org/10.1080/17568947.2022.2019645

Smith A, Jones B, Johnson C (2023b) Challenges and future directions in sustainable biomaterials for tissue engineering. J Biomater Sci Polym Ed. https://doi.org/10.1080/09205063.2022.2068417

Thompson P, Garcia R (2019) Regulatory landscape for sustainable medical materials. Regul Aff J 27:223–230

Wang X, Zhang H (2019) Applications of cellulose-based hydrogels in biomedical engineering. Biotechnol Adv 37:107452

Wang Y et al (2018) Mechanical and biological properties of cellulose nanofiber reinforced gelatin hydrogels for potential use in meniscus regeneration. J Biomed Mater Res A 106:309–318. https://doi.org/10.1002/jbm.a.36293

Wang X et al (2020a) Utilization of wood pulp-derived cellulose in hydrogel synthesis. Aust J Biol Sci 42:567–578
Wang Q et al (2020b) Microwave-assisted synthesis of cellulose-based hydrogels for tissue engineering applications. Mater Sci Eng C. https://doi.org/10.1016/j.msec.2020.110890
Wang J et al (2020c) Mechanical properties of cellulose-based hydrogels for biomedical engineering applications: a review. Carbohydr Polym. https://doi.org/10.1016/j.carbpol.2019.115641
Wang Q et al (2020d) Recent advances in biodegradable cellulose-based hydrogels for biomedical applications. Front Chem. https://doi.org/10.3389/fchem.2020.00081
Wang X, Liu Q, Gao P (2021a) Biocompatibility and safety of natural polymers in tissue engineering. Adv Healthc Mater 10:2001234
Wang X et al (2021b) Versatility of cellulose-derived materials for various applications. J Adv Mater 34:1234–1247
Wang X et al (2022) Eco-friendly synthesis of biomaterials. J Green Chem 14:1234–1247
Wang S, Wei Y, Wang Y, Cheng Y (2023) Cyclodextrin regulated natural polysaccharide hydrogels for biomedical applications—a review. Carbohydr Polym 313:120760
Yang J et al (2019) Mechanical properties of cellulose-based hydrogels: an updated review. Biopolymers. https://doi.org/10.1002/bip.23245
Yue K, Trujillo-de Santiago G, Alvarez MM, Tamayol A, Annabi N, Khademhosseini A (2015) Synthesis, properties, and biomedical applications of gelatin methacryloyl (GelMA) hydrogels. Biomaterials 73:254–271. https://doi.org/10.1016/j.biomaterials.2015.09.041
Zhang S et al (2017) Hydroxyethyl cellulose-based hydrogels for tissue engineering. Biomater Sci 23:345–356
Zhang H et al (2019) Bamboo as a source of cellulose for hydrogel synthesis. J Sustain Mater 28: 789–802

Ainil Hawa Jasni, a former graduate of Monash University (MU) and National Defence University of Malaysia (NDUM). She completed her PhD journey at Faculty of Engineering, International Islamic University Malaysia. She obtained her Bachelor of Science in Biotechnology from MU and Master of Science in Biology from NDUM. She had been working under the area of nanotechnology, material science, biotechnology in military applications including microbiology, biosensor, proteomic, and electrospinning since 2013.

Azlin Suhaida Azmi, a former graduate of the Widener University, Pennsylvania, USA started her career as a Chemist at First Malaysia Coating Sdn. Bhd. on April 1999. On September 1999, she received an offer from UniversitiTeknologi PETRONAS (UTP) and working as trainee lecturer. She continued her study in MSc in Process Integration at University of Manchester Institute of Science and Technology (UMIST), Manchester, UK in 2001 for 1 year. Early 2002 she was back to UTP and being appointed and served as Lecturer until October 2006. On November 2006, she moved to International Islamic University Malaysia (IIUM) as lecturer before she pursued her study at University of Malaya (UM), Malaysia on December 2007. She obtained her PhD in Bioprocess in Chemical Engineering on September 2012. Currently she is attached as an Associate Professor to the Department of Chemical Engineering & Sustainability, International Islamic University Malaysia, Kuala Lumpur.

Noor Illi Puad Mohamad Puad graduated with B. Eng (Biochemical-Biotechnology) (Honors) from International Islamic University Malaysia (IIUM) in 2007. She was then appointed as an Assistant Lecturer at the Calatrava Department of Biotechnology Engineering, IIUM in the same year. Later in 2011, she obtained her PhD in Chemical Engineering and Analytical Science from The University of Manchester, UK. Her research interest is mainly on Plant Cell Culture Technology, Flux Balance Analysis, Kinetic Modelling and Simulation, Plant Secondary Metabolite and Natural Products, Bioprocess and Renewable Energy. Presently, she is an Assistant Professor at the Department of Chemical Engineering & Sustainability, Faculty of Engineering, International Islamic University Malaysia (IIUM).

Fathilah Ali is an Assistant Professor at the Department of Biotechnology Engineering, Faculty of Engineering, International Islamic University Malaysia (IIUM), Gombak 50,728, Kuala Lumpur. She obtained her PhD in Chemical and Biomolecular Engineering from Korea Advanced Institute of Technology (KAIST) in 2013. Her research interest is in the area of Polymer Blend and Composites, Polymer Nanocomposites, Synthesis of polymers, block copolymers and polyurethane, Synthesis of Nanoparticles using natural sources, Polymers in Sensors, Polymers for packaging materials, Functional Polymers for lithography, Polymerization of biopolymer from the monomers produced from fermentation process, and Separation and Purification Techniques. She has nearly 2 years' experience in teaching on Process Plant and Design, Biopharmaceutical Engineering, Seminar for undergraduates and Fluid Mechanics. Dr. Fathilah also serves as a reviewer for polymer related journals. She published 11 papers in National and International conferences, 6 papers in SCOPUS and ISI journals and 2 book chapters. She was secretary for Nano Research Group and currently the Deputy Dean of Student Affairs (Jan 2015—present).

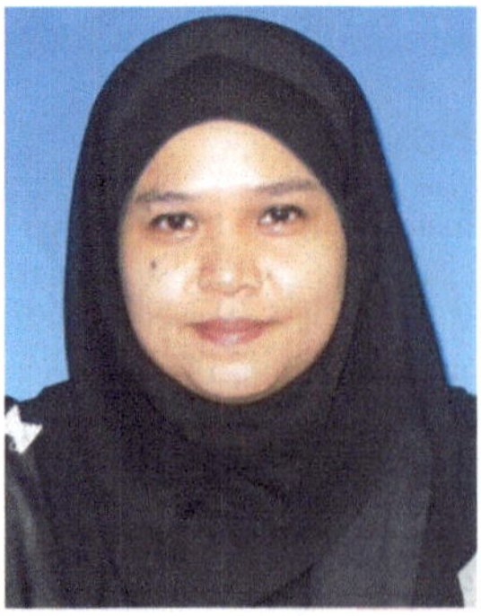

Yusilawati Ahmad Nor obtained her qualifications of Chemical and Biological Science (Nanotechnology) for her Doctor of Philosophy from University of Queensland and she acquired her Biotechnology Engineering Masters Degree from the International Islamic University Malaysia (IIUM). She is now working as an Assistant Professor at the Department of Chemical Engineering & Sustainability, Faculty of Engineering IIUM.

Hydroxyapatite-Starch-Based Sustainable Biomaterials

15

Christian Chapa

Abstract

In this chapter, key aspects of biomaterials composed of hydroxyapatite in combination with starch are discussed with emphasis on nanomedicine applications, especially drug delivery. Both biomaterials, hydroxyapatite and starch, present advantages considering sustainability aspects. In the introduction broadly discussing the potentiality and outstanding advantages of these biomaterials for such an application, the structure, properties and biomedical applications of hydroxyapatite as well as the structure and pharmaceutical preparations of starch are mentioned. In addition, a rationale for the combination of hydroxyapatite and starch for the creation of sustainable green biomaterials with enhanced functionalities will be highlighted. Also, separate hydroxyapatite-based, and starch-based drug delivery systems will be explored, and then we will highlight the properties that the biomaterial seeks to be a drug delivery system and that this combination could satisfy. Next, a review of hydroxyapatite-starch based drug delivery systems found in formulations reported in the literature is presented, followed by some strategies for drug incorporation into hydroxyapatite-starch composites. Finally, the main challenges faced by this system are addressed, e.g., choosing the routes of administration according to the idea that hydroxyapatite is unlikely to degrade under physiological conditions. Undoubtedly, fabrication techniques will improve by adopting environmentally sustainable methods, so that in the future we will have options for effective and affordable drug delivery to solve complex medical challenges.

C. Chapa (✉)
Grupo de investigación en Nanomedicina, Instituto de Ingeniería y Tecnología, Universidad Autónoma de Ciudad Juárez, Ciudad Juárez, Mexico
e-mail: christian.chapa@uacj.mx

R. Malviya, S. Sundram (eds.), *Sustainable Green Biomaterials As Drug Delivery Systems*, Biomaterials, Bioengineering and Sustainability 1,
https://doi.org/10.1007/978-3-031-79062-1_15

Keywords

Biomaterials · Sustainability · Drug delivery · Biodegradability · Nanomedicine · Hydroxyapatite · Electrospinning · Controlled release · Tissue engineering

Abbreviations

BMPs	bone morphogenic proteins
ClAlPc	chloroaluminum phthalocyanine
ECM	extracellular matrix
HA	hydroxyapatite
NAD+	nicotinamide adenine dinucleotide
NHAp	nanohydroxyapatite
NMN	nicotinamide mononucleotide
PCL	poly ε-caprolactone
PLA	polylactic acid
SBF	simulated body fluid
TIPS	thermally induced phase separation technique
β-CD	cyclodextrin
β-TCP	β-tricalcium phosphate

15.1 Introduction

To discuss about hydroxyapatite compounds combined with starch is to talk about innovative materials that represent a sustainable approach to various areas, including nanomedicine, particularly drug delivery. It is undeniable that biomaterials in nanomedicine have become the main avenues for researching and developing advanced drug delivery systems. These nanomedicine systems seek to improve therapeutic efficacy or minimize the side effects of current treatments, something that the state of the art of medicine has not achieved. Quite simply, hydroxyapatite and starch-based compounds are two of the materials that have attracted the most attention. On the other hand, hydroxyapatite, which is a natural mineral form of calcium apatite, is mainly found in tissue engineering applications, because its excellent biocompatibility and osteoconductive properties have been demonstrated. However, many researchers have worked to demonstrate its potential beyond bone regeneration, consequently the material has been explored, like many others, in drug delivery systems. It is true that the materials that have been explored for this application are too many, many and all of them have their advantages and disadvantages, but the ability of hydroxyapatite to adsorb and release various therapeutic agents is remarkable, there are not many biomaterials that have this ability. On the other hand, starch, a glucose homopolymer, which is naturally found in bread, rice, pasta, cereals, potatoes, peas, corn and is also found as a binder, disintegrator and excipient in pharmaceutical formulations, has numerous properties that have been amply demonstrated such as biocompatibility, biodegradability and ease of modification, qualities that are extremely useful for nanomedicine. Therefore, when hydroxyapatite and starch are combined, they would form a composite material that takes advantage of the strengths of both components, and we would be in the presence of a biomaterial for sustainable drug delivery systems. We say sustainable for reasons that are addressed later in this chapter.

Unquestionably, the field of drug delivery is interdisciplinary and multidisciplinary because it encompasses principles of chemistry, biology, materials science and engineering to obtain and study systems capable of releasing therapeutic agents, bioactive compounds, etc., in a precise and controlled manner. Moreover, sustainability is currently in vogue in scientific research on this topic, which is why we see unique advantages in hydroxyapatite compounds in combination with starch. It should be noted that the structure of hydroxyapatite would provide a robust matrix for the encapsulation of drugs, and its large area-to-surface ratio would theoretically improve the carrying capacity of drugs while its biocompatibility is a guarantee that the body does not reject it due to immunological reactions. At the same time, starch would be contributing to being degraded by the action of body fluids or by specialized enzymes in their hydrolysis that would contribute to the adaptation or control of the kinetics of release of the drugs carried.

In this book we discuss sustainable green biomaterials as drug delivery systems, the reader will not find a combination that promises to be more effective but also more sustainable, than the combination of hydroxyapatite with starch. Keep in mind that starch is mostly derived from plants, it is renewable and biodegradable. While it has been shown that hydroxyapatite can be synthesized from various renewable resources, such as eggshells and animal bones, which are often considered waste products and which we will recover later in this chapter. Renewable sources are the best weapon to ensure that these materials provide a sustainable way to produce the biomaterial, and even add value to materials that would otherwise contribute to environmental pollution. Sustainability is based on the principle of ensuring the needs of the present without compromising the needs of future generations, always without sacrificing environmental protection, economic growth and social development, therefore, this combination is aligned with the emphasis of green chemistry and sustainable development in biomedical research. Both materials and their combination will have biocompatibility, biodegradability and efficacy properties in the encapsulation and delivery of drugs, all together it is an undeniable potential to advance advanced and environmentally responsible drug delivery systems.

15.2 Hydroxyapatite

It is well documented in the scientific literature that hydroxyapatite (HA) is a compound that mimics the structure and chemical composition of natural bone. What's more, probably the vast majority, if not all scientific publications, highlight this quality. Indeed, the stoichiometry of calcium and phosphorus in this biomaterial offers a rigid crystalline structure in the shape of a double pyramid with six sides (hexagonal bipyramidal) whose formula is represented as $Ca_{10}(PO_4)_6(OH)_2$ or $Ca_5(PO_4)_3OH$. The first formula option indicates the presence of ten calcium ions (Ca^{2+}) in each unit cell, while the second is the reduction to the minimum expression but always keeps a precise ratio of calcium and phosphorus atoms (Ca/P) within the crystalline unit cell, which in this case is 1.67 (10:6). If this ratio deviates from the ideal value, the crystal structure becomes less stable and dissolves more easily. Interestingly, hydroxyapatite with a slight calcium deficiency (Ca/P ratio of 1.60) shows slightly better bioactivity compared to the perfect ratio material (Ca/P 1.67). The basic composition of the mineral component of bone is hydroxyapatite whose formula $Ca_{10}(PO_4)_6(OH)_2$. In addition to containing calcium and phosphate, it also contains minimal proportions of sodium, chlorine, carbonates, and magnesium, which act in the remodeling of the bone. Hydroxyapatite is part of the apatite family, has a hexagonal structure, with a space group P63/m and has a Ca:P ratio of 1.67. The crystal structure of the material has a six-fold symmetry along a specific axis (c-axis) and three identical axes (a-axes) arranged perpendicularly at 120 degrees to each other. Unit cell dimensions for this crystal structure are reported to be 0.9422 nm for both a and b axes, and 0.688 nm for the c-axis. There are several common methods used to create this material, including wet chemical precipitation, sol-gel, hydrothermal, and microwave irradiation.

15.2.1 Biomedical Applications of Hydroxyapatite

Bone tissue is a specialized connective tissue that, unlike other tissues, is mineralized. This mineralization gives the bone the ability to provide support and protection. The extracellular matrix of the tissue is composed of the organic matrix and the inorganic matrix. The organic matrix represents 90% of type I collagen and to a lesser extent type IV collagen. The other 10% is represented by the fundamental substance, which are other non-collagen proteins such as: proteoglycan macromolecules such as hyaluronan, chondroitin sulfate and keratan sulfate; multi-adhesive glycoproteins such as osteonectin, sialoproteins; vitamin K-dependent proteins such as osteocalcin, and finally, growth factors and cytokines, the most important being bone morphogenic proteins (BMPs).

There are four types of cells that can be found in bone. Osteoprogenitor cells are cells derived from mesenchymal stem cells that give rise to osteoblasts. Osteoblasts are derived from mesenchymal stem cells whose function is to secrete the extracellular matrix that, when surrounded by it, is called osteocyte and stops secreting matrix, becoming one of those responsible for the viability of bone tissue. Osteoclasts are bone resorption cells present on bony surfaces where bone is being removed or remodeled (rearranged) or where bone has been injured. One tissue is classified into two distinct structural organizations: compact and spongy. Compact bone tissue is a dense, compact layer, which forms the outer surface of the bone. This comprises 80% of the bone mass and is covered by the periosteum and lined up by the endosteum. Cancellous bone tissue comprises 20% of the bone mass and forms the inner part of the bone with thin web-like tissue spicules and is lined only by the endosteum. Compact and spongy bone tissue, in turn, presents a structural organization composed of osteons on a micrometer scale that have a central duct called Havers' duct that contains vessels and nerves where the extensions of the osteocytes arrive to carry out an exchange of substances. Each osteon is made up of concentric lamellae and these in turn are made up of collagen fibers on a nanometer scale. Each collagen fiber is in turn made up of collagen fibrils where specific attachment points will be made that will serve as nucleation sites for the hydroxyapatite crystals.

Bioactivity pertains to the capability of bone scaffolds to establish a direct bond with the surrounding bone, avoiding the formation of fibrous tissue. The evaluation of bioactivity commonly involves testing the biomaterial's potential to form apatite in a simulated body fluid (SBF), which has ion concentrations similar to human blood plasma. A material is considered bioactive if it enhances apatite crystallization in a hydroxyapatite-supersaturated solution. However, the reliability of using in vitro SBF tests to predict a material's bioactivity *in vivo* has been debated, as Bohner and Lemaitre showed that "*in vitro* bioactivity" tests in SBF solutions cannot be used to predict the in vivo bone bonding ability of a material.

Scientists can produce HA using various methods, including chemical precipitation and electrospinning. This unique material holds promise for several biomedical applications due to its biocompatible and bioactive properties. It can function as an implant material to promote bone regeneration, a carrier for delivering drugs, and even a system for gene delivery. When processed into tiny particles, hydroxyapatite can be used to create materials for bone tissue engineering. These particles degrade slowly within the body (biodegradable) and have excellent properties for encouraging bone growth (osteoconductive and osteoinductive). HAnanoparticles modified with a special compound (poly(sodium 4-styrene sulfonate)) can act as carriers for antibiotics like vancomycin. This allows for controlled release of the medication after implantation of a bone scaffold. Another environmental contribution of hydroxyapatite microspheres is that they have a high capacity to absorb heavy metals thanks to their porosity. For this reason, several researchers use them to study and apply them in waters contaminated with these harmful elements and other pollutants.

15.2.2 Hydroxyapatite Composites and Their Application in Drug Delivery

Its biocompatibility and ability to induce bone formation has already been more than demonstrated. However, hydroxyapatite has been emerging as a strategic component in some drug delivery systems and we are seeing it alone or in combination with polymers. Recent research promotes various methodologies where hydroxyapatite is used and that is related to the controlled administration of drugs. Researchers take advantage of its properties to propose solutions to improve the efficacy and reduce the side effects of some pharmacological treatments.

Where we see a very significant advance is the use of particle formulations for dental disinfectants. Recently, the manufacture of calcium phosphate particles with immobilized ciprofloxacin was carried out by a coprecipitation process. In this process, the aging time significantly influences the physicochemical properties of the resulting particles. Interestingly, particles with higher ciprofloxacin content showed a slower release of the drug in physiological saline solutions in a less soluble crystalline matrix of hydroxyapatite, compared to amorphous particles. Both types of particles demonstrated antibacterial and acid-neutralizing activities against important oral bacteria such as *Streptococcus mutans* and *Porphyromonas gingivalis* (Pal et al. 2024). In the same sense, hydroxyapatite microspheres exhibit defects in mechanical properties, such as low resistance and brittleness, but have a large specific surface area and good injection properties. In addition, these microspheres can carry proteins and bioactive factors, as well as drugs to facilitate bone or dental regeneration. Consequently, it is foreseeable that the preparation of microspheres composed in this way will be suitable for such clinical applications (Wang et al. 2024).

Hydroxyapatite is prepared in other presentations, for example, in another study hydroxyapatite nanorods were prepared for the release of bumetanide and meloxicam, these active ingredients are poorly soluble in water. The method used on this occasion was known as hydrothermal synthesis. Compared to compounds alone, hydroxyapatite helped improve drug dissolution, solubility, and wettability rates. Precisely this hybrid system is a good example of how hydoxoapatite improves desirable properties in drug delivery (Friuli et al. 2024). In addition, hydroxyapatite was shown to be effective in improving the bioavailability and tissue distribution of nicotinamide mononucleotide (NMN) that was incorporated into the delivery system. In the same way as an example described above, here the chemical precipitation method was used to obtain nanoparticles and NMN was incorporated by physical adsorption. Again, the hydroxyapatite system showed significantly slower release compared to free NMN. This time, hydroxapatite extended circulation time, improved bioavailability, and raised plasma levels of NMN, and NAD+, with specific distribution in tissues such as the brain and liver in *in vivo* studies (Zhang et al. 2023).

Another compound that has been involved in demonstrating that hydroxyapatite contributes in the field of nanomedicine is chloroaluminum phthalocyanine (ClAlPc). On this occasion, the method for synthesizing hydroxyapatite microspheres was using vaterite as a template and the drug was encapsulated by mixing the components. Notably, the encapsulation efficiency was confirmed with photophysical and photobiological studies. The results showed a significant reduction in cell viability of gliosarcoma cells treated with photodynamic therapy (Ambrosio et al. 2023). Even bioactive compounds such as allicin, derived from garlic, which has shown beneficial effects on dyslipidemia, obesity, endothelial dysfunction, hypertension, heart attack, heart disease and even on cardiac arrhythmia, can also be loaded on hydroxyapatite. In one study, the biomaterial was shown to contribute to the improved antibacterial properties and controlled release of allicin (Bose et al. 2022).

As can be seen, hydroxyapatite, in its various forms, is being consolidated as a component in the formulation of nanomedicine systems for drug delivery. Its biocompatibility and ability to improve the bioavailability of medicines, along with its applicability in bone and tooth regeneration, make it a versatile and effective option in modern medicine. Recent studies highlight its potential in creating sustained-release systems, offering new prospects for the development of more effective and targeted treatments. When combined with polymeric materials, such as starch, it gives rise to potential applications ranging from tissue engineering, photodynamic therapy, to enhanced drug delivery (Mahdavinia et al. 2019; Ahamed et al. 2022).

15.3 Starch

15.3.1 Sources, Structure and Properties of Starch

Plants store energy in the form of complex carbohydrates, particularly starch. These starch molecules are very attractive to water because they have numerous hydroxyl groups that can form hydrogen bonds with water molecules. Starch is especially abundant in underground stems (tubers) like potatoes and in seeds (Mweta et al. 2010; Hong et al. 2011). Starch itself is a combination of two different types of glucose polymers: amylose and amylopectin. Amylose is a long, unbranched chain of sugar molecules (glucose) linked together in a specific way. These chains can vary greatly in size. Amylopectin, on the other hand, is much more complex. It's also a large molecule, but unlike amylose, it's highly branched. The branches occur at regular intervals along the main chain. The macromolecules that make up starch grains are arranged in layers. Amylose and amylopectin are two distinct polymer structures that make up starches. Most starches contain 80% amylopectin and 20% amylose. Amylose molecules, found in the inner layers, are helically wound chains, made up of between 200 and 20,000 moles of glucose linked by glycosidic bonds 1–4. In the outer layer is amylopectin, which is structured differently from amylose. The amylopectin molecule contains diglycan bonds 1–4 and 1–6. The glucose molecules in the amylopectin skeleton are connected to each other by glycosidic bonds. Backbone branching is common and arises from the coupling of 1–6-glycosides with additional glucose molecules. In branches, endpoints account for 4–5% of total links.

15.3.2 Biomedical Applications of Starch

The pharmaceutical preparations use starch as an excipient in a variety of forms, including as a disintegrant, binder, absorbent, viscosifier, diluent, and slipping agent. It is commonly used as a thinner in face powders, as an absorbent in talcum powders, and as an emulsion stabilizer in the cosmetic industry. Starch is also being studied as a food packaging material (García-Guzmán et al. 2022) and derivatives (Azfaralariff et al. 2020; Chen et al. 2020; Sathyan and Nisha 2022). The search for new sources of this carbohydrate that allow it to achieve different functional qualities is inspired by the varied properties of starch. Starch has become a starter material for the creation of excipients with new capabilities due to its biodegradability, biocompatibility, and bioavailability.

The use of starches as carriers of active molecules has been studied to find alternative approaches to the problems of minimizing the side effects of certain active molecules, protecting them from harmful environmental factors, transporting them to the site of action and doing so in a regulated manner, as well as meeting the need to use biodegradable and biocompatible materials as excipients. Starches have been used in the development of encapsulation techniques for this purpose.

15.4 Rationale for Combining HA and Starch: Creation of Sustainable Green Biomaterials with Enhanced Functionalities

HA in combination with starch has emerged with the aim of creating sustainable green biomaterials with enhanced functionalities that fulfill the functions of bone both at the cellular and structural level for bone replacement or repair (Miculescu et al. 2017; Koski et al. 2018; Ordon et al. 2019; Mohd Roslan et al. 2021; You et al. 2022; Le et al. 2023; Murugan and Akshata 2023), and for other important biomedical applications such as drug delivery. Nanomedicine, as a multi- and interdisciplinary research area, has made it necessary to search for materials and techniques that meet the different requirements for the creation of a sustainable biomaterial, so the following points need to be considered:

(a) Biocompatibility. Materials that do not trigger an immune or cytotoxic response are needed.
(b) Porosity. A porous structure promotes tissue growth and nutrient exchange. Pore size of 50–100 microns is essential for the growth of bone tissue.
(c) Osteoinductivity. The support must be able to allow the process of recruitment and differentiation of stem and osteoprogenitor cells.
(d) Biodegradability. Materials that have resorption rates that go hand in hand with bone formation should be used.
(e) Mechanical properties. The support must offer adequate mechanical strength to maintain the integrity of the support, and it has been shown that trying to mimic the mechanical properties offers better differentiation of osteogenic cells.
(f) Surface properties. There are two that are fundamental for cells: chemical and morphological. Having similar morphological properties will allow the migration process of osteogenic cells to the new matrix. The chemical properties will allow the cells to adhere to the matrix.

All these demanding requirements for a sustainable biomaterial that is used in applications as sophisticated as drug delivery are met by designing and obtaining biomaterials that combine hydroxyapatite with starch.

HA-starch composites exhibit excellent biocompatibility, making them suitable for biomedical applications. HA, being similar to the mineral component of bone, is inherently biocompatible and supports bone cell attachment and proliferation. Starch, being a natural polysaccharide, is also biocompatible and biodegradable. The combination of these materials creates a composite that integrates well with human tissues, promoting cellular activities essential for bone healing and regeneration. Moreover, the degradation products of starch are non-toxic and can be metabolized by the body, further enhancing the composite's compatibility.

15.5 Synthesis Techniques for HA-Starch Composites

Sustainable biomaterial manufacturing techniques constitute a significant factor in the biomedical function. The morphology of the material is essential for the release of drugs because each tissue has different characteristics: porosity, degradation, structure, mechanical properties, etc. There is a wide area of techniques used for the creation of biomaterials, therefore it is necessary to pay attention to the characteristics of the biomaterial and understand the interaction that you want it to acquire, in order to select the appropriate technique. In the field of nanomedicine, the methods more common used for the development of biomaterials that combine HA with starch are electrospinning, sol-gel, reaction-diffusion systems, solvent casting (Beh et al. 2021a), thermally induced phase separation (Beh et al. 2021b), 3D printing (Koski and Bose 2019), among others (Hosseinzadeh and Ramin 2018).

15.5.1 Electrospinning

Electrospinning has proven to be an effective technique in the production of supports as the diameter of their fibers can be controlled, as well as their spatial orientation. It is generally defined as the process by which fibers are made using an electric field acting on a polymeric solution. Electrospinning begins with a drop of polymer at the tip of a needle, which is stimulated by an electric field until it forms a surface tension in the droplet that begins to give way, resulting in the deformation of the droplet until it acquires a conical shape called the Taylor cone. This stage is critical since the final morphology of the fibers depends on the correct formation of this cone. Once the cone is formed, the surface tension exerted by the force of the electric field on the polymer causes a deformation on the surface of the polymer, generating fibers. These fibers do not follow a straight path but undergo a series of electrodynamic instabilities, resulting in a trajectory in concentric circles until they reach the collector.

The fibers obtained through the Taylor cone are captured on different types of collectors, depending on the spatial arrangement of the nanofibers in the support.

Fixed collectors are used to obtain random or undirected nanofibers, while rotating collectors are used for the generation of directed nanofibers. The matrix of the support created by electrospun nanofibers not only provides a defined area but also offers a spatial arrangement that is extremely important. It has been discovered that supports with directed nanofibers are more similar to a natural 3D extracellular matrix (ECM) than supports with undirected nanofibers, thereby enhancing the development and proliferation of cells in these materials. This is a crucial characteristic for the development of a support. However, biocompatible polymeric materials with both natural and synthetic materials are being investigated for applications in tissue engineering, drug delivery and other biomedical applications (Villarreal-Gómez et al. 2016; Khalf and Madihally 2017; Naveenkumar et al. 2023; Tören et al. 2024).

Electrospinning can be utilized to create composite fibers that combine the biocompatibility and bioactivity of hydroxyapatite with the biodegradable and renewable properties of starch. The incorporation of hydroxyapatite into electrospun starch fibers can enhance the mechanical properties and biological functionality of the resulting biomaterial, making it suitable for applications in bone tissue engineering. Characteristically, this electrospinning technique allows the creation of fibers, although it is known that the ability to control the diameter and orientation of the fiber requires the design of experiments to test different variables ranging from the voltage, the gauge of the syringe, to the viscosity of the polymer solution. But once controlled, materials are obtained that even closely mimic the structure and function of natural tissues, particularly in terms of cell adhesion, proliferation and differentiation.

15.5.2 Sol-Gel

If we could address a method that would be versatile and effective for creating hydroxyapatite and starch-based biomaterials with potential applications in tissue engineering and regenerative medicine, we would be talking about the sol gel. The method classically occurs in the following stages: hydrolysis of precursor (sol formation), i.e. the hydrolysis of metal alkoxides to produce hydroxyl groups in the presence of stoichiometric amounts of water and acid or basic catalyst, then comes the polycondensation (gelation), here occurs the polycondensation of the resulting hydroxyl groups and residual alkoxy groups to form three-dimensional networks, then aging, drying and finally, the calcination which is performed at increased temperatures varying within the range of 400–800 °C. During the calcination stage the dry gel structure is dehydrated. Volatile organics are also removed in this stage. Although some authors claim to obtain polymer gels using the sol-gel method, sometimes they do not comply with the classically involved stages, especially the calcination, therefore these methods could not be called sol-gel.

However, the method is traditionally associated with obtaining ceramic oxides, but researchers have adapted it to obtain calcium phosphates as well, including hydroxyapatite. Thus, it is said that hydroxyapatite can be prepared using the sol-gel technique and that this method is able to improve the chemical homogeneity of the resulting hydroxyapatite, especially when it is used as a coating. The method is simple and more accessible than the plasma spray method, currently used for biomedical applications. Among its advantages, it has the ability to precisely adjust the properties of the material but with an experimental design that allows control of the variables, so we could be facing the development of new generation biomaterials. Undoubtedly, the sol-gel method is a widely used process, especially to produce ceramics that are then used as coatings or catalyst supports that provide stability and regulated physicochemical properties. It could be said that the method was developed as an alternative for the preparation of glass and ceramics at low temperatures. The final properties of the material obtained by this method depend on conditions such as temperature and pressure during the drying process. Critical conditions can yield an aerogel, while ambient conditions typically produce a xerogel. The sol-gel process offers several advantages for the synthesis of hydroxyapatite-starch-based biomaterials:

(a) **Homogeneity:** Ensures uniform distribution of hydroxyapatite within the starch matrix.
(b) **Control of Composition:** Allows precise control over the ratio of hydroxyapatite to starch.
(c) **Low-Temperature Process:** Suitable for incorporating temperature-sensitive biological molecules.
(d) **Reduction of Impurities:** The use of pure precursors minimizes contamination.
(e) **Encapsulation Capability:** Enables the incorporation of elements such as enzymes or growth factors.
(f) **Cost-Effective Equipment:** The process does not require expensive apparatus.
(g) **Optical Properties:** Can be advantageous for developing optical and biological sensors.

15.5.3 Thermally Induced Phase Separation

The thermally induced phase separation technique (TIPS) consists of reducing the temperature of a polymeric solution to induce a phase separation, what happens is that they are separated into two phases, one with a high concentration of polymer and the other with a low concentration. When the solvent is separated, it is removed, leaving a space that becomes the pores of the material. Thermally induced phase separation can be said to have two stages: phase separation and solvent removal.

Solid-liquid phase separation occurs when low temperature leads to solvent crystallization inducing phase separation of the polymer solution. It is important to say that the morphology that will be obtained by this separation will change with the type of solvent, the crystallization temperature, and the polymer concentrations. For the removal of the solvent, the freeze-drying process is used. Freeze-drying is considered a technique in which a material is frozen and then dehydrated by sublimation without losing the shape of the material. This entire process is carried out through the freeze-dyer machine, which has the corresponding sections to carry out the process.

The stages of the process are:

(a) Freezing. The material is put in low temperature conditions.
(b) Vacuum. Once frozen, the material is placed in a section where it will be emptied. This allows the solvent to evaporate without passing through the liquid phase.

15.6 Starch-Based Drug Delivery Systems

Interest in biopolymers as agents in drug delivery has grown considerably in recent years. We can list materials such as polylactic acid (PLA), chitosan, cellulose, alginate and of course starch. All these biopolymers have properties that make them ideal for biomedical applications. Focusing on starch, this glucose polymer found in the diet of virtually all humans is, obviously, biocompatible, but also biodegradable, which gives it the stamp of sustainability. In the laboratory, it is used to form gelled matrices capable of encapsulating drugs (Lukova et al. 2023). Consider another example, polylactic acid is known for its biomedical applications. It also provides us with the ability to form nanoparticles that can release drugs in a controlled manner (Tudorachi et al. 2017; Pariy et al. 2022; Taib et al. 2022; Diaz Varela et al. 2024). Certainly, chitosan, which is a chitin derivative, is also biocompatible and biodegradable, and has been widely documented and demonstrated to have antimicrobial properties and to improve mucosal permeability, facilitating drug release (Alvarez-Barreto et al. 2017; Marouf et al. 2020; Kedir et al. 2022). Cellulose and its derivatives, indeed hydroxyapatite-cellulose, is an example in drug delivery due to its ability to form films and membranes with adequate mechanical and release properties (Morán et al. 2013). Biopolymers also include alginate, a polysaccharide derived from algae that forms gels in the presence of calcium ions, allowing drugs to be encapsulated and released in a controlled manner (Huq et al. 2017; Song et al. 2018; Rosch et al. 2019; Thomas et al. 2021; Abka-khajouei et al. 2022; Shaikh et al. 2022). In view of the above, there is no doubt that biopolymers give us a great diversity of options for use in the formulation of these drug delivery systems and, in fact, their study encompasses various active substances, routes of administration and methods of preparation. (Kamaly et al. 2016; Barclay et al. 2019; Alvarez-Barreto et al. 2021; Valdez et al. 2022; Gaona et al. 2023).

Emphasis should be placed on starch-based nanomedicine systems, as they offer properties that facilitate topical drug delivery by promoting controlled release (Mao et al. 2004; Rydell et al. 2005; Mundargi et al. 2008; Kim et al. 2015; Okunlola and Ghomorai 2018; Yasar et al. 2018; de Oliveira Cardoso et al. 2020; Marto et al. 2020; Ali et al. 2021; Lemos et al. 2021; Obireddy and Lai 2021; Ren et al. 2021; Choi et al. 2022; Ab'lah et al. 2023). The properties that make starch an excellent candidate for the development of nanoparticulate drug carriers in the area of nanomedicine lies in its natural polymeric nature of glucose monomers. The reader can find recent exhaustive reviews based on natural polymers being employed in drug delivery systems elsewhere (Fazal et al. 2023), but without a doubt starch-based products and their derived compounds are remarkable. For example, hydroxyethyl starch is combined with curcumin (Acevedo-Guevara et al. 2018), with Fe_3O_4 for a magnetic carrier (Castrejón-Parga et al. 2015), and with alginate to deliver medications such as metoprolol tartrate, tolbutamide, ibuprofen, and metformin HCl. Sustainability, as mentioned above, is due to the fact that starch sources are merely natural.

Previously, we have talked about hydroxyapatite without emphasizing the most recommended form for the application we have been discussing, nanohydroxyapatite (nHAp). It is well known that nanoparticles have distinct and outstanding properties compared to their bulk counterparts, well, the same is true for nHAp and it is becoming a material of great interest in drug delivery systems due to its distinctive properties (Ghorbani et al. 2019; Gui et al. 2022; Munir et al. 2022; Asghar et al. 2023) such as high surface area, porosity and inherent biocompatibility as a calcium phosphate that mimics the mineral composition of human bone (Pandey et al. 2018; Santillana-Marín et al. 2018; Singh et al. 2020). Of particular note, porosity facilitates drug loading and release by allowing a high adsorption capacity of molecules, consequently, what is being studied are the conditions to achieve sustained and controlled drug release. nHAp can successfully load and release various therapeutic agents in a controlled manner while at the same time it can be also functionalized with various chemical groups to enhance its potential affinity for certain drugs without compromising biocompatibility so that it does not cause adverse responses in the body. In addition, several studies have demonstrated their capability to enhance the antibacterial and antioxidant properties of some compounds when nHAp is the carrier, further underscoring their multifunctionality in biomedical applications (Ragab et al. 2014; Pandey et al. 2018; Fatimah et al. 2021; González-Torres et al. 2021; Asghar et al. 2023). For example, zinc-doped hydroxyapatite nanoparticles loaded with ciprofloxacin have shown increased drug loading and a tunable controlled release profile significantly improving antibacterial and antioxidant activities (Devanand Venkatasubbu et al. 2011).

Without neglecting the extensively reported properties such as efficacy and biodistribution, the property of composite materials for drug delivery that should be gaining more relevance is the sustainability of the manufacturing processes. Researchers currently developing such nanomedicine systems should take a growing interest in embracing green methods for the synthesis of these materials by looking to minimize the use of toxic solvents, reduce energy consumption and employ renewable raw materials. Fortunately, the green synthesis of nHAp, for example, can be carried out using plant extracts (Averous and Halley 2014; Anaya-Barajas et al. 2019; Sriprapha et al. 2024). Similarly, modification of biopolymers such as starch by environmentally friendly techniques can result in materials with properties suitable for selective, controlled, prolonged or targeted drug delivery (Morán et al. 2013; Alvarez-Barreto et al. 2021). This book is about sustainable biomaterials for drug delivery. Therefore, it should be emphasized that hydroxyapatite can be derived from natural sources such as eggshells, animal bones and seashells, which is where sustainability lies. While these are not chemical synthesis methods, they can be treated by chemical methods such as grinding to fine particles. You can also choose to adapt traditional methods to move towards greener chemistry. In this same sense, we can highlight that starch is a biodegradable polymer that is obtained in a sustainable way from various plants, this polymer is a great example of sustainable methods of obtaining, because it can be obtained even from vegetable waste product of industrial processes of food production, in addition to aligning with environmental objectives is obtained as a benefit interesting drug delivery system (Kurczewska 2022). Accordingly, the combination of these materials using environmentally friendly methods would result in highly sustainable solutions. Methodologies can even be designed to combine them where properties like crystallinity, particle size, porosity, morphology, among others that are determinants of drug release efficacy are controlled, given that pharmaceutical results require their precise control.

15.7 Properties That the Biomaterial Seeks to Be a Drug Delivery System

It is well recognized that individual materials can act collaboratively, and the properties of the combined material can act in concordance. In that sense, it is expected that hydroxyapatite combined with starch will offer synergistic properties for drug delivery systems whose final properties will be a result of a mutual interaction of both materials. In principle, both materials can present to a greater or lesser extent crystallinity that would affect the adsorption capacity or encapsulation efficiency and the rate of drug release or release kinetics. We could be in the presence of a hydroxyapatite with high crystallinity that presents a slower and sustained release, which is not necessarily detrimental if what is sought is a therapeutic application by sustained release action over time. Particle size is also a property that should not be left to chance because it plays an important role. In general, it has been accepted that smaller particles have a larger surface area, which can be used to increase the absorption rate of the drug. As for starch, we cannot fail to mention that its structure is anything but consistent between sources, let alone the degree of cross-linking, which can influence the release kinetics. However, by predictably characterizing the starch to be worked with, we could know if it has a denser and more cross-linked structure which would be beneficial in slowing down the release of the drug for sustained release (Lukova et al. 2023).

Therefore, to achieve drug encapsulation in hydroxyapatite in combination with starch matrices requires control of several other important parameters such as size and morphology, which ultimately affect the encapsulation efficiency and drug release kinetics. Particle size and morphology are both parameters that can be tunable during synthesis and can definitely impact on the biodistribution and bioavailability of the drug (Ren et al. 2021). To be precise, encapsulation efficiency refers to the amount of drug that can be incorporated into the matrix and is a key factor in ensuring adequate therapeutic doses, while release kinetics, on the other hand, determines the rate at which the drug is released from the material. All reports of release systems using this combination, and others, should report these parameters and relate them to the properties we have mentioned, in order to be able to compare and make decisions to obtain the biomaterial that is most appropriate for the treatment of chronic diseases that require sustained release.

Consequently, among the properties that the combination of these biomaterials should be carefully observed are:

(a) **Crystallinity**: To call crystallinity in both hydroxyapatite and starch is to refer to one of the properties that both shares. Crystallinity, as is well known, affects mechanical strength, degradation rate, interaction with surrounding water molecules. On the one hand, high crystallinity in hydroxyapatite would result in high stability and bioactivity, although it is unlikely to relate bioactivity to drug release, it is plausible that crystallinity also determines prolonged drug release. On the other hand, crystallinity in starch is associated with amylopectin containing and the amorphous phase with amylose. And given that the content of both polymers depends on the nature of the source, this will have a decisive influence on the ability to form stable and uniform matrices for drug encapsulation.

(b) **Particle size**: Particle size is closely related to drug loading capacity and release kinetics. Smaller particles typically have a larger surface area for drug adsorption, leading to higher drug loading efficiency. Even the particle size distribution influences the release profile of encapsulated drugs. In the case of hydroxyapatite nanoparticles, they can be obtained in various sizes and sieved to reduce their size distribution, aiming for smaller particle sizes to be associated with higher drug loading and capacity for controlled release. It is difficult to find reports of starch particle sizes in the nanometer scale when they are obtained from their sources, regularly the materials are microstructured. But through physical or chemical processes, particle sizes in the nanometer scale can be achieved.

(c) **Porosity**: Matrices formed by polymers that produce porous structures in which the drug is disseminated allow the drug to infiltrate into the pores of the matrix system, followed by a slow diffusion step of the drugs in the channels. The rate-limiting step in drug release from this type of system is the penetration of the dissolution liquid into the matrix, with water permeation promoted by the addition of wetting agents. The increased surface area resulting from the porosity of the nanohydroxyapatite may contribute to the ability of a matrix to load drugs. As for starch, in addition to the structural and biological properties already mentioned, it is also possible to functionalize its surface, making it more efficient and selective as a drug-carrying matrix, since the anchoring of functional groups on porous matrices causes changes in pore size and drug-substrate interaction, resulting in lower drug loading and a slower release rate, which is desirable. Not to be omitted that the release of the active ingredient is essentially controlled by diffusion mechanisms through the pores or by erosion, in addition to hydrolysis in the case of hydrophilic polymers with one or the other mechanism prevailing depending on the properties of the drug.

15.8 Hydroxyapatite-Starch Based Drug Delivery Systems

In the existing literature there are formulations that already incorporate hydroxyapatite and starch for drug delivery. Table 15.1 presents a summarized overview of recent publications where several formulations incorporating hydroxyapatite-starch based materials were developed to assess their efficacy in drug delivery systems covering from antibiotics to glucocoticoids. As can be seen, the frameworks were designed to improve specific properties of the drug release studies, in general the release profiles were better in the systems that took advantage of the unique properties of the composites.

It is clear from the data presented in Table 15.1 that a ternary mixture of polylactic acid (PLA), starch and poly ε-caprolactone (PCL) was used alongside nanohydroxyapatite (nHA) by fusion. Although the resulting nanocomposite does not exclusively combine hydroxyapatite with starch, its mention in this discussion is not misplaced, as it should be noted that the composite with 3% nHA demonstrated better hydrolytic degradation, higher hydrophilicity, greater antibacterial activity and optimized triclosan release profiles (Davachi et al. 2017). Triclosan is an antiseptic agent which belongs to the chemical group of polychlorinated phenoxyphenols. It is effective against gram-negative and gram-positive bacteria, fungi and molds. It is a regulated ingredient commonly found in medicines, soaps, lotions, deodorants and toothpastes.

Yet another formulation hydroxyapatite was combined with β-CD to carry dexamethasone, a hydrophobic drug. Again, hydroxyapatite is not directly combined with starch, but this example is particularly interesting because cyclodextrins are produced from starch by enzymatic conversion, which is why they are sometimes

Table 15.1 Formulations of drug delivery system based on hydroxyapatite combined with starch

Formulation	Drug	Results	Ref.
Polylactic acid/starch/poly ε-caprolactone, mixed with nano hydroxyapatite (nHA) via melt blending	Triclosan	3% nHA increases material's degradation, antibacterial action, and drug release.	Davachi et al. (2017)
Hydroxyapatite (HA)/ Cyclodextrin (β-CD) nanocomposite	Dexamethasone	The release of dexamethasone from β-CD/HA was about 4.6% at pH 5.3 and 18.7% at pH 7.4.	Salimi and Molaei (2021)
Dispersion of hydroxyapatite into urea and thermoplastic starch at nanoscale	Urea (fertilizer)	Study shows breaking up hydroxyapatite particles and dispersing them in water-soluble materials improves how well phosphate releases	Giroto et al. (2015)
HA and pure β- tri calcium phosphate based porous scaffolds formed by applying together starch consolidation with foaming method	Ceftriaxone	HAp exhibited better drug release profile than β- tri calcium phosphate when drug was used alone	Kundu et al. (2010)

also called "cycloamylases". These compounds are used in food, pharmaceuticals, drug delivery systems. Moreover, in this work the nanocomposite showed pH-sensitive drug release, with a release of 4.6% at pH 5.3 and 18.7% at pH 7.4, consequently, it is highlighted that the system functions as a pH-sensitive sustainable release carrier for the hydrophobic Drug (Salimi and Molaei 2021). Dexamethasone is a drug belonging to the glucocorticoid class. It mainly can treat diseases of rheumatic, immune, skin, ocular, endocrine, pulmonary, blood, gastrointestinal, neurological and neoplastic origin. Dexamethasone has an anti-inflammatory action due to the inhibition of phospholipase A2, which causes the inhibition of the entire arachidonic acid cascade, impeding the formation of prostaglandin, thromboxane and leukotrienes. In addition, they also act by reducing the migration and adhesion of leukocytes, thus reducing the release of cytokines at the site of inflammation, resulting in immunosuppressive activity.

In a different approach, the incorporation of hydroxyapatite nanoparticles into thermoplastic starch matrices was explored by the authors. The thermoplastic starch has an amylose content higher than 70% and are gelatinized vegetable starch based. With the use of specific plasticizing solvents, can produce thermoplastic materials with good performance properties and inherent biodegradability. The starch is typically plasticized, unstructured or blended with other materials to form useful mechanical properties. Although generally, the applications of thermoplastic starch polymers are generally films, such as shopping bags, bread bags, bait bags, wrapping, "washable" medical device backing material and mulch film, their inclusion in a delivery system, not for drugs, but for compounds of interest is noteworthy as well as sustainable. The biodegradation of starch-based polymers is the result of an enzymatic attack on the glycosidic bonds between the carbohydrate groups leading to a reduction in chain length and the cleavage of the carbohydrate units into monosaccharides, disaccharides and oligosaccharides, which are readily utilized in biochemical pathways. The work aimed to improve the solubility of poorly soluble phosphate phases. It was revealed that the dispersion of HAp within the matrices improved phosphate release and the influence of hydroxyapatite particles and their dispersion in water soluble materials was demonstrated to significantly improve phosphate release. This improved release profile heralds that the combined material increases fertilizer efficacy by achieving a more sustained release of nutrients (Giroto et al. 2015).

Previously, the use of porous hydroxyapatite (HAp) and β-tricalcium phosphate (β-TCP) scaffolds fabricated by a starch consolidation and foaming method had already been reported. For that reason, this study is also incorporated in this analysis, since it combines both materials that we have been mentioning, this time the study was about the local administration of ceftriaxone in the treatment of osteomyelitis As mentioned before, the porosity of the materials is one of the main properties to be controlled and in this study the importance of high porosity and interconnectivity in the release of the drug was highlighted. Ceftriaxone was incorporated into the scaffolds and the results showed that hydroxyapatite provided a better drug release profile than β-TCP when used alone (Kundu et al. 2010). Ceftriaxone is a broad-spectrum injectable antibiotic used to eliminate excess bacteria in the body and can therefore be used to treat various types of infections, from sexually transmitted infections to skin infections, pneumonia and even meningitis.

15.9 Strategies for Drug Incorporation in Hydroxyapatite-Starch Compounds

Studies where drug incorporation into hydroxyapatite (HA)-starch composites will certainly continue to appear in the literature in the coming years. Conducting these investigations would involve various strategies designed to optimize drug loading and release profiles that take into account the properties of the composite. For example, the common approach of co-precipitation would incorporate drugs during the synthesis of the starch and HA composite. With this method, a uniform distribution of the drug within the composite matrix would be sought. Perhaps the most obvious or most commonly used strategy is surface adsorption, whereby drugs are adsorbed on the surface of the preformed hydroxyapatite particles combined with starch through weak bonds. This technique would be particularly useful for drugs that have a strong affinity for the starch or hydroxyapatite surface. On the other hand, encapsulation techniques should be considered, the starch polymer lends itself to form a layer around the hydroxyapatite particles containing adsorbed drug or the two separately. This method of encapsulation can protect drugs from premature degradation and allow a sustained release over time, the role of hydroxyapatite would also play to that end.

As in most, if not all, nanomedicine drug delivery systems, the kinetics of drug release from HA starch composites will be influenced by several factors, including as mentioned above the porosity and crystallinity of the composite, the properties of the drug and the interaction between the drug and the composite matrix. Drug release is anticipated to occur through a combination of diffusion, degradation and dissolution mechanisms. Initially, a burst release would occur due to the release of adsorbed drug at or near the surface of the composite, this is usually seen by a steep slope in graphs plotting the amount of drug released as a function of time. This first phase is usually followed by a more sustained release phase in which the drug encapsulated within the matrix is gradually released as the compound degrades, keep in mind that it is the starch that is expected to degrade, the hydroxyapatite less often. Again it is important to emphasize that higher porosity allows for higher drug loading and facilitates faster drug diffusion, but it will also depend on several properties such as the molecular mass of the drug and stereochemistry. Returning to the rate of starch degradation, it must be said that this will affect drug release as the starch degrades in compartments where amylase enzymes are present, when degraded it would create channels that allow drug diffusion, contributing to the sustained release phase, these release profiles could present interesting kinetics.

15.10 Future Perspectives and Challenges

Despite their promising potential, hydroxyapatite composites in combination with starch face several limitations in drug delivery applications. An important limitation is the chemical resistance of hydroxyapatite. While hydroxyapatite provides some structural integrity, routes of administration must be chosen in accordance with the idea that hydroxyapatite is unlikely to degrade under physiological conditions. While the incorporation of starch may provide elements of degradability, the overall strength of the composite in particular would be determined by the persistence of the hydroxyapatite, which is perhaps why studies are focusing on bone or dental applications. In any case, achieving a balance between the degradation of the starch component and the stability of the hydroxyapatite is a fundamental limitation to be overcome in the future to advance this topic. To address these limitations, researchers can resort to surface modification techniques, size control, etc. to promote better properties.

15.11 Conclusion

In conclusion, hydroxyapatite composites combined with starch definitely represent a sustainable approach to drug delivery. However, their potential application as a nanomedicine system for controlled drug release is still not being harnessed for the potentially wide range of medical applications that could benefit from the combination of biocompatibility, bioactivity and other capabilities. What it is clear is that innovations in composite formulation, manufacturing techniques will further enhance their functionality and versatility by adopting green methods so that, as these materials continue to evolve, sustainable options with great potential for drug delivery will offer efficient and affordable solutions to complex medical challenges.

References

Ab'lah NN et al (2023) Reinvention of starch for oral drug delivery system design. Int J Biol Macromol 241. Available at: https://doi.org/10.1016/J.IJBIOMAC.2023.124506

Abka-khajouei R et al (2022) Structures, properties and applications of alginates. Mar Drugs 20(6). Available at: https://doi.org/10.3390/MD20060364

Acevedo-Guevara L et al (2018) Development of native and modified banana starch nanoparticles as vehicles for curcumin. Int J Biol Macromol 111:498–504. Available at: https://doi.org/10.1016/J.IJBIOMAC.2018.01.063

Ahamed AF, Manimohan M, Kalaivasan N (2022) Fabrication of biologically active fish bone derived hydroxyapatite and montmorillonite blended sodium alginate composite for in-vitro drug delivery studies. J Inorg Organomet Polym Mater 32(10):3902–3922. Available at: https://doi.org/10.1007/S10904-022-02401-1/METRICS

Ali KA et al (2021) Starch-based nanomaterials in drug delivery applications. In: Biopolymer-based nanomaterials in drug delivery and biomedical applications. Academic Press, London, pp 31–56. Available at: https://doi.org/10.1016/B978-0-12-820874-8.00023-3

Alvarez-Barreto J et al (2017) Mesenchymal stem cell culture on composite hydrogels of hydroxyapatite nanoparticles and photo-crosslinking chitosan. Revista Mexicana de Ingenieria Biomedica 38(3):524–536. Available at: https://doi.org/10.17488/RMIB.38.3.2

Alvarez-Barreto JF et al (2021) Development of films, based on oxidized Ipomea Batatas starch, with protein encapsulation. Revista Mexicana de Ingenieria Biomedica 42(2):119–131. Available at: https://doi.org/10.17488/RMIB.42.2.10

Ambrosio JAR et al (2023) Hydroxyapatite microspheres used as a drug delivery system for gliosarcoma strain 9l/Lacz treatment by photodynamic therapy protocols. Photodiagn Photodyn Ther 44. Available at: https://doi.org/10.1016/J.PDPDT.2023.103830

Anaya-Barajas D et al (2019) Biomaterials for bone tissue regeneration extracted from fish wastes. Revista Mexicana de Ingenieria Biomedica 40(3):1–10. Available at: https://doi.org/10.17488/RMIB.40.3.13

Asghar MS et al (2023) In vitro controlled drug delivery of cationic substituted hydroxyapatite nanoparticles; enhanced anti-chelating and antibacterial response. Kuwait J Sci 50(2):97–104. Available at: https://doi.org/10.1016/J.KJS.2023.02.014

Averous L, Halley PJ (2014) Starch polymers: from the field to industrial products. In: Starch polymers: from genetic engineering to green applications. Elsevier, Burlington, pp 3–10. Available at: https://doi.org/10.1016/B978-0-444-53730-0.00018-X

Azfaralariff A et al (2020) Food-grade particle stabilized pickering emulsion using modified sago (Metroxylon sagu) starch nanocrystal. J Food Eng 280. Available at: https://doi.org/10.1016/J.JFOODENG.2020.109974

Barclay TG et al (2019) Review of polysaccharide particle-based functional drug delivery. Carbohydr Polym 221:94–112. Available at: https://doi.org/10.1016/J.CARBPOL.2019.05.067

Beh CY, Cheng EM, Mohd Nasir NF, Eng SK et al (2021a) Dielectric and material analysis on physicochemical activity of porous hydroxyapatite/cornstarch composites. Int J Biol Macromol 166:1543–1553. Available at: https://doi.org/10.1016/J.IJBIOMAC.2020.11.034

Beh CY, Cheng EM, Mohd Nasir NF, Khor SF et al (2021b) Low frequency dielectric and optical behavior on physicochemical properties of hydroxyapatite/cornstarch composite. J Colloid Interface Sci 600:187–198. Available at: https://doi.org/10.1016/J.JCIS.2021.03.158

Bose S et al (2022) Allicin-loaded hydroxyapatite: enhanced release, Cytocompatibility, and antibacterial properties for bone tissue engineering applications. JOM 74(9):3349–3356. Available at: https://doi.org/10.1007/S11837-022-05366-1

Castrejón-Parga KY et al (2015) Chitosan-starch film reinforced with magnetite-decorated carbon nanotubes. J Alloys Compd 615(S1):S505–S510. Available at: https://doi.org/10.1016/j.jallcom.2013.12.269

Chen J et al (2020) Preparation, characterization, physicochemical property and potential application of porous starch: a review. Int J Biol Macromol 148:1169–1181. Available at: https://doi.org/10.1016/j.ijbiomac.2020.02.055

Choi I, Li N, Zhong Q (2022) Enhancing bioaccessibility of resveratrol by loading in natural porous starch microparticles. Int J Biol Macromol 194:982–992. Available at: https://doi.org/10.1016/J.IJBIOMAC.2021.11.157

Davachi SM et al (2017) Interface modified polylactic acid/starch/poly ε-caprolactone antibacterial nanocomposite blends for medical applications. Carbohydr Polym 155:336–344. Available at: https://doi.org/10.1016/J.CARBPOL.2016.08.037

de Oliveira Cardoso VM et al (2020) Insights into the impact of cross-linking processes on physicochemical characteristics and mucoadhesive potential of gellan gum/retrograded starch microparticles as a platform for colonic drug release. J Drug Deliv Sci Technol 55. Available at: https://doi.org/10.1016/J.JDDST.2019.101445

Devanand Venkatasubbu G et al (2011) Nanocrystalline hydroxyapatite and zinc-doped hydroxyapatite as carrier material for controlled delivery of ciprofloxacin. 3 Biotech 1(3):173–186. Available at: https://doi.org/10.1007/S13205-011-0021-9

Diaz Varela JY et al (2024) The role of multi-walled carbon nanotubes in enhancing the hydrolysis and thermal stability of PLA. Sci Rep 14(1):1–11. Available at: https://doi.org/10.1038/s41598-024-58755-8

Fatimah I et al (2021) Biosynthesized gold nanoparticles-doped hydroxyapatite as antibacterial and antioxidant nanocomposite. Mater Res Express 8(11). Available at: https://doi.org/10.1088/2053-1591/AC3309

Fazal T et al (2023) Recent developments in natural biopolymer based drug delivery systems. RSC Advances 13(33):23087. Available at: https://doi.org/10.1039/D3RA03369D

Flores Valdez JD et al (2022) Hydroxyapatite and biopolymer composites with promising biomedical applications. Revista Mexicana de Ingenieria Biomedica 43(2):6–23. Available at: https://doi.org/10.17488/RMIB.43.2.1

Friuli V et al (2024) Hydroxyapatite nanorods based drug delivery systems for Bumetanide and Meloxicam, poorly water soluble active principles. Nanomaterials (Basel, Switzerland) 14(1). Available at: https://doi.org/10.3390/NANO14010113

Gaona CGC et al (2023) Novel studies in the designs of natural, synthetic, and compound hydrogels with biomedical applications. Revista Mexicana de Ingenieria Biomedica 44(2):74–96. Available at: https://doi.org/10.17488/RMIB.44.2.6

García-Guzmán L et al (2022) Progress in starch-based materials for food packaging applications. Polysaccharides 3(1):136–177. Available at: https://doi.org/10.3390/POLYSACCHARIDES3010007

Ghorbani F et al (2019) A facile method to synthesize mussel-inspired polydopamine nanospheres as an active template for in situ formation of biomimetic hydroxyapatite. Mater Sci Eng C 94: 729–739. Available at: https://doi.org/10.1016/J.MSEC.2018.10.010

Giroto AS, Fidélis SC, Ribeiro C (2015) Controlled release from hydroxyapatite nanoparticles incorporated into biodegradable, soluble host matrixes. RSC Advances 5(126):104179–104186. Available at: https://doi.org/10.1039/C5RA17669G

González-Torres V et al (2021) Antibacterial activity analysis of hydroxyapatite based materials with fluorine and silver. Revista Mexicana de Ingenieria Biomedica 42(2):49–57. Available at: https://doi.org/10.17488/RMIB.42.2.4

Gui X et al (2022) Synthesis and application of nanometer hydroxyapatite in biomedicine. Nanotechnol Rev 11(1):2154–2168. Available at: https://doi.org/10.1515/NTREV-2022-0127

Hong Y et al (2011) Study on physicochemical characteristics of waxy potato starch in comparison with other waxy starches. Starch Stärke 63(12):754–759. Available at: https://doi.org/10.1002/STAR.201100013

Hosseinzadeh H, Ramin S (2018) Fabrication of starch-graft-poly(acrylamide)/graphene oxide/hydroxyapatite nanocomposite hydrogel adsorbent for removal of malachite green dye from aqueous solution. Int J Biol Macromol 106:101–115. Available at: https://doi.org/10.1016/J.IJBIOMAC.2017.07.182

Huq T et al (2017) Alginate based nanocomposite for microencapsulation of probiotic: effect of cellulose nanocrystal (CNC) and lecithin. Carbohydr Polym 168:61–69. Available at: https://doi.org/10.1016/j.carbpol.2017.03.032

Kamaly N et al (2016) Degradable controlled-release polymers and polymeric nanoparticles: mechanisms of controlling drug release. Chem Rev 116(4):2602–2663. Available at: https://doi.org/10.1021/ACS.CHEMREV.5B00346

Kedir WM et al (2022) Pharmaceutical and drug delivery applications of chitosan biopolymer and its modified nanocomposite: a review. Heliyon 8(8). Available at: https://doi.org/10.1016/J.HELIYON.2022.E10196

Khalf A, Madihally SV (2017) Recent advances in multiaxial electrospinning for drug delivery. Eur J Pharm Biopharm 112:1–17. Available at: https://doi.org/10.1016/J.EJPB.2016.11.010

Kim HY, Park SS, Lim ST (2015) Preparation, characterization and utilization of starch nanoparticles. Colloids Surf B: Biointerfaces 126:607–620. Available at: https://doi.org/10.1016/J.COLSURFB.2014.11.011

Koski C, Bose S (2019) Effects of amylose content on the mechanical properties of starch-hydroxyapatite 3D printed bone scaffolds. Addit Manuf 30:100817. Available at: https://doi.org/10.1016/J.ADDMA.2019.100817

Koski C et al (2018) Starch-hydroxyapatite composite bone scaffold fabrication utilizing a slurry extrusion-based solid freeform fabricator. Addit Manuf 24:47–59. Available at: https://doi.org/10.1016/J.ADDMA.2018.08.030

Kundu B et al (2010) Development of porous HAp and β-TCP scaffolds by starch consolidation with foaming method and drug-chitosan bilayered scaffold based drug delivery system. J Mater Sci Mater Med 21(11):2955–2969. Available at: https://doi.org/10.1007/S10856-010-4127-0

Kurczewska J (2022) Recent reports on polysaccharide-based materials for drug delivery. Polymers 14(19). Available at: https://doi.org/10.3390/POLYM14194189

Le TDH et al (2023) Hydroxyapatite−loaded starch/polyvinyl alcohol scaffold for bone regeneration application: preparation and characterization. J Sol-Gel Sci Technol 107(2):441–451. Available at: https://doi.org/10.1007/S10971-023-06137-3

Lemos PVF et al (2021) Starch chemical modifications applied to drug delivery systems: from fundamentals to FDA-approved raw materials. Int J Biol Macromol 184:218–234. Available at: https://doi.org/10.1016/J.IJBIOMAC.2021.06.077

Lukova P, Katsarov P, Pilicheva B (2023) Application of starch, cellulose, and their derivatives in the development of microparticle drug-delivery systems. Polymers 15(17):3615. Available at: https://doi.org/10.3390/POLYM15173615

Mahdavinia GR et al (2019) In vitro evaluation of sustained ciprofloxacin release from κ-carrageenan-crosslinked chitosan/hydroxyapatite hydrogel nanocomposites. Int J Biol Macromol 126:443–453. Available at:. https://doi.org/10.1016/J.IJBIOMAC.2018.12.240

Mao S et al (2004) Intranasal administration of melatonin starch microspheres. Int J Pharm 272(1–2):37–43. Available at: https://doi.org/10.1016/J.IJPHARM.2003.11.028

Marouf N, Nojehdehian H, Ghorbani F (2020) Physicochemical properties of chitosan–hydroxyapatite matrix incorporated with Ginkgo biloba-loaded PLGA microspheres for tissue engineering applications. Polym Polym Compos 28(5):320–330. Available at: https://doi.org/10.1177/0967391119874972

Marto J, Ribeiro HM, Almeida AJ (2020) Starch-based nanocapsules as drug carriers for topical drug delivery. In: Smart nanocontainers: micro and nano technologies. Springer, Singapore, pp 287–294. Available at: https://doi.org/10.1016/B978-0-12-816770-0.00017-4

Miculescu F et al (2017) Progress in hydroxyapatite-starch based sustainable biomaterials for biomedical bone substitution applications. ACS Sustain Chem Eng 5(10):8491–8512. Available at: https://doi.org/10.1021/ACSSUSCHEMENG.7B02314

Mohd Roslan MR et al (2021) The state of starch/hydroxyapatite composite scaffold in bone tissue engineering with consideration for dielectric measurement as an alternative characterization technique. Materials 14(8). Available at: https://doi.org/10.3390/MA14081960

Morán JI, Vázquez A, Cyras VP (2013) Bio-nanocomposites based on derivatized potato starch and cellulose, preparation and characterization. J Mater Sci 48(20):7196–7203. Available at: https://doi.org/10.1007/S10853-013-7536-X

Mundargi RC et al (2008) Formulation and in-vitro evaluation of novel starch-based tableted microspheres for controlled release of ampicillin. Carbohydr Polym 71(1):42–53. Available at: https://doi.org/10.1016/J.CARBPOL.2007.05.013

Munir MU et al (2022) Synthesis, characterization, functionalization and bio-applications of hydroxyapatite nanomaterials: an overview. Int J Nanomedicine 17:1903–1925. Available at: https://doi.org/10.2147/IJN.S360670

Murugan E, Akshata CR (2023) Dextrose, maltose and starch guide crystallization of strontium-substituted hydroxyapatite: a comparative study for bone tissue engineering application. Int J Biol Macromol 248:125927. Available at: https://doi.org/10.1016/J.IJBIOMAC.2023.125927

Mweta DE et al (2010) Isolation and physicochemical characterisation of starch from cocoyam (Colocasia esculenta) grown in Malawi. J Sci Food Agric 90(11):1886–1896. Available at: https://doi.org/10.1002/JSFA.4029

Naveenkumar R, Senthilvelan S, Karthikeyan B (2023) A review on the recent developments in electrospinned nanofibers for drug delivery. Biomed Mater Device 2(1):342–364. Available at: https://doi.org/10.1007/S44174-023-00121-9

Nur-Azzah ABT et al (2022) A review on poly lactic acid (PLA) as a biodegradable polymer. Polym Bull 80(2):1179–1213. Available at: https://doi.org/10.1007/S00289-022-04160-Y

Obireddy SR, Lai WF (2021) Preparation and characterization of 2-hydroxyethyl starch microparticles for co-delivery of multiple bioactive agents. Drug Deliv 28(1):1562–1568. Available at: https://doi.org/10.1080/10717544.2021.1955043

Okunlola A, Ghomorai T (2018) Development of ibuprofen microspheres using acetylated plantain starches as polymer for sustained release. J Pharm Investig 48(5):551–564. Available at: https://doi.org/10.1007/S40005-017-0345-5

Ordon K et al (2019) Examining the effect of starch and hydroxyapatite crosslinking on the thermal properties of polyurethane-based biomaterials. Thermochim Acta 682:178414. Available at: https://doi.org/10.1016/J.TCA.2019.178414

Pal A et al (2024) Fabrication of ciprofloxacin-immobilized calcium phosphate particles for dental drug delivery. Materials (Basel, Switzerland) 17(9). Available at: https://doi.org/10.3390/MA17092035

Pandey A et al (2018) Antioxidant and antibacterial hydroxyapatite-based biocomposite for orthopedic applications. Mater Sci Eng C 88:13–24. Available at: https://doi.org/10.1016/j.msec.2018.02.014

Pariy IO et al (2022) Hybrid biodegradable electrospun scaffolds based on poly(l-lactic acid) and reduced graphene oxide with improved piezoelectric response. Poly J 54(10):1237–1252. Available at: https://doi.org/10.1038/s41428-022-00669-1

Ragab HS et al (2014) Synthesis and in vitro antibacterial properties of hydroxyapatite nanoparticles. IOSR J Pharm Biol Sci 9(1):77–85. Available at: https://doi.org/10.9790/3008-09167785

Ren N et al (2021) Preparation of rutin-loaded microparticles by debranched lentil starch-based wall materials: structure, morphology and in vitro release behavior. Int J Biol Macromol 173:293–306. Available at: https://doi.org/10.1016/J.IJBIOMAC.2021.01.122

Rosch JG et al (2019) Inverse-micelle synthesis of doxorubicin-loaded alginate/chitosan nanoparticles and in vitro assessment of breast cancer cytotoxicity. Colloid Interface Sci Commun 28:69–74. Available at: https://doi.org/10.1016/j.colcom.2018.12.002

Rydell N, Stertman L, Sjöholm I (2005) Starch microparticles as vaccine adjuvant. Expert Opin Drug Deliv 2(5):807–828. Available at: https://doi.org/10.1517/17425247.2.5.807

Salimi E, Molaei MJ (2021) Preparation and in-vitro evaluation of β-CD/HA nanocomposite as a potential carrier for hydrophobic drugs. J Biomater Appl 36(2):246–251. Available at: https://doi.org/10.1177/08853282211012289

Santillana-Marín RA et al (2018) Bioactivity and hemolysis test of hydroxyapatite/zirconia composites (HAp/ZrO2-8Y2O3) for bone implants. Revista Mexicana de Ingenieria Biomedica 39(3):272–280. Available at: https://doi.org/10.17488/RMIB.39.3.5

Sathyan S, Nisha P (2022) Optimization and characterization of porous starch from corn starch and application studies in emulsion stabilization. Food Bioprocess Technol 15(9):2084–2099. Available at: https://doi.org/10.1007/S11947-022-02843-Y

Shaikh MAJ et al (2022) Sodium alginate based drug delivery in management of breast cancer. Carbohydr Polym 292. Available at: https://doi.org/10.1016/J.CARBPOL.2022.119689

Singh G, Singh RP, Jolly SS (2020) Customized hydroxyapatites for bone-tissue engineering and drug delivery applications: a review. J Sol-Gel Sci Technol 94(3):505–530. Available at: https://doi.org/10.1007/S10971-020-05222-1

Song W et al (2018) Magnetic alginate/chitosan nanoparticles for targeted delivery of curcumin into human breast cancer cells. Nano 8(11):907. Available at: https://doi.org/10.3390/nano8110907

Sriprapha P et al (2024) Induced porous structure with a slight change in mechanical properties of hydroxyapatite-based nanocomposites synthesized from waste bovine bone and their

bioactivity. Mater Today Sustain 26:100710. Available at: https://doi.org/10.1016/J.MTSUST.2024.100710
Thomas D, Mathew N, Nath MS (2021) Starch modified alginate nanoparticles for drug delivery application. Int J Biol Macromol 173:277–284. Available at: https://doi.org/10.1016/J.IJBIOMAC.2020.12.227
Tören E et al (2024) Recent advances in biopolymer based electrospun nanomaterials for drug delivery systems. Polym Adv Technol 35(3):e6309. Available at: https://doi.org/10.1002/PAT.6309
Tudorachi N et al (2017) Studies on the nanocomposites based on carboxymethyl starch-g-lactic acid-co-glycolic acid copolymer and magnetite. J Therm Anal Calorim:1–14. Available at: https://doi.org/10.1007/s10973-017-6682-9
Villarreal-Gómez LJ et al (2016) Electrospinning as a powerful technique for biomedical applications: a critically selected survey. J Biomater Sci Polym Ed 27(2):157–176. Available at: https://doi.org/10.1080/09205063.2015.1116885
Wang Z, Shang J, Zhang Z (2024) Composite or modified hydroxyapatite microspheres as drug delivery carrier for bone and tooth tissue engineering. Curr Med Chem 31. Available at: https://doi.org/10.2174/0109298673303632240320073606
Yasar H et al (2018) Starch-chitosan polyplexes: a versatile carrier system for anti-infectives and gene delivery. Polymers 10(3). Available at: https://doi.org/10.3390/POLYM10030252
You BC et al (2022) Dielectric and biodegradation properties of biodegradable nano-hydroxyapatite/starch bone scaffold. J Mater Res Technol 18:3215–3226. Available at: https://doi.org/10.1016/J.JMRT.2022.04.014
Zhang D et al (2023) Hydroxyapatite-based nano-drug delivery system for nicotinamide mononucleotide (NMN): significantly enhancing NMN bioavailability and replenishing in vivo nicotinamide adenine dinucleotide (NAD+) levels. J Pharm Pharmacol 75(12):1569–1580. Available at: https://doi.org/10.1093/JPP/RGAD090

Dr. Christian Chapa is a leading researcher at the Institute of Engineering and Technology of the Autonomous University of Ciudad Juarez (Mexico), and head of the NANOMEDICINE research group. His research interests focus on nanomedicine, physico-chemical properties of biomaterials and nanomaterials where teragnostic approach is the main approach. In 2012, he formed the Nanomedicine Research Group for the proposal and execution of research projects related to the preclinical investigation of interactions between biomaterials or nanomaterials with biomolecules, clinical analytes or drugs to develop nanomedicine systems proposed for the identification of molecular targets for the diagnosis and treatment of a diverse number of human diseases and develop activities focused on student learning and innovative application of knowledge. In 2020, he formed the Laboratory for Integration of Data and Evidence in Health and Science Reviews, LIDERSC, to conduct critical reviews of the scientific literature with the goal of providing robust and reliable evidence to answer research questions and support decision making in health and science. Christian Chapa has published more than 30 scientific articles in scientific journals of international recognition, 7 book chapters and has more than 50 presentations in scientific congresses. Christian Chapa was president of the Scientific Committee of the Mexican Society of Biomedical Engineering in 2018–2019 and received recognition from the International Federation for Medical and Biological Engineering for his outstanding work as vice president of the Scientific Committee for the VIII Latin American Congress of Biomedical Engineering. He is a researcher of the National Council of Humanities, Science and Technology of Mexico, participates in programs to promote scientific vocations, such as the Summer of Research. Currently, he collaborates with other research groups ByNEF and BIOCIM to propose and execute academic and research projects with social impact in the areas of nanotechnology and bioengineering.

Surfactant-Free Synthesis of Metal and Metal Oxide Nanomaterials: Sustainable and Eco-Synthesis Methods

16

Sayali Churi, Ushasi Das, Popat Mohite, Sanchita Mandal, and Sudarshan Singh

Abstract

Surfactant free synthesis of metallic nanomaterials offers multiple benefits in medical, pharmaceutical, biological, biomedical, and catalytic application due to their high surface-to-volume ratio and fascinating physicochemical properties. Though there are several techniques with different advantageous and challenges in synthesis metal oxide or metallic nanomaterials have been investigated, the biocompatibility, stability, and purity of such nanomaterials also controllability with reproducibility are most challenging issue faced by researcher and scientist. Moreover, the metal oxide and metallic nanoparticles due to variation in size and shape and percentage of reduction are prone to lose reactivity and ultimately therapeutic efficacy, as this can precipitate out or form aggregate as bulk metals; therefore, various stabilizers briefly surfactant or allied materials are typically required. Plant-based is an alternative technique to synthesize stable metal oxide or metallic nanoparticles, however bench to market of such techniques is not yet well-setup, thus several surfactant-free strategies such as solvent mediated synthesis, ion-mediated synthesis, and several physical processes have been introduced for fabrication of oxide and reduced metallic nanoparticles as

S. Churi (✉) · P. Mohite
Department Pharmaceutical Chemistry, AETs St. John Institute of Pharmacy and Research, Palghar, Maharashtra, India
e-mail: sayalic@sjipr.edu.in

U. Das · S. Mandal
Department of Pharmaceutical Technology, Jadavpur University, Kolkata, West Bengal, India
e-mail: smandal.pharmacy@jadavpuruniversity.in

S. Singh
Office of Research Administration, Chiang Mai University, Chiang Mai, Thailand

Faculty of Pharmacy, Chiang Mai University, Chiang Mai, Thailand

R. Malviya, S. Sundram (eds.), *Sustainable Green Biomaterials As Drug Delivery Systems*, Biomaterials, Bioengineering and Sustainability 1,
https://doi.org/10.1007/978-3-031-79062-1_16

eco-friendly techniques by avoiding use of toxic chemicals. Thus, this chapter provides a brief on the utilization of surfactant free synthesis process for manufacturing of metallic nanomaterials with their associated applications and challenges.

Keywords

Nanoparticles · Surfactant-free synthesis · Sonochemical · Catalysis · Metal oxides

16.1 Introduction

The typical size range for metal nanoparticles is 1–100 nm. Metal nanoparticles exhibit distinct physiochemical features concerning optics, magnetism, and specific chemical reactions as they have large great surface energy, better surface area, and quantum confinement (Syukri et al. 2020, 2021; Nwabor et al. 2020, 2021a, b; Ontong et al. 2020; Jayeoye et al. 2021a, b; Kumar et al. 2023; Puri et al. 2023; Nagime et al. 2023) confinement;.A large surface: volume ratio and intriguing physicochemical features of metal and metal oxide nanoparticles have generated great interest in pharmaceutical, biological, medical, biomedical, and catalytic purposes (Singh et al. 2022; Syukri et al. 2024; Kumar et al. 2024; Jayeoye et al. 2024; Shah et al. 2024). Synthesis of nanomaterials including metal and metal-oxide, numerous methods have been presented to date, each with unique benefits, drawbacks, and difficulties. Important and difficult problems in this context are the stability, controllability, and repeatability of synthesis methods in addition to the purity, stability, and corrosion resistance of nanomaterials. Because metal and metal oxide nanoparticles can precipitate out or form aggregate as bulk metals, they are susceptible to losing their reactivity. As a result, various stabilizers, including functionalized dendrimers, ligands, polymers, inorganic materials (such as carbon, zeolites, metal oxides, and sol–gel clays), or ionic surfactants, can be used to prevent it from occurring. In order to synthesize metal and metal oxide nanoparticles with the advantages of affordability, ease of use, and environmentally benign qualities, a number of surfactant-free strategies have been introduced recently, such as micro plasma based, laser synthesis, the Co_4Cat process, and mono-alcohol fabrication, techniques. These strategies avoid the use of toxic additives or surfactants (Jiang et al. 2013).

In comparison to their bulk counterparts, these nanoparticles also exhibit lower melting points and electrical/thermal conductivity. As metal nanoparticles get smaller and have a measurable number of atoms on their surfaces, their physicochemical characteristics alter. The morphology (i.e., shape and dimensionality), composition (i.e., whether the nanoparticles are of an alloy or a metal), and agglomeration of the nanoparticles are important in addition to these size-based effects, as are the physical and chemical properties of the nanoparticles.

Scientific interest in metal nanoparticle research is growing as a result of the discovery of controllable-solution-based methods for producing metal nanoparticles (Khan et al. 2019).

16.2 Metal and Metal Oxide Nanomaterial Synthesis Devoid of Surfactants

In colloidal synthesis without surfactants, the solvent functions as a stabilizer, reducing agent, and reaction media. Forming valuable metallic nanoparticles (NPs) that don't include any molecules with a molar mass higher than 100 g mol^{-1} by this process shows promise. Numerous surfactant-free methods for formulating NPs have been documented, such as those utilizing alkaline mono-alcohols, plasma, alkaline polyols, DMF, or lasers. Although most catalysis research still involves surfactants with/without direct deposition of NPs on confirms in one-pot syntheses, these approaches provide control over the chemical structure and size of surfactant-devoid NPs. The greatest source for examining and explaining the role of ligands and surfactants in NP formation and catalysis are those that are surfactant-free.

16.2.1 Solvent-Assisted Synthesis

16.2.1.1 N,N-Dimethylformamide (DMF) Assisted Synthesis

Dimethylamine or DMF is a polar, aprotic, organic solvent that was originally produced in 1893 by the French scientist Albert Verley. It is generated either by treating dimethylamine coupled with carbon monoxide or by distilling a combination of dimethylamine hydrochloride and potassium formate (Amooref and Hautala 1983) Pastoriza-Santos's team (Pastorizo-Santos and Liz-Marzán 2009) published the first study on the metal nanoparticles synthesis using DMF in 1999. Since then, it has been demonstrated that this technique is adaptable for producing different kinds of metallic nanoparticles with monitored surface treatment, dimension, and structure. DMF plays three distinct functions as a reductant, solvent, and protectant. Strong and adaptable, DMF finds usage in a versatile range of processes, including the production of block copolymers, the fabrication of colloids, and organic synthesis including non-solvent- hydrolysis, allylation, decarboxylation, and polymerization (Muzart 2009). Moreover, it finds application in the production of films, fibers, adhesives, surface coatings, and synthetic leathers. DMF is utilized as an initial component utilized for reactions in the dye making process. Additionally, it serves as a catalyst in significant organic processes such as the Beckmann rearrangement, Friedel–Crafts reaction, and Vilsmeier–Haack reaction (Kumar et al. 2010). DMF serves as an origin of important precursors in a variety of processes because it has the ability to react as an nucleophilic or electrophilic mediator (Kawasaki 2013; Heravi et al. 2018). Distinct from surfactant-free metallic nanoparticles made using methyl isobutyl ketone and ethylene glycol as reductants, the DMF-reduction technique yields thermally stable metal nanoparticles or nanoclusters with a limited variability

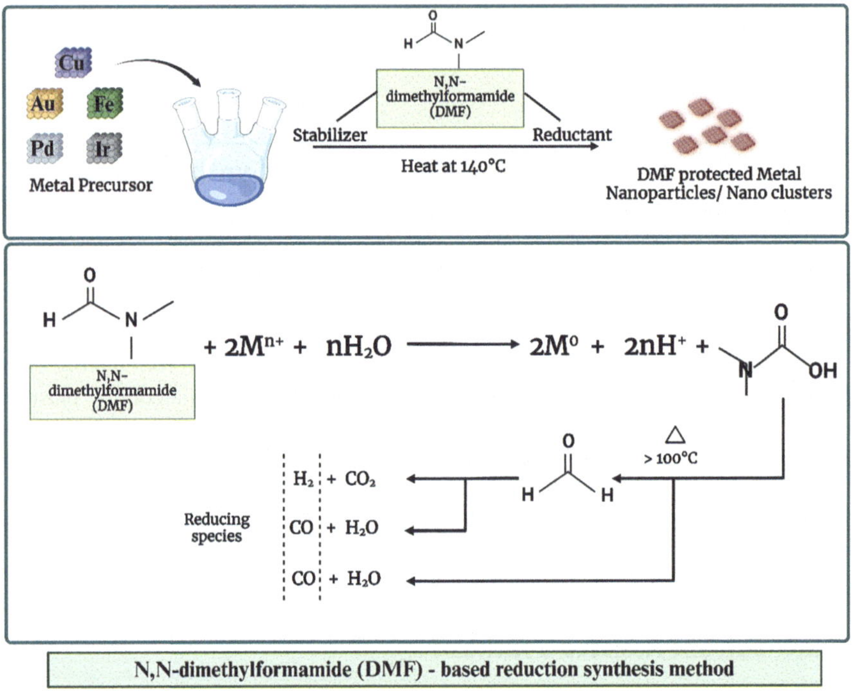

Fig. 16.1 N, N-dimethylformamide (DMF) based reduction synthesis method and reaction pathway

in size (<7 nm) (Nagata and Obora 2020). When synthesizing acyl halides and acyl chlorides from carboxylic acids, this chemical compound is frequently utilized as a catalyst (Dubey et al. 2010; Nagata and Obora 2020). Imidoyl chloride is reversibly formed as part of its catalytic process. In addition, it helps in the synthesis of azepine, condensation processes, cyclo-addition reactions, cyclic carbonates, reductive ring-cleavage, allylation of aldehydes, alcohol conversion to bromides, and acylation of aromatic compounds. Various processes have been suggested to explain the methods by which DMF reduces gold and silver ions. Nonetheless, the production of H^+ ions and carbamic acid is a component of each of these reactions (Fig. 16.1). In DMF (Nagata and Obora 2020), the resulting carbamic acid is weak and quickly breaks down into $(CH_3)_2NH$ and CO_2. Several metals and bimetallic compounds may be synthesized as quantum dots, clusters, and nanoparticles using the surfactant-free form of the ingredient. Since DMF protection reduces considerably at elevated temperatures, DMF-stabilized clusters are the perfect choice for unconventional catalysts composed of untreated NPs. In DMF, colloids an remain stable for several weeks, even after being redispersed in a variety of solvents, such as water-based solutions (Quinson et al. 2023).

Reduction of a metal source in dimethyl acetamide (DMF) in the atmosphere and reflux reaction yields colloids. The presence of DMF on the NP surface is confirmed by experimental data, and the stabilization is achieved by the interaction with the amide groups of DMF. Because of the tiny size of these produced clusters, size management might be difficult. As the reaction time rises, Ag nanoparticles with a size control of 4–10 nm may be generated. An extensive investigation of the effects of reaction duration, temperature, water content, and metal precursor salt were conducted for Pt nanoparticles. By adjusting reactivity parameters with autoclave synthesis, cubic or truncated octahedral nanoparticles successfully synthesized (Kawasaki et al. 2010; Da Costa et al. 2011).

The optical characteristics of Au nanoparticles (AuNPs), such as fluorescence and plasmon resonance that are formed by DMF synthesis have been the subject of investigations. The impact of ligand functionalizations on these characteristics was assessed by the investigators using surfactant-free Au NPs as an instance. Several ligands, such as 11-mercaptoundecanoic acid, thioctic acid, 1-dodecanethiol, PVP, dodecylamine, poly(amido) amine, triphenylphosphine, and thiocholine bromide, were used to functionalize the same batch of Au NPs. PVP was added, and this had a considerable impact on the emission wavelengths, whereas the absorption wavelengths did not alter at varied concentration levels. According to the authors, the ligand concentration impact is a result of cross-linking (Liu et al. 2008; Duchesne and Zhang 2012; Maman et al. 2019).

Capping agents not only stabilize nanoparticle dispersion but also influence particle morphology, as seen with ethyleneglycol or higher polyols. Although PVP is frequently used to modify the shape of nanoparticles, other substances are equally interesting. This is focuses on synthesizing nanoparticles using other capping agents and discusses the use of PVP as a stabilizer and morphology control, both in the presence and absence of preformed seed. DMF is a flexible catalyst that may be used to create Au and Ag nanoparticles with specific optical characteristics. As different capping agents offer different surface functions, they can be utilized as stabilizers during particle synthesis. PVP enables adaptable particle size and shape adjustment. The primary element affecting the ultimate particle shape is the crystallinity of the original seeds, perhaps in combination with PVP. PVP modifies the development process by preferentially interacting using the seeds' crystallographic features. The benefit of DMF is that it may provide a variety of particle morphologies in a broad size range and with narrow size distributions, including triangles, decahedra, octahedrons, and spheres. This method provides a more adaptable way to synthesize metallic particles (Pastorizo-Santos and Liz-Marzán 2009).

DMF synthesis has opened up new perspectives for electrocatalysis and heterogeneous catalysis, especially uncontrolled heterogeneous catalytic reactions. This creates the opportunity to investigate further reactions using supported catalysts, such electrocatalytic processes. Strong metallic surface contact, the non-ecofriendly solvent DMF, the high boiling point, and the restricted potential for scale-up are some of the constraints of the synthesis. A lengthy synthesis duration and an elevated temperature are necessary for the reaction, and a metal quantity of approximately 1 mM is the ideal (Prat et al. 2015).

16.2.1.2 Ethylene Glycol Assisted Synthesis

Charles-Adolphe Wurtz, a French scientist, discovered ethylene glycol (EG), the most basic diol, in 1859 by saponifying ethylene glycol diacetate with potassium hydroxide.Ethylene glycol serves as a reducing ingredient in methods as well as a solvent for precursor molecules.Large-scale nanostructure synthesis may be easily accomplished at an affordable price and with great versatility using the ethylene glycol-mediated synthesis methodology, also known as the polyol process (Yue et al. 2012; Soni et al. 2020). Fiévet et al. (2018) initially presented the polyol process in the late 1980s. It is a liquid-phase synthesis method that yields finely split metals from their salts in polyalcohols, hydroxides, or oxides. Polyalcohols like α-diols and ethylene glycols are examples of liquid organic compounds that may be used in this procedure as both a reducing substance and a solvent. Copper, electropositive metals like nickel and cobalt, and noble metals can all be reduced to zero-valent state ions by polyols. For the purpose of forming metal and alloy powders with non-agglomerated nanoparticles that have a precise form, a restricted variation in size, and a monitored size in the micrometer or submicrometer range, the polyol process was thought to be a flexible and intriguing technique (Li et al. 2007, 2008). Because ethylene glycol-mediated synthesis works in an enclosed framework and reduces metallic prerequisites to their zero-valence state, it is a sustainably beneficial and energy-conscious approach (Fig. 16.2).

A class of glycols known as polyols is chelation-based and water-comparable, providing adequate command on size of particles, dispersity, and variations. Propanediol, butanediol, pentanediol, glycerol, pentaerythritol, and certain carbohydrates are representatives of this group of compounds. A viable method for producing homogeneous magnetic iron oxide nanoparticles that may be applied to magnetic resonance imaging is polyol. The polyol approach performs because of their capacity to interact with the nuclei developed, which makes post-synthesis intervention simple. Thermal annealing, carboxylates, amines, and other coordination exchangers, as well as repeated washing with plain water are effective post-synthesis treatments for eliminating polyols (Watt et al. 2017; Deshmukh and Niederberger 2017; Gul et al. 2019). Polyol functions as a stabilizing agent, reducing agent, and solvent that manages particle development and inhibits aggregation. A polyol is used to dissolve or suspend precursor chemicals such as hydroxides, oxides, nitrates, sulfates, and acetates. The resulting product is subsequently heated to reflux. In order to generate polyols, a soluble metal precursor must first be reduced to metal nuclei, which then nucleate into metal nanoparticles. This process provides an intermediate molecule. The reagent polyvinylpyrrolide (PVP) functions as a shape-controlling and stabilizing substance facilitating both heat degradation and diverse nucleation (Bhattarai et al. 2015). Alkaline EG was used by Wang et al. to decrease metal ions such as ruthenium, rhodium, and platinum in order to formulate metal nanoparticles without incorporating surfactants. They synthesized robust noble metal nanoparticles (Pt, Rh, and Ru) with an average diameter of 1–2 nm by using metal solutions in EG that contained NaOH and had concentrations ranging from 0.3 to 3.7 g/l. Despite the use of natural stabilizing agents, the stability of these nanoparticles was ascribed to the adsorption of EG and OH-ions on their surfaces,

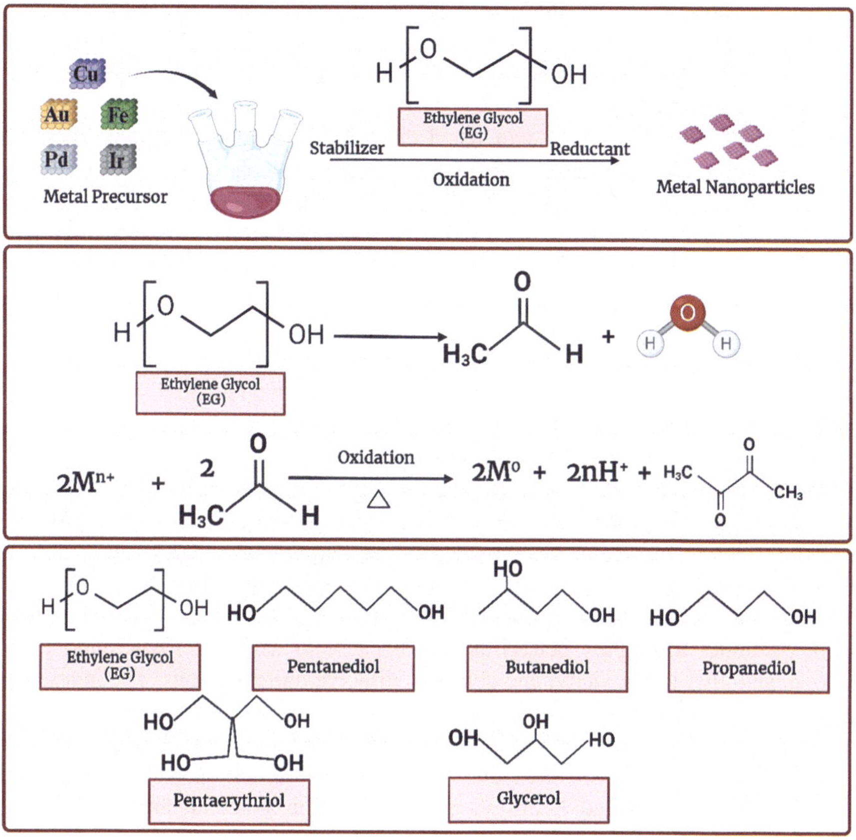

Fig. 16.2 Ethylene glycol mediated synthesis pathway

which prevented aggregation. Pt/Ru and Pt/Rh bimetallic nanoparticles have also been synthesized using the alkaline EG synthesis technique (Wang et al. 2000).

16.2.1.3 Benzyl Alcohol Assisted Synthesis

Benzyl alcohol is a colorless, transparent liquid with minimal vapor pressure, excellent polarity, solubility in water and solvents, and a subtle aromatic fragrance. It also has mild toxicity. It is frequently employed as a solvent and textile dye assistance, as well as in industrial processes and perfumes. In nature it is noticed in plants, fruits, tea, and wines, it is a constituent in commercial and pharmaceutical products (Corcoran and Ray 2014). Without the use of surfactants, binary, ternary, and doped metal oxide nanocrystals as well as oxide-based hybrid materials may be produced by the benzyl alcohol approach. Using an autoclave set to 180–230 °C, benzyl alcohol is reacted with precursors such as metal chloride, alkoxide, acetate, or acetylacetonate. Excellent quality crystallized metal oxide nanoparticles may be continuously synthesized using this easy, time- and energy-efficient approach. The initial concentration of the precursor and the duration of the irradiation process control the crystal size. Tert-butanol can serve as a substitute if oxygen particles are being targeted without chelating surface ligands (Bilecka and Niederberger 2010; Mutin and Vioux 2016).

A typical alcohol to react with metal chlorides was benzyl alcohol (BZOH). Because of the p-π conjugation, BZOH is an aromatic alcohol with a benzene ring on the α-carbon, which efficiently distributes the positive charge of the carbocation onto the big benzyl group. This renders it more vulnerable to ionization and nucleophilic agent damage. Because metal chlorides are inexpensive, oxygen-free precursors that supply metal resources and prevent oxygen from interfering with the production of oxides, they enable a deeper understanding of reaction processes (Hu et al. 2013). Because the carbon–oxygen dissociative interaction is endothermic, raising the temperature can increase the reactivity of benzyl alcohol. Based on this property, it was discovered that several metal chlorides might react with benzyl alcohol by varying the reaction temperature.

In this study, metal chlorides, metallic alkoxides, acetates, acetylacetonates, and other metal organic groups are reacted using benzyl alcohol as a solvent to investigate the synthesis of various metal oxides (Niederberger et al. 2002; Ba et al. 2005). Prior research has mostly focused on high-valence metal chlorides, with little attention paid to low-valence metal chlorides that include benzyl alcohol. This work continues to employ benzyl alcohol because to its distinct reactivity, great thermal stability, low toxicity, and ability to dissolve inorganic ions. The investigation intends to further understand and categorize the reaction process in addition to broadening the reaction's field of use. Metal chloride-based benzyl alcohol-related interactions are known to be one type of alcohol-driven synthesis that uses alcoholic solvents with an elevated hydroxyl group ionization potential and can readily provide oxygen supplies. It has also been discovered that tertiary alcohol can supply oxygen donors for the manufacture of metal oxides through studies into tertiary alcohol-derived reactivity with metal chlorides. The E1 reaction, however, is the main reaction that drives the production of metal oxides in tertiary alcohol-based synthesis because tertiary carbocation is easier to eradicate a proton from and convert to an alkene. For benzyl alcohol, the SN1 reaction pathway is the one that matters. The purpose of this work is to extend the idea of organic to inorganic synthesis and demonstrate the validity and application of the SN1 reaction principle in explaining this benzyl alcohol-based synthesis (Pinna et al. 2004; Bilecka et al. 2009; Szeifert et al. 2010).

16.2.1.4 Methyl Isobutyl Ketone Assisted Synthesis

Acetone is an outcome of the manufacturing of cumene. Because of its poor miscibility with water, it may be converted into methyl isobutyl ketone (MIBK), a solvent used in paints, rubbers, medicines, and liquid-liquid extraction of valuable metals. The three stages involved in producing MIBK in a batch reactor are acetone coupling to make diacetone alcohol, diacetone alcohol dehydration to form mesityl oxide, and mesityl oxide hydrogenation to form MIBK via a metal catalyst. This process is the conventional method. Recent studies, however, have demonstrated that MIBK may be made more effectively in a single step with a metal-supported catalyst. Acetone condensation can occur via either an enol or an enolate process, depending on the acidity/basicity of the support. The intensity of the acid/base sites also influences the pace of conversion (Sifniades et al. 2010; Diallo-Garcia et al. 2011; Shylesh et al. 2015).

In this investigation, Esumi and associates investigated the production of palladium nanoparticles by the thermal breakdown of bis(2,4-pentanedionato) palladium (II) Pd(acac) in a range of organic solvents without the use of additives, including DMSO, p-xylene, oxylene, bromobenzene, and chloroform. Pd(acac) had to be dissolved in an organic solvent and refluxed at boiling point for a duration of 20–24 h as part of the preparation process. A stable dispersion of Pd nanoparticles was initially generated by MIBK, which served as simultaneously the solvent and the stabilizing ingredient (Esumi et al. 1989). The mean diameter of the as-prepared Pd nanoparticles was 8–10 nm, and their size distributions were mostly unaffected by the quantity of precursor Pd ions. Solvents that display high dipole moments have the tendency to dissolve the palladium particles, stabilizing the Pd nanoparticles in MIBK. Chen and colleagues employed microwave irradiation to break down Pd acetate in a MIBK solution with KOH, resulting in the synthesis of solvent-stabilized Pd nanoparticles with approximate diameters of 33 nm. The solvent-stabilized Pd nanoparticles were shown by TEM and X-ray diffraction studies to be composed of clusters formed from numerous tiny 3–4 nm palladium nanocrystals.

16.2.2 Simple Ion Assisted Synthesis

16.2.2.1 Citrate Assisted Synthesis

A typical stabilizing compound for metal nanoparticles is citrate, which offers a negatively charged surface that may be moved by ligands or other molecules (Fig. 16.3). Due to their lack of steric stability, noble metal nanoparticles are prone to aggregation in non-aqueous solvents and elevated salt solutions despite having good stability in water and weakly-buffered solutions. Its tiny molecule forms weak associations with the surface of the particles due to the numerous carboxylic acid groups. Because of its great movability, the surface can interchange ligands with other ligands or proteins. Molecules containing amines or thiols can

Fig. 16.3 Citrate surface

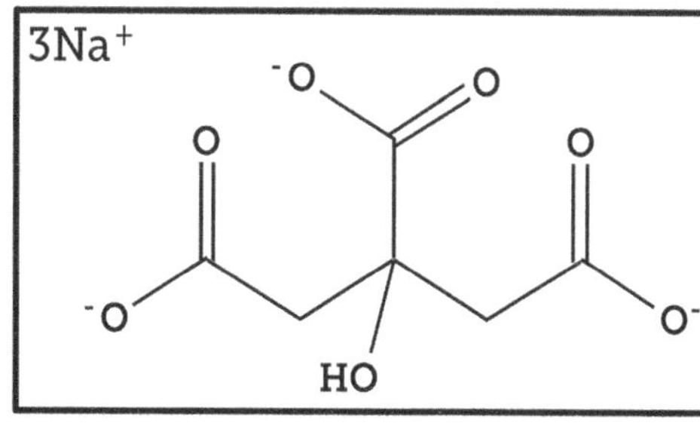

readily replace citrate and form a strong bond with surfaces that are gold or silver. The hydrodynamic diameter and the TEM measured diameter are almost same. Citrate ions undergo interactions with the surfaces of gold or silver nanoparticles, as reported by Turkevich et al., when utilized as a reductant in the formation of metal colloidal materials (John Turkevich et al. 1941). In the citrate reduction method used to produce gold colloids, citrate ions serve as simultaneously a reducing agent and a stabilizing factor. These ions form a charged coating on the nanoparticle surfaces that generates electrical repulsive forces in an aqueous environment. The interaction of the amino or thiol group with the gold nanoparticles displaces poorer bound citrate ions. The size of the gold nanoparticles is dependent on the ratio of gold ions to the reducing agent. Turkevich et al.-based approaches have been developed to alter the sizes and shapes of the nanoparticles. A variety of sizes, from 9 to 120 nm, may be produced in Au nanoparticles using the citrate-mediated synthesis method. For stable nanoparticles, the lowest possible actual amount of the gold ions is under two millimeters. Elevated concentrations cause precipitation on vessel walls, but suspensions at concentrations below 1 mm remain for several months. Utilizing revised Turkevich strategies, monodispersed sub-10 nanometer silver and gold nanoparticles have been generated subsequently (Pillai and Kamat 2004; Uppal et al. 2010).

Turkevich strategy is used to study the processes of nucleation of nanoparticles, conversion of silver ions into atoms, and citrate oxidation. Citrate is not as powerful a reductant as the oxidized radical Cit(−H)•. The generation ratios of CO_2 and dicarboxy acetone (DCA) during Cit(H)• oxidation validate the decarboxylation mechanism (Al Gharib et al. 2019). Pulse radiolysis of solutions containing silver perchlorate and sodium citrate from 20 ps to 800 ms exhibits transient spectra and kinetics. The mechanism initiates the nucleation and growth of reduced silver oligomers by transferring one electron from citrate radicals to silver ions. OH• radicals sequentially abstract H from citrate after this. The reduction potential of citrate and Cit(-H)• can account for a portion of the antioxidant properties of the compound.

16.2.2.2 Amino Acid Assisted Synthesis

Due to their zwitterionic behavior, lack of a tightly linked bond, and structural chirality, amino acids are significant compounds (Nidya et al. 2015). Because they may be utilized as capping agents for the formation of nanoparticles (Vemula et al. 2007) despite having any negative effects, amino acids and vitamins are environmentally safe substances. Protein building components, amino acids, are the perfect biodegradable capping agents for fabricating metal nanoparticles without the need for an extra surfactant. Amino acids are used in this approach as a reductant and stabilizer, much like citrate is. For the formation of metal nanoparticles, a variety of amino acids can be employed, including glutamic acid, histidine, aspartic acid, tryptophan, phenylalanine, cysteine, and methionine. These amino acids have stabilizing and reductant properties (Shao et al. 2004; Ma and Han 2008; Liu et al. 2010). The quickest reducing agent among the twenty amino acids is tryptophan. Using histidine as a reducing agent and a protective ligand, Yang et al. produced water-soluble Au 10 NCs that fluoresced high-intensity bluish-green light, as demonstrated by a straightforward process (Yang et al. 2011).

Reducing stability and managing the size and shape of nanoparticles (NPs) are problems for green chemical techniques. One way to address this issue is to use eco-friendly reagents such as α-amino acids (α-AA) as capping and reducing agents. In order to conjugate important biomolecules, improve biocompatibility, and minimize problems in biomedical applications, α-AA provides amino and carboxy groups. It has been demonstrated to be effective in producing homogeneous, quasi-spherical gold nanoparticles by reducing gold salts. The benefits of α-AA are its availability, low cost, and nontoxicity. Numerous research works have documented the use of one, several, or all 20 proteinogenic α-AA in the synthesis of AuNPs. Nonetheless, the majority of research uses distinct reaction parameters for various amino acids. For contrast, only three papers employed all 20 proteinogenic α-AA for AuNPs production under comparable or identical reaction circumstances. Gold complexes can be reduced by α-AA in a concentration-dependent way, where smaller particles result from greater concentrations and vice versa. A reaction technique for converting gold complexes with amino acids has been suggested in a short number of works (Berghian-Grosan et al. 2014; Chakraborty et al. 2018; Sun et al. 2018).

The study used various reagents, including l-glycine, l-alanine, l-valine, l-leucine, l-isoleucine, dl-phenylalanine, l-tryptophan, l-methionine, l-proline, l-cysteine, l-glutamine, l-asparagine, l-tyrosine, l-serine, dl-threonine, l-(4)-hydroxyproline, l-glutamic acid, l-aspartic acid, l-histidine, l-lysine, and l-arginine, to investigate the physicochemical properties of α-amino acids (Figat et al. 2023). These properties, including solubility, stability, pK, and pI, were expected to influence the reactions' outcomes. The pK is particularly important as it is known that the form of gold salts and reducing agents affects the reactions' results. The study aimed to understand the unique physicochemical properties of these amino acids (Das and Raj 2011).

16.2.2.3 Iodide Assisted Synthesis

When forming metal nanoparticles, such anisotropic Au nanoparticles, iodide is a commonly employed surfactant. Raj et al. showed that single-crystalline polyhedral Au nanostructures may be synthesized via iodide-mediated reduction of $AuCl_4$-ions. This is accomplished by disproportioning Au (I) to metallic Au (0) and reducing an Au (III) complex to a metastable Au (I) complex. The quantity of iodide ions and the existence of Ag ions determine the form and surface structure of Au nanoparticles. Higher concentrations of KI (10 μ m) lead to uncertainty, whereas lower quantities protect the Au nanoparticles. Dendritic Au nanostructures develop more quickly when silver ions are present (Rashid et al. 2006).

Metal nanocrystals (NCs) based on rhodium have been shown to be efficient catalysts for a number of processes, such as hydroformylation, CO oxidation, NO reduction, hydrogenations, and electro-oxidations. However, in comparison with palladium, platinum, silver, and gold, their molecular-level control has received less research attention. A research employing foreign metal crystal surfaces as substrates found Rh epitaxial overgrowth initiated by iodide. Different metal substrates can support epigrowth, and their catalytic characteristics were studied in the electro-oxidation of formic acid and CO stripping. In the overgrowth process, iodide is essential, even on gold surfaces with a significant lattice mismatch. Selected etching of core substrates has also been used to create hollow Rh nanostructures. This new role of iodide in the overgrowth and high control for Rh could be crucial for future nanoscale control of this important metal's architecture for heterogeneous catalysis (Sneed et al. 2012).

16.2.2.4 Buffer Assisted Synthesis

Sodium borohydride ($NaBH_4$) and/or sodium citrate are used as reducing agents in the chemical transformation of chloroauric acid ($HAuCl_4$) to create gold nanoparticles (AuNPs). To prevent therapeutic side effects, safe, biocompatible NPs with improved furtivity and biocompatibility are needed for AuNP uses in medicine. Natural macromolecules have garnered attention as an eco-friendly method of producing AuNPs has emerged in recent years. There have been reports of natural substances such as chitosan, cellulose, and spider-silk fiber acting as Au^{3+} reducing agents. For highly biocompatible AuNPs, proteins which are currently utilized in nanoparticle synthesis make excellent candidates. Based on drug-gold complexes, Moustaoui et al. suggested a novel technique for producing AuNPs as nanotherapeutic agents (Arib et al. 2021).

Xie et al. used 2-[4-(2-Hydroxyethyl)-1-piperazinyl] ethanesulfonic acid (HEPES) as a relatively mild reducing, protecting, and shape-directing agent for the beneficial surfactant-free manufacture of Au nanoflowers. The process involved mixing 100 mM HEPES with 9 mL of deionized water, then adding 250 μl of 20 mM $HAuCl_4$ solution. The resulting Au nanoflowers, also called gold nanoflowers, had more than ten points and were almost spherical in shape. They also included three-dimensionally expanding nanomaterials. Three stages are involved in producing gold nanoflowers: reducing Au(III) ions to basic Au nanocrystals; agglomerating initial Au nanocrystals into intermediary agglomerates; and developing agglomerates anisotropically into flower-like nanostructures (Xie et al. 2008).

16.2.2.5 Tetrakis (Hydroxymethyl)Phosphonium Chloride (THPC) Assisted Synthesis

THPC is used in the production of silver, gold, platinum, and palladium metal nanoparticles as a reducing and stabilizing agent. It makes it easier for metal ions to be reduced under control to create nanoparticles with the correct sizes and characteristics. THPC was employed in the study as a reducing agent to create silver, platinum, and palladium seed nanoparticles, which had a diameter of around 4 nm. Afterwards, on silica cores, these nanoshells were utilized to create pure silver, platinum, and palladium nanoshells as well as hybrid platinum/silver nanoshells. Transmission electron microscopy (TEM), selected area electron diffraction, X-ray photoelectron spectroscopy, FTIR, and ultraviolet-visible spectroscopy were among the methods used to describe the nanoshells. A variety of metal nanoshells, including very uniform platinum and palladium nanoshells, were quickly seeded to develop using the THPC-mNPs. The researchers also optimized the nanoshell syntheses for generating metallic shells with continuous, smooth surfaces (Bryan et al. 2016). With potential uses in catalysis (Hueso et al. 2013), tetrakis-(hydroxymethyl)-phosphonium chloride (THPC) has been employed as a stabilizing ligand and reducing agent in the one-step production of monometallic nanoparticles and bi−/tri-metallic nanoalloys comprising noble metals. These colloidal suspensions exhibit excellent stability in aqueous solutions and mean sizes of less than 4 nm.

16.2.3 Physical Process-Mediated Synthesis

16.2.3.1 Photochemically-Mediated Synthesis

Recent advances in photoredox catalysis have led to a significant rise in the use of synthetic processes mediated by photochemistry. But problems with photochemical reaction scaling, mainly with light entering bigger reaction vessels, secondary photoreactions at longer reaction times, and thermal effect control, have forced continuous flow photochemical reactors to be widely used in labs and development settings. Segmented flow is frequently needed for multiphasic systems (liquid/liquid, gas liquid), yet the majority of reactor designs are straightforward tubular reactors that manage homogenous reactions. Using patterns or oscillations, mixing is frequently utilized to increase mass transfer rates. According to recent reports, there is a need for photochemical reactors that can handle solid/liquid biphasic systems. These reactors can be continuous stirred tank reactors (CSTRs), plate flow reactors that combine oscillatory flow and static mixers, or immersion well reactors operating in semi-continuous recirculatory mode. The 'fReactor' is a compact, easily reconfigurable, modular CSTR platform with strong mass transfer characteristics that can handle multi-phasic mixes of gas, liquid, and solid. This technology, which combines strong mass transport with a high surface area to volume ratio for good light penetration, has demonstrated first uses as a continuous flow photochemical reactor (Grimm et al. 2020; Bonfield et al. 2020).

McGilvray et al. described the photochemical synthesis of stable, unprotected gold nanoparticles without the need for conventional stabilizing compounds. They subjected it to radiation at a wavelength of 350 nm using 1-[4-(2-hydroxyethoxy) phenyl]-2-hydroxy-2-methyl-1-propane-1-one (Irgacure-2959). The reducing agents called ketyl radicals changed Au^{3+} into Au 0 to develop Au nanoparticles. Over several months, the synthesis process proceeds quickly and steadily. To modify the particle size, the illuminating radiation's intensity can be adjusted. Particles with a diameter of 150–300 nm were formed after 3 days of exposure to room lighting, whereas Au nanoparticles with sizes of 40 ± 10, 12 ± 3, and 8 ± 2 nm were created by irradiation with light intensities of 7, 40, and 100 W/m^2. Water-soluble thiols such as 3-mercapto-1-propane sulfonic acid (MPSA) can be used to functionalize these Au nanoparticles (McGilvray et al. 2006).

16.2.3.2 Sonochemically Assisted Synthesis

For the synthesis of noble metal nanoparticles (NPs) like Au, Ag, Pt, and Pb, the sonochemical process is an intriguing approach (Bang and Suslick 2010; Kumar et al. 2020). The underlying process is called cavitation, which produces gaseous bubbles of μm sizes that burst and collapse inside the medium. As a result, there is sufficient chemical susceptibility to decrease noble metals, including silver, since there is enough energy, temperature, and pressure to change solvent molecules into highly reactive free radicals. Sonochemical processes provide several benefits, including as spherical and uniform variations, symmetric distributions, and superior purity of nanomaterials, rapid reaction speeds, adjustable settings, clarity, and stability. Sonochemical techniques are not dependent on high temperatures, extended reaction durations, pH regulation, shape control agents, or steric capping agents like NP wet synthesis techniques are. Sonochemistry is a promising method for producing AgNPs in the sub-15 nm range with specified dimensional features and possibly stable metallic nuclei in aqueous conditions for an extended length of time since it also provides the ability to produce AgNPs with particular particle sizes (Bang and Suslick 2010; Xu and Suslick 2010).

The primary process for the sono-chemical production of metallic nanoparticles is the cavitation phenomena. It produces gaseous bubbles with μm diameters that burst and collapse inside the medium, producing enough heat, pressure, and energy to change solvent molecules into reactive species' free radicals, which may lower noble metals like silver (Restrepo and Villa 2021; Calderón-Jiménez et al. 2022).

Utilizing ultrasonic irradiation as a green reducing agent, a green sono-chemical process was devised to manufacture silver nanoparticles (Ag-NPs) at various concentrations of kappa carrageenan (κ-carrageenan). As the concentration of κ-carrageenan increased, so did the amount of Ag-NPs. Ag/κ-carrageenan was formed, as demonstrated by UV-visible spectroscopy, and X-ray diffraction showed a face-centered cubic structure. The TEM and SEM pictures revealed a spherical form, and the green and economical characteristic was provided by light irradiation (Elsupikhe et al. 2015).

16.2.3.3 Laser Ablation-Mediated Synthesis

Colloidal metallic nanoparticles such as silver and gold are produced by laser ablation of a solid noble metal target in a liquid media. There are several solvents utilized, including n-hexane, ethanol, and water. Larger clusters of metal are formed when smaller metal clusters are ablated from the solid metal. Even in the absence of surfactants, these metal nanoparticles remain stable against aggregation in water. It was discovered that the partly oxidized surfaces of Au nanoparticles created by laser ablation in water were negatively charged. The polarity of the solvent, processing duration, and laser intensity may all be adjusted to regulate the formation of nanoparticles in the absence of surfactants (Sylvestre et al. 2005; Muto et al. 2007).

Wet chemical techniques provide higher productivity and more control over laser-ablated nanoparticles. Colloid concentration has grown along with productivity thanks to recent developments in dynamic flow-based laser ablation technology. By adjusting solvent flow and flow conditions, Freeland et al. (2021) were able to accomplish effective silicon nanoparticle ablation (Freeland et al. 2021; Balachandran et al. 2022). Chemically pure silver-109 nanoparticles were obtained by Aneta Plaza et al. using a 2D galvanometer scanner in conjunction with a pulsed fiber laser. Marc Labusch et al. achieved steady ablation processes by use of acoustic emission monitoring. The colloid quality and environmental friendliness of LASiS are making them increasingly appealing for the production of conductive inks and nanoparticles. This overview covers the variables that influence the characteristics and yield of LASiS nanoparticles, how printed electronics use them, and the problems that need to be solved to enhance performance (Labusch et al. 2019).

16.3 Application of Metallic Nanoparticles Synthesized Using Surfactant Free Solutions

Ultrahigh cooling performance is constrained by a multitude of industrial technologies to maintain the instrumentation cool for efficient operation. Heat-transfer efficiency of the base fluids is improved by the formation of nanomaterial suspension (Wan et al. 2012). Numerous research teams have created nanofluids using various nanomaterials, such as metal and metal oxide, and they have demonstrated improved thermal conductivity (Baby and Ramaprabhu 2011). For appropriate dispersion, nanoparticles are typically either dispersed in surfactants or functionalized by a capping agent. The thermal conductivity of nanofluids may be reduced by the addition of surfactants and/or capping agents.

Using no capping agent, nickeloxide (NiO) and cobaltoxide (CoO) nano-particles were created in the first stage. The resulting nanoparticles were then disseminated in base-fluid using gentle ultrasonic mixing technique in the second stage. In reality metal salts are precipitated with hydroxide ions and then dehydrated to create metal oxides. Particle sizes can vary when water is utilized, and this might be challenging to regulate the process when using organic solvents. Ethanol, serves as a solvent along with source of OH ions. At elevated temperatures, a slow release of OH ions can be achieved. This method gives you control over particle size while also reducing the quantity of additional additives you employ in the reaction (Parashar et al. 2010).

Rashmi Parashar and colleagues presented a novel and simple method for creating scattered metal oxide nanoparticles without the use of capping ligands or inorganic bases. A separate 0.2 M solution of Cobalt (II) chloride hexahydrate and Nickel (II) acetate tetrahydrate was produced in 25 ml of HPLC grade water for the production of CoO and NiO nanoparticles. After adding 1.5 ml of 0.2 M Cobalt (II) chloride hexahydrate aqueous solution and 1.5 ml HPLC grade water, 24 ml of room-temperature ethanol solvent was added. For 20 h, the reaction was stirred at 80 °C while the container was maintained tightly closed. After that, the first step's reaction-mix was transferred to a 25 ml autoclave along with a teflon lining so that it could undergo a hydrothermal reaction for 3 h at 160 °C.

The identical procedures used to make cobalt oxide nanoparticles were used to prepare nickel oxide nanoparticles. The thermal conductivity of nanofluids was determined using the Hot Disk Thermal Constants Analyzer. The device measures the thermal conductivities of nanofluids using the transient plane source method. Since the current synthesis technique does not allow for the filtering or separation of the nanoparticles of extremely small sizes from the suspension product, they utilized the nanoparticles as liquid suspensions whose portion added to base fluids, such as ethylene glycol. Heat conduction observed in NO-EG nanofluids is lower than in CO-EG (Parashar et al. 2014).

16.3.1 Synthetic Catalysis

The existence of strongly bound organic layers which is protective on action and surrounding metal nanoparticles might render it unfit for catalytic applications, since this process occurs on surfaces of metal. Hence for catalysis, the metal nanoparticles produced by manufacturing techniques which devoid use of surfactants will be appropriate. Studies on supported metal nanoparticles, or those utilized in heterogeneous catalysis, done extensively (Zhang et al. 2012; Wu et al. 2012) however they are not within the purview of this review. Only unsupported metal nanoparticles, or colloidal metal nanoparticles, that are produced as synthetic catalysts without the use of surfactants are covered in this review. The section on electrocatalysis will cover heterogeneous catalysis employing supported metal nanoparticles made without the need of surfactants.

The surfactant devoid synthesis of Pd-nanoclusters for use in catalytic cross-coupling reactions was described by Obora and colleague (Hyotanishi et al. 2011). The DMF-based reduction process was used to create the Pd-nanoclusters, which were 1–1.5 nm in size. Because the DMF solvents stabilized the Pd-nanoclusters, they displayed good dispersion in a variety of polar solvents. During organic synthesis, the Suzuki Miyaura and Mizoroki Heck cross-coupling reactions are two very efficient ways to create carbon-carbon bonds. The DMF-stabilized Pd-nanoclusters investigated as catalysts in these processes. In Suzuki-Miyaura and Mizoroki-Heck cross-coupling reactions, the Pd-nanoclusters exhibited strong catalysis capability. In the presence of 10^{-7} mol% of Pd-nanoclusters, the turnover number (TN) was as high as $6.0 \times 10^{8.}$ The Pd-nanoclusters exhibited a greater catalytic activity compared to the subnanometer Pd-nanoclusters (about 0.7 nm) supported on polymer micelles, demonstrating a high TN of around 2.8×10^{5} (Okamoto et al. 2005). Additionally, by employing a liquid-liquid extraction technique, the researchers devised a way to recycle this catalyst during Suzuki-Miyaura cross coupling reaction at least five times.

Precious metals are expensive, which restricts the industrial uses for them. Because copper is a cheap metal, it is frequently utilized in industry as an alternative. Significant research has been done to determine whether copper may be used in catalysis (Evano et al. 2008; Rao and Fu 2011). Obora and associates used the DFM-based reduction approach to create Cu nanoparticles having size roughly 2 nm, and they did it without the need for surfactants. In case of Ullmann-coupling reaction, these Cu nanoparticles served as catalysts (Isomura et al. 2012). As aryl halides and phenols were cross-coupled as per Ullmann reaction in the absence of ligand, Cu nanoparticles showed an elevated catalytic ability with a TN of 2.2×10^{-4}.

In a well-known model reaction that is catalysed by metals (Yamamoto et al. 2012; Hervés et al. 2012) 4-nitrophenol is reduced to 4-aminophenol by $NaBH_4$. Kawasaki and colleagues examined catalytic characteristics of DMF-stabilized Au NCs. Even at low concentrations (10^{-7} g), the DMF-stabilized Au NCs had strong catalytic ability; at 25 °C, the pseudo-first-order rate constant (k app) was 1.0×10^{-2} for 1.5 μm of the gold catalyst. The idea was that the key to obtaining a high catalytic activity exhibiting DMF-stabilized Au NCs was the reorganization of the DMF layer, which is really an activation process.

Mandal and colleagues presented a wet-chemical approach of preparing metal nanoparticles without the use of surfactants. These nanoparticles were also employed by them as artificial catalysts. The production of Au nanoparticles ranging in size from 20 to 140 nm was achieved through utilizing ammonium bismuth citrate as stabilizing and reducing agent. With a great surface area 42 m^2/g, the Au nanoparticles resembled sponges and bearing pores with sizes 2.4–6.0 nm. With a rate constant $k = 2.1 \times 10^{-3}\ s^{-1}$, these sponge-like Au nanoparticles demonstrated a strong catalysis in the reduction of PNP using $NaBH_4$ (Rashid et al. 2006). Moreover, ferric ammonium citrate utilized as reducing agent to create polygonal Au nanoparticles without need for a polymeric template or surfactant (Rashid and Mandal 2008). When it came to the reduction of nitrophenol, polygonal Au nanoparticles' catalysis outperformed that of the spherical Au nanoparticles made using the Turkevich approach by a factor of 300–1000. Additionally, an increase in activity comparable to this was noted in the aerobic oxidation of several d-hexoses. The polygonal gold nanoparticles enhanced catalytic activity was ascribed to their increased number of sharp corners and edges.

The ability of citrate stabilized Au nanoparticles to catalyze the reaction of thiosulfate ions with ferrercyanide was reported by Freund and colleagues (Freund and Spiro 1985). The catalytic rate rise as the quantity of Au nanoparticles increased, indicating the catalytic rate proportional to the surface area of Au nanoparticles. The reaction was proposed to have happened on the surfaces of the Au nanoparticles and not been restrained by diffusion. According to research by Rossi and colleagues (Comotti et al. 2004; Welch and Compton 2006), Under strongly basic conditions, bare Au nanoparticles with a mean diameter of 3.6 nm were employed to enhance aerobic oxidation of glucose in an aqueous solution. This process allowed 21% conversion of glucose within the 200 s. The researchers discovered that the diameter of Au nanoparticles was inversely correlated with their catalytic activity, utilizing Au nanoparticles ranging in size from 3 to 6 nm. Au nanoparticles bigger than 6 nm size, on other hand, didn't exhibit a linear rise in catalysis; instead, sudden change was noted at about 10 nm in diameter. This demonstrated the catalytic activity's discontinuity at the nanoscale.

The same quantity of gold was used to examine the differences in the catalysis of Au nanoparticles supported on carbon versus those that are not. In first 100 s of the oxidation of glucose, the two types of nanoparticles' initial response rates were rather equal. The inclination of Au nanoparticles to expand and forms agglomerates in the medium, however, meant that this colloidal Au-nanoparticles were unstable and eventually experienced deactivation.

16.3.2 Electrocatalysis

Main goal of research is creating an effective metal nanoparticle electrocatalysts that may transfer electrons between reactants and electrodes. In order to attain a current density that is near to equilibrium potential, these catalysts transfer electrons. Electrode-supported metal nanoparticles have been used to exhibit electro-catalytic reactions for example H_2O_2 reduction, CO oxidation, oxidation, and oxygen reduction. However, the use of surfactants-polymers, which hindered catalytic sites and reduce catalytic activity, is frequently necessary for the production of materials with predictable sizes and shapes (Welch and Compton 2006; Hu and Dong 2008).

According to Zhou et al. (2003), a polyol method free of surfactants was used to create Pt nanoparticles the mean diameter of roughly 2.9 nm. Carbon was used to support the nanoparticles, and up to 40% weight percentage of Pt was loaded. In comparison to commercial catalysts, the as-synthesised Pt/C electrocatalyst demonstrated superior electrocatalysis as cathode in course of reaction for oxygen reduction in direct methyl fuel cells (DMFC).

A study by Li et al. (Kiadó and Vol 2004) described the creation of Pt/MWNT nanocomposites, which are supported by multiwalled carbon nanotubes. The Pt salt was reduced using an aqueous solution (HCHO reduction) and a Ption salt was reduced in an EG solution. In DMFC, those nanocomposites served as cathode catalysts. The Pt nanoparticles, ranging in size from 2 to 5 nm, were uniformly distributed through the utilization of the surfactant-free polyol technique in the context of Pt/MWNTs. With a Brunauer-Emmett-Teller (B-E-T) having surface area as 237 m^2/g, the Pt/MWNT nanocomposites exhibited a much greater catalytic performance than the commercial catalyst Pt/XC-72. As a result, the surfactant-free polyol method of producing tiny metal nanoparticles provides an appealing means of obtaining heterogeneous catalyst (Wang et al. 2005).

Using a surfactant-free, microwave-assisted polyol synthesis process, Kundu et al. (2011) developed Pt nanocomposites supported by reduced graphene oxide (rGO) with a diameter of 2–3 nm that can serve as electrocatalysts in the oxidation of methanol and conversion of H_2.

The rGO/Pt nanocomposites' tolerance, methanol oxidation potential, and current were measured; the findings demonstrated that the catalysts had high H_2 conversion and strong electrocatalytic activity. By combining a $HauCl_4$ solution with rGO, Yin et al. created a unique method for growing ultrafine Au NCs on rGO sheets without the need for an extra protective molecule or reductant (Yin et al. 2012). The graphene oxide sheets had Au NCs with an average diameter of 1.8 nm and a limited dispersion of 0.2 nm.

This method might also be used to get Pt and Pd agglomerates of other metals on the rGO sheets. When utilized as the cathode in fuel cells, the synthesised Au

NC/rGO hybrids demonstrated outstanding performance as catalysts in the oxygen reduction reaction. The hybrids showed good stability, better methanol tolerance, and great onset potential.The reduction of the precursors and the subsequent annealing of the reduced precursors in glycerol at 300 °C are the two phases of the one-pot technique of synthesising AuCu intermetallic nanoparticles (iNPs) supported on high-surface-area carbon, as reported by Wang et al. (2012). Toto avoid blocking the AuCuiNPs' active sites, no surfactant was used throughout the synthesis procedure.

When utilized in oxygen reduction in an alkaline media, it's discovered that AuCu-iNPs exhibited a catalysis greater than that of the regular Au-nanoparticles. Anisotropic Pt nanostructures, including nanorods, nanodendrites, nanowires, and nanotubes, have also been reported to demonstrate improved catalytic performance, in addition to single-nanosized Pt-NCs (<3 nm) (Subhramannia and Pillai 2008). By avoiding the use of surfactants, templates, or seeds, Jia et al. (2011) created Pt nanostructure networks with superior electrochemical activity and durability by chemically reducing chloroplatinic acid with benzyl alcohol under microwave irradiation. To assess the Pt nanostructured networks' suitability for use in DMFCs, their catalytic activity during the oxidation of methanol was measured.

According to the quantity of methanol oxidized on the Pt network, the peak anodic current magnitude during the forward scan at roughly 0.65 V was proportionate. When used as an electrode, Pt nanostructured networks had a mass current density of 0.32 A mg^{-1}, which was 2.7 times higher than that of a Pt/C electrode found in stores (0.12 A mg^{-1}). Without the need of a template or surfactant, Sun et al. (2008) showed how to manufacture 3D flower-like platinum nano-structures on a large scale by reducing hexachloro-platinic acid with formic acid at ambient conditions. Following the mixing of those two solutions, the Pt nanoflowers grew progressively. The massive amounts of single-crystal nanowires that were created by the growth of planes made up the as-prepared Pt nanoflowers. Pt nanoflowers that were deposited on carbon paper were subjected to cyclic voltammetry (CV), with the paper being submerged in H_2SO_4 at a pH of 1. This showed that the surface of the Pt nanoflower-carbon paper had more Pt accessible electrochemically than that of marketed electrode.

16.3.3 Surface-Enhanced Raman Scattering

Recent developments in analytics applied as chemosensors-biosensors for various domains of life sciences, environmental supervising, and food safety are currently being invented as the outcome of improvements in the production of metal nanoparticles. One such innovative method that utilizes metal nanoparticles is surface plasmon resonance (SERS), which offers the advantages of label-free investigation, surface selectivity, and multiplexing and high sensitivity (Álvarez-Puebla and Liz-Marzán 2010; Fan et al. 2011; Zhang et al. 2011).

The enhancement factor (EF) in SERS can be as high as 10^{14} or 10^{15} when Au/Ag nanoparticles utilized that it is much greater than the conditions when we don't use metal nanoparticles. A variety of technologies, such as complex lithographic techniques, electrochemical roughening, colloidal lithography, deposition techniques, and solution-phase synthesis techniques, can be used to produce SERS substrates. It would be advantageous if the nanoparticles were to develop without the utilizing surfactant and an extra-capping agent because the layers of such adsorbed organic species would not interfere as much with the SERS analysis.

A simplistic surfactant-free technique employing a suspension of zinc microparticles to serve as heterogeneous-reducing factor produced a great yield (>85%) of Ag-dendrites, according to Wen et al. (2006). 5–50 μ m in length and having 20–30 nm in diameter made up the Ag-dendritic stems. The electrochemical glucose biosensor's sensitivity could rise by one or two orders of magnitude when silver nano-dendrites were incorporated into it. Room temperature, acetone-based mixed solvent method without surfactants was used by Han et al. to create hierarchically ordered silver nanostructures (Han et al. 2009). The key to managing the Ag nanoparticles' nucleation, development, conversion, and assembly in this surfactant-free synthesis technique was acetone.

Acetone was essential in regulating the Ag nanoparticles' nucleation, development, conversion, and assembly in this surfactant-free production technique. When employing the hierarchical Ag nanostructure, adenine's SERS performance was comparable to that of triangular Ag nanoplates. Furthermore, compared to polyhedral Ag nanoparticles, the hierarchical Ag nanostructure produced a significantly better SERS performance. For SERS, Xie et al. employed flower-shaped Au nanoparticles made using the Good's buffer (HEPES) mediated reaction approach. When compared to spherical Au nanoparticles, the flower-shaped nanoparticles showed substantial SERS affects, produced a >10 times rise in SERS intensity (Xie et al. 2008). In the RAW264.7 cell lines, use of Raman-active tags in living cells were experimented.

It has been shown that nanostructured metals, such as copper, nickel, and palladium, work well as SERS substrates same as nanoparticles of gold and silver. The SERS activity of crystalline copper nanostructures made in benzyl alcohol via microwave-surfactant synthesis was assessed by Dar et al. (2012). As a model analyte, 4-mercaptobenzoic acid was used at concentrations as 10^{-3} to 10^{-6} m. This showed that produced stable copper nanostructures were efficient SERS substrates. Pd nanowire (Feng et al. 2008) networks produced using a surfactant-free approach in an EG/DMSO combination were reported by Feng et al. to be effective SERS materials as well, with a lower detection limit 10^{-8} m for 4-mercaptopyridine.

By reducing nickel chloride in EG with a concentrated hydrozine hydrate solution serving as the reducing agent, Krishnadas et al. (2011) produced nickel nanowires. No surfactantsor external magnetic field were used in the process. SERS activity was seen in the nickel nanowires where crystal-violet were employed as analyte. An estimate of EF found as 1.7×10^{-4}.

16.4 Challenges, Limitation, and Future Perspective

Surfactant-free synthesis of metallic nanomaterials offers several advantages but also poses significant challenges and limitations. One of the primary advantages is the elimination of surfactants, which can simplify downstream processing and reduce potential environmental impact. This approach often relies on the use of capping agents or stabilizers such as polymers or biomolecules to control particle size and morphology. However, the absence of surfactants can lead to difficulties in achieving uniform particle size distribution and controlling the nucleation and growth kinetics, which are crucial for obtaining desired properties. Moreover, the synthesis conditions must be precisely controlled to prevent particle aggregation and ensure stability. Scaling up surfactant-free methods to industrial levels presents another challenge due to the need for consistent quality and reproducibility. Future perspectives include advancing the understanding of nucleation and growth mechanisms in surfactant-free environments, exploring novel stabilizers or capping agents to enhance control over nanomaterial properties, and integrating these materials into practical applications such as catalysis, sensing, and biomedical devices. Addressing these challenges and expanding the scope of surfactant-free synthesis could unlock new opportunities for tailoring nanomaterials with enhanced functionalities and sustainability in various technological domains.

16.5 Conclusions

In conclusion, surfactant-free synthesis of metallic nanomaterials holds incredible promise across various fields including medical, pharmaceutical, biological, biomedical, and catalytic applications due to their unique physicochemical properties and high surface-to-volume ratio. Despite the numerous advantages, challenges such as biocompatibility, stability, purity, and reproducibility remain acceptable research area by for researchers and scientists. The variation in size, shape, and reduction percentage of metal oxide and metallic nanoparticles can significantly impact their reactivity and therapeutic efficacy, which requires the use of stabilizers or surfactants to maintain stability and prevent aggregation. Plant-based synthesis offers a sustainable alternative but requires further development to bridge the gap from bench to market. Emerging surfactant-free strategies, such as solvent-mediated synthesis, ion-mediated synthesis, and various physical processes, are being actively explored as eco-friendly approaches that mitigate the use of toxic chemicals and enhance the sustainability of nanomaterial production. These advancements underscore the ongoing efforts to improve synthesis techniques and optimize the properties of metallic nanomaterials for diverse applications. Continued research into understanding nucleation, growth mechanisms, and the influence of synthesis parameters will be crucial for advancing the field towards scalable and reproducible manufacturing processes. Ultimately, integrating these innovations into practical applications has the potential to revolutionize industries by offering enhanced materials with tailored properties for specific biomedical, catalytic, and environmental challenges.

References

Al Gharib S, Marignier JL, El Omar AK et al (2019) Key role of the oxidized citrate-free radical in the nucleation mechanism of the metal nanoparticle Turkevich synthesis. J Phys Chem C 123: 22624–22633. https://doi.org/10.1021/acs.jpcc.9b06090

Álvarez-Puebla RA, Liz-Marzán LM (2010) Environmental applications of plasmon assisted Raman scattering. Energy Environ Sci 3:1011–1017. https://doi.org/10.1039/c002437f

Amooref JE, Hautala E (1983) Odor as an aid to chemical safety: odor thresholds compared with threshold limit values and volatilities for 214 industrial chemicals in air and water dilution. J Appl Toxicol 3(6):272–290

Arib C, Spadavecchia J, de la Chapelle ML (2021) Enzyme mediated synthesis of hybrid polyedric gold nanoparticles. Sci Rep 11. https://doi.org/10.1038/s41598-021-81751-1

Ba J, Polleux J, Antonietti M, Niederberger M (2005) Non-aqueous synthesis of tin oxide nanocrystals and their assembly into ordered porous mesostructures. Adv Mater 17:2509–2512. https://doi.org/10.1002/adma.200501018

Baby T, Ramaprabhu S (2011) Synthesis and nanofluid application of silver nanoparticles decorated. J Mater Chem 21:9702–9709

Balachandran A, Sreenilayam SP, Madanan K et al (2022) Nanoparticle production via laser ablation synthesis in solution method and printed electronic application – a brief review. Results Eng 16. https://doi.org/10.1016/j.rineng.2022.100646

Bang JH, Suslick KS (2010) Applications of ultrasound to the synthesis of nanostructured materials. Adv Mater 22:1039–1059

Berghian-Grosan C, Olenic L, Katona G et al (2014) L-Leucine for gold nanoparticles synthesis and their cytotoxic effects evaluation. Amino Acids 46:2545–2552. https://doi.org/10.1007/s00726-014-1814-z

Bhattarai N, Khanal S, Velazquez-Salazar JJ, Jose-Yacaman M (2015) Advanced electron microscopy in the study of multimetallic nanoparticles. In: Advanced transmission electron microscopy: applications to nanomaterials. Springer International Publishing, Cham, pp 59–92

Bilecka I, Niederberger M (2010) New developments in the nonaqueous and/or non-hydrolytic sol-gel synthesis of inorganic nanoparticles. Electrochimica Acta 55:7717–7725

Bilecka I, Elser P, Niederberger M (2009) Kinetic and thermodynamic aspects in the microwave-assisted synthesis of ZnO nanoparticles in benzyl alcohol. ACS Nano 3:467–477. https://doi.org/10.1021/nn800842b

Bonfield HE, Knauber T, Lévesque F et al (2020) Photons as a 21st century reagent. Nat Commun 11

Bryan WW, Jamison AC, Chinwangso P et al (2016) Preparation of THPC-generated silver, platinum, and palladium nanoparticles and their use in the synthesis of Ag, Pt, Pd, and Pt/Ag nanoshells. RSC Adv 6:68150–68159. https://doi.org/10.1039/c6ra10717f

Calderón-Jiménez B, Montoro Bustos AR, Pereira Reyes R et al (2022) Novel pathway for the sonochemical synthesis of silver nanoparticles with near-spherical shape and high stability in aqueous media. Sci Rep 12. https://doi.org/10.1038/s41598-022-04921-9

Chakraborty A, Boer JC, Selomulya C, Plebanski M (2018) Amino acid functionalized inorganic nanoparticles as cutting-edge therapeutic and diagnostic agents. Bioconjug Chem 29:657–671

Comotti M, Della Pina C, Matarrese R, Rossi M (2004) The catalytic activity of "naked" gold particles. Angew Chem Int Ed 43:5812–5815. https://doi.org/10.1002/anie.200460446

Corcoran GB, Ray SD (2014) Benzyl alcohol. In: Encyclopedia of toxicology: third edition. Elsevier, pp 429–432

Da Costa LP, Formiga ALB, Mazali IO, Sigoli FA (2011) Spontaneous formation of highly dispersed spheroidal metallic silver nanoparticles in surfactant-free N,N-dimethylacetamide. Synth Met 161:1517–1521. https://doi.org/10.1016/j.synthmet.2011.04.018

Dar MI, Sampath S, Shivashankar SA (2012) Microwave-assisted, surfactant-free synthesis of air-stable copper nanostructures and their SERS study. J Mater Chem 22:22418–22423. https://doi.org/10.1039/c2jm35629e

Das AK, Raj CR (2011) Iodide-mediated reduction of AuCl4- and a new green route for the synthesis of single crystalline au nanostructures with pronounced electrocatalytic activity. J Phys Chem C 115:21041–21046. https://doi.org/10.1021/jp207233b

Deshmukh R, Niederberger M (2017) Mechanistic aspects in the formation, growth and surface functionalization of metal oxide nanoparticles in organic solvents. Chem Eur J 23:8542–8570

Diallo-Garcia S, Laurencin D, Krafft JM et al (2011) Influence of magnesium substitution on the basic properties of hydroxyapatites. J Phys Chem C 115:24317–24327. https://doi.org/10.1021/jp209316k

Dubey A, Upadhyay AK, Kumar P (2010) Pivaloyl chloride/DMF: a new reagent for conversion of alcohols to chlorides. Tetrahedron Lett 51:744–746. https://doi.org/10.1016/j.tetlet.2009.11.131

Duchesne PN, Zhang P (2012) Local structure of fluorescent platinum nanoclusters. Nanoscale 4: 4199–4205. https://doi.org/10.1039/c2nr30500c

Elsupikhe RF, Shameli K, Ahmad MB et al (2015) Green sonochemical synthesis of silver nanoparticles at varying concentrations of κ-carrageenan. Nanoscale Res Lett 10. https://doi.org/10.1186/s11671-015-0916-1

Esumi K, Tano T, Meguro K (1989) Preparation of Organopalladium particles from thermal decomposition of its organic complex in organic solvents. Langmuir 5:268–270

Evano G, Blanchard N, Toumi M (2008) Copper-mediated coupling reactions and their applications in natural products and designed biomolecules synthesis. Chem Rev 108:3054–3131

Fan M, Andrade GFS, Brolo AG (2011) A review on the fabrication of substrates for surface enhanced Raman spectroscopy and their applications in analytical chemistry. Anal Chim Acta 693:7–25

Feng C, Zhang R, Yin P et al (2008) Direct solution synthesis of Pd nanowire networks and their application in surface-enhanced Raman scattering. Nanotechnology 19. https://doi.org/10.1088/0957-4484/19/30/305601

Fievet F, Ammar-Merah S, Brayner R et al (2018) The polyol process: a unique method for easy access to metal nanoparticles with tailored sizes, shapes and compositions. Chem Soc Rev 47: 5187–5233. https://doi.org/10.1039/c7cs00777a

Figat AM, Bartosewicz B, Liszewska M et al (2023) α-Amino acids as reducing and capping agents in gold nanoparticles synthesis using the Turkevich method. Langmuir 39:8646–8657. https://doi.org/10.1021/acs.langmuir.3c00507

Freeland B, Mccann R, O'neill P et al (2021) Real-time monitoring and control for high-efficiency autonomous laser fabrication of silicon nanoparticle colloids. Int J Adv Manuf Technol 114:1–14. https://doi.org/10.1007/s00170-021-06772-6/Published

Freund PL, Spiro M (1985) Colloidal catalysis: the effect of sol size and concentration. J Phys Chem 89:1074

Grimm I, Hauer ST, Schulte T et al (2020) Upscaling photoredox cross-coupling reactions in batch using immersion-well reactors. Org Process Res Dev 24:1185–1193. https://doi.org/10.1021/acs.oprd.0c00070

Gul S, Khan SB, Rehman IU et al (2019) A comprehensive review of magnetic nanomaterials modern day theranostics. Front Mater 6

Han Y, Liu S, Han M et al (2009) Fabrication of hierarchical nanostructure of silver via a surfactant-free mixed solvents route. Cryst Growth Des 9:3941–3947. https://doi.org/10.1021/cg900066z

Heravi MM, Ghavidel M, Mohammadkhani L (2018) Beyond a solvent: triple roles of dimethylformamide in organic chemistry. RSC Adv 8:27832–27862

Hervés P, Pérez-Lorenzo M, Liz-Marzán LM et al (2012) Catalysis by metallic nanoparticles in aqueous solution: model reactions. Chem Soc Rev 41:5577–5587. https://doi.org/10.1039/c2cs35029g

Hu X, Dong S (2008) Metal nanomaterials and carbon nanotubes – synthesis, functionalization and potential applications towards electrochemistry. J Mater Chem 18:1279–1295. https://doi.org/10.1039/b713255g

Hu M, Xu J, Gao J et al (2013) Benzyl alcohol-based synthesis of oxide nanoparticles: the perspective of SN1 reaction mechanism. J Chem Soc Dalton Trans 42:9777–9784. https://doi.org/10.1039/c3dt50680k

Hueso JL, Sebastián V, Mayoral Á et al (2013) Beyond gold: rediscovering tetrakis-(hydroxymethyl)-phosphonium chloride (THPC) as an effective agent for the synthesis of ultra-small noble metal nanoparticles and Pt-containing nanoalloys. RSC Adv 3:10427–10433. https://doi.org/10.1039/c3ra40774h

Hyotanishi M, Isomura Y, Yamamoto H et al (2011) Surfactant-free synthesis of palladium nanoclusters for their use in catalytic cross-coupling reactions. Chem Commun 47:5750–5752. https://doi.org/10.1039/c1cc11487e

Isomura Y, Narushima T, Kawasaki H et al (2012) Surfactant-free single-nano-sized colloidal Cu nanoparticles for use as an active catalyst in Ullmann-coupling reaction. Chem Commun 48: 3784–3786. https://doi.org/10.1039/c2cc30975k

Jayeoye TJ, Eze FN, Olatunde OO et al (2021a) Multifarious biological applications and toxic Hg^{2+} sensing potentiality of biogenic silver nanoparticles based on securidaca inappendiculata hassk stem extract. Int J Nanomedicine 16:7557–7574. https://doi.org/10.2147/IJN.S325996

Jayeoye TJ, Eze FN, Singh S et al (2021b) Synthesis of gold nanoparticles/polyaniline boronic acid/sodium alginate aqueous nanocomposite based on chemical oxidative polymerization for biological applications. Int J Biol Macromol 179:196–205. https://doi.org/10.1016/j.ijbiomac.2021.02.199

Jayeoye TJ, Nwude EF, Singh S et al (2024) Sustainable synthesis of gold nanoparticles for drug delivery and cosmeceutical applications: a review. BioNanoScience. https://doi.org/10.1007/s12668-024-01436-7

Jia F, Wang F, Lin Y, Zhang L (2011) Microwave-induced formation of platinum nanostructured networks with superior electrochemical activity and stability. Chem Eur J 17:14603–14610. https://doi.org/10.1002/chem.201101179

Jiang L, You T, Yin P et al (2013) Surface-enhanced Raman scattering spectra of adsorbates on Cu_2O nanospheres: charge-transfer and electromagnetic enhancement. Nanoscale 5:2784–2789. https://doi.org/10.1039/c3nr33502j

John Turkevich B, Cooper Stevenson P, Hillier J (1941) A study of the nucleation and growth processes in the synthesis of colloidal gold. Turkevich and Hillier

Kawasaki H (2013) Surfactant-free solution-based synthesis of metallic nanoparticles toward efficient use of the nanoparticles' surfaces and their application in catalysis and chemo–/biosensing. Nanotechnol Rev 2:5–25

Kawasaki H, Yamamoto H, Fujimori H et al (2010) Surfactant-free solution synthesis of fluorescent platinum subnanoclusters. Chem Commun 46:3759–3761. https://doi.org/10.1039/b925117k

Khan T, Ullah N, Khan MA et al (2019) Plant-based gold nanoparticles; a comprehensive review of the decade-long research on synthesis, mechanistic aspects and diverse applications. Adv Colloid Interf Sci 272

Kiadó A, Vol B (2004) Multi-walled carbon nanotubes supported Pt-Fe cathodic catalyst for direct methanol fuel cell. Kluwer Academic Publishers

Krishnadas KR, Sajanlal PR, Pradeep T (2011) Pristine and hybrid nickel nanowires: template-, magnetic field-, and surfactant-free wet chemical synthesis and Raman studies. J Phys Chem C 115:4483–4490. https://doi.org/10.1021/jp110498x

Kumar R, Wadhwa D, Prakash O (2010) Beckmann Rearrangement of 2-Hydroxy-5-Methylacetophenone Oxime using Vilsmeier-Haack Reagent (POCI3/DMF): synthesis of some new heterocycles. Heterocycl Commun 16(2–3):201–205

Kumar A, Mishra B, Tripathi BP (2020) Polydopamine assisted synthesis of ultrafine silver nanoparticles for heterogeneous catalysis and water remediation. Nano-Struct Nano-Objects 23. https://doi.org/10.1016/j.nanoso.2020.100489

Kumar A, Shah SR, Jayeoye TJ et al (2023) Biogenic metallic nanoparticles: biomedical, analytical, food preservation, and applications in other consumable products. Front Nanotechnol 5

Kumar A, Jayeoye TJ, Mohite P et al (2024) Sustainable and consumer-centric nanotechnology-based materials: an update on the multifaceted applications, risks and tremendous opportunities. Nano-Struct Nano-Objects 38:101148. https://doi.org/10.1016/j.nanoso.2024.101148

Kundu P, Nethravathi C, Deshpande PA et al (2011) Ultrafast microwave-assisted route to surfactant-free ultrafine Pt nanoparticles on graphene: synergistic co-reduction mechanism and high catalytic activity. Chem Mater 23:2772–2780. https://doi.org/10.1021/cm200329a

Labusch M, Cunha APA, Wirtz SF et al (2019) Acoustic emission control avoids fluence shifts caused by target runaway during laser synthesis of colloids. Appl Surf Sci 479:887–895. https://doi.org/10.1016/j.apsusc.2019.02.080

Li C, Shuford KL, Park Q-H et al (2007) High-yield synthesis of single-crystalline gold Nano-octahedra. Angew Chem 119:3328–3332. https://doi.org/10.1002/ange.200604167

Li C, Shuford KL, Chen M et al (2008) A facile polyol route to uniform gold octahedra with tailorable size and their optical properties. ACS Nano 2:1760–1769. https://doi.org/10.1021/nn800264q

Liu X, Li C, Xu J et al (2008) Surfactant-free synthesis and functionalization of highly fluorescent gold quantum dots. J Phys Chem C 112:10778–10783. https://doi.org/10.1021/jp8028227

Liu Z, Zu Y, Fu Y et al (2010) Hydrothermal synthesis of histidine-functionalized single-crystalline gold nanoparticles and their pH-dependent UV absorption characteristic. Colloids Surf B Biointerfaces 76:311–316. https://doi.org/10.1016/j.colsurfb.2009.11.010

Ma Z, Han H (2008) One-step synthesis of cystine-coated gold nanoparticles in aqueous solution. Colloids Surf A Physicochem Eng Asp 317:229–233. https://doi.org/10.1016/j.colsurfa.2007.10.018

Maman MP, Gratious S, Aparna RK, Mandal S (2019) A N, N-dimethylformamide protected rhodium nanocluster with engrossing optical properties and its application. Dalton Trans 48: 6522–6526. https://doi.org/10.1039/c9dt01036j

McGilvray KL, Decan MR, Wang D, Scaiano JC (2006) Facile photochemical synthesis of unprotected aqueous gold nanoparticles. J Am Chem Soc 128:15980–15981. https://doi.org/10.1021/ja066522h

Mutin PH, Vioux A (2016) Non-hydrolytic Sol–Gel routes. In: Reference module in materials science and materials engineering. Elsevier

Muto H, Yamada K, Miyajima K, Mafuné F (2007) Estimation of surface oxide on surfactant-free gold nanoparticles laser-ablated in water. J Phys Chem C 111:17221–17226. https://doi.org/10.1021/jp075582m

Muzart J (2009) N,N-Dimethylformamide: much more than a solvent. Tetrahedron 65:8313–8323

Nagata T, Obora Y (2020) N, N-Dimethylformamide-protected single-sized metal nanoparticles and their use as catalysts for organic transformations. ACS Omega 5:98–103

Nagime PV, Singh S, Shaikh NM et al (2023) Biogenic fabrication of silver nanoparticles using Calotropis procera flower extract with enhanced Biomimetics attributes. Materials 16. https://doi.org/10.3390/ma16114058

Nidya M, Umadevi M, Rajkumar BJM (2015) Optical and morphological studies of L-histidine functionalised silver nanoparticles synthesised by two different methods. J Exp Nanosci 10: 167–180. https://doi.org/10.1080/17458080.2013.812810

Niederberger M, Bartl MH, Stucky GD (2002) Benzyl alcohol and transition metal chlorides as a versatile reaction system for the nonaqueous and low-temperature synthesis of crystalline nano-objects with controlled dimensionality. J Am Chem Soc 124:13642–13643. https://doi.org/10.1021/ja027115i

Nwabor OF, Singh S, Paosen S et al (2020) Enhancement of food shelf life with polyvinyl alcohol-chitosan nanocomposite films from bioactive eucalyptus leaf extracts. Food Biosci 36. https://doi.org/10.1016/j.fbio.2020.100609

Nwabor OF, Singh S, Ontong JC et al (2021a) Valorization of wastepaper through antimicrobial functionalization with biogenic silver nanoparticles, a sustainable packaging composite. Waste Biomass Valoriz 12:3287–3301. https://doi.org/10.1007/s12649-020-01237-5

Nwabor OF, Singh S, Wunnoo S et al (2021b) Facile deposition of biogenic silver nanoparticles on porous alumina discs, an efficient antimicrobial, antibiofilm, and antifouling strategy for functional contact surfaces. Biofouling 37:538–554. https://doi.org/10.1080/08927014.2021.1934457

Okamoto K, Akiyama R, Yoshida H et al (2005) Formation of nanoarchitectures including subnanometer palladium clusters and their use as highly active catalysts. J Am Chem Soc 127:2125–2135. https://doi.org/10.1021/ja047095f

Ontong JC, Singh S, Nwabor OF et al (2020) Potential of antimicrobial topical gel with synthesized biogenic silver nanoparticle using Rhodomyrtus tomentosa leaf extract and silk sericin. Biotechnol Lett 42:2653–2664. https://doi.org/10.1007/s10529-020-02971-5

Parashar V, Pandey SK, Pandey AC (2010) Low-temperature synthesis of quantum size gadolinium monosulfide (GdS) nanoparticles and their pathogen capture efficiency. Chem Commun 46: 3143–3145. https://doi.org/10.1039/c000553c

Parashar R, Wan M, Yadav RR et al (2014) Surfactant free synthesis of metal oxide (Co and Ni) nanoparticles and applications to heat propagation in nanofluids. Mater Lett 132:440–443. https://doi.org/10.1016/j.matlet.2014.06.126

Pastorizo-Santos I, Liz-Marzán LM (2009) N,N-dimethylformamide as a reaction medium for metal nanoparticle synthesis. Adv Funct Mater 19:679–688. https://doi.org/10.1002/adfm.200801566

Pillai ZS, Kamat PV (2004) What factors control the size and shape of silver nanoparticles in the citrate ion reduction method? J Phys Chem B 108:945–951. https://doi.org/10.1021/jp037018r

Pinna N, Neri G, Antonietti M, Niederberger M (2004) Nonaqueous synthesis of nanocrystalline semiconducting metal oxides for gas sensing. Angew Chem Int Ed 43:4345–4349. https://doi.org/10.1002/anie.200460610

Prat D, Wells A, Hayler J et al (2015) CHEM21 selection guide of classical- and less classical-solvents. Green Chem 18:288–296. https://doi.org/10.1039/c5gc01008j

Puri A, Mohite P, Patil S et al (2023) Facile green synthesis and characterization of Terminalia arjuna bark phenolic–selenium nanogel: a biocompatible and green nano-biomaterial for multifaceted biological applications. Front Chem 11. https://doi.org/10.3389/fchem.2023.1273360

Quinson J, Kunz S, Arenz M (2023) Surfactant-free colloidal syntheses of precious metal nanoparticles for improved catalysts. ACS Catal 13:4903

Rao H, Fu H (2011) Copper-catalyzed coupling reactions. Synlett 745–769. https://doi.org/10.1055/s-0030-1259919

Rashid MH, Mandal TK (2008) Templateless synthesis of polygonal gold nanoparticles: an unsupported and reusable catalyst with superior activity. Adv Funct Mater 18:2261–2271. https://doi.org/10.1002/adfm.200800085

Rashid MH, Bhattacharjee RR, Kotal A, Mandal TK (2006) Synthesis of spongy gold nanocrystals with pronounced catalytic activities. Langmuir 22:7141–7143. https://doi.org/10.1021/la060939j

Restrepo CV, Villa CC (2021) Synthesis of silver nanoparticles, influence of capping agents, and dependence on size and shape: a review. Environ Nanotechnol Monit Manag 15. https://doi.org/10.1016/j.enmm.2021.100428

Shah DD, Chorawala MR, Mansuri MKA et al (2024) Biogenic metallic nanoparticles: from green synthesis to clinical translation. Naunyn Schmiedeberg's Arch Pharmacol. https://doi.org/10.1007/s00210-024-03236-y

Shao Y, Jin Y, Dong S (2004) Synthesis of gold nanoplates by aspartate reduction of gold chloride. Chem Commun 4:1104–1105. https://doi.org/10.1039/b315732f

Shylesh S, Hanna D, Gomes J et al (2015) The role of hydroxyl group acidity on the activity of silica-supported secondary amines for the self-condensation of n-butanal. ChemSusChem 8: 466–472. https://doi.org/10.1002/cssc.201402443

Sifniades S, Levy AB, Bahl H (2010) Acetone. In: Ullmann's encyclopedia of industrial chemistry. Wiley-VCH Verlag GmbH & Co. KGaA, Weinheim

Singh S, Chunglok W, Nwabor OF et al (2022) Hydrophilic biopolymer matrix antibacterial Peel-off facial mask functionalized with biogenic nanostructured material for cosmeceutical applications. J Polym Environ 30:938–953. https://doi.org/10.1007/s10924-021-02249-5

Sneed BT, Kuo C-H, Brodsky CN, Tsung C-K (2012) Iodide-mediated control of rhodium epitaxial growth on well-defined noble metal nanocrystals: synthesis, characterization, and structure-dependent catalytic properties. J Am Chem Soc 134:18417

Soni J, Sahiba N, Sethiya A, Agarwal S (2020) Polyethylene glycol: a promising approach for sustainable organic synthesis. J Mol Liq 315

Subhramannia M, Pillai VK (2008) Shape-dependent electrocatalytic activity of platinum nanostructures. J Mater Chem 18:5858–5870. https://doi.org/10.1039/b811149a

Sun S, Yang D, Villers D et al (2008) Template- and surfactant-free room temperature synthesis of self-assembled 3D Pt nanoflowers from single-crystal nanowires. Adv Mater 20:571–574. https://doi.org/10.1002/adma.200701408

Sun J, Liu G, Fu S et al (2018) Gold nanoparticles of multiple shapes synthesized in l-Tryptophan aqueous solution. Trans Tianjin Univ 24:401–414. https://doi.org/10.1007/s12209-018-0141-y

Sylvestre JP, Kabashin AV, Sacher E, Meunier M (2005) Femtosecond laser ablation of gold in water: influence of the laser-produced plasma on the nanoparticle size distribution. Appl Phys A Mater Sci Process 80:753–758. https://doi.org/10.1007/s00339-004-3081-4

Syukri DM, Nwabor OF, Singh S et al (2020) Antibacterial-coated silk surgical sutures by ex situ deposition of silver nanoparticles synthesized with Eucalyptus camaldulensis eradicates infections. J Microbiol Methods 174. https://doi.org/10.1016/j.mimet.2020.105955

Syukri DM, Nwabor OF, Singh S, Voravuthikunchai SP (2021) Antibacterial functionalization of nylon monofilament surgical sutures through in situ deposition of biogenic silver nanoparticles. Surf Coat Technol 413. https://doi.org/10.1016/j.surfcoat.2021.127090

Syukri DM, Singh S, Nwabor OF et al (2024) Prevention of post-operative bacterial colonization on mice buccal mucosa using biogenic silver nanoparticles-coated nylon sutures. Regen Eng Transl Med 10:1–15

Szeifert JM, Feckl JM, Fattakhova-Rohlfing D et al (2010) Ultrasmall titania nanocrystals and their direct assembly into mesoporous structures showing fast lithium insertion. J Am Chem Soc 132: 12605–12611. https://doi.org/10.1021/ja101810e

Uppal MA, Kafizas A, Ewing MB, Parkin IP (2010) The effect of initiation method on the size, monodispersity and shape of gold nanoparticles formed by the Turkevich method. New J Chem 34:2906–2914. https://doi.org/10.1039/c0nj00505c

Vemula PK, Aslam U, Mallia VA, John G (2007) In situ synthesis of gold nanoparticles using molecular gels and liquid crystals from vitamin-C amphiphiles. Chem Mater 19:138–140. https://doi.org/10.1021/cm062464n

Wan M, Yadav RR, Yadav KL, Yadaw SB (2012) Synthesis and experimental investigation on thermal conductivity of nanofluids containing functionalized Polyaniline nanofibers. Exp Thermal Fluid Sci 41:158–164. https://doi.org/10.1016/j.expthermflusci.2012.03.030

Wang Y, Ren J, Deng K et al (2000) Preparation of tractable platinum, rhodium, and ruthenium nanoclusters with small particle size in organic media. Chem Mater 12:1622–1627. https://doi.org/10.1021/cm0000853

Wang Y, Zhang J, Wang X et al (2005) Metal nanoclusters stabilized with simple ions and solvents – promising building blocks for future catalysts. Top Catal 35:35–41. https://doi.org/10.1007/s11244-005-3811-7

Wang G, Xiao L, Huang B et al (2012) AuCu intermetallic nanoparticles: surfactant-free synthesis and novel electrochemistry. J Mater Chem 22:15769–15774. https://doi.org/10.1039/c2jm32264a

Watt J, Bleier GC, Austin MJ et al (2017) Non-volatile iron carbonyls as versatile precursors for the synthesis of iron-containing nanoparticles. Nanoscale 9:6632–6637. https://doi.org/10.1039/c7nr01028a

Welch CM, Compton RG (2006) The use of nanoparticles in electroanalysis: a review. Anal Bioanal Chem 384:601–619

Wen X, Xie YT, Mak MWC et al (2006) Dendritic nanostructures of silver: facile synthesis, structural characterizations, and sensing applications. Langmuir 22:4836–4842. https://doi.org/10.1021/la060267x

Wu J, Li P, Pan YT et al (2012) Surface lattice-engineered bimetallic nanoparticles and their catalytic properties. Chem Soc Rev 41:8066–8074. https://doi.org/10.1039/c2cs35189g

Xie J, Zhang Q, Lee JY, Wang DIC (2008) The synthesis of SERS-active gold nanoflower tags for in vivo applications. ACS Nano 2:2473–2480. https://doi.org/10.1021/nn800442q

Xu H, Suslick KS (2010) Sonochemical synthesis of highly fluorescent Ag nanoclusters. ACS Nano 4:3209–3214. https://doi.org/10.1021/nn100987k

Yamamoto H, Yano H, Kouchi H et al (2012) N,N-Dimethylformamide-stabilized gold nanoclusters as a catalyst for the reduction of 4-nitrophenol. Nanoscale 4:4148–4154. https://doi.org/10.1039/c2nr30222e

Yang X, Shi M, Zhou R et al (2011) Blending of HAuCl4 and histidine in aqueous solution: a simple approach to the Au10 cluster. Nanoscale 3:2596–2601. https://doi.org/10.1039/c1nr10287g

Yin H, Tang H, Wang D et al (2012) Facile synthesis of surfactant-free au cluster/graphene hybrids for high-performance oxygen reduction reaction. ACS Nano 6:8288–8297. https://doi.org/10.1021/nn302984x

Yue H, Zhao Y, Ma X, Gong J (2012) Ethylene glycol: properties, synthesis, and applications. Chem Soc Rev 41:4218–4244. https://doi.org/10.1039/c2cs15359a

Zhang Y, Hong H, Myklejord DV, Cai W (2011) Molecular imaging with SERS-active nanoparticles. Small 7:3261–3269

Zhang Y, Cui X, Shi F, Deng Y (2012) Nano-gold catalysis in fine chemical synthesis. Chem Rev 112:2467–2505

Zhou Z, Wang S, Zhou W et al (2003) Novel synthesis of highly active Pt/C cathode electrocatalyst for direct methanol fuel cell. Chem Commun 3:394–395. https://doi.org/10.1039/b211075j

Ms. Sayali Churi is currently working as an Assistant professor at St. John Institute of Pharmacy and Research, Palghar. She has a total of 2.8 years' experience in teaching and 1 year Industry experience in analytical research and development department. She has presented research papers and published abstracts at 2 international conferences. She won and represented University of Mumbai in Avishkar Research Convention at both UG and PG level. She is a lifetime member of IPA and ISTE. Her research areas are phytochemical analysis, extraction techniques, herbal product development, analytical method development and validation.

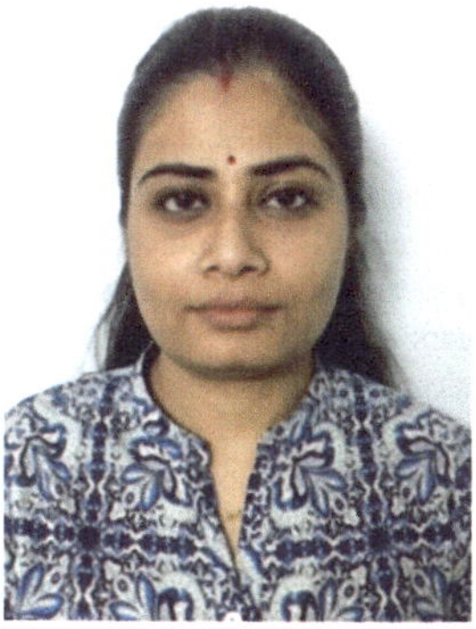

Ushasi Das Presently a PhD Research Scholar in the Department of Pharmaceutical Technology at Jadavpur University, Kolkata. Her academic journey commenced with an undergraduate degree in Pharmacy from NSHM College of Pharmaceutical Technology, NSHM knowledge campus, Kolkata in 2011, followed by post-graduate studies M. Pharm in Pharmaceutics at NSHM College of Pharmaceutical Technology, NSHM knowledge campus, affiliated to West Bengal University of Technology, culminating in 2013. Over the past 10 years, she has made significant contributions to various institutions as an Analytical Chemist, including Joy Beauty Care Pvt. Ltd. and Central Drugs Laboratory, Kolkata. She has authored over 14 research and review articles, edited two books, and six book chapters.

Dr. Popat Mohite is currently working as an Associate professor and Head of the Department at St. John Institute of Pharmacy and Research, Palghar. He has a total of 16 years' experience in teaching. He has several National and International Journals publications to his credit and Research grants from the University of Pune. He has presented research papers at several national and international conferences. He has guided more than 35 PG students. He has published 85 research and review articles in various national and international journals of repute. He has authored 03 books in Lambert publishing house. He has granted 01 and published 01 patents. He is a reviewer and member of the editorial board of various journals. He is a lifetime member of IPA, APTI and ISTE. He was awarded with Young Researcher Award, Young Scientist and Young Faculty award. His research areas are synthesis and biological evaluation, QSAR study formulation, surface functionalization of polymers and analytical method development and validation. His current Google Scholar h-Index is 18; i10-Index is 28 with citation of 1075.

Dr. Sanchita Mandal is currently working as an assistant professor in the Department of Pharm Tech. at Jadavpur University. She had qualified in GATE 2004, was Awarded UGC JRF in 2007 and the CSIR SRF (senior research fellowship) in 2010. She was awarded her PhD from Jadavpur University in 2012. She attended two oral presentations in Thailand and NTU Singapore. Stood 2nd in the International Workshop-Seminar on 'Liposome: From Theory to Practice' 2010. She has also been an Approved Mumbai University faculty member since 2012 and has worked at various colleges in Mumbai. Recently, she has published four books between 2021 and 24 and many Book Chapters. She also published an AU Patent, in December 2021, Australian Government IP Australia. She has joined Jadavpur University in 2021. She is also guiding four M. Pharm Student and three PhD students and working with silicone nanoparticles, combating anti-microbial resistance with hybrid nanogel, graphene and metal nano composites, novel wound healing enzymes. So far she has gained total experience of 18 years including Research and Teaching.

Dr. Sudarshan Singh presently working as a Frontier Proactive Researcher in Chiang Mai University, Thailand. He has authored several technical books and books chapters and more than two hundreds of research/review papers in reputed peer-reviewed journals. He has registered five patents on his credit. He is working as reviewer of various international journals. He has eleven years of teaching experience in undergraduate and postgraduate courses of pharmacy with more than six years of postdoctoral research experience. He has guided several undergraduate and postgraduate students in their project work. Moreover, during his postdoctoral research at Prince of Songkhla University, Walailak University, and Chiang Mai University, Thailand, co-supervised doctorate students and won award of top researcher for the year of 2020–2021. He is a life member of professional bodies such as the Society of Pharmacovigilance India and a registered pharmacist at the State Pharmacy Council of Chhattisgarh, India.

Index

R. Malviya, S. Sundram (eds.), *Sustainable Green Biomaterials As Drug Delivery Systems*, Biomaterials, Bioengineering and Sustainability 1,
https://doi.org/10.1007/978-3-031-79062-1

The manufacturer's authorised representative in the EU is Springer
Nature Customer Service Centre GmbH, Europaplatz 3, 69115 Heidelberg,
Germany. If you have any concerns regarding our products, please
contact ProductSafety@springernature.com

Printed and bound by CPI Group (UK) Ltd, Croydon, CR0 4YY
01/10/2026
02225949-0017